Hilbert's Programs

and

Beyond

Wilfried Sieg

Hilbert's Programs
and Beyond

OXFORD
UNIVERSITY PRESS

Oxford University Press is a department of the University of Oxford. It furthers the University's objective of excellence in research, scholarship, and education by publishing worldwide.

Oxford New York
Auckland Cape Town Dar es Salaam Hong Kong Karachi
Kuala Lumpur Madrid Melbourne Mexico City Nairobi
New Delhi Shanghai Taipei Toronto

With offices in
Argentina Austria Brazil Chile Czech Republic France Greece
Guatemala Hungary Italy Japan Poland Portugal Singapore
South Korea Switzerland Thailand Turkey Ukraine Vietnam

Oxford is a registered trade mark of Oxford University Press in the UK and certain other countries.

Published in the United States of America by
Oxford University Press
198 Madison Avenue, New York, NY 10016

© Oxford University Press 2013

First issued as an Oxford University Press paperback, 2019

All rights reserved. No part of this publication may be reproduced, stored in a retrieval system, or transmitted, in any form or by any means, without the prior permission in writing of Oxford University Press, or as expressly permitted by law, by license, or under terms agreed with the appropriate reproduction rights organization. Inquiries concerning reproduction outside the scope of the above should be sent to the Rights Department, Oxford University Press, at the address above.

You must not circulate this work in any other form
and you must impose this same condition on any acquirer.

Library of Congress Cataloging-in-Publication Data
Sieg, Wilfried, 1945-
Hilbert's programs and beyond / Wilfried Sieg.
p. cm.
ISBN 978-0-19-537222-9 (hardback : alk. paper) — ISBN 978-0-19-970715-7 (e-book) —
ISBN 978-0-19-094819-1 (pbk.)
1. Mathematics—Philosophy. 2. Hilbert, David, 1862-1943. I. Title.
QA8.4.S545 2013
510.1—dc23
2012022356

To Gail, Emily & Clara

Preface for the Paperback Edition

Hilbert's Programs and Beyond was published in 2013. The book's title indicates that Hilbert, in his foundational reflections stretching from the early 1890s to the early 1930s, pursued a number of *different* programs; consequently, attention should not be exclusively restricted to his finitist consistency program of the 1920s. This remark concerns the first part of the title, "Hilbert's Programs". Its second part, "Beyond", points to continued proof theoretic work,[1] but also to my ways of providing contexts for these programs. On the one hand, I analyzed their *mathematical roots* in the 19th century transformation of mathematics, especially those laid bare in Dedekind's foundational work. On the other hand, I explored the *philosophical horizons* for the expanding work in proof theory, especially those revealed by Bernays' methodological perspective.

Both ways have been pursued in subsequent work and, as it turned out, they have reinforced each other. As to the *mathematical roots*, I have deepened my understanding of the logical framework for Dedekind's mathematical structuralism and connected them to contemporaneous philosophical developments, e.g., in Lotze's very influential *Logik* where mathematics is viewed to be a part of logic.[2] As to the *philosophical horizons*, I have expanded the reach of the mathematical core of Bernays' methodological frames that are so crucial for proof theoretic work. That has been at the center of my own considerations ever since I introduced *accessible domains* in 1990. However, a fully unified perspective emerged only slowly. It is articulated in my recent essay with the fitting title, *Methodological frames: Paul*

---

[1]Rathjen & Sieg's *Proof Theory*, a contribution to the Stanford Encyclopedia of Philosophy, presents additional and important parts of proof theoretic work that go beyond what is described in this volume. Gentzen's work, very much in the foreground of my essays and the SEP article, is deeply rooted in prior work of Hilbert and Bernays; see, *In the shadow of incompleteness: Hilbert and Gentzen* (essay II.3 in this volume).

[2]The deeper understanding has been brought out in two papers with Morris, respectively Schlimm: *Dedekind's Structuralism: Creating concepts and deriving theorems*, in: Logic, Philosophy of Mathematics and their History, E. Reck (ed.), College Publications, 2018, pp. 251–301; *Dedekind's Abstract Concepts: Models and Mappings*, Philosophia Mathematica 25 (3), 2017, pp. 292–317. The first essay articulates in section C.2 and the Concluding Remarks Dedekind's connection to Lotze, in particular, sharing the view of mathematics as part of logic and adopting his concept of *abstraction*. The anti-Kantianism of Dedekind (and Hilbert in his foundational work around the turn of the century) is exposed in my *On Tait on Kant and Finitism*, The Journal of Philosophy, CXIII (5/6), 2016, pp. 274–285.

Bernays, mathematical structuralism and proof theory; its basis is found in the sharp juxtaposition of *structural* and *formal* axiomatics.[3]

These complementary ways of Hilbert's axiomatic method have shaped proof theoretic work that was hinted at in the last essay of this volume, *Searching for proofs (and uncovering capacities of the mathematical mind)*. This work exploits contemporary computer technology and, at the same time, returns to the classical goal of considering mathematical proofs as objects of theoretical study. Formal representation of proofs and their meta-mathematical investigation are important, but—from this standpoint—subservient to the examination of what Hilbert called, in his Zürich talk of 1917, "the notion of the specifically mathematical proof".[4] Nearly twenty years later, in exactly this spirit, Gentzen asserted, "The objects of proof theory shall be the *proofs* carried out in mathematics proper" (Gentzen 1936, 499).

Hilbert made in (1927, 475) a remarkably broad claim concerning the significance of formalized mathematics when carried out "according to certain definite rules, in which the *technique of our thinking* is expressed":

> These rules form a closed system that can be discovered and definitively stated. The fundamental idea of my proof theory is none other than to describe the activity of our understanding, to make a protocol of the rules according to which our thinking actually proceeds.... If any totality of observations and phenomena deserves to be made the object of a serious and thorough investigation, it is this one.

Such "a serious and thorough investigation" can, and indeed must, begin in our days with a *computer-based formal and faithful reconstruction* of parts of mathematics. For that purpose, it refines methods of *natural deduction* and expands them to an interactive process of *natural formalization*; that process is embedded in a detailed conceptual framework and is structured for human intelligibility and discovery.[5] To take the further step toward *automated* theorem proving, we must isolate creative elements in proofs, implement them as guiding ideas for proof search, and carry out computational experiments to determine the efficacy of particular search procedures. In this way, we will undoubtedly uncover sophisticated capacities of the mathematical mind and arrive at a deepened understanding of the human mind.

Pittsburgh, January 6, 2019

---

[3] I am alluding in this sentence to my paper, *The ways of Hilbert's axiomatics: structural and formal*, that was published in Perspectives of Science 22 (1), 2014, pp. 133–157. The more recent essay is to appear in: E. Reck and G. Schiemer (eds), *The Pre-History of Mathematical Structuralism*, Oxford University Press (in press).

[4] Hilbert's (1918) was given as a talk to the Swiss Mathematical Society in Zürich in September of 1917. In celebration of its 100th anniversary, conferences were organized in Zürich and Lisbon. In my contribution to these conferences, *Proofs as objects: Hilbert's pivotal thought*, I focused on this very idea and connected it to Hilbert's 24th problem.

[5] See my paper with Walsh, *Natural Formalization: Deriving the Cantor-Bernstein Theorem in ZF* (under submission). It contains a thorough discussion of a related, but quite different approach, namely, Ganesalingam and Gowers' "human-centered automatic theorem-proving".

Preface

As the plural in *Hilbert's Programs and Beyond* suggests, the essays I selected for this volume go beyond the analysis of the finitist consistency program Hilbert pursued during the 1920s. They show how Hilbert, since the early 1890s, attempted in different ways to come to grips with a changed and changing subject, i.e., with mathematics in its modern abstract form as it began to evolve during the 19th century. They also argue that Hilbert's quest can be continued by broadening his methodological perspective, refining the mathematical investigations, and expanding proof theoretic tools. The search for finitist consistency proofs of *the* Hilbert Program is discussed in detail, but it is seen in an intellectual and scientific context with deep roots in the past and with remarkable future impact transcending logic and the foundations of mathematics.

The essays have been published over a period of 25 years and are presented here in a uniform format and with a common bibliography.[6] I corrected typos silently and addressed, in explicit notes, the real mistakes I am aware of. Some of the essays overlap, as they were written for different audiences and were to be self-contained. My philosophical views, as well as my understanding of mathematical and historical facts, have evolved: the *Brief Guides* that precede each of the four groups of essays indicate connections and changes; the *Introduction* to the volume articulates a broad perspective for Hilbert's Programs. I am aware of the shortcomings implicit in the fact that the essays were written over a quarter century; but I think that they illuminate, coherently, central themes in the philosophy of mathematics, significant aspects of the history of mathematical logic, and the foundational goals of proof theory.

I feel fortunate to have been drawn to the mathematical developments, proof theoretic investigations, and methodological reflections that are richly represented not only in Hilbert's papers and lectures, but also in the essays of Paul Bernays, the two volumes of *Grundlagen der Mathematik*, and the work in the tradition of the Hilbert School. Three university teachers deeply influenced my mathematical and logical education, namely, Karl Peter Grotemeyer in Berlin, Dieter Rödding in Münster, and Solomon Feferman at Stanford. Grotemeyer introduced me to structuralist

---

[6]They are reprinted here with the contractual or explicit permission of the copyright holders; the two papers that were co-authored with Charles Parsons, respectively Dirk Schlimm, appear also with their permissions.

mathematics, and Rödding to the craft of logical work. Feferman, most importantly, guided me in the crucial step to independent research in proof theory; he has been mentor, collaborator, and friend. At the institutions where I have worked, a number of colleagues influenced me deeply: Dagfinn Føllesdal, Georg Kreisel, and Patrick Suppes at Stanford University; Isaac Levi, Sidney Morgenbesser, and Charles Parsons at Columbia University; Clark Glymour, Dana Scott, and Teddy Seidenfeld at Carnegie Mellon University. Three colleagues, who do not fall fully into these institutional categories, have stimulated my work for a long time and continue to do so. They are Martin Davis, Howard Stein, and especially William Tait. I am also indebted to many other colleagues, friends, and students for keen insights and observations; they are mentioned in the individual essays.

Early in our academic careers, Wilfried Buchholz, Wolfram Pohlers, and I did complementary proof theoretic work on theories of generalized inductive definitions. That work, together with papers by and with Feferman, was published as volume 897 of the Springer Lecture Notes in Mathematics; it was also the starting-point for my philosophical and historical explorations. The latter have profited, for much longer than a decade now, from working with William Ewald, Michael Hallett, and Ulrich Majer on the edition of Hilbert's unpublished lecture notes on logic and arithmetic. Those notes have opened a new view on the development of proof theory and mathematical logic that is reflected in my essays published after 1999. I am grateful for the permission to use source material from the Niedersächsische Staats- und Universitätsbibliothek in Göttingen pertaining not only to Hilbert, but also to Dedekind; Dr. Helmut Rohlfing has been extremely helpful.

Academic work is done in institutional settings, and I have had the good fortune to work at three great universities. The departments of philosophy at Stanford and Columbia provided rich, stimulating environments. Special appreciation goes to Carnegie Mellon, which has become "my place". While interdisciplinary work is important at many universities, at Carnegie Mellon it is so important as to be ordinary, i.e., it is part of the regular order of things. I treasure that aspect of the university and thrive participating in its vigorous intellectual life. More specific thanks are owed to another institution, the Swedish Collegium for Advanced Study (SCAS) in Uppsala; my stay as a Fellow at that scholars' utopia allowed me to focus on this volume in spring and early summer of 2009. Since then I have written two additional essays, (II.3) and (III.3), as well as the long *Introduction*. Last, but certainly not least, my thanks go to Dawn McLaughlin. My essays would hardly have seen the light of day in this collected form, had it not been for her excellent editorial and TEX- nical support, patiently sustained over almost five years.

Wilfried Sieg
July 2012
Pittsburgh

Published with Permission

I.1 Richard Dedekind. Wikimedia. "Richard Dedekind," http://commons.wikimedia.org/wiki/File:Dedekind.jpeg.

I.2 Leopold Kronecker. Wikimedia. "Leopold Kronecker," http://commons.wikimedia.org/wiki/File:Leopold_Kronecker.jpg.

I.3 David Hilbert. Courtesy of Niedersächsische Staats- und Universitätsbibliothek Goettingen, Abteilung Spezialsammlungen und Bestandserhaltung.

II.1 David Hilbert. Attributed to Elisabeth Reidemeister. Courtesy of the Emilio Segrè Visual Archives, Nina Courant Collection, American Institute of Physics.

II.2 Johan von Neumann. Courtesy of Los Alamos National Laboratory Archives. Unless otherwise indicated, this information has been authored by an employee or employees of the Los Alamos National Security, LLC (LANS), operator of the Los Alamos National Laboratory under Contract No. DE-AC52-06NA25396 with the U.S. Department of Energy. The U.S. Government has rights to use, reproduce, and distribute this information. The public may copy and use this information without charge, provided that this Notice and any statement of authorship are reproduced on all copies. Neither the Government nor LANS makes any warranty, express or implied, or assumes any liability or responsibility for the use of this information.

II.3 Kurt Gödel. Photographer unknown. From The Shelby White and Leon Levy Archives Center, Institute for Advanced Study, Princeton, NJ, USA.

II.4 Gerhard Gentzen. Courtesy of Eckart Menzler.

III.1 Paul Bernays. Courtesy of Wilfried Sieg.

Contents

Introduction   1
    In.1   A Perspective on Hilbert's Programs   3
    In.2   Milestones   29

I Mathematical roots   33
    I.1   Dedekind's analysis of number: systems and axioms   35
    I.2   Methods for real arithmetic   73
    I.3   Hilbert's programs: 1917–1922   91

II Analyses
                        Historical   129

    II.1   Finitist proof theory: 1922–1934   131
    II.2   After Königsberg   145
    II.3   In the shadow of incompleteness: Hilbert and Gentzen   155
    II.4   Gödel at Zilsel's   193
    II.5   Hilbert and Bernays: 1939   215

                        Systematical   227

    II.6   Foundations for analysis and proof theory   229
    II.7   Reductions of theories for analysis   263
    II.8   Hilbert's program sixty years later   281
    II.9   On reverse mathematics   291
    II.10   Relative consistency and accessible domains   299

III Philosophical horizons   327
    III.1   Aspects of mathematical experience   329

III.2	Beyond Hilbert's reach?	345
III.3	Searching for proofs (and uncovering capacities of the mathematical mind)	377
Bibliography		403
Index		435

Introduction

A brief note (to the reader)

This *Introduction* consists of two main parts, an expansive *Perspective* and related *Milestones*. The *Perspective* describes a web of connections between Hilbert's Programs and isolates essential themes that are, from a contemporary standpoint, a brilliant and permanent part of the intellectual history of the 20th century. The *Milestones* form a short list of Hilbert's significant lecture courses, important addresses, and crucial publications — all tied to his work on the foundations of mathematics. Even a glimpse at the milestones shows that characterizing modern abstract mathematics and "securing" it from contradictions were life-long preoccupations of Hilbert. For him, the character of modern mathematics was distinctly tied to the axiomatic method and its use to introduce structural definitions; it had raised, already for Dedekind, the methodologically central consistency problem. Hilbert took on that problem in different programmatic ways and formulated in 1922 his finitist consistency program, *the* Hilbert Program as it is usually seen.

When examining the intricate web of connections and essential themes, starting in section 4 of the *Perspective*, I take for granted some familiarity with mathematical logic and related foundational work. I want to emphasize, however, that sections 4 through 8 of the *Perspective* are by no means a "prerequisite" for reading the essays with understanding. The essays were written, after all, as individual papers and are standing on their own feet in circumscribed contexts. The *Perspective*, by contrast, draws on the essays to give a unified view of Hilbert's Programs. Thus, dear reader, a non-linear approach to engaging with the core material recommends itself. One might read, for example, sections 1 through 3 of the *Perspective* first and survey the *Milestones* second, next read essay (III.1) in its entirety, and finally return to the *Perspective* and read sections 4 through 8. After that follow your curiosity: enjoy exploring Hilbert's Programs and the work that goes beyond them.

In.1

A Perspective on Hilbert's Programs

David Hilbert was among the great mathematicians who expounded the centrality of mathematics in human thought. Late in 1930, he gave a lecture in his native Königsberg, which was also Kant's town, and addressed the "vexed question about the share which thought, on the one hand, and experience, on the other, have in our knowledge". To illuminate his view on an implicitly Kantian question, Hilbert asserts that mathematics is "the instrument that mediates between theory and practice, between thought and observation" and that it "builds the connecting bridges and makes them ever sounder". The very next sentence formulates a far-reaching consequence of this instrumental view of mathematics and its use in the sciences: "... our entire modern culture, in so far as it rests on the penetration and utilization of nature, has its foundation in mathematics." (Hilbert 1930, p. 1163)

Hilbert's claims for mathematics hold in particular for those parts of the subject he helped to create for foundational studies, namely, mathematical logic and proof theory. Whereas classical analysis and its abstract extensions have been crucial for physics and resultant technological advances, logic and proof theory have proven indispensible for computer science and resultant computational developments. Both kinds of developments have had a revolutionary impact on our social and economic life, while retaining a reflective, philosophical side. After all, we have been exploring, in the first case, the physical universe and our place in it and, in the second case, intelligent machinery and the human mind, i.e., aspects of our own intelligence.

Hilbert made his observations for a subject that had been dramatically expanded, indeed transformed, since its classical beginnings as a systematic science. The character of this distinctly *modern mathematics* is indicated by its use of *abstract* concepts. Such concepts do not exclusively apply to privileged objects like numbers or geometric magnitudes; rather, they are introduced through *structural definitions* and apply to many different classes of objects with appropriate relational connections. This transformation took place in the second half of the 19th century and is reflected in Hilbert's *Grundlagen der Geometrie* (1899a). It presented difficult methodological challenges, which Hilbert addressed in a variety of ways over the next thirty years. The intellectual journey Hilbert and his collaborators took between 1899 and 1930 is remarkable; equally remarkable is the journey we have taken since. My essays try to convey a sense of the importance and sheer excitement of both journeys.

This introductory essay depicts systematic directions of and historical contexts for Hilbert's investigations. At the same time, it attempts to illuminate relations with the papers collected here. Section 1 distinguishes between two directions of Hilbert's work and argues for the importance of historical contexts. The rationale for the overall organization of the papers is delineated in section 2, and from this discussion emerge methodological themes with four associated core concepts. These themes have their roots, as section 3 describes, in foundational views represented by Dedekind and Kronecker; they are explored in sections 4 through 7. The last theme is focused on computational procedures and has added significance, as it opens possibilities for probing the mathematical mind. Section 8 gives an outlook on such possibilities together with a unified perspective on Hilbert's Programs.

1 Directions & contexts

In order to address the methodological challenges of modern mathematics, Hilbert formulated particular programmatic approaches. He strove to reach *metamathematical* goals and, in particular, to establish in a *mathematically conclusive* way that the abstract concepts of mathematics are free from internal contradictions. This central goal was pursued, as Hilbert's collaborator Bernays expressed it later, within different methodological frames (methodische Rahmen). Initially, around the turn from the 19th to the 20th century, it was to be achieved within a logicist frame inspired by Dedekind; later, in the 1920s, after structural definitions had given way to strictly formal theories, Hilbert sought a solution within a finitist frame influenced by Kronecker. In either case, the reach and significance of the methodological frame must be critically examined.

Such an examination is a philosophical task and also an integral part of Hilbert's Programs. Thus, the latter are located properly between mathematics and philosophy. At first sight the programs seem to be important only for mathematics and the philosophy thereof. This impression, I argue, should be corrected by the fact that Hilbert emphasized a broad role for mathematics throughout his career. That role is sometimes hidden in aphoristic remarks. Here is a striking example:

Arithmetic symbols are written diagrams, and geometric figures are graphic formulae.

This remark is concise and sharply to the point, but is also enigmatic and open to conflicting interpretations. Hilbert made it in the *Introduction* to his

Introduction: A Perspective

Paris address from 1900 just after having characterized "as entirely erroneous" the opinion of some eminent men that only arithmetic concepts "are susceptible of a fully rigorous treatment". In contrast, he believes mathematics should also investigate ideas coming from other fields, be that epistemology or geometry or the natural sciences, and it should take on the concrete task of formulating the basic principles in a "simple and complete set of axioms, so that the exactness of the new ideas and their applicability to deduction shall be in no respect inferior to those of the old arithmetic concepts". Signs are then introduced for the new concepts and chosen so as "to remind us of the phenomena which were the occasion for the formation of the new concepts". That is especially true for geometric figures; but how can they be used as a means of strict proof? — Here is the central point:

> The use... presupposes the exact knowledge and mastery of the axioms that underlie those figures; and in order that these geometric figures may be incorporated in the general treasure of mathematical signs, a rigorous axiomatic investigation of their conceptual content is necessary. (Hilbert 1900a, p. 1100)

These observations take for granted, as Hilbert puts it, the "agreement between geometric and arithmetic thought" underlying Minkowski's *Die Geometrie der Zahlen*. The group of eminent men whose opinion concerning the rigorous treatment of non-arithmetic concepts Hilbert had just chastised includes Gauss and Kronecker; they would certainly not agree that there is such an agreement.

Hilbert holds his contrary ecumenic view on account of mathematics' character as *mathesis universalis*: it is at the center of the "interplay of thought and experience" and is to contribute to the solution of problems, be they internal to the subject or derived from external sources. Solutions of problems must be established "by means of a finite number of steps based upon a finite number of hypotheses which are implied in the statement of the problem and which must always be exactly formulated". Hilbert continues:

> This requirement of logical deduction by means of a finite number of processes is simply the requirement of rigor in reasoning. Indeed, the requirement of rigor, which has become proverbial in mathematics, corresponds to a universal philosophical necessity of our understanding; and, on the other hand, only by satisfying this requirement do the problem's thought content and suggestiveness attain their full effect. (Hilbert 1900a, p. 1099)

The attempt of gaining a contextual understanding of an aphoristic remark led us from arithmetic symbols and geometric figures to a glimpse of Hilbert's view of the character of mathematics and the fundamental role of the *axiomatic method*. That method emerged from the transformation of mathematics in the 19th century and is at the subject's core, not as a rigid mechanical procedure or rigmarole, but rather as a flexible tool for conceptual clarification and systematic exploration in the small and the large. In his talk *Axiomatisches Denken*, given in September of 1917, Hilbert describes its centrality as follows:

> ... Anything at all that can be the object of scientific thought becomes dependent on the axiomatic method, and thereby indirectly on mathematics, as soon as it is ripe for the formation of a theory. By pushing ahead to ever deeper layers of axioms ... we also gain ever deeper insights into the nature of scientific thought itself, and we become ever more conscious of the unity of our knowledge. Under the banner of the axiomatic method, mathematics appears destined to play a leading role in science. (Hilbert 1918, p. 1115)

To use the method creatively means to think critically. Hilbert's examination of the foundations of geometry is a paradigm of such a use that lives up to new principles of logical rigor and completeness, but also pursues novel metamathematical studies of geometry itself and its relationship to analysis.

2 Connections & themes

My views on Hilbert's Programs have evolved along different, sometimes meandering paths that do not provide vistas of all aspects that might be deemed relevant. For example, as to the programs' historical framing, there is no sustained discussion of Cantor or Frege; as to the mathematical development of proof theory, I have mostly side-stepped the discussion of systems of ordinal notations as well as that of functional interpretations; as to the philosophical analysis of the finitist program, there is no detailed discussion connecting finitism with earlier philosophical, in particular, Kantian views. At first I thought that an *Introduction* should address these and other missing topics, but there are already excellent discussions in the literature, and I realized that such an undertaking would require at the very least an additional volume. Thus, the draft of a long *Introduction* transformed itself into a briefer *Perspective*. This refocused *Introduction* is to explain, first of all, why the essays of this volume are organized in the way they are, and to bring out, secondly, overarching themes that emerged from Hilbert's Programs and are relevant for the metamathematical as well as philosophical work that has been inspired by them.

A glance at the Table of Contents shows that sixteen essays are presented in three parts entitled *Mathematical roots*, *Analyses*, and *Philosophical horizons*. The ten analyses of Part II are subdivided into five *historical* and five *systematical* ones; Parts I and III consist of three essays each. The chronologically first essay (II.6) examines the foundational point of then recent proof theoretic research and its relevance for extended versions of the finitist program. In order to grasp the finitist program's intellectual core, one has to understand the evolution of Hilbert's foundational investigations and their roots in 19th century mathematics. For his Paris lecture in 1900, Hilbert had singled out the *arithmetization of analysis* and the *discovery of Non-Euclidean geometries* as the most

Introduction: A Perspective

important foundational advances of the 19th century. These advances and underlying methodological problems shaped Hilbert's foundational work, as he had to come to grips with the abstract character of modern mathematics reflected especially in Dedekind's work. This is the topic of essays (I.1) and (I.2), the second of which emphasizes also Hilbert's ways of incorporating aspects of Kronecker's constructive approach to mathematics. Essay (I.3) describes the significance of logical developments and the impact of Whitehead and Russell's *Principia Mathematica* on the foundational studies in Göttingen.

The chronologically last essay (II.3) was completed in the spring of 2011. It is concerned with the pivotal period in the history of proof theory between 1930 and 1934. It details, on the one hand, the brief evolution of Gödel's results in 1930 and reactions from members of the Hilbert School, which are also the topic of (II.2); it describes, on the other hand, the evolution of Gentzen's work leading to his first consistency proof for arithmetic. On closer inspection, it examines the probing questions of my first essay through the eyes of the pioneers. It reveals also that Gentzen was deeply influenced by the proof theoretic work of the 1920s (discussed in II.1) and was directly motivated by Hilbert's last and neglected 1931 paper, *Der Beweis des tertium non datur*. Gentzen's investigations, in turn, influenced Gödel's and Bernays's reflections on the prospect for proof theory in the late 1930s; those reflections are discussed in (II.4) and (II.5).

Two directions of proof theoretic research, roughly fifty years later, are detailed in (II.7) through (II.9); they both exploit Gentzen's sequent calculi and the crucial technique of cut-elimination. One direction was pursued successfully for a subsystem of second-order arithmetic. That subsystem suffices to formalize classical analysis as developed in Supplement IV of (Hilbert and Bernays 1939); it was shown to be consistent relative to an intuitionist theory of constructive ordinals. The other direction isolates weak formal theories that are sufficient and necessary for particular theorems; such theories are used to obtain mathematically interesting information by characterizing their provably recursive functions. The sense in which work along the first direction was philosophically successful is brought out in (II.10). I also discuss there, for the first time, the deep conceptual connections between Dedekind and Hilbert.

If Hilbert's foundational work is seen, as I think it should be, in this broad historical and systematic context, it opens a wide philosophical horizon and continues to raise fascinating mathematical as well as methodological questions. Essay (III.1) sets the stage for the pursuit of extended Hilbertian themes by describing two aspects of mathematical experience. The first aspect is related to finitist mathematics and is explored in (III.2); it concerns classes of uniformly generated, or inductively defined, mathematical objects, *i.d. classes*. The second aspect is central to an emerging part of proof theory and is examined in essay (III.3); it concerns a theory of *conceptually structured* proofs. These aspects highlight two capacities of the mathematical mind, namely, (i) to recognize the correctness of principles for i.d. classes and (ii) to find (and refine) proofs once suitable concepts have been introduced. These capacities are seen against the backdrop of precise formalisms and an associated concept of effective procedures.

Four notions are implicitly appealed to in these orienting remarks: *structural definition, rigorous proof, accessible domain*, and *mechanical procedure*. They shape the context of proof theoretic investigations, but are independent of any particular program and have to be taken into account by *any* informed philosophy of mathematics. The first notion highlights a dramatic shift in the evolution of mathematics; the second turns proofs in their formal variety into objects of investigation; the third isolates a central aspect of generated mathematical objects; the fourth connects to the intellectual core of both the practical impact of computers and the theoretical approach of cognitive science. These notions are discussed, respectively, in sections 4 through 7 below.

The century bounded by 1854 and 1954 could be called *centuria mirabilis*; it spans an extraordinary period in the history of mathematics, a period that is also remarkable through its self-reflection via foundational studies.[1] Hilbert is not the only prominent mathematician who took foundational problems seriously, but he is the one who aimed wholeheartedly to infuse their investigation with mathematical rigor. In the 1920s, it appeared as if philosophy of mathematics could become a part of mathematics. Bernays wrote in late 1921 on the emerging finitist program:

> This is precisely the great advantage of Hilbert's proposal, that the problems and difficulties arising in the foundations of mathematics are transferred from the epistemological-philosophical to the genuinely mathematical domain.[2] (Bernays 1922b, p. 19)

The real story is more intricate and illuminates the complex interaction of philosophy and mathematics; it is definitely more intricate than the disputes of the 1920s and their continued discussion suggest.

3 Foundational views

If there was a *Grundlagenstreit*, i.e., a dispute on the foundations of mathematics, between intuitionism and formalism around 1920, it could have ended in September 1922 when Hilbert presented the finitist program

---

[1] These years are associated with striking essays of Dedekind and Turing. The essays are not in the standard canon of the foundational literature, but express deep methodological problems informally and with unguarded candor. In addition, the first connects our discussion even farther to the past, whereas the second points into the future, indeed, to us. The essays are (Dedekind 1854) and (Turing 1954). — *Centuria mirabilis* is to playfully resonate with Newton's *annus mirabilis*. It is an anachronistic "translation" of the English phrase "century of miracles" and perhaps should be more fully "annorum centuria mirabilis". (See the Oxford English Dictionary for details about the etymology of "century".)

[2] Von Neumann shared this view as late as September 1930; see the first paragraph of his (von Neumann 1931, p. 116).

Introduction: A Perspective

in a talk at Leipzig; that talk was published as (Hilbert 1923). It could have been recognized that there was no controversy about the *Grundlagen* or rather the contentual part of mathematics in which "classical" mathematics was to be grounded.[3] There was a *Streit*, but its roots went deeper than the polemical remarks from either side reveal; the fundamental differences are reflected in essays that were published almost simultaneously more than thirty years earlier, Kronecker's *Über den Zahlbegriff* (1887) and Dedekind's *Was sind und was sollen die Zahlen?* (1888).

Dedekind's essay appeared in the very year in which Hilbert took his *Rundreise* from Königsberg to the German centers of mathematical research. In December of 1930, Hilbert gave a talk in Hamburg and remarked about this trip:

> At my first stop, in Berlin, I heard people in all mathematical circles, both young and old, discuss Dedekind's essay *Was sind und was sollen die Zahlen?*, which had just been published — their remarks were mostly antagonistic. This essay, together with Frege's investigation, is the most important first and profound attempt to ground elementary number theory. (Hilbert 1931a, p. 487)

Among the people Hilbert visited was (the excellent mathematician) Paul du Bois-Reymond who told him that Dedekind's essay was horrifying to him. That comment is not based, as the longer remark I just quoted clearly is, on remembrances of events that took place a long time ago; rather, it is recorded in Hilbert's diary of his *Rundreise*.[4]

For Dedekind, the essay (1888) was the last step in attaining an arithmetic foundation for analysis; the first step had been taken in his (1872c) where he showed that the system of all cuts of rational numbers constitutes a complete ordered field. In this early work, the system of natural numbers and its expansion to the systems of whole and rational numbers are taken for granted. Dedekind, as a matter of fact, thought he knew how to achieve such an expansion and considered the characterization of the natural numbers as the central remaining task. From manuscripts (see section 3.2 of essay (I.1)), it is clear that he was already concerned with this task in 1872, but its completion took a long time and required significant methodological innovations.

The steps of the expansion from natural to real numbers are arithmetic ones for Dedekind, and he views it as "something obvious and not at all novel that every theorem of algebra and higher analysis, no matter how remote, can be expressed as a theorem about natural numbers". He reports that Dirichlet had made that claim repeatedly and continues:

> But I don't view it as something meritorious — and nothing was further from Dirichlet's thought — to actually carry out this wearisome circumlocution and to insist that no other than natural numbers are to be used and recognized. (1888, p. 792)

---

[3] A related and striking remark Brouwer made in 1928 is discussed in (III.2, section 1).
[4] Cod. Ms. Hilbert 741, 1/5, and quoted in (Dugac 1976, p. 203).

The expanding steps are only presented in manuscripts and are discussed in sections 3 and 4 of essay (I.1). In his (1888), Dedekind develops the theory of natural numbers and attempts to ground it in logic.

Kronecker, in contrast to Dedekind, takes for granted the natural numbers as finite ordinals, perhaps not directly given to us through God's good work, but as abstracted from or identified with notation systems that start out with an initial symbol and proceed uniformly to generate further notations:

> The stock of notations we possess in the ordinal numbers is always adequate, because it is not so much an actual but an ideal stock. In the laws of *formation* for our words or numerals, which refer to numbers, we have the true "capacity" to satisfy every demand. (1887, p. 950)

Kronecker views such studies of the number concept as preceding the properly mathematical work in arithmetic. He considers, on this *open field of philosophical groundwork*, arithmetic operations and develops *particular* laws for them, e.g., commutativity of addition. However, and here is a second contrast to Dedekind, he does not formulate general principles like those of definition by recursion or proof by induction.[5] In the extended version of his essay, Kronecker focuses on the *elimination* of all concepts that are "foreign to true arithmetic". The list of such notions includes the concepts of negative, rational, as well as real and imaginary algebraic numbers. This is a third contrast to Dedekind, who strives to *introduce* mathematical concepts in a methodologically justified way.

Kronecker made restrictive demands on the formation of mathematical concepts, e.g., he insisted on their decidability.[6] Being well aware of these demands, Dedekind attached the following footnote to his remark that a system S, as an object of our thinking, is completely determined as soon as "of each thing it is determined, whether or not it is an element of S" (Dedekind 1888, p. 2):

> How this determination is brought about, and whether we know of a way of deciding upon it, is a matter of utter indifference for all that follows; the general laws to be developed in no way depend upon it; they hold under all circumstances. I mention this expressly because Kronecker not long ago (in *Kronecker 1886*) has endeavored

---

[5]Two different ways of treating natural numbers are discussed in Chapter 2 of (Schröder 1873). Schröder contrasts the "independente Behandlungsweise" with the "recurrente", which he associates with Grassmann. The latter treatment, Schröder claims on p. 51, can hardly be surpassed in terms of thoroughness and foundational rigor; the former is more directly accessible and proceeds with shorter inference sequences (Schlussreihen), but it makes greater demands on "inner intuition" (inneres Anschauungsvermögen). Kronecker can be viewed as pursuing the "independente Behandlungsweise" and Dedekind the "recurrente". Even at the very beginning of his work on natural numbers in manuscripts from the 1870s, Dedekind dramatically sharpened the "recurrente" treatment by separating the generation of natural numbers by a successor operation from the recursive definition of addition. This is discussed in sections 3.1 and 5 of essay (I.1).

[6]There are other demands as well: for example, calculability of functions and constructivity of existence proofs. The latter demand was the central point of Hilbert's discontent with the Kroneckerian perspective.

Introduction: A Perspective

to impose certain limitations upon the free formation of concepts in mathematics, which I do not believe to be justified; . . . .

This footnote is directed against one in (Kronecker 1886), where Kronecker argues not only against Dedekind's concepts like module or ideal, but also against a *general concept of irrational number*. Explaining his rejection, Kronecker writes:

> Even the *general* concept of an infinite series ... is in my opinion ... only admissible on condition that in every special case, on the basis of the arithmetical law for the formation of the terms (or of the coefficients), certain presuppositions are shown to be satisfied, which permit the series to be applied like finite expressions, and which consequently make it really unnecessary to go beyond the concept of a *finite* series. (Kronecker 1886, p. 947)

The effect of such a finiteness condition on the concept of real number is stated in a letter Kummer wrote to Schwartz on 15 March 1872. Kummer remarks that he and Kronecker share the conviction that "the effort to create enough individual points to fill out a continuum, i.e., enough real numbers to fill out a line, is as vain as the ancient efforts to prove Euclid's parallel postulate".[7]

If *individual points* can be created only in accord with Kronecker's finiteness demand, then Kummer's observation is provable. One first notices that the system of real numbers thus created is countable. Next one has to address the question, what is the geometric line that cannot be filled by the individually created points? In 1872, the very year of Kummer's letter, Dedekind had characterized an *arithmetic continuum* as the system of all cuts of rational numbers.[8] As that system is isomorphic to the continuous geometric line and is uncountable, Kummer's observation has been established. The argument I just sketched is, of course, anachronistic, but brings out the strikingly different approaches to the arithmetization of the geometric continuum. In this way, it makes evident the impact of broader foundational views on mathematical practice. At the heart of the difference between these foundational positions is the freedom of introducing abstract concepts — given by *structural definitions*.

4 Structural definitions

The discovery of Non-Euclidean geometries was for Hilbert, as mentioned at the beginning of section 2, the second major advance of foundational work in the 19th century. It pressed for a new way of conceiving the role of mathematics in conceptually structuring our physical experience. Kronecker was led, as Gauss had been, to a radical separation of

---

[7] This remark is found in (Edwards 2009, p. 14); I learned of it from Ivahn Smadja.

[8] The system of cuts is motivated by a comparison with the geometric line that is complete by *axiomatic fiat*: "The assumption of this property of the line [its completeness] is nothing but an axiom which attributes to the line its continuity, by which we think continuity into the line." (Dedekind 1872c) in (Ewald 1996, p. 771).

geometry and arithmetic: geometry is viewed as deeply and inextricably connected to empirical space, whereas the theory of natural numbers is taken to be *a priori*.[9] Dedekind shared this perspective to some extent, but was led to a more abstract view of mathematics grounded in arithmetic and, ultimately, in logic. Already in (Dedekind 1872c, p. 328), he remarks that "real" space might be discontinuous, but that — even if we knew for certain that it is discontinuous — nothing could prevent us from completing it "in thought". He describes in his (1888, p. vii) a mathematical, indeed Euclidean space that is everywhere discontinuous.[10] The arithmetic continuum is central, in Dedekind's view, for investigating real space, as this pure number-domain is the *only* means "to render the notion of continuous space clear and definite." (1888, p. 340)

Dedekind's insistence that nothing can prevent us from completing space *in thought* raises the question of what we *can* think mathematically, i.e., what kind of notions we can freely introduce and what safeguards have to be kept. The question of safeguards was systematically addressed only in Dedekind's (1888) and (1890a), but it is already hinted at in the essay on continuity and irrational numbers. It is emphasized in a long note of (Dedekind 1877, p. 269), where it is made perfectly clear, with detailed references to (1872c), that these methodological considerations are important for the foundations of mathematics as well as for mathematical practice, e.g., in algebraic number theory when introducing ideals. Indeed, his general attitude towards the "creation of concepts" is expressed in his *Habilitationsrede* from 1854. Dedekind refers to it in the Preface to (1888), and I will come back to it in section 8 below.

The mathematical notions Dedekind is increasingly concerned with in the 1870s are ultimately given by *structural definitions* and include *simply infinite systems* as well as *complete ordered fields*. As an example, let me consider the first notion: a system N is *simply infinite* if a distinguished element 1 of N and a mapping f on N exist that satisfy the following conditions: (i) $f[N]$, the image of N under f, is a subset of N, (ii) N is the chain of the system $\{1\}$, i.e., is the intersection of all systems that contain 1 as an element and are closed under f, (iii) 1 is not an element of $f[N]$, and (iv) f is injective. The foundation for such notions is provided by *logic*; indeed, Dedekind views arithmetic, but also algebra and analysis, as a part of logic. This is to imply, he emphasizes in the Preface to (1888), that the

---

[9] Weierstrass had a similar view. In (Petri and Schappacher 2007, p. 353) one finds this remark of Weierstrass: "For the foundation of pure analysis all that is required is the concept of number, while geometry has to borrow many notions from experience. We will try here to construct all of analysis from the concept of number."

[10] The space is characterized by Dedekind as follows: Select three arbitrary non-collinear points A, B, C such that the ratios of the distances AB, AC, BC are algebraic numbers; then let the space consist of all points M for which the ratios of AM, BM, CM to AB are also algebraic numbers. Dedekind asserts that this space is everywhere discontinuous. Hilbert's corresponding model A in (1899a) is defined inductively as the class of algebraic numbers that contains 1 and is closed under addition, subtraction, multiplication, division as well as under the operation of taking the square root of $1 + t^2$, where t is already an element of A.

Introduction: A Perspective

number-concept is not dependent on "the notions or intuitions of space and time" but is rather "an immediate product of the pure laws of thought".

Dedekind finds the reason for the logical origin of the number-concept in a unique and fundamental capacity of the human mind, namely, the ability "to relate things to things, to let a thing correspond to a thing, or to represent a thing by a thing". This capacity is reflected, in mathematics with perfect generality, by the notion of *mapping* (Abbildung) between arbitrary systems. Two brief remarks on this notion are in order. First, mappings are not defined as particular systems as is done in set theory but constitute a separate category of basic logical things. Second, mappings are understood as "laws" that associate with every element of some system S a determinate thing, its image.[11] Dedekind made the conceptual shift from a restricted notion of *function* to mappings around 1872; see section 5.1 of essay (I.1). One restrictive feature of functions was due to the requirement that their domains and co-domains had to be identical, indeed, consist of numbers.[12]

The notion of function, not depending on analytic expressions relating arguments and values, is important for the structural definition of a simply infinite system. The further step of freeing it from the requirement that domain and co-domain be identical is crucial for the metamathematical investigation of these systems. Only with the notion of mapping can Dedekind's *theorem of definition by induction* be formulated in full generality and, thus, be used to obtain the existence of a canonical isomorphism between arbitrary simply infinite systems. The latter fact is fundamental for Dedekind's *complete justification* of his definition of natural numbers given in section 73 of his (1888): the elements of *any* simply infinite system can be called (or viewed as) natural numbers, when "we entirely neglect the special character of the elements, simply retaining their distinguishability and taking into account only the relations to one another in which they are placed by the ordering mapping" of that system. This reflects Dedekind's *indifference to* any *identification* of natural numbers with particular kinds of mathematical or logical objects.[13]

This attitude of *indifference to identification* also impacts Dedekind's view of the expansions of number concepts. Originally he thought of them as involving, in a first step, the creation of new mathematical objects and, in a second step, the extension of mathematical laws from the smaller to the larger domain. He later reshaped his view: laws are first articulated as characteristic conditions of a structural definition; then a system of objects satisfying them is obtained through logical construction. (This shift is

---

[11](Dedekind 1888, p. 799).

[12]Even during the first decades of the 20th century, this was a restrictive condition for functions in complex analysis, a subject called in German simply *Funktionentheorie*; the condition originally indicated an implicit contrast to operations on geometric magnitudes when, say, multiplication led from lengths to areas.

[13]See (Kanamori 2012, p. 47); the phrase is attributed to (Burgess 2009). — Dedekind would indeed object to any specific identification; cf. section 4.3 of essay (I.1).

discussed in essay (I.1), sections 3 and 4.) In fact, Dedekind's ways are those one would follow in any contemporary course on number systems.

Hilbert stands in the Dedekindian tradition of what he and Bernays called, in the 1920s, *existential axiomatics*. He presented his axiomatizations of geometry and of analysis in (1899a) and (1900b); they are not modern formal theories, but structural definitions assuming the existence of a domain or, in the case of geometry, several domains that satisfy the characteristic conditions. Hilbert sharply expresses indifference to identification when he gives an analytic model for geometry and a geometric model for analysis in (1899a). For grasping the relation between Dedekind and Hilbert's considerations, two synonymies should be kept in mind: Hilbert's axioms are Dedekind's characteristic conditions of a structural definition; Hilbert's models of axioms are Dedekind's systems falling under the corresponding structural definition.[14] There is one additional commonality between the two mathematicians, namely, their insistence on proving all that can be proved in a rigorous way — in finitely many logical steps from clearly articulated starting-points, thus respecting the nature of, what Dedekind called, our *Treppenverstand*, and what Hilbert saw as corresponding to "a universal philosophical necessity of our understanding". (Hilbert's phrase is from his (1900a); see the longer quotation in section 1 above.)

5 Rigorous proofs

The systematic development of logic and mathematics in truly *formal* theories made proof theoretic investigations possible. Bernays described this connection in the following way:

It [proof theory] is the systematic investigation of the various applications and effects of logical reasoning in those mathematical disciplines, in which concept formations and assumptions are fixed in such a way that a *strict formalization of proofs* [my emphasis] is possible using the expressive means of symbolic logic. (Bernays 1954, p. 9)

---

[14]This is expressed explicitly in Hilbert's letter to Frege when he writes on 29 December 1899, that he has nothing against renaming his axioms as characteristic conditions (Merkmale); indeed, coming back to the issue a little later, he emphasizes that it is a mere formality and just a matter of taste. See (Frege 1980, p. 12). — It is worth mentioning that Zermelo's work in his (1908c) extends Dedekind's methodological approach considering, as Hilbert had done, a domain of individuals that is taken to satisfy the characteristic conditions of a structural definition or axioms. The subject is viewed as constituting an indispensable component of the science of mathematics: "Set theory is that branch of mathematics whose task is to investigate mathematically the fundamental notions 'number', 'order', and 'function', taking them in their pristine, simple form, and to develop thereby the logical foundations of all of arithmetic and analysis; . . . " Zermelo's axiomatization and systematic development of set theory parallels the analysis and synthesis of Dedekind's work on number theory. The metamathematical side of (Dedekind 1888) had to wait until the brilliant (Zermelo 1930), where it is shown that suitable segments of the cumulative hierarchy are models of the Zermelo-Fraenkel axioms (plus large cardinal axioms) and are unique up to ϵ-isomorphism, when the size of the set of *urelements* is given and the height of the hierarchy is suitably fixed.

Introduction: A Perspective 15

The idea of formalizing arguments was of course not invented in the Hilbert School. It had a long tradition in logic and, separately, in mathematics; its origins go back to the logic of Aristotle and the Stoics and to the geometry of Euclid. Leibniz envisaged logical calculi based on a universal language or *characteristica universalis*. Influenced by Hobbes, he thought, "every work of the human mind consists in computation" and every problem, once formulated in the *characteristica universalis*, can be decided by calculation; see (Leibniz 1666). The central components of these considerations, symbolic languages and mathematical calculations, emerged ever more sharply. The further developments responded in part also to a *normative* demand for radical intersubjectivity grounded in those minimal cognitive capacities that are needed to carry out *algorithmic* or *mechanical* procedures.[15]

Frege viewed his *Begriffsschrift* as standing in the Leibnizian tradition with the more modest goal of creating a *characteristica mathematica* and tools for the construction of *formal* proofs in mathematics. To point out in what ways such proofs differ from ordinary mathematical proofs, he compares them in his *Grundgesetze* with the kind of proofs given in Euclid's *Elements*. The propositions that are used without proof should, of course, be listed as it is done in the *Elements*, but Frege demands furthermore — and in this he sees himself as going beyond the ancient geometer — "all the methods of inference used must be specified in advance. Otherwise, it is impossible to ensure satisfaction of the first demand [of proving everything that can be proved]." In addition, such proofs are epistemologically significant:

Since there are no gaps in the chains of inferences, each axiom, assumption, hypothesis, or whatever you like to call it, upon which a proof is founded, is brought to light; and so we gain a basis for deciding the epistemological nature of the law that is proved.[16]

Contentual considerations are not allowed to impact proof steps. So it must be possible to recognize the application of an inference as correct on account of the form of the sentences occurring in it and that has to be done in a perfectly algorithmic way. Indeed, Frege claims that in his logical system "inference is conducted like a calculation" and observes:

I do not mean this in a narrow sense, as if it were subject to an algorithm the same as ... ordinary addition or multiplication, but only in the sense that there is an algorithm at all, i.e., a totality of rules which governs the transition from one

---

[15] In the case of mathematics there were also *practical* demands, e.g., in navigation to determine positions and in the sciences, more generally, not only to develop comprehensive mathematical models, but also to make predictions effectively. - The efforts for joining logical and mathematical arguments in the second half of the 19th century are shown in Mugnai's (2010) to have significant and intriguing roots in the 17th century. I mention that Jacob Bernoulli in his (1685) compares algebraic with logical reasoning and claims that algebra is the true logic.

[16] That fundamental point is also made in the Preface to *Principia Mathematica* on p. vi: "We have found it necessary to give very full proofs, because otherwise it is scarcely possible to see what hypotheses are really required, or whether our results follow from our explicit premises."

sentence or from two sentences to a new one in such a way that nothing happens except in conformity with these rules. (Frege 1984, p. 237)

Frege's work concerning logical calculi influenced Hilbert mostly in an indirect way through Whitehead and Russell's *Principia Mathematica*. That *opus magnum* was decisive for Hilbert's way of transforming logic as developed there into *mathematical logic* as presented with stunning novelty in his lectures (*1917/18); see essay (I.3). Hilbert emphasized the significance of the formal features on every occasion; for example, when presenting his most famous lecture from the 1920s, *On the Infinite*, he asserted in June of 1925:

> In the logical calculus we possess a sign language that is capable of representing mathematical propositions in formulas and of expressing logical inference through formal processes. . . . Certain of these formulas correspond to the mathematical axioms, and to contentual inference there correspond the rules according to which the formulas follow each other; hence contentual inference is replaced by manipulation of signs [äußeres Handeln] according to rules, . . .

Hilbert reiterates this basic perspective in a talk of July 1927 in Hamburg[17] and actually goes further by claiming that the formal rules express "the technique of our thinking"; he continues:

> These rules form a closed system that can be discovered and definitively stated. The fundamental idea of my proof theory is none other than to describe the activity of our understanding, to make a protocol of the rules according to which our thinking actually proceeds. (Hilbert 1927, p. 475)

The logical inference principles given in this talk are axiomatic versions of the introduction and elimination rules Gentzen formulated in 1932. Details of this development are given in essay (II.3) and, in particular, the methodological and pragmatic considerations guiding Hilbert and Bernays's formulation are exposed there.

Hilbert's goal was the formal representation of mathematics, and *Principia Mathematica* provided the basis for it. A streamlined version of the system of *Principia* was used to develop analysis in (*1917/18). This formal theory was understood as having a properly foundational, logicist interpretation with the generated or constructed ramified hierarchy of types as its domain. However, the axiom of reducibility was indispensible for the mathematical development. Its detailed examination in (*1920b) led to the conclusion that the domain had to satisfy an axiomatic existential closure

---

[17]The longer quotation is found in (van Heijenoort 1967, p. 381) and the reiteration from 1927 (l.c., p. 467) is given here: "For in my theory contentual inference is replaced by manipulation of signs [äußeres Handeln] according to rules; in this way the axiomatic method attains that reliability and perfection that it can and must reach, if it is to become the basic instrument of all theoretical research." — In this quotation, but also that in the main text, "äußeres Handeln" might be more directly translated by "external operations". The remark in the Hamburg quotation is reminiscent of Leibniz's *calculus ratiocinator* that was "to bring under mathematical laws human reasoning, which is the most excellent and useful thing we have". (Leibniz's observation is quoted in (Parkinson 1966, p. 105).)

condition undermining the logical constructibility of its elements. Hilbert and Bernays viewed that as a return to (the Dedekind-Hilbert) existential axiomatics and its central foundational problem.

Dedekind had addressed in (1888) that central problem for the notion of a simply infinite system by showing that a particular system falls under it, thus guaranteeing quasi-semantically its consistency. His "logical proof" of the existence of a simply infinite system assumed that his *Gedankenwelt*, i.e., the system of everything thinkable, exists as a system; however, the notion of such a universal system is inconsistent as Cantor pointed out. Hilbert became aware of this inconsistency in 1897 through his correspondence with Cantor, but did not abandon the Dedekind-style axiomatic method. Instead, he took a step from Dedekind's quasi-semantic notion of consistency to a quasi-syntactic one, calling axioms *consistent* if no contradiction can be derived from them in finitely many logical steps (without specifying what these steps are).

The further development of Hilbert's ideas is complex and is discussed in a number of essays, for example, (I.3) and (II.2). Here I jump directly to the finitist consistency program that started in early 1922. It was to achieve a *structural reduction* for ever more encompassing parts of mathematics to a fixed, elementary, and meaningful part of itself; see section 3.1 of essay (II.10). The guiding "geometric" idea was this: *project* abstract structures into the domain of finitist mathematics by first using formal axiomatic theories instead of structural definitions and then formalizing mathematical proofs. In this way one arrives at combinatorial objects that can be investigated from the finitist standpoint with the goal of establishing the syntactic consistency of the formal theories.

6 Accessible domains

With its clear reductive goal, the finitist program was attractive to talented mathematicians like Ackermann, von Neumann, and Herbrand. Even Gödel admitted, in his 1938 lecture at Zilsel's, that Hilbert's original program would have been of "enormous epistemological value" if it could have been carried out. The main reason: "Everything would really have been reduced to a concrete basis, on which everyone must be able to agree." The goal of mediating between contrary perspectives and finding "a concrete basis on which everyone must be able to agree" was indeed Hilbert's. However, the metamathematical way of achieving it was blocked by Gödel's second incompleteness theorem, i.e., the "absolute" reductive goal turned out to be utterly elusive for a finitist.

Proof theory did not fade away, neither as a mathematical-logical discipline nor as a subject with foundational ambitions. The absolutely central idea of proving the consistency of *formal theories* through mathematical means that are in some sense more elementary than those expressed in the theories under investigation, that idea has been sustained. The methods permitted in consistency proofs have been traditionally categorized as *constructive*, because processes are carried out effectively and mathematical objects are obtained by construction. Bernays highlights these features in his (1954), as the axiomatic standpoint ultimately must be grounded in contentual knowledge or, to phrase it differently, be embedded in a *methodological frame*. To be suitable for proof theoretic work, such a frame must satisfy at least one condition: "The objects (making up the intended model of the theory) are not taken from the domain as being already given, but are rather constituted by generative processes."

When Bernays reflects in his (1937a) on constraints for frames, he takes the *arithmetic perspective in the strict sense* as central. In the background is the distinction between arithmetic and geometric knowledge:

Accordingly, arithmetical is the representation (Vorstellung) of a figure that is composed of discrete parts, in which the parts themselves are considered either only in their relation to the whole figure or according to certain coarser distinctive features that have been specially singled out; arithmetical is also the representation of a formal process that is performed with such a figure and that is considered only with regard to the change that it causes. By contrast, geometrical are the representations of continuous change, of continuously variable magnitudes, moreover topological representations, like those of the shapes of lines and areas. (p. 107)

These considerations underlie the requirement that frames must have domains of objects that are constituted through generative *arithmetic* processes. Contrasting this special form of contentual axiomatics with its general existential form, Bernays calls it *sharpened axiomatics* (verschärfte Axiomatik). The philosophical significance of consistency proofs is thus to be seen and assessed in terms of the objective underpinnings of the frames within which reductions are achieved.

A *crucial question* is, consequently, what kind of procedures can be viewed as generative arithmetic ones. The (clauses of) elementary inductive definitions of syntactic notions, like formula or proof, were initially viewed in that light. Now we have an extremely general articulation for the rule-governed or inductive generation of mathematical objects; see (Aczel 1977). It is so general that it encompasses elementary i.d. classes and constructive number classes, but also segments of the cumulative hierarchy of sets. Such i.d. classes obey the principle of proof by induction and, if they are deterministic, also that of definition by recursion.[18] I call such i.d. classes *accessible domains*; they are introduced in (II.10) and provide an opportunity for significant philosophical and mathematical work. I will return to that opportunity at the end of section 8 below.

The crucial tool for achieving a structural reduction, to reemphasize, is formalization together with the effective, finitist description of a formalism. This effective *external* description of a formalism is complemented, in a subtle and important way, by an *internal* feature concerned with the evaluation of number theoretic functions that are included in the formalism;

---

[18]The elements of deterministic i.d. classes have unique and well-founded construction trees. It should be noted that (the graphs of) recursive functions can also be defined inductively.

Introduction: A Perspective 19

the functions may have been specified by recursion. Such evaluations are crucial steps in the consistency proofs for fragments of arithmetic given in early 1922; see essays (I.3) and (II.1). They systematically use the equational rules of the fragments and are finitistically shown to terminate for any set of arguments. This was *the* fundamental idea of the new proof theory, but its reach was limited. The reason for its limitation was uncovered by Gödel: as soon as a consistent formalism is strong enough to internalize its effective description, it is incomplete and requires, for a proof of its consistency, means that are stronger than those that can be formalized in it.

Gödel established these claims specifically for (a version of) the system of *Principia Mathematica*. To prove them in the generality indicated in the last sentence of the previous paragraph, one has to characterize the very notion of formalism or formal system in a rigorous way. In the early 1950s, Gödel started to point to the work in (Turing 1936) as providing "a precise and unquestionably adequate definition of the general concept of formal system" and claimed that it secures the most general formulation of his incompleteness theorems:

. . . the existence of undecidable arithmetical propositions and the non-demonstrability of the consistency of a system in the same system can now be proved rigorously for *every* consistent formal theory containing a certain amount of finitary number theory. (Gödel 1964, p. 369)

Our understanding of the strict regimentation of (proofs in) a formal system is informed by the notion of "mechanical procedure" to be explored next. This exploration begins with a brief remark about Herbrand's 1931 paper, *Sur la non-contradiction de l'arithmétique*, and then sketches the developments surrounding the notion of effectively calculable function in the 1930s (before turning to Turing's work).

7 Mechanical procedures

Herbrand characterized in his (1931) classes of finitistically calculable functions in a form that also covers non-primitive recursive functions, like the Ackermann function. Building on Herbrand's formulation, Gödel arrived in his (1934) at the notion of *general recursive function* in two steps.[19] The first step isolates two simple calculation rules in a purely equational calculus (the rule of replacing variables by numerals and that of substituting complex closed terms by their numerical values). The second, very radical step replaces the demand that the termination of a calculation procedure be finitistically provable by the requirement that the value of a function, as a matter of classical fact, be calculable in finitely many steps (for any set of arguments). Gödel gives a much-simplified version of his equational calculus in the manuscript *Undecidable Diophantine propositions* (Gödel 193?) and argues there that arithmetic calculations are

---

[19] As to the connection between Herbrand's and Gödel's definitions, see their correspondence in volume V of Gödel's *Collected Works*.

captured correctly and adequately by deductions in his calculus. Thus, he implicitly accepted at that point the identification of effectively calculable functions with general recursive ones, an identification that had already been urged by Church in his abstract (1935).

In section 7 of his classical paper (1936), Church tried to give a convincing argument, if not outright proof, that all effectively calculable functions are indeed general recursive because their values can be calculated "in a logic". Gödel, in a very similar spirit, claimed in the Postscriptum to his note (1936) that general recursive functions are *absolute*, i.e., if their values can be calculated in a theory of finite or transfinite types, then they can be calculated already at the lowest level, namely, in arithmetic. Church's and Gödel's observations can be rigorously proved when the deductive steps (in a logic for Church and in type theories for Gödel) are not only elementary in some informal sense, but general recursive. Thus, there is a subtle circularity when insisting, at the same time, that these observations establish the adequacy of general recursiveness to capture the informal notion of effective calculability.

Hilbert and Bernays analyzed in *Supplement II* of their (1939) the concept of *reckonable function* (regelrecht auswertbare Funktion) via "deductive formalisms" that satisfy *recursiveness conditions*; the crucial condition requires the proof predicate of such formalisms to be primitive recursive. Their concept of reckonable function is shown to be co-extensional with that of general recursive function. This was a most fitting closure for one *significant* way of thinking about calculations, namely, as deductions in a formalism using steps of restricted complexity. It avoids, in a way, Church's and Gödel's subtle circularity by explicitly imposing the restrictive recursiveness conditions on formalisms; however, Hilbert and Bernays gave no intrinsic reasons why these conditions should be adopted.[20]

Though a full, rigorous characterization of mechanical procedures is important for a *general* formulation of the incompleteness theorems, significant cases can be proved without it; Gödel's (1931d) does exactly that for his version of the system of *Principia Mathematica*. For a negative solution of the *Entscheidungsproblem*, however, an adequate notion is essential.[21] After all, any mathematical argument, appealing to a rigorously defined class of mechanical procedures, establishes only that no element *of this particular class* solves the problem. To justify the *further claim* that a negative solution of the decision problem has been obtained, good reasons have to be given that the defined class encompasses *all* mechanical procedures.

---

[20] Gödel generalized the observation on the absoluteness of general recursive functions in his (1946) by considering instead of type theories *any* formal theory extending arithmetic. The subtle circularity is even more obvious here, as the character of *formal* theories has to be exploited in any proof of the observation. For a detailed account of these and further developments, see my paper *On Computability*, i.e., (Sieg 2009a).

[21] The *Entscheidungsproblem* was viewed by Hilbert as one of the most important problems of mathematical logic. Its positive solution would have provided a mechanical procedure for deciding, in finitely many steps, whether a formula in the language of predicate logic is provable or not.

Turing was inspired by lectures of Newman to tackle the *Entscheidungsproblem*; he introduced in his (1936) idealized computing machines and proved that procedures executable by them do not solve the decision problem. He also attempted to provide the required good reasons in section 9 of his paper, but acknowledges that his considerations are not a proof in a strictly mathematical sense.

From a philosophical perspective, however, Turing does something very important: he takes on the challenge of characterizing mechanical procedures that can be carried out by a (human) computer.[22] It is the dramatic novelty of Turing's approach that he brings the computing agent, a human being, into the analysis; that is exactly right, as we are to carry out decision procedures. The human agent's sensory and memory limitations allow Turing to argue that the elementary steps taken in mechanical procedures must be *bounded* and *local* in a strict sense. When the basic configurations on which computers operate are strings of symbols, one can *prove* that the restricted steps are recursive. For the development of computability theory and its use as the basis for the emerging subject of computer science, it was crucial that Turing had introduced a theoretical notion that could serve as the model of "machine computations".[23] The very model allowed, as Turing had shown, the construction of a "universal machine"; that was of extreme practical importance as its architecture was adopted for actual digital computing machines.

Towards the end of the *centuria mirabilis*, in February of 1954, Turing's paper *Solvable and unsolvable problems* was published.[24] It describes the mathematical-logical result Turing had obtained in 1936 and analyzes its conceptual foundation in terms of procedures on possibly three-dimensional discrete symbolic configurations, "puzzles", and their reduction to operations on strings, "substitution puzzles". Substitution puzzles are a form of Post's production systems. Already in the 1920s, Post had used such systems and their canonical versions to make precise the concept of a mechanically generated set (of strings of symbols). Post's approach naturally dovetails with Turing's considerations in (section 9 of) his 1936-paper, as the analysis of human calculability ends there quite literally with a production system, which is then easily seen to be executable by a Turing machine.[25] Conversely, Post presents in his (1947) Turing machines via

---

[22] That is Turing's term, in accord with usage in the 1930s, for a human being carrying out an algorithmic procedure. Computers in our sense are always called machines.

[23] Interestingly, Turing's basic considerations can be carried over to machines, when the latter are conceived as "discrete mechanical devices". The psychological limitations of human computers are replaced by physical limitations of mechanical devices; see (Gandy 1980) and (Sieg 2009a).

[24] 1954 is also the year of Turing's tragic death; he committed suicide on June 7.

[25] By 1954, Turing had learned of Post's work: after all, he expanded the result of (Post 1947) in his own (1950b); there he expressed his appreciation of Post's conceptual work by referring in the introductory paragraph, p. 491, to the "logical computing machines introduced by Post and the author". — This connection is described in (Sieg 2012); cf. also (Stillwell 2004) and (de Mol 2006).

production systems in a mathematically fruitful and elegant way. This Post-Turing approach to effectiveness is in full harmony with the syntactic descriptive one of Hilbert's metamathematics, more so than that based on the arithmetization of syntax in the tradition of Gödel's 1931-paper.[26]

In the next section, I will appeal to the original target of Turing's analysis and to the limitative results in order to elucidate aspects of the mathematical mind. To reemphasize, Turing's work successfully analyzed the notion of human calculability and, for that reason, is so pertinent and important for the task ahead.

8 Outlook: Formality & mind

When Gödel, in his (1964), sees Turing's analysis as grounding the general formulation of the incompleteness theorems, he also remarks that these results "do not establish any bounds for the powers of human reason, but rather for the potentialities of pure formalism in mathematics". That is in line with his reflections in the manuscript (193?) where he agrees with Hilbert's claim (oft repeated since 1900) that *all* precisely formulated mathematical problems are in principle solvable. Although the claim seems to be in conflict with his theorems, Gödel argues that it is not: his theorems only imply that even in number theory it is not possible "to formalize mathematical evidence". "The conviction about which Hilbert speaks", he continues, "remains entirely untouched." But how is it that, in mathematics, the human mind can transcend the limits of pure formalism?

In the earlier sections 3 and 4, I pointed to elements of an innermathematical change that expanded and indeed transformed the subject in the 19th century. This transformation revealed, as Stein observes in his (1988), a *capacity of the human mind* that had been discovered already in Greece between the 6th and 4th century BC. Stein emphasizes that the 19th century rediscovery of this capacity teaches us something new about its nature and claims, furthermore, that what has been learned "constitutes one of the greatest advances in philosophy". In order to grasp the nature of the philosophical advance (which is not made explicit by Stein), we have to deepen our understanding of the mind's *mathematical* capacity or, rather, of the mind's capacities as they come to light in mathematics and its use. Hilbert addressed both these facets in his own ways: he asserted in the 1920s, as mentioned in section 5 above, that the formal representation of mathematics uncovers the rules according to which our thinking actually proceeds, and he emphasized throughout his career that mathematics bridges thought and experience.

The undecidability and incompleteness results bound the reach of mathematical methods, when the latter are constrained to be mechanical procedures codified by Turing's machines. They also limit Hilbert's assertion

---

[26] Gödel's original proof of a version of his first incompleteness theorem did not involve an arithmetization of syntax; see essay (II.3).

concerning actual mathematical thinking and, at the same time, they seem to open the door to searching for "transcendent" capacities of the mind in mathematics. Understanding by "reason" something close to Dedekind's *Treppenverstand*, Turing attests to such an opening in the last sentence of his (1954): "These [unsolvable problems] may be regarded as going some way towards a demonstration, within mathematics itself, of the inadequacy of 'reason' unsupported by common sense." Turing explored ways of extending formal reason by common sense in his manuscripts and papers on intelligent machinery from the late 1940s, and so did, in a certain way, Gödel and Bernays. Whereas Turing emphasized individual initiative and cultural context as crucial for creative mathematical work, Gödel focused on a deeper understanding of abstract notions (in particular of the concept of set), and Bernays pointed to dynamic aspects of the human mind. Such issues and related directions for research are discussed in essay (III.3); here I want to remain focused on the connection to the 19th century transformation of mathematics.

Links between (a more adequate) understanding of a field of inquiry and dynamic aspects of the human mind are explicitly made in Dedekind's writings. In the Preface to his (1888), he refers to his *Habilitationsrede* from 1854. There he had claimed that the imperfection of our intellectual powers necessitates the introduction of pertinent notions in the sciences, as well as in mathematics. After all, we frame the object of a science by introducing concepts and in this way formulate hypotheses on its inner nature. How well our concepts capture that nature is determined by their usefulness for scientific developments; in mathematics, their efficacy is shown by the possibility and ease of constructing proofs. Dedekind's foundational essays can be taken to formulate hypotheses on the nature of the continuum and of natural numbers by introducing, respectively, the notion of a complete ordered field and that of a simply infinite system. The characteristic conditions of these concepts are the starting-points for stepwise arguments with our *Treppenverstand*: structural notions and detailed proofs are tools for widening the reach of our intellectual powers. Ironically then, the human mind has developed effective approaches to partially overcome its own imperfection.

Hilbert's axiomatic method reflects these central elements of Dedekind's thought. That can be seen in many of Hilbert's writings, but perhaps most clearly and extensively in his lectures (*1921/22). The axiomatic method, he emphasizes there, does not attempt to gain absolute certainty that is transferred from axioms to theorems, but rather separates the investigation of logical connections from the question of factual truth. The construction of an axiomatic theory is viewed as a procedure of mapping a field of inquiry to a *Begriffsfachwerk*[27] in such a way that "the concepts correspond to

---

[27]*Fachwerk* is usually taken as part of the compound word *Fachwerkhaus*, i.e., timber-frame or half-timbered house; however, it has a more general meaning in engineering: it is a construction consisting of several rods that are connected to each other at both ends. Apart from its instantiation in a *Fachwerkhaus*, it can also be seen in bridges and cranes.

the objects of the domain of knowledge and the logical relations between the concepts correspond to the statements concerning the objects". This mapping allows the purely logical investigation of connections in a theory; Hilbert asserts:

> The theory has nothing to do anymore with the real objects or with the intuitive content of knowledge (Erkenntnis); it is a pure thought construct (Gedankengebilde) of which one cannot say that it is true or false. And yet, this *Begriffsfachwerk* is significant for gaining knowledge of reality, as it constitutes a possible form of real connections.[28]

Mathematics has the task of developing such *Begriffsfachwerke* independently of whether they have been motivated by experience or by systematic speculation.[29] In this way Hilbert deepens one of the important themes expressed already in his Paris talk; cf. section 1 above.

Hilbert had used the very term already early on, for example in his letter of 29 December 1899 to Frege; he writes there (Frege 1980, p. 13): "Ja, es ist doch selbstverständlich eine jede Theorie nur ein Fachwerk von Begriffen oder ein Schema von Begriffen nebst ihren notwendigen Beziehungen zu einander, und die Grundelemente können in beliebiger Weise gedacht werden." Here is an English translation: Well, any theory is evidently only a *Fachwerk* of concepts or a schema of concepts together with their necessary relations to each other, and the basic elements can be thought in arbitrary ways.

From Dirk Schlimm I learned that Hilbert had used *Fachwerk* already in his (*1894a). The remarks Hilbert made in those lectures are close in spirit to Dedekind's in (1854). Hilbert asserts that every science aims to "order" the facts in its domain or to "describe" its phenomena and continues: "Dieses Ordnen oder Beschreiben geschieht vermittels gewisser Begriffe, die durch die Gesetze der Logik unter sich zu verknüpfen sind.

Die Wissenschaft ist desto fortgeschrittner, d.h. das Fachwerk der Begriffe ist desto vollkommner, je leichter jede Erscheinung oder Tatsache untergebracht wird. Die Geometrie ist eine Wissenschaft, welche im Wesentlichen so weit fortgeschritten ist, dass alle ihre Thatsachen bereits durch logische Schlüsse aus früheren abgeleitet werden können."

Schlimm's translation, slightly modified, is given now: "This ordering or describing is done with certain concepts that are to be interconnected by the laws of logic.

A science is the more advanced, i.e., the *Fachwerk* of concepts is the more complete, the easier it is to accommodate any phenomenon or fact. Geometry is a science that has essentially advanced so far that all of its facts can already be deduced from earlier ones by logical inferences."

[28]These remarks are quoted in their German original in Note 50 of essay (III.2). — In his (1922a), Bernays describes these connections for Hilbert's thoughts on geometry; in his (1930b), he does so generally.

[29]The task of developing *Begriffsfachwerke* has to be complemented by methodological reflections that connect them to "reality" and thus analyze more deeply the role mathematics plays in the sciences. This is explicit in Dedekind's remarks about geometry, and clearly at the core of the modern "structuralism in the sciences" as investigated by Suppes, van Fraassen and others.

Hilbert's Programs should be triangulated with mathematics, philosophy and the sciences. Here is a brief indication of why; for a fuller account see (Hilbert 2009) and (Corry 2004). Hilbert insisted, in a talk he gave in Copenhagen in March of 1921, that mathematics has two faces: one is directed towards philosophy, the other to the sciences. These two faces are indicated when Hilbert answers the question in what ways a physicist uses mathematical analysis: "... erstens zur *Klärung und Formulierung* seiner Ideen und zweitens als *Instrument* der Rechnung zur raschen und sicheren Gewinnung numerischer Resultate, durch die er die Richtigkeit seiner Ideen prüft." (Hilbert 1921, Cod. Ms. 589, p. 29) Here is an English translation: ... first of all, in order to *clarify and formulate* his ideas and secondly as a computational *instrument* to quickly and safely obtain numerical results by means of which he checks the correctness of his ideas.

Introduction: A Perspective

For *Begriffsfachwerke* possibly to serve as images (Abbilder) of real connections, a minimal condition has to be satisfied, namely, "The provable claims of the theory must not contradict each other, i.e., the theory must be possible in itself; thus, the consistency problem arises." In a full circle, we are back at the beginning of Hilbert's foundational investigations and their connection to Dedekind's, when Hilbert viewed in 1899 the consistency problem for the arithmetic of real numbers as absolutely central. Now in late 1921, he still considers the consistency issue "as the most important and most difficult problem for the investigation of axiom systems" and is about to confront it for elementary number theory. He and Bernays tackle, in early 1922, this restricted problem in a completely new way from within the methodological frame of finitist mathematics.

When focusing on the foundational problems for 19th century mathematics and viewing them in light of the experience we have gained, one is naturally led to a unified perspective on Hilbert's Programs. That perspective takes a more abstract stance when reflecting on the "crucial question" of section 6 concerning generative arithmetic procedures. Is that question, prominently raised by Bernays, tied to the opposition of geometry and arithmetic? I think not: the *systematic issue* is broader and directly connected to two aspects of mathematical experience. There is, on the one hand, the *conceptual* aspect; it involves abstract notions like that of a topological space or a group, which have many different interpretations and can be used to articulate analogies. There is, on the other hand, the *constructive* one; it refers to particular kinds of objects like natural numbers and ordinals that are uniquely generated in accord with rules (understood very broadly as indicated in section 6). The first aspect allows us to shape proofs with *conceptual understanding*, whereas the second aspect allows us to consider mathematical objects that obey *principles* grounded in their uniform generation.

This broader systematic issue raises mathematical and philosophical questions, some of which are indicated in essay (III.2). It remains a topic for continued work under the heading *reductive structuralism*.[30] Based on the two aspects of mathematical experience just described, reductive structuralism distinguishes between two types of structural definitions and, correspondingly, two types of structures, namely, abstract ones whose domains contain "structureless" points and accessible ones whose domains contain objects with an intrinsic "internal structure". The "internal structure" of the

---

[30] The contemporary discussion of structuralism has two quite separate forums: mathematics and philosophy. In mathematics it goes back to Dedekind, Hilbert, Noether, and many other mathematicians in the 1920s, e.g., van der Waerden and Hasse. It led to Bourbaki's structuralism and, ultimately, to category theory; see section 6 of essay III.1. Bernays, already in the 1930s, stated this view clearly, when considering mathematics as "the science of idealized structures". My considerations try to bridge the gap between the *structuralism of mathematical practice* and the *structuralism of philosophical analysis*. The latter is almost exclusively concerned with structures that are categorically fixed; indeed, it mostly deals with epistemological and metaphysical issues for natural numbers. Let me note that models of complete ordered fields ("real numbers") are also unique up to isomorphism; but their domains are not accessible. Thus, the notion of a complete ordered field is an abstract one.

objects in such domains reflects their deterministic generation and guarantees that accessible domains are unique up to a canonical isomorphism. Accessible domains are fruitful targets of mathematical study and systematic theoretical development, but they are also used to define models of abstract notions and guarantee their consistency. In a deeply related way, they can provide a unifying frame for mathematics.[31] Hilbert emphasized this latter point with respect to set theory when remarking in his lecture course (*1920b):

> Set theory encompasses all mathematical theories (like number theory, analysis, geometry) in the following sense: the relations that hold between the objects of one of these mathematical disciplines are represented in a completely corresponding way by relations that obtain [between objects] in a sub-domain of Zermelo's set theory.

Once abstract structures and accessible domains have been distinguished, the philosophy of mathematics is confronted with two central tasks, namely,
(i) to investigate the function of abstract structures in mathematical practice and the role they play for mathematical understanding,
and
(ii) to analyze the function of accessibility notions in foundational theories and the role they play in grounding principles.
The position between mathematics and philosophy is the perfect place for mathematically informed and philosophically sophisticated reflection.[32] Here we also have a concrete starting-point for exploring the philosophical advance Stein saw as having been brought about by the transformative development of mathematics; it is rooted in and generalizes the approach Hilbert and Bernays took through their structural reduction. In essay (III.2), I quote a comment of Stein's on Hilbert. Having gently complained about Hilbert's insistence, in his later foundational writings, that the statements of ordinary mathematics are meaningless and only finitist statements have meaning, Stein points out:

> Hilbert certainly never abandoned the view that mathematics is an organon for the sciences . . . ; and he surely did not think that physics is meaningless, or its discourse a play with "blind" symbols. His point is, I think, this rather: that the mathematical *logos* has no responsibility to any imposed *standard* of meaning: not to Kantian or Brouwerian "intuition", not to finite or effective decidability, not to anyone's

---

[31]Both types of considerations cut across foundational standpoints. As to the first type, see the discussion in essay (III.2) at the very end of section 5 and in footnote 69; as to the second, one can make for, e.g., Martin-Löf's type theory and constructive mathematics, a similar claim as Hilbert does, in the next quotation, for Zermelo's set theory and classical mathematics.

[32]The converse of a remark of Poincaré's should guide such reflection. The soundness of that remark, used as the motto of essay (II.6), still impresses me as it did when I read it for the first time many years ago. Here is his remark: "Douter de tout ou tout croire, ce sont deux solutions également commodes qui l'une et l'autre nous dispensent de réfléchir." The meaning of the elegant French phrase is less elegantly given by: "To doubt all or to believe all, these are two equally facile solutions that both dispense us from reflecting."

Introduction: A Perspective

metaphysical standards for "ontology"; its *sole* "formal" or "legal" responsibility is to be consistent (of course, it has also what one might call a "moral" or "aesthetic" responsibility: to be useful, or interesting, or beautiful; but to this it cannot be constrained — poetry is not produced through censorship). (Stein 1988, p. 255)

The mathematical and philosophical challenges are, of course, to gain a more profound understanding of how mathematics is integrated with our cognitive capacities, thus, can be *useful*, and to analyze on the basis of what we can actually live up to the responsibility of being *consistent*. These are exactly the central issues for reductive structuralism.[33]

---

[33] This *Perspective* was "in the works" for a long time. Several colleagues and students made valuable suggestions, offered gentle encouragement, and gave critical advice. I am grateful to C. Benzmüller, W. Buchholz, I. Cervesato, P. Corvini, S. Feferman, J. Floyd, A. Kanamori, E. Livnat, R. Lupacchini, U. Majer, P. Martin-Löf, E. Menzler, R. Morris, C. D. Parsons, V. Peckhaus, M. Rathjen, D. Schlimm, O. Shagrir, W. W. Tait, and M. van Lambalgen.

In.2

Milestones

Hilbert's life and work are portrayed in Constance Reid's *Hilbert*. This excellent biography from 1970 gives a vivid impression of the influences on Hilbert; it depicts his wide-ranging interactions with colleagues across fields and nationalities, and it describes his impact on a remarkable group of doctoral students. This group included Otto Blumenthal and Hermann Weyl, who deepen the appreciation of Hilbert's mathematical and scientific work with their (1935), respectively (1944). A concise discussion of Hilbert's work on the foundations of mathematics is given in (Bernays 1967); contemporary accounts of that work are, of course, woven into all the essays. As a complement to the broad *Perspective*, I am listing here the *Milestones* in Hilbert's foundational work: significant lecture courses, important addresses, and crucial publications. The list provides a skeletal historical structure. (The bibliographical details are found in the *Bibliography* to this volume.)

1889-1900: Lectures on geometry, quadrature of the circle, and the concept of number; among them are (*1894a), (*1894b), (*1897b), and (*1898/99). This sequence of lectures culminated in two publications, namely, *Grundlagen der Geometrie* (1899a) and *Über den Zahlbegriff* (1900b). The intricate background for (1900b) is examined in essay (I.2); Hilbert's paper articulates forcefully the consistency problem for the arithmetic of real numbers. The Second Problem in (1900a) expands on this formulation.

1900 Mathematische Probleme, address to the International Congress of Mathematicians, Paris; (Hilbert 1900a).

1904 Über die Grundlagen der Logik und der Arithmetik, address to the International Congress of Mathematicians, Heidelberg; (Hilbert 1905a).

1904-1917: In his Heidelberg talk, Hilbert suggested a new way of addressing the consistency problem and exemplified it for a small fragment of the arithmetic of natural numbers. That syntactic, "proof theoretic" approach was not pursued again until 1920, most likely on account of Poincaré's pertinent criticism. There is no relevant publication until 1918, in spite of the fact that Hilbert continued to be engaged with foundational issues. In the summer semester of 1905, he gave a broad and reflective course with the title *Logische Prinzipien des mathematischen Denkens*, (*1905b). The concern with the foundations of mathematics is further witnessed by his lectures (*1908), (*1910), (*1913), and (*1914/15). In the summer semester of 1917, he gave a course on set theory and indicated at its end that he would "examine more closely a foundation for logic" in the following term. His pivotal Zürich talk preceded the promised new course.

1917 Axiomatisches Denken, talk to the Swiss Mathematical Society, Zürich; (Hilbert 1918).

1917/18 Prinzipien der Mathematik, lectures given in the winter semester 1917/18.

1917-1922: *Prinzipien der Mathematik* was the course Hilbert had announced at the end of (*1917); its content and context are described in essay (I.3). It was the very first course in properly mathematical logic and marked the beginning of Hilbert's collaboration with Bernays. *Prinzipien der Mathematik* was followed by (*1920a), (*1920b), (*1921/22), and (*1922/23); these courses allow us to see the internal development toward proof theory and the finitist consistency program. The initial part of the metamathematical development toward proof theory was indicated in talks Hilbert gave in Copenhagen and Hamburg in the spring and summer of 1921, published as (Hilbert 1922). The dramatic methodological step to the finitist consistency program was taken only in February of 1922. The full program was then presented, for the first time outside of Göttingen, in Hilbert's Leipzig address of 1922. Bernays remarked in his (1935a, p. 204): "With the presentation of proof theory as given in the Leipzig talk the principled form of its structure had been reached."

1922 Die logischen Grundlagen der Mathematik, address to the German Society of Natural Scientists, Leipzig; (Hilbert 1923).

1925 Über das Unendliche, talk to the Westphalian Mathematical Society, Münster; (Hilbert 1926).

1927 Die Grundlagen der Mathematik, talk to the Mathematical Seminar, Hamburg; (Hilbert 1927).

1928 Probleme der Grundlegung der Mathematik, address to the International Congress of Mathematicians, Bologna; (Hilbert 1928).

The Bologna address articulated, in an impressively clear way, central problems of mathematical logic, from the consistency of analysis to the semantic completeness of first-order logic. In that year, the book with Ackermann,

Grundzüge der theoretischen Logik was also published; it is a modestly expanded version of the lecture notes (*1917/18) incorporating also parts of (*1920a) and (*1921/22). In the summer of 1929, Hilbert gave a course on set theory that builds on (*1917), but includes in its last third a detailed discussion of proof theoretic investigations; it has been preserved in (*1929). The nexus of the following three talks with Gödel's incompleteness theorems is fascinatingly complicated and is discussed in essay (II.3). The very last talk (1931b) had a decisive effect on Gentzen's proof theoretic investigations, which began in late 1931 and ultimately led to his consistency proof for elementary arithmetic, (Gentzen 1936).

1930 Naturerkennen und Logik, address to the Society of German Scientists and Physicians, Königsberg; (Hilbert 1930).

1930 Die Grundlegung der elementaren Zahlenlehre, talk to the Hamburg Philosophical Society; (Hilbert 1931a).

1931 Beweis des tertium non datur, talk to the Göttingen Society of Science; (Hilbert 1931b).

Finally, there are two volumes of the superb book with Bernays, *Grundlagen der Mathematik*, (Hilbert and Bernays 1934 and 1939). The first volume together with sections 1 through 3 of the second volume give a refined presentation of the proof theoretic work that had actually been done in the 1920s and early 1930s. It is discussed in essay (II.1). The remainder of the second volume describes the changed situation after Gödel's and Gentzen's work; see essay (II.5).

The problem of rigorously proving the consistency of a mathematical theory holds Hilbert's foundational work together. In contrast to Dedekind and logicians of the 19th century, Hilbert viewed consistency early on, definitely since 1899, as a quasi-combinatorial notion: no contradiction can be obtained from the axioms in *finitely many logical steps*. In his persistent attempt to solve the consistency problem, Hilbert held *prima facie* dramatically different positions. First, around 1900, he pursued a kind of *Dedekindian logicism* and, almost 20 years later, he took quite seriously Russell's attempt of founding mathematics in logic. Second, in 1920 and 1921, he adopted *Kroneckerian constructivism* for a restricted mathematical base theory that was then to be expanded stepwise to fully cover mathematical practice; each step was to be secured through a consistency argument. (His (1931b) can be viewed as taking up that approach.) Third, *Hilbertian finitist proof theory* was adumbrated in 1904 without a comprehensive framework for formalizing mathematical practice and without a clear sense for the need to restrict metamathematical means. The finitist approach was articulated in early 1922 and underlies of course what is usually called *Hilbert's Program*. We are dealing with rich and fascinating intellectual issues — not only as historical phenomena, but also as inspiring sources for contemporary mathematical and philosophical work.

I Mathematical roots

A brief guide

In 1980, I began writing my first essay, (Sieg 1984), on proof theory and the foundations for analysis. Hilbert's work in proof theory was seen at the time as being centered on the *Grundlagenstreit* with Brouwer and Weyl. To my astonishment I discovered soon that his foundational work was deeply rooted in 19th century mathematics. Those roots went deeper than establishing a connection to the paradox of Russell and Zermelo, which had prompted Hilbert's Heidelberg talk of 1904. That talk actually foreshadowed the later proof theoretic approach, and it seemed to be just a part of the broad dialectical context for the debate in the 1920s.

Naturally, I was led to Hilbert's second problem in his Paris lecture of 1900. It called for a consistency proof of arithmetic (i.e., here, analysis) and appeared to be the culmination of the 19th century quest for rigor and secure foundations through the "arithmetization of analysis". For the arithmetic axioms Hilbert referred to his paper *Über den Zahlbegriff* that had been completed only a few months earlier. This brief, and in many respects mysterious, paper challenged my perspective. Why was the focus on real numbers, conceived as elements of a complete ordered field? Why was completeness formulated as it was? Why was the genetic method so sharply opposed to the axiomatic? How, finally, did Hilbert envision proving the consistency of these axioms?

Mysteries remained then or opened up later, when I attempted to gain a deeper understanding of central problems and developments. My 1984 essay, which is reprinted as (II.6), focused on arithmetization as *the* foundational issue for Hilbert. The more radical methodological changes in 19th century mathematics became topical only in my (1990a), where I pointed out the connection between Dedekind's essays (1872c) and (1888), but also his deep influence on Hilbert's axiomatic method. Essay (I.1), written with Dirk Schlimm, uses material from the Dedekind Nachlass to grasp better the stunning novelty of his reflections. Some of the earlier mysteries are resolved in Essay (I.2), whereas material from the Hilbert Nachlass is used in essay (I.3) to throw a surprising light on the origins of mathematical logic. The latter essay provides in detail the background for finitist proof theory as articulated for the first time in 1922.

Richard Dedekind

Leopold Kronecker

David Hilbert

I.1

Dedekind's analysis of number: systems and axioms

(with Dirk Schlimm)

> Treated philosophically,
> it [mathematics] becomes a part of philosophy.
> Herbart*

1 Introduction

In 1888 Hilbert made his *Rundreise* from Königsberg to other German university towns. He arrived in Berlin just as Dedekind's *Was sind und was sollen die Zahlen?* had been published. Hilbert reports that in mathematical circles everyone, young and old, talked about Dedekind's essay, but mostly in an opposing or even hostile sense.[1] A year earlier, Helmholtz and Kronecker had published articles on the concept of number in a *Festschrift* for Eduard Zeller. When reading those essays in parallel to Dedekind's and assuming that they reflect accurately more standard contemporaneous views, it is easy to understand how difficult it must have been to grasp and

---

*Herbart, as quoted in (Scholz 1982, p. 437).

[1] In (Hilbert 1931a, p. 487): "Im Jahre 1888 machte ich als junger Privatdozent von Königsberg aus eine Rundreise an die deutschen Universitäten. Auf meiner ersten Station, in Berlin, hörte ich in allen mathematischen Kreisen bei jung und alt von der damals eben erschienenen Arbeit Dedekinds *Was sind und was sollen die Zahlen?* sprechen — meist in gegnerischem Sinne. Die Abhandlung ist neben der Untersuchung von Frege der wichtigste erste tiefgreifende Versuch einer Begründung der elementaren Zahlenlehre." On this trip Hilbert visited also Paul du Bois-Reymond who told Hilbert "die dedekindsche Arbeit 'Was sollen Zahlen' sei ihm grässlich". In Hilbert's report, Cod. Ms. 741, 1/5 and also mentioned in (Dugac 1976, p. 203).

In the preface to the second edition of his (1888), Dedekind reports: "Die vorliegende Schrift hat bald nach ihrem Erscheinen neben günstigen auch ungünstige Beurteilungen gefunden, ja es sind ihr arge Fehler vorgeworfen. Ich habe mich von der Richtigkeit dieser Vorwürfe nicht überzeugen können und lasse jetzt die seit kurzem vergriffene Schrift, zu deren öffentlicher Verteidigung es mir an Zeit fehlt, ohne jede Änderung wieder abdrucken, indem ich nur folgende Bemerkungen dem ersten Vorwort hinzufüge." The preface was written in August 1893.

appreciate Dedekind's remarkably novel and thoroughly abstract approach. This is true even for people sympathetic with Dedekind's ways. Consider, for example, the remark Frobenius made in a letter of 23 December 1893 to Dedekind's collaborator and friend Heinrich Weber who was planning to write a book on algebra:

> I hope you often walk on the paths of Dedekind, but avoid the too abstract corners, which he now likes so much to visit. His newest edition contains so many beauties, §173 is highly ingenious, but his permutations are too disembodied, and it is also unnecessary to push abstraction so far.[2]

This remark was made by someone who refers to Dedekind as "our admired friend and master". The use of permutations, i.e., isomorphisms, in Dedekind's algebraic investigations is systematically related to the use of similar mappings in *Was sind und was sollen die Zahlen?* (The introduction of the general concept of mapping and its structure-preserving variety for mathematical investigations is perhaps the methodologically most distinctive and most radical step in Dedekind's work.)

Dedekind was well aware that such difficulties would arise. In the preface to the first edition of his (1888) he writes that anyone with sound common sense can understand his essay and that philosophical or mathematical school knowledge is not needed in the least. He continues, as if anticipating the reproach of having pushed mathematical abstraction and logical analysis too far:

> But I know very well that many a reader will hardly recognize his numbers, which have accompanied him as faithful and familiar friends all his life, in the shadowy figures I present to him; he will be frightened by the long series of simple inferences corresponding to our step-by-step understanding, by the sober analysis of the sequence of thoughts on which the laws of numbers depend, and he will become impatient at having to follow proofs for truths which to his supposed inner intuition seem evident and certain from the very beginning.[3]

---

[2]Frobenius refers to Dedekind's investigations in Supplement XI of (Dirichlet 1863). The letter is found in (Dugac 1976, p. 269). Here is the German text: "Hoffentlich gehen Sie vielfach die Wege von Dedekind, vermeiden aber die gar zu abstrakten Winkel, die er jetzt so gern aufsucht. Seine neueste Auflage enthält so viele Schönheiten, der §173 ist hochgenial, aber seine Permutationen sind zu körperlos, und es ist doch auch unnöthig, die Abstraktion so weit zu treiben."

[3](Ewald 1996, p. 791). "Aber ich weiß sehr wohl, daß gar mancher in den schattenhaften Gestalten, die ich ihm vorführe, seine Zahlen, die ihn als treue und vertraute Freunde durch das ganze Leben begleitet haben, kaum wiedererkennen mag; er wird durch die lange, der Beschaffenheit unseres Treppenverstandes entsprechende Reihe von einfachen Schlüssen, durch die nüchterne Zergliederung der Gedankenreihen, auf denen die Gesetze der Zahlen beruhen, abgeschreckt und ungeduldig darüber werden, Beweise für Wahrheiten verfolgen zu sollen, die ihm nach seiner vermeintlichen inneren Anschauung von vornherein einleuchtend und gewiß erscheinen." Dedekind expressed such sentiments also in a letter to Klein written on 6 April 1888; the letter is contained in Appendix XXV of (Dugac 1976, pp. 188–189). Even a quite positive review like that by Meyer remarks: "Der Verfasser sieht bei seinen Darlegungen von specifischen mathematischen Kenntnissen völlig ab, er wendet sich demgemäss an jeden Gebildeten. [. . .] Für unsere Vorstellung allerdings sinken die gemeinhin Zahlen genannten Dinge vermöge der erwähnten Abstractionen zu blossen Schatten herab, dafür sind sie aber

Dedekind arrived at his approach only after protracted labor as he emphasized in his letter to Keferstein dated 27 February 1890; in this letter Dedekind defended his essay against Keferstein's critical review (1890). Indeed, Dedekind had started to develop his views concerning numbers in a manuscript, or rather a sequence of manuscripts, written during the period between 1872 and 1878.[4] These intellectual developments are not isolated foundational ruminations, but have to be seen in the context of Dedekind's contemporaneous work on algebraic number theory; cf. Section 4.

The publication of the essays by Helmholtz and Kronecker moved Dedekind finally to sharpen, complete, and publish his considerations. He characterized his views as "being in some respects similar [to those of Helmholtz and Kronecker], but through their grounding essentially different".[5] This is a gentle formulation of sharp mathematical and philosophical differences. The differences emerged in Dedekind's reflections slowly and over a long period of time, but they ultimately resulted in a dramatic shift. The latter can be understood, or so we will argue more explicitly below, as articulating an *axiomatic* approach that is joined with a *genetic* one in a methodologically coherent way. In his essay *Über den Zahlbegriff* Hilbert distinguished sharply between the axiomatic and genetic method, but did not recognize then the complementary roles they play for the foundations of arithmetic. Dedekind's and Hilbert's investigations have to be seen against the backdrop of the arithmetization of analysis, that is, of the reduction of analysis to number theory. Dedekind's approach is associated with a novel structuralist perspective on mathematics and is grounded in *logic broadly conceived*. Hilbert sustains this general perspective in what he later calls *existential axiomatics*, but he gives up a logicist in favor of a finitist grounding of mathematics; of course, that presupposes the formalistic sharpening of the axiomatic method and the syntactic formulation of the consistency problem. For that development, see (Sieg 1999, Sieg 2002).

By tracing its development we provide a view of Dedekind's evolving foundational position that apparently differs from Hilbert's: to our knowledge, Hilbert never considered Dedekind as having used the axiomatic method. The view we provide is definitely in conflict with that of contemporary writers like Ferreirós, Corry, and McCarty. According to Ferreirós, Dedekind is non-modern in logical matters, as he can be viewed as "anti-axiomatic"; according to Corry, Dedekind is non-modern in mathematical matters, as he can't be taken to be a mathematical structuralist; finally, according to McCarty, Dedekind is non-modern in philosophical matters,

---

auch aller subjectiven Willkür entzogen, und, strengen rein logischen Regeln unterworfen, bieten sie für den Arithmetiker völligen Ersatz für jene populären Zahlen." (Cf. (Dugac 1976) on Meyer, pp. 93, 176.)

[4] These manuscripts are analyzed in Section 5; their dating is Dedekind's own.

[5] These observations are made in the first note to the preface of the first edition: "Das Erscheinen dieser Abhandlungen [i.e., the essays by Helmholtz and Kronecker] ist die Veranlassung, die mich bewogen hat, nun auch mit meiner, in mancher Beziehung ähnlichen, aber durch ihre Begründung doch wesentlich verschiedenen Auffassung hervorzutreten, die ich mir seit vielen Jahren und ohne jede Beeinflussung von irgendwelcher Seite gebildet habe."

as he is a thoroughgoing Kantian.[6] The reason for Ferreirós's and Corry's judgments is rooted, ultimately, in a particular understanding of the foundational essays (1872c) and (1888). (That understanding is made explicit by Ferreirós on pp. 119–124 and by Corry on pp. 71–75.) Our paper should make it very clear that their understanding of Dedekind's (methodology for the) treatment of real and natural numbers is inadequate. This also holds for McCarty. When contrasting (1872c) and (1888) in his paper (1995), McCarty points out that the essay from 1888 contains a categoricity result, whereas that from 1872 does not. McCarty asks on p. 81, why Dedekind does not establish that the geometric straight line and the system of rational cuts are isomorphic. He continues, "To this the short—but by my lights correct—answer is: Dedekind thinks that such an isomorphism would be impossible to establish." Our contrary answer is indicated in the Concluding Remarks, Section 7.

Our general views are informed by the work of Belna, Dugac, Gray, Mehrtens, Noether, Parsons, Stein, and Tait. However, they have been shaped most importantly by a close reading of manuscripts in Dedekind's Nachlass. It is fair to say that these manuscripts—including *Arithmetische Grundlagen*, the three drafts listed in Section 3.3, (1871/1872), (1872/1878), and also (1887a)—have not yet been taken into account properly for a detailed analysis of the development of Dedekind's foundational views and its intimate connection to the evolution of his mathematical work. Our essay continues and deepens earlier work in (Sieg 1990a, Sieg 2000) and (Schlimm 2000), but focuses almost exclusively on the systematic development of Dedekind's approach to the foundations of the theory of numbers. And what a stunning development it is! In a second essay, Dedekind's general methodological concerns will take center stage.

---

[6] As to Ferreirós, we refer to the discussion, *Dedekind's deductive method*, in section 5.3 of his book (1999), where it is claimed on p. 247, that ". . . Dedekind's deductive method seems rather strange, and could be even called anti-axiomatic. . . . The underlying elementary logic [in (Dedekind 1888)] — although transparently employed — is not made explicit, and above all arithmetic is understood as requiring no axiom. All of this places Dedekind's contribution in a peculiar historical position, as an intermediate step that would quickly be abandoned (or, if you wish, superseded)." That perspective is taken then in Section 6.3 to judge Dedekind's influence on Hilbert and his school with the central question formulated on p. 246: ". . . why Dedekind's strong deductivism did not lead to an axiomatic approach". *As to Corry*, we refer to the discussion of Dedekind's and Hilbert's influence "on the rise of the structural approach to algebra" in his book (1996) as expressed, for example, on pp. 170–171; though both Dedekind and Hilbert introduced "a kind of axiomatic analysis when dealing with their algebraic entities" and in their algebraic works "displayed and promoted" structural features, they did so "independently of any adoption of the modern axiomatic approach". For Dedekind, in particular, Corry notes on p. 129 in a section entitled "Dedekind and the structural image of algebra": "Although Dedekind used many of the concepts that were later to become the hard core of structural algebra, these concepts play very different roles in those of his works in which they appear. Therefore, they cannot be identified with the notion of an algebraic structure." Finally, *as to McCarty*, he argues that the solutions to the "mysteries" in Dedekind's thought are "to be found in the doctrines of Kant's Transcendental Dialectic". (p. 70) These points of difference are taken up also below.

Let us give a brief orientation of this paper. Section 2 is concerned with the important *Habilitationsrede* of 1854, as it reveals Dedekind's perspective on the classical number systems and some broad methodological issues. We expose a *subtle, but pervasive circularity* in Dedekind's considerations, when he connects the creation of numbers beyond the naturals with the *extension* of operations. This subtle circularity is addressed, fully and satisfactorily, through the developments described next. Section 3 presents Dedekind's more systematic treatment of numbers around 1872. He introduces the successor function for natural numbers and takes a dramatic step of "analysis" toward their coherent extension to integers and rationals. This step is complemented by a "synthesis" described in Sections 4.1 and 4.2; the central demands underlying these extensions—together with a quite new aspect of abstraction—are emphasized, when we consider the *free creation* of irrational numbers in Section 4.3. This leaves open in (1872c) the question how the natural numbers can be characterized. The evolution of Dedekind's theory of chains and the formulation of the Dedekind-Peano axioms for natural numbers are described in Section 5, based on a detailed analysis of (1872/1878).

Remarkable metamathematical investigations of this axiom system are presented in (1888). Dedekind's attempt to establish its consistency and his proof of its categoricity are of course crucial here; they are discussed in Section 6. The categoricity result allows him to justify the claim that the numbers can be called a *free creation* of the human mind. "Free creation" is understood here in a different, but related way from "free creation" in (1872c); however, the meaning is radically different from "creation" in (1854). This central part of Dedekind's foundational work, elucidated by some general reflections, is in accord with the spirit of Herbart's remark we quoted as the motto for our essay: the mathematical work opens up a completely novel and distinctive philosophical perspective on the nature of number and, indeed, of mathematics.

2 Extending operations

Richard Dedekind, born in 1831 as a citizen of Braunschweig, finished his dissertation under Gauss in 1852 and gave a talk on the occasion of his Habilitation only two years later. The talk was entitled *Über die Einführung neuer Funktionen in der Mathematik* and was presented on 30 June 1854 to an audience that included Gauss, the classical philologist Hoeck, the historian Waitz, and the physicist Weber. Dedekind had chosen to talk about the general way, "in which new functions, or, as one might also want to say, new *operations*, are added to the chain of already existing ones in the progressive development of this science (i.e., mathematics)".[7] For Dedekind in (1854), the introduction of new functions was an extremely

---

[7](Dedekind 1854, p. 428). The German text: "Diese Vorlesung hat nicht etwa . . . die Einführung einer bestimmten Klasse neuer Funktionen in die Mathematik, sondern vielmehr allgemein die Art und Weise zum Gegenstande, wie in der fortschreitenden Entwicklung

important component in the development of mathematics; for us now, Dedekind's observations reveal general aspects of his intellectual approach as well as special features of his understanding of the classical number systems.

2.1 Systematic reflections. The concrete analyses of the introduction of some functions are preceded by expansive remarks about the role of functions and concepts in organizing a body of knowledge, in "shaping a system". That role pertains to the law as well as to the sciences and, in particular, to mathematics. Dedekind made these remarks at the age of twenty-three for a particular occasion. Nevertheless, they bring out striking characteristics of his way of thinking and, consequently, of his later mathematical work. Their intrinsic significance is underlined by the fact that he returned to them in (1888).

In the preface to (1888) Dedekind mentions with some satisfaction that the purpose of his *Habilitationsrede* had been approved by Gauss; he characterizes it then and there as defending the claim that the most significant and most fruitful advances in mathematics and other sciences have been made "by the creation and introduction of new concepts, rendered necessary by the frequent recurrence of complex phenomena, which could be controlled only with difficulty by the old ones".[8] This need to introduce new and more appropriate notions arises for Dedekind, in (1854), from the fact that human intellectual powers are imperfect; their limitation leads us to frame the object of a science in different forms or systems. To introduce a concept, "as a motive for shaping the system", means in a certain sense to formulate an hypothesis concerning the inner nature of a science, and it is only the further development that determines the real value of such a notion by its efficacy in recognizing general truths. These truths, in turn, affect the formulation of definitions. Dedekind summarizes his considerations in a most revealing way:

So it may very well happen that the concepts, introduced for whatever motive, have to be modified, because they were initially conceived either too narrowly or too broadly; they will require modification so that their efficacy, their import, can be extended to a larger domain. The greatest art of the systematizer lies in carefully turning over definitions for the sake of the discovered laws or truths in which they play a role.[9]

---

dieser Wissenschaft (i.e., der Mathematik) neue Funktionen, oder, wie man ebensowohl sagen kann, neue *Operationen* zu der Kette der bisherigen hinzugefügt werden."

[8](Dedekind 1888, VI). The German text: "... die größten und fruchtbarsten Fortschritte in der Mathematik und anderen Wissenschaften sind vorzugsweise durch die Schöpfung und Einführung neuer Begriffe gemacht, nachdem die häufige Wiederkehr zusammengesetzter Erscheinungen, welche von den alten Begriffen nur mühselig beherrscht werden, dazu gedrängt hat."

[9](Ewald 1996, p. 756) [4]. The German text — (Dedekind 1854, p. 430) — is as follows: "So zeigt sich wohl, daß die aus irgendeinem Motive eingeführten Begriffe, weil sie anfangs zu beschränkt oder zu weit gefaßt waren, einer Abänderung bedürfen, um ihre Wirksamkeit, ihre Tragweite auf ein größeres Gebiet erstrecken zu können. Dieses Drehen und Wenden der

Dedekind turns his attention then to mathematics. Definitions in mathematics are initially of a restricted form, but their generalizations are determined without arbitrariness. Indeed, Dedekind asserts, "they follow with compelling necessity from the earlier narrower ones". I.e., they do follow with necessity, if one applies the principle that some laws holding for the initial definitions are viewed as generally valid. These laws become consequently the source of the generalized definitions, when one asks, "How must the general definition be formulated such that the found characteristic law is always satisfied?" Dedekind views this as the distinctive feature of mathematical definitions, and the feature by which mathematics is distinguished from the other sciences. This claim will be taken up below; here we just note that in mathematics the *creation* of new objects may be involved, whereas the objects of the other sciences are presumably given. In order to illustrate this general point, we consider one of Dedekind's mathematical examples—an example that provides furthermore a real insight into his contemporaneous understanding of the classical number systems.

2.2 Generally valid, subtly circular. Dedekind describes elementary arithmetic as being "based on the formation of ordinal and cardinal numbers" and continues, "the successive progress from one member of the series of the absolute whole numbers to the next is the first and simplest operation of arithmetic; all other operations rest on it".[10] Addition, multiplication, and exponentiation are obtained by iterating "the first and simplest operation", addition, and multiplication, respectively, and then joining these iterations into single acts. For the further development of arithmetic these definitions of the basic operations are insufficient as they are restricted to the very small domain of the positive integers. The demand that one should be able to carry out the inverse operations of subtraction, division, etc., without any restrictions leads to the creation of "the negative, fractional, irrational and finally also the so-called imaginary numbers". Indeed,

---

Definitionen, den aufgefundenen Gesetzen oder Wahrheiten zuliebe, in denen sie eine Rolle spielen, bildet die größte Kunst des Systematikers." In the Introduction to the second edition of (Dirichlet 1863) he emphasized this general aspect for the particular mathematical work. He presented in the tenth supplement his general theory of ideals in order, as be put it, "to cast, from a higher standpoint, a new light on the main subject of the whole book". In German, "Endlich habe ich in dieses Supplement eine allgemeine Theorie der Ideale aufgenommen, um auf den Hauptgegenstand des ganzen Buches von einem höheren Standpunkte aus ein neues Licht zu werfen;" he continues, "hierbei habe ich mich freilich auf die Darstellung der Grundlagen beschränken müssen, doch hoffe ich, daß das Streben nach charakteristischen Grundbegriffen, welches in anderen Teilen der Mathematik mit so schönem Erfolg gekrönt ist, mir nicht ganz mißglückt sein möge." (Dedekind 1932, pp. 367 and 397).

[10] (Ewald 1996, p. 757ff). The German text, (Dedekind 1854, pp. 430–431), is as follows: "Die Elementararithmetik geht aus von der Bildung der Ordinal- und Kardinalzahlen; der sukzessive Fortschritt von einem Gliede der Reihe der absoluten ganzen Zahlen zu dem nächstfolgenden ist die erste und einfachste Operation der Arithmetik; auf ihr fußen alle andern. Faßt man die mehrere Male hintereinander wiederholte Ausführung dieser Elementaroperation in einem einzigen Akt zusammen, so gelangt man zum Begriff der Addition. Aus diesem bildet sich auf ähnliche Weise der der Multiplikation, aus diesem der der Potenzierung."

Dedekind views this last demand as another formulation of the demand "to create anew by each of these operations the whole given number domain".

Having expanded the domain of numbers by means of the inverse operations, a crucial question arises, namely, how to extend the definitions of the fundamental operations so that they are applicable to the newly created numbers. Here Dedekind joins the above general reflections and considers in detail the extension of multiplication from the natural numbers to all integers. The extension of the definition of multiplication is non-arbitrary, Dedekind asserts, if one follows his principle of the general validity of laws as the source for deriving "the meaning of the operations for the new number domains". This source cannot be exploited without a *subtle circularity* for addition itself: the new numbers are generated by the unrestricted inverse of a restricted operation, which is then extended to this generated broader domain! In spite of Dedekind's protestation—that the definition of the extended operation "involves an a priori complete arbitrariness" (see next paragraph)—he appeals to the very character of that generation. The intricate dependency can be observed most clearly also in Dedekind's considerations for the extension of multiplication from the natural numbers to all integers.

As noted earlier, Dedekind defines multiplication for the natural numbers as joining the iteration of addition into one single act, and it is of course assumed now that addition and subtraction are already available for all integers. Prima facie, the definition of multiplication via iteration makes sense only if the multiplicator is positive; the multiplicator is the number which indicates how often one has to iterate the addition of the multiplicand. The multiplicand can be positive or negative. Dedekind asserts:

> A special definition is therefore needed in order to admit negative multiplicators as well, and to liberate in this way the operation from the initial restriction; but such a definition involves an *a priori* complete arbitrariness, and it would only later be determined whether this arbitrarily chosen definition would bring any real advantage to arithmetic; and even if this succeeded, one could only call it a lucky guess, a happy coincidence—the sort of thing a scientific method ought to avoid.[11]

What considerations might provide grounds for a principled definition of the extended operation of multiplication?—"One has to investigate", Dedekind demands, "which laws govern the product, if the multiplicator is successively subjected to the same changes by which the series of negative numbers is generated from the series of the absolute whole numbers in the first place." The broader domain is obtained, of course, by the unrestricted

---

[11](Ewald 1996, p. 758) [8]. The German text — (Dedekind 1854, pp. 431–432) — is as follows: "Es bedarf daher einer besonderen Definition, um auch negative Multiplikatoren zuzulassen, und auf diese Weise die Operation von der anfänglichen Beschränkung zu befreien; eine solche involviert aber eine a priori vollständige Willkürlichkeit, und es würde sich erst später entscheiden, ob denn die so beliebig gewählte Definition der Arithmetik einen wesentlichen Nutzen brächte; und glückte es auch, so könnte man dies doch immer nur ein zufälliges Erraten, ein glückliches Zutreffen nennen, von welchem eine wissenschaftliche Methode sich frei halten soll."

Dedekind's analysis of number

inversion of addition, i.e., by considering $(m - n)$ for arbitrary natural numbers m and n. Dedekind observes that $a \times (m + 1) = a \times m + a$, which yields the "addition theorem for the multiplicator" $a \times (m + n) = a \times m + a \times n$. From this follows the "subtraction theorem" $a \times (m - n) = a \times m - a \times n$, but only as long as the minuend m is greater than the subtrahend n. Taking this law as valid also for the case that the difference representing the multiplicator is negative, one obtains the definition of multiplication for the generated new numbers.[12] Thus, Dedekind concludes, "It is no longer an accident that the general law for multiplication is in both cases exactly the same." Dedekind obtains in a similar way the generalized definition of exponentiation for rational numbers.

2.3 Imaginary and real problems.

The extension of the basic operations to the real and imaginary numbers is only alluded to. Dedekind claims, "These advances [obtained by creating the new classes of numbers] are so immense that it is difficult to decide which of the many paths that are opened up here one should follow first." So much is clear, however, that the operations of arithmetic have to be extended to these new classes and that no extension is possible along the lines sketched above without grasping the "generation" of the real and imaginary numbers. Here, "at least with the treatment of the imaginary numbers", the main difficulties for the systematic development of arithmetic begin. Dedekind ends the discussion of the number systems in a very surprising way:

> However, one might well hope that a truly solid edifice of arithmetic will be attained by persistently applying the principle not to permit ourselves any arbitrariness, but always to be led on by the discovered laws. Everybody knows that until now, an unobjectionable theory of the imaginary numbers, not to mention those newly invented by Hamilton, does not exist, or at any rate has not been published yet.[13]

Four years later, in the fall of 1858, Dedekind lectured on the infinitesimal calculus at the "Eidgenössisches Polytechnikum" in Zürich. He reports in (1872c) that he was motivated—by the "overwhelming feeling of dissatisfaction" with the need to appeal to geometric evidences when discussing certain limit considerations—to search for "a purely arithmetic and completely rigorous foundation of the principles of infinitesimal analysis". He

---

[12] Notice a most interesting feature: a negative number can be taken to be represented, for the purpose of defining the extended operation, by a pair of positive ones. If the multiplicand a were to be represented in the same way, this approach is technically very close to the later one discussed in Section 4.1.

[13] (Ewald 1996, p. 759) [9]. The German text — (Dedekind 1854, p. 434) — is as follows: "Indessen ist wohl zu hoffen, daß man durch beharrliche Anwendung des Grundsatzes, sich auch hier keine Willkürlichkeit zu erlauben, sondern immer durch die gefundenen Gesetze selbst sich weiterleiten zu lassen, zu einem wirklich festen Gebäude der Arithmetik gelangen wird. Bis jetzt ist bekanntlich eine vorwurfsfreie Theorie der imaginären, geschweige denn der neuerdings von Hamilton erdachten Zahlen entweder nicht vorhanden, oder doch wenigstens noch nicht publiziert." To see why Gauss's geometric interpretation of complex numbers did not satisfy Dedekind's purely arithmetic ambitions, it is instructive to read Gauss's defense of the use of complex numbers in his (1831), in particular, pp. 310–313.

found it in his examination of continuity and the resulting definition of real numbers as, or rather through, cuts of rationals.

Dedekind discussed the solution with his friend Heinrich Durège at the time and presented the material to the "Wissenschaftlicher Verein" in Braunschweig on 11 January 1864, but also in some of his lectures on the differential and integral calculus.[14] Already in 1870 he had the intention of publishing his theory of continuity according to a letter from his friend Adolf Dauber.[15] We have the extended draft (1871/1872) of the essay *Stetigkeit und irrationale Zahlen*; that was seemingly written in late 1871 and early 1872. We should notice that by this time Dedekind had isolated the concept of a *field* (Körper) in Supplement X of (Dirichlet 1863). This concept plays a significant role in (1872c); the novel and careful definition of the arithmetic operations on rational cuts via their definition on rationals will be discussed in Section 4.2. This fits marvelously with Dedekind's evolving view of natural numbers and their extensions to integers and rationals around 1872.

Assuming that the difficulty mentioned explicitly in the *Habilitationsrede* (to obtain an "irreproachable theory of imaginary numbers") has been resolved[16] and that the definition of real numbers in terms of cuts answers Dedekind's concerns for a rigorous foundation of analysis, two questions are clearly implicit in the above and remain open for Dedekind in 1872: (i) What are (the principles for) natural numbers? and (ii) How are the integers and rational numbers obtained, or how are they created, starting with the natural numbers? In the *Habilitationsrede* Dedekind takes for granted that the new mathematical objects (the negative and fractional numbers) have been obtained already from the natural numbers; the central issue is there, how to extend the basic arithmetic operations to the wider number systems. Question (ii) is addressed in a sequence of manuscripts contained in Cod. Ms. Dedekind III, 4, and it seems that the issues were settled to Dedekind's satisfaction before the essay on continuity and irrational numbers was completed. Question (i) was not settled at that time; on the contrary, Dedekind struggled with it intermittently over the next six years. The intense work is reflected in the manuscript (1872/1878); it served as the very first draft for the 1888 essay on the nature and meaning of numbers and is published as Appendix LVI in (Dugac 1976) with the title *Gedanken über die Zahlen*.

---

[14]See, for example, the outline for such a course in the winter semester of 1862/1863, published in (Dugac 1976) as Appendix IV.

[15]Dauber asks Dedekind in a letter of 20 June 1871, whether Dedekind had come closer to realizing his plans for publishing his theory of continuity, and remarks that Dedekind had written him about such plans a year earlier. The letter is part of Appendix XXVI in (Dugac 1976); the remark can be found on p. 192.

[16]Ferreirós reports on p. 220 of his (1999) that Dedekind borrowed in 1857 Hamilton's *Lectures on Quaternions* from the Göttingen Library. Hamilton gives in the Preface to his book the definition of complex numbers as pairs of reals. Pairs are viewed there as genuine mathematical objects for which operations can be defined appropriately; see (Hamilton 1853, pp. 381–385). "Thus", Ferreirós concludes convincingly, "Dedekind could regard the problem of complex numbers as satisfactorily solved, . . . " Indeed, Dedekind uses Hamilton's way later; cf. Section 4.1.

Dedekind's analysis of number 45

We will analyze that work in Section 5.1, whereas the next two sections are devoted to turning the puzzle of manuscripts on extensions into an informative mosaic that answers question (ii).

3 Extending domains

In this section two important steps are described and analyzed: (i) the successor operation is separated clearly from the other arithmetic operations as the one that generates the domain of natural numbers (above, and even in (1872c), *all* the arithmetic operations are on a par); (ii) basic domains of integers and rationals are characterized axiomatically (guaranteeing invertibility of addition and multiplication, but also providing without subtle circularity the basis for extending the arithmetic operations). The axiomatic "analysis" is complemented by a "synthesis" in Section 4.1: Dedekind gives an explicit definition of appropriate domains that form models of those axioms.[17] These considerations foreshadow the broad methodological moves in (1888); namely, an axiomatic characterization of simply infinite systems and the explicit definition of a model. To obtain an appropriate axiomatic analysis of the natural numbers as a simply infinite system will take significantly more work. That is obtained in the manuscript (1872/1878), at the end of which we find the first formulation of the Dedekind-Peano axioms. But one step at a time!

3.1 Analyzing naïvely.
The manuscript (1872/1878) has the subtitle *Attempt to analyze the number concept from the naïve point of view* (Versuch einer Analyse des Zahlbegriffs vom naiven Standpuncte aus). Is it in the "naïve" approach to the topic that Dedekind sees, as he does in (1888), a certain similarity between his view and that of Helmholtz and Kronecker? What did Dedekind have in mind, when calling his approach naïve? An answer to the second question seems to be given in his letter to Keferstein by the remark addressing the rhetorical question, "How did my essay come into being?"

Surely not all at once, rather it is a synthesis constructed after protracted labor, which is based on a preceding analysis of the sequence of natural numbers as it presents itself, in experience so to speak, to our consideration.[18]

A thoroughgoing analysis of the data of ordinary mathematical experience, free from philosophical preconceptions, is fundamental for Dedekind. Such an analysis, as Dedekind demanded already in (1854), should lead

---

[17]Dedekind uses the concepts *analysis* and *synthesis* here. Dedekind's use illuminates, but is also illuminated by, that of the ancient geometers, in particular, as formulated by Pappus; see (Beaney 2003).

[18](van Heijenoort 1967, p. 99). The German text is: "Gewiss nicht in einem Zuge, sondern sie ist eine nach langer Arbeit aufgebaute Synthesis, die sich auf eine vorausgehende Analyse der Reihe der natürlichen Zahlen stützt, so wie diese sich, gewissermassen erfahrungsmässig, unserer Betrachtung darbietet."

to notions that reflect the nature of the subject and prove their efficacy in its development. The independence from traditional philosophical preconceptions is brought out clearly, when Dedekind at the very beginning of (1872/1878) writes that the notions he uses for the foundation of the number concept "remain necessary for arithmetic even when the notion of cardinal number is assumed as immediately evident ('inner intuition')".[19]

Recall that in (1854) elementary arithmetic begins with the formation of ordinal and cardinal numbers. Dedekind views the "successive progress from one member of the sequence of positive integers to the next" as "the first and simplest operation of arithmetic" on which all other operations rest. Addition is obtained by joining iterations of this "first and simplest operation" into a single act; in completely parallel ways one obtains multiplication from addition and exponentiation from multiplication. This standpoint concerning the character of natural numbers is hardly changed, when Dedekind expresses his views in Section 1 of (1872c). There he uses *chain*, the central notion of (1888), not yet in the precise sense of the later work, but rather as a fitting informal notion to capture the structural character of the domain that has been obtained by successively generating its objects through the "simplest arithmetical act":

I regard the whole of arithmetic as a necessary, or at least natural, consequence of the simplest arithmetical act, that of counting, and counting itself is nothing other than the successive creation of the infinite series of positive integers in which each individual is defined by the one immediately preceding; the simplest act is to pass from an already-created individual to its successor that is to be newly created. The chain of these numbers already forms in itself an exceedingly useful instrument for the human mind; it presents an inexhaustible wealth of remarkable laws, which one obtains by introducing the four fundamental operations of arithmetic.[20]

One should notice that Dedekind speaks of counting as "nothing other" than the *successive creation* of the individual positive integers.

An elementary and restricted development of arithmetic is given in the contemporaneous manuscript *Arithmetische Grundlagen*; this manuscript is found in three distinct versions in Dedekind's Nachlass (Cod. Ms. Dedekind III, 4, II).[21] The development uses only the definition principle

---

[19](Dugac 1976, p. 293). The German text is included in footnote 33.

[20](Ewald 1996, p. 768). The German text — (Dedekind 1872c, pp. 5–6) — is as follows: "Ich sehe die ganze Arithmetik als eine notwendige oder wenigstens natürliche Folge des einfachsten arithmetischen Aktes, des Zählens, an, und das Zählen selbst ist nichts anderes als die sukzessive Schöpfung der unendlichen Reihe der positiven ganzen Zahlen, in welcher jedes Individuum durch das unmittelbar vorhergehende definiert ist; der einfachste Akt ist der Übergang von einem schon erschaffenen Individuum zu dem darauffolgenden neu zu erschaffenden. Die Kette dieser Zahlen bildet an sich schon ein überaus nützliches Hilfsmittel für den menschlichen Geist, und sie bietet einen unerschöpflichen Reichtum an merkwürdigen Gesetzen dar, zu welchen man durch die Einführung der vier arithmetischen Grundoperationen gelangt."

[21]Ferreirós discusses *Arithmetische Grundlagen* on p. 218 and, more extensively, on pp. 222–224. Our perspectives are different on the dating of the manuscript and on the "rational reconstruction" of the mathematical content and context. Our reasons for differing are

Dedekind's analysis of number 47

by recursion and the proof principle by induction. The first version starts out in the following way:

§1
Act of creation 1; $1+1 = 2$; $2+1 = 3$; $3+1 = 4 \ldots$ numbers (ordinal).

§2
Definition of addition by $a+(b+1) = (a+b)+1$. After this, consequences are according to the nature of the subject—always to be deduced by complete induction.[22]

This is only slightly modified in the second version that reads:

§1
Creation of the numbers: 1; $1+1 = 2$; $2+1 = 3$; $3+1 = 4 \ldots$ from each number a the *following* number $a + 1$ is formed by the act $+1$. Therefore, everything by complete induction.

§2
Definition of addition: $a + (b + 1) = (a + b) + 1$.[23]

In both versions elementary arithmetic is then briefly and very thoroughly developed. Dedekind establishes (in different ways) associativity and commutativity of addition and multiplication and ends with a proof of the distributive law $a \times (b + c) = a \times b + a \times c$. In the second version, he remarks on the margin that this law can be obtained much more directly from the definition of multiplication and the associativity of addition. Such a more direct argument is indeed presented in the third version.

Most remarkable about the third version of *Arithmetische Grundlagen* is the fact that Dedekind separates the generating "successor operation" from addition, i.e., the sequence of numbers is now indicated by 1, $\varphi(1) = 2$, $\varphi(2) = 3$, $\varphi(3) = 4, \ldots$, and the recursive definition of addition is given by the two equations $a + \varphi(b) = \varphi(a+b)$ and $a + 1 = \varphi(a)$ instead of just by the single equation $a + (b+1) = (a+b) + 1$. This notational change to the unary successor operation indicates the beginning of a quite dramatic conceptual shift that finds its systematic expression in the manuscript (1872/1878) and provides one solid reason for thinking that *Arithmetische Grundlagen* was completed in (early) 1872.

3.2 Creating in circles. The third version makes also quite clear that Dedekind is trying to use these foundations for constructing the extended number systems, here, of all integers. Dedekind defines subtraction by

---

presented with the detailed discussion of the manuscript below. On one crucial issue we do agree with Ferreirós, namely, that the introduction of the successor operation in (what we take to be) the third version of the manuscript is of utmost significance and a central result of the informal analysis.

[22]§1 Schöpfungsakt 1; $1+1 = 2$; $2+1 = 3$; $3+1 = 4 \ldots$ Zahlen (Ordinal). §2 Erklärung der Addition durch $a + (b + 1) = (a + b) + 1$. Hiernach Folgerungen, der Natur der Sache nach [,] immer durch die vollständige Induktion abzuleiten.

[23]§1 *Erschaffung der Zahlen*: 1; $1+1 = 2$; $2+1 = 3$; $3+1 = 4 \ldots$ aus jeder Zahl a wird durch den Act $+1$ die *folgende* Zahl $a + 1$ gebildet. — Deshalb alles durch vollständige Induction. §2 Erklärung der Addition: $a + (b + 1) = (a + b) + 1$.

$a - b = c$, in case $a = b + c$; this is taken, implicitly, as the motivation for considering an extension of the positive integers that contains 0 (zero) and the negative numbers 1^*, 2^*, 3^*, etc. The successor operation is suitably extended by setting, in particular, $0 + 1 = 1$, $1^* + 1 = 0$, $2^* + 1 = 1^*$, $3^* + 1 = 2^*$, etc. Having defined the predecessor operation $b = a - 1$, in case $b + 1 = a$, he considers $1 - 1$, $(1 - 1) - 1$, etc. as the *new numbers*.

Together with the systematic development up to the distributive law (central for restricting the possible extensions of multiplication in (1854) and called there the "addition theorem for the multiplicator") this sets the stage for a development along the lines suggested in his *Habilitationsrede*. Indeed, it sets the stage in a much more refined way, but it leaves in place the subtle circularity we diagnosed in (1854); it is now directly visible through the juxtaposition of the non-positive numbers 0, 1^*, 2^*, etc., and the new numbers $1 - 1$, $(1 - 1) - 1$, $((1 - 1) - 1) - 1$, etc. Thus, Dedekind assumes here a domain containing also zero and the negative numbers in order to define the extended successor operation. That allows him, in turn, to define the general predecessor operation and to describe the desired extension of the system of natural numbers by the *new numbers*.[24] But of what objects does the first extension really consist? What are the negative numbers? (Dedekind's answers to these questions are discussed fully in Section 4.)

There is no indication on the manuscript itself as to when *Arithmetische Grundlagen* was written. We conjecture, for three reasons, that it was completed in early 1872. The first reason is simply the fact that the beginnings of the various versions are in accord with the informal description in (1872c). The second reason was mentioned already, when we looked at the third version and noticed an important and rather unique overlap with (1872/1878). Finally, the third reason is provided by the systematic context of creating the system of rational numbers on these arithmetic foundations. In (1871/1872) Dedekind emphasizes that the rational numbers are a *free creation*. He also claims that the "instrument mathematicians have constructed by creating the rational numbers" has to be refined by the creation of the irrational numbers in a purely arithmetic way.

Just as negative and fractional rational numbers are formed by a free creation, and just as the laws of operating with these numbers are reduced to the laws of

---

[24]That way of proceeding was not uncommon at the time; indeed, Heine pursues a similar route in his *Elemente der Functionenlehre*. Though Heine's is a natural way of proceeding, Dedekind must have found it (and his own approach) quite unsatisfactory at this juncture. Heine answers the general question "*What are numbers?*" not by a conceptual definition, but rather by taking a purely formal standpoint (acerbically criticized by Frege): "In the definition [of numbers] I adopt the purely formal standpoint, *by calling certain tangible marks numbers*, such that the existence of these numbers is not in question." Dedekind received Heine's paper, when working on the draft of (1872c). Heine describes his way of introducing the negative numbers on pp. 173–174 of his essay. As to possible precedents of Dedekind's way of proceeding cf. (Ferreirós 1999, p. 219, note 1).

Dedekind's analysis of number 49

operating with positive integers (at least it *should* be done in this way), in the same way the irrational numbers must also be defined by means of the rational numbers.[25]

This long sentence is repeated almost verbatim in the publication (1872c). Here it is (and we urge readers to notice the italicized replacement for the parenthetical remark in the above quotation):

Just as negative and fractional rational numbers are formed by a free creation, and just as the laws of operating with these numbers *must and can* be reduced to the laws of operating with positive integers, in the same way the irrational numbers must also be completely defined by means of the rational numbers alone.[26]

What is the mathematical substance that allows us to understand the shift from *should* to *must and can*?

We conjecture that the material contained in Cod. Ms. Dedekind III, 4 provides the answer: having established proper *arithmetic foundations*, Dedekind convinces himself in detail that the system of rational numbers can be created, and that the laws for calculating with these numbers can be reduced to those for calculating with the positive whole numbers. This is done, however, in a completely novel axiomatic way.

3.3 Analyzing axiomatically. Dedekind's *Nachlass* contains several manuscripts dealing with the extension of the natural numbers to the integers and rational numbers. Particular ways of extending the number concept are pursued in the following manuscripts: (i) Cod. Ms. Dedekind III, 4, I, pp. 1–4, entitled *Die Schöpfung der Null und der negativen ganzen Zahlen*, (ii) Cod. Ms. Dedekind III, 4, I, pp. 5–7, without title, but we will refer to it as *Ganze und rationale Zahlen*, and (iii) Cod. Ms. Dedekind III, 2, I, entitled *Die Erweiterung des Zahlbegriffs auf Grund der Reihe der natürlichen Zahlen*. The first two manuscripts, we conjecture, were written in 1872.[27] The third one was written after 1888, as it refers explicitly to the essay (1888); it gives an altogether modern approach. In this subsection we give a detailed account of the first manuscript.

---

[25] In (Dugac 1976, p. 205). "So wie die negativen und gebrochenen rationalen Zahlen durch eine freie Schöpfung hergestellt, und wie die Gesetze der Rechnungen mit diesen Zahlen auf die Gesetze der Rechnungen mit ganzen positiven Zahlen zurückgeführt werden (so *sollte* es wenigstens geschehen), ebenso müssen auch die irrationalen Zahlen durch die rationalen Zahlen definiert werden." The emphasis of "sollte" is Dedekind's.

[26] (Ewald 1996, p. 771). The German text is: "So wie die negativen und gebrochenen rationalen Zahlen durch eine freie Schöpfung hergestellt, und wie die Gesetze der Rechnungen mit diesen Zahlen auf die Gesetze der Rechnungen mit ganzen positiven Zahlen zurückgeführt werden *müssen und können*, ebenso hat man dahin zu streben, daß auch die irrationalen Zahlen durch die rationalen Zahlen allein vollständig definiert werden."

[27] To be more precise, we conjecture that the first manuscript was written in 1872, whereas the second one was written much later, but that its essential content goes back to 1872. (The evidence for the conjecture that the second manuscript was written later is quite direct: one part of the detailed calculations is written on the back of a receipt for a journal subscription - from 1907.)

Our description in Section 3.2 of how to generate the integers from the natural numbers is based on remarks in the third version of *Arithmetische Grundlagen*. The generation proceeds essentially in two steps, the creation of the negative numbers motivated by the demand of the general invertibility of addition and the creation of the new numbers by the generalized predecessor operation. In *Die Schöpfung der Null und der negativen ganzen Zahlen* a beautifully detailed presentation of the first step of those considerations is given. That is *one* way of describing it; more accurately, however, Dedekind separates cleanly the discussion of the general invertibility of addition, extension of operations and the permanence of laws from the generation of mathematical objects satisfying those laws.

The first manuscript formulates at the outset basic facts regarding the series of natural numbers N: (1) N is closed under addition; addition is (2) commutative and (3) associative; (4) if $a > b$, then there exists one and only one natural number c, such that $b + c = a$, whereas in the opposite case, when $a \leq b$, no such number c exists. Dedekind notes that the fourth condition states a certain *irregularity* and raises the crucial question, whether it is possible to extend the sequence N to a system M (by the addition of elements or numbers to be newly generated) in such a way that M satisfies conditions (1)–(3) and also (4'), i.e., for any two elements a and b from M, there exists exactly one element c, such that $b + c = a$. And he asks, how rich must the *smallest* such system M be.

In the following *Investigation*, which is also called *Analysis*, Dedekind assumes the existence of such a system M. He reasons that M must contain a unique element 0 (called zero), such that $a + 0 = a$; furthermore, for every element a in N there must be a new element a^* in M, such that $a + a^* = 0$. Thus, any system M satisfying (1)-(4') must contain in addition to the elements of N the new element zero and all the different new elements a^*. Dedekind considers now the system P consisting of just N together with these new elements and shows that P has already the completeness expressed by conditions (1)-(4'); P is obviously the smallest such system, as it must be contained in any complete system M. The investigation is carried out in exemplary mathematical clarity, but it assumes quite explicitly the existence of a suitable M. This methodologically crucial issue is presumably addressed in the second, and unfortunately incomplete, section of the manuscript that is entitled *Synthesis*. Here is the full text of that section:

From the sequence N of natural numbers a is to be created a system P, which contains in addition to the elements a also an element 0 and for each a a corresponding element a^*, with the stipulation that all these elements in P are *different* from each other (easy to formulate more precisely; on the *possibility* of such a creation, see farther below).[28]

---

[28]The German text on p. 4 of Cod. Ms. Dedekind III, 4, 1, is this: "Man erschaffe aus der Reihe N der natürlichen Zahlen a ein System P, welches außer den Elementen a noch ein Element 0, und zu jedem a ein entsprechendes Element a^* enthält, mit der Festlegung, daß alle diese Elemente in P von einander *verschieden* sind (leicht genauer auszudrücken; über die *Möglichkeit* einer solchen Schöpfung weiter unten)."

Dedekind's analysis of number

There is no "farther below" and thus no discussion of the *possibility of such a creation*. The manuscript ends abruptly on page 4 with the remark just quoted. The folder contains, however, additional material that was written at a later date (as argued above), but its substance was undoubtedly clear to Dedekind in 1872 and can be understood as realizing such a creation.

4 Creating models

The systematic considerations are continued in *Ganze und Rationale Zahlen*. This manuscript has two main parts: the first deals with the *extension of the domain* of all natural numbers to that of all integers; the second is concerned with the "transition from the domain G of all whole numbers to the field R of all rational numbers". The first part consists of three handwritten pages together with a few *Zettel* filled with detailed calculations concerning integers; the second part sketches very briefly similar considerations for the rationals on just one page. We describe the first part in detail, despite the fact that the steps are routine for a modern reader: Dedekind has finally found a way out of the subtle circularity involved in his earlier considerations of the various number systems (and their creation from the natural numbers).

4.1 Pairs as numbers.

Dedekind starts out with the domain N of all natural numbers together with the operations of addition and multiplication. Both operations satisfy the commutative and associative laws, and the distributive law connects them. The domain G of all whole numbers is then formed from N, as Dedekind puts it, *by extension*: "Any two numbers m, n in N generate a number (m, n) in G." Dedekind defines two pairs of numbers (m, n) and (m', n') as *identical* when $m + n' = m' + n$ and verifies that this relation is symmetric and transitive. As it is obviously also reflexive, it is an equivalence relation. Then he defines *addition* on pairs by letting the sum of (m, n) and (m', n') be identical to the pair $(m + m', n + n')$. Having checked that the defined addition yields identical results when applied to identical pairs, he verifies easily the associative and commutative laws. *Multiplication* for pairs (m, n) and (m', n') is given by $(mm'+nn', mn'+m'n)$ and is treated in a completely parallel way: uniqueness is checked (that is actually a quite lengthy argument and spills over onto the *Zettel*) and laws are verified; the final step is the verification of the distributive law.

This is the central part of constructing the integers as pairs of natural numbers that represent positive and negative numbers, but of course also zero. It is reminiscent of the very early considerations in (1854), when Dedekind extends subtraction from the natural numbers to the integers and, in essence, uses differences between natural numbers to represent negative numbers. Thinking of the pairs (m, n) as differences $m - n$ and using the ordinary calculation rules, the operations are obtained in a direct way and obey the standard laws. A parallel construction is sketched in the second part of this manuscript to obtain the rationals R from the integers

G: for pairs (m,n) and (m',n')—where m, n, m' and n' are in G, but n and n' are different from zero—"identity" is defined by $mn' = m'n$; this is again an equivalence relation. Thinking of pairs (m,n) as fractions m/n, addition and multiplication are defined via the ordinary calculation rules as $(mn' + nm', nn')$, respectively (mm', nn'). The various laws can be verified. It is also clear, though Dedekind does not prove it, that the inverted operations can be performed without any restriction.

We emphasize that this manuscript is in very rough form and indicates only the bare minimum of the needed considerations. But even so, it does provide a quite novel way in which to ensure the permanence of laws. Dedekind does not create—out of thin air—new individual elements: he rather obtains by pairing natural numbers, respectively integers, new systems of genuine mathematical objects. The arithmetic operations are then defined in terms of the operations on natural numbers, respectively integers. These systems satisfy the laws or axiomatic conditions for integers and rationals, i.e., Dedekind exhibits models for these laws. In fact, the models presented are exactly the ones that are still being employed today: except that in a modern exposition one would deal with equivalence classes of pairs.

That is done very beautifully in the final and later manuscript concerned with the extension of the number systems, *Die Erweiterung des Zahlbegriffs auf Grund der Reihe der natürlichen Zahlen*. It should be noticed clearly, however, that Dedekind could have taken this last step in 1872. There was no ideological reason for avoiding infinite mathematical objects; indeed, he had used such objects in the ideal theoretic investigations of Supplement X for the second edition of Dirichlet's *Zahlentheorie* of 1871, but also in the 1872 essay on continuity and irrational numbers. Yet there is one question that is left open: The rational numbers, "are" they these specific infinite objects? — A pertinent answer can be extracted from (1872c), as Dedekind's essay answers an analogous question for the reals. In the introduction to (1888), Dedekind situates his treatment of the natural numbers in the general context of providing, as he puts it, a completely clear picture of the science of numbers. He refers to the example of the real numbers presented in (1872c) and remarks that the other classes of numbers can be treated easily in a *quite similar fashion*. It is this observation, made also very clearly in Dedekind's letter to Weber dated 24 January 1888, that allows us to use the methodological considerations concerning the reals for our present context of the rational numbers. What has Dedekind to say about the question, what the reals really "are"? (The reader should look also at the related considerations in Section 6.2 and note 62.)

4.2 Systems as numbers. In his considerations of this very question Dedekind heeds, first of all, his own later warning in a letter to Lipschitz of 27 July 1876 that "nothing is more dangerous in mathematics than to make existence assumptions without sufficient proof". This refers to the definition of the system of real numbers. Recall that the system of reals

Dedekind's analysis of number

is to allow us to pursue all phenomena of the geometric line in a purely arithmetic way. Thus, it has to be defined by means of rational numbers and (the laws for) the arithmetic operations have to be reduced to (those for) the operations on rational numbers. The construction has to be done in such a way that the resulting system has the same kind of continuity or completeness as the geometric line. We will emphasize, on the one hand, the considerations involved in extending the system of rationals to that of the reals and bring out, on the other hand, the new answer to the question that parallels the above for rationals: "Are" the constructed objects, i.e., the cuts, *really* the real numbers? (The central issues are discussed in almost identical ways in (1871/1872) and (1872c).)

Cuts are partitions (A_1, A_2) of the system of rationals with the property that all a_1 in A_1 are less than all a_2 in A_2; they are viewed extensionally: $(A_1, A_2) = (B_1, B_2)$ if and only if A_1 and A_2 have the same members as B_1 and B_2, respectively. If A_2 contains a smallest element a', then the cut (A_1, A_2) is said to have been engendered by a'; the fact that not all cuts are engendered by rationals constitutes the incompleteness or discontinuity of the domain of rationals.[29] Dedekind continues, in the section entitled *Creation of irrational numbers*:

Thus, whenever we have a cut (A_1, A_2) produced by no rational number, we *create* a new number, an *irrational* number α, which we regard as completely defined by this cut (A_1, A_2); we shall say that the number α corresponds to this cut, or that it produces this cut. From now on, therefore, to every definite cut there corresponds a definite rational or irrational number, and we regard two numbers as *different* or *unequal* if and only if they correspond to essentially different cuts.[30]

The system of real numbers consists thus of all rational numbers (corresponding of course to the cuts engendered by them) together with these newly *created* irrational ones or, to put it in other words, the system of rationals has been extended by these irrational numbers. The crucial point is this: reals are not identified with cuts, but rather "correspond" to cuts; the latter are for Dedekind genuine mathematical objects, and the relations between reals and operations on them are defined in terms of the corresponding cuts.

The ordering between two reals α and β corresponding to the cuts (A_1, A_2) and (B_1, B_2) is defined as follows: $\alpha < \beta$ if and only if $A_1 \subset B_1$

---

[29] There is a simple issue of whether the partition (A', B') that is exactly like (A, B) except that b' is no longer the smallest element of B but the largest element in A' should also be a cut or not; Dedekind discusses these matters in (1871/1872) on p. 11, i.e., on p. 207 in (Dugac 1976). For his own presentation, he decides to consider such cuts as not *essentially different*.

[30] (Ewald 1996, p. 773). "Jedesmal nun, wenn ein Schnitt (A_1, A_2) vorliegt, welcher durch keine rationale Zahl hervorgebracht wird, so *erschaffen* wir eine neue, eine *irrationale* Zahl α, welche wir als durch diesen Schnitt (A_1, A_2) vollständig definiert ansehen; wir werden sagen, daß die Zahl α diesem Schnitt entspricht, oder daß sie diesen Schnitt hervorbringt. Es entspricht also von jetzt ab jedem bestimmten Schnitt eine und nur eine rationale oder irrationale Zahl und wir sehen zwei Zahlen stets und nur dann als *verschieden* oder *ungleich* an, wenn sie wesentlich verschiedenen Schnitten entsprechen."

(if, for any rational cut, the rational that engenders the cut is always, say, in the right part of the cut). Addition and multiplication of reals is defined in terms of the corresponding operations for the rationals. Consider two reals α and β that correspond to the cuts (A_1, A_2) and (B_1, B_2); the sum $\alpha + \beta$ corresponds to the cut (C_1, C_2), where C_1 consists of all c that are smaller than $a_1 + b_1$ for some a_1 in A_1 and b_1 in B_1, and C_2 consists of the remaining rational numbers. Multiplication can be defined in a similar way, and it is not difficult to verify the arithmetic laws for a field. Dedekind verifies also the order laws and proves that the system of reals is continuous. The system of reals or, more directly, the system of all cuts has been recognized as a complete ordered field.

It should be noticed that Dedekind uses "creation" here with a different sense than in the early discussions: for one, not individual mathematical objects are created, but rather systems thereof; in addition, the elements of those systems correspond to the elements of an already established system. (In the case of the reals, they correspond to rational cuts.) Perhaps to emphasize this new sense, Dedekind speaks in both (1871/1872) and (1872c) of free creation.

4.3 Free creation. Dedekind had excellent reasons for not identifying the real numbers with cuts of rationals. He articulated them very clearly in his early correspondence with Lipschitz already in 1876 and, as we mentioned above, in his letter to Weber dated 24 January 1888. The correspondence with Lipschitz was partially stimulated by the preparation of Dedekind's essay *Sur la théorie des nombres entiers algébriques*, published in 1877 in the Bulletin des sciences mathématiques. Lipschitz had actually suggested that Dedekind be invited to report on his work in algebraic number theory.[31] The resulting attempt by Dedekind to present his work (essentially contained in Supplement X of the second edition of Dirichlet's *Zahlentheorie*) in a new and possibly more accessible way contains in the *Introduction* a long methodological note; it is attached to remarks about Kummer's ideal numbers and his own ideals. In that note he points to (1872c) as making even more evident—for the case of introducing the irrational numbers and defining the arithmetic operations on them—the "legitimacy, or rather necessity, of such demands, which must always be imposed with the introduction or creation of new arithmetic elements". He refers here to the demands concerning the precise definition of new mathematical objects in terms of already existing ones and the general definition of operations on them in terms of the given ones. In contemporary language, the structures of pairs and cuts provide models of the axioms for integers, rationals, and reals; the particular elements of these structures are not identified with the respective numbers, but the latter are specifically obtained by an *abstracting* free creation.

---

[31]Cf. the letter from Lipschitz to Dedekind dated 11 March 1876; in (Lipschitz 1986, pp. 47–48).

Dedekind's analysis of number

If we think of the genetic method as underlying the construction of mathematical objects, systems of which are models of appropriate axiom systems, we can see very clearly how it complements in Dedekind's hands an axiomatic approach. However, an arithmetization of analysis that satisfies Dedekind's methodological demands for creating the irrational numbers has not been achieved yet: for that it is essential to characterize the very basis of the construction, the natural numbers. First steps beyond *Arithmetische Grundlagen* are taken in the manuscript (1872/1878) for (1888) that was written, modified, and extended between 1872 and 1878. At the end of this period Dedekind must have thought about publishing a booklet with the very title of (1888), as Heinrich Weber writes in a letter of 13 November 1878:[32] "I am awaiting your book *Was sind und was sollen die Zahlen* with great anticipation." In the Introduction to (1888) on page IV, the earlier manuscript is said to contain "all essential basic thoughts of my present essay". Dedekind mentions as the main points the "sharp distinction between the finite and the infinite", the concept of cardinal, the justification of proof by induction and definition by recursion.

The emphasis in the draft is, however, almost exclusively on the proof principle; there are some very brief, almost cryptic hints concerning definition by recursion. From a modern perspective there is so much more to the final essay; for one, the detailed metamathematical considerations and Dedekind's reflections based on them. In the letter to Keferstein they are properly emphasized, and we will discuss them in Section 6, in particular, the existence and uniqueness, up to isomorphism, of simply infinite systems. That will be the background for discussing the free creation of numbers with a more systematically founded perspective.

5 Chain of a system

Weber's "great anticipation" was more than justified already in 1878, as Dedekind's reflections had led him to a novel conceptualization of natural numbers within, what he viewed as, a *logical* framework using the fundamental concepts of *system* and *mapping*. Indeed, in the manuscript (1872/1878) Dedekind writes:

If one accurately tracks what we are doing when we count a set or a number of things, one is necessarily led to the concept of correspondence or mapping. The concepts of system, of mapping, which shall be introduced in the following in order

---

[32]The letter is found in Appendix L of (Dugac 1976, p. 272). In German the remark is: "Deinem Buch *Was sind und was sollen die Zahlen* sehe ich mit grosser Spannung entgegen." Dedekind responded on 19 November 1878, saying: "Du fragst auch nach meiner Untersuchung über den Uranfang der Arithmetik: "Was sind und was sollen die Zahlen?" Sie ruht und ich zweifle, ob ich sie je publiciren werde; sie ist auch nur in rohem Entwurfe aufgeschrieben, mit dem Motto: "Was beweisbar ist, soll in der Wissenschaft nicht ohne Beweis geglaubt werden." Die Hauptsache ist die Unterscheidung des Zählbaren vom Unzählbaren, und der Begriff der Anzahl, und die Begründung der sog. vollständigen Induction." (Dedekind 1932, p. 486).

to ground the concept of number, cardinal number, remain indispensable for arithmetic even if one wants to assume the concept of cardinal number as being immediately evident ("inner intuition").[33]

This is the basis for the *radical break* with the considerations in (1854) and the description of the positive integers in (1872c), a break that was hinted at by the notational change from the creative act $+1$ to the successor operation φ in the third version of *Arithmetische Grundlagen*. However, the facts one is forced to accept from an informal analysis of number using these new conceptual tools "are still far from being adequate for completely characterizing the nature of the number sequence N";[34] for that the general notion of the *chain of a system A* is introduced. The specialization of A to the system $\{1\}$ leads to the "complete" characterization of N as a simply infinite system.

5.1 Mappings (between 2 and 3).

We do not mean to discuss mappings between the integers two and three, but rather emphasize the significance of the notion of mapping that emerged in Dedekind's work between the publication of the second and third edition of Dirichlet's *Zahlentheorie* (1863) in 1871 and 1879, respectively. This was a fruitful and important period in Dedekind's work on algebraic number theory: he published the essay *Sur la théorie des nombres entiers algébriques* and worked intermittently, but strenuously, on a proper formulation of his *Gedanken über Zahlen*. The broad considerations, which were central for the mathematical and the foundational work, are highlighted in the *announcement* of the third edition and in a footnote to that very work. Indeed, Dedekind refers back to these considerations in (a note to §161 of) the fourth edition of (1863) indicating very clearly, how important those reflections were for him:

It is stated already in the third edition of the present work (1863, 3rd edition, footnote on p. 470) that the entire science of numbers is also based on this intellectual ability to compare a thing a with a thing a', or to relate a to a', or to let a correspond to a', without which no thinking at all is possible. The development of this thought has meanwhile been published in my essay "Was sind und was sollen die Zahlen?" (Braunschweig, 1888); ...[35]

---

[33] From (1872/1878), printed in (Dugac 1976, p. 293): "Verfolgt man genau, was wir beim Abzählen der Menge oder Anzahl von Dingen thun, so wird man nothwendig auf den Begriff der Correspondenz oder Abbildung geführt.
Die Begriffe des Systems, der Abbildung, welche im Folgenden eingeführt werden, um den Begriff der Zahl, der Anzahl zu begründen, bleiben auch dann für die Arithmetik unentbehrlich, selbst wenn man den Begriff der Anzahl als unmittelbar evident ("innere Anschauung") voraussetzen wollte."

[34] Dedekind in his letter to Keferstein, in (van Heijenoort 1967, p. 100). The German text is: "Aber ich habe in meiner Entgegnung ... gezeigt, daß diese Tatsachen noch lange nicht ausreichen, um das Wesen der Zahlenreihe N vollständig zu erfassen."

[35] "Schon in der dritten Auflage dieses Werkes (1879, Anmerkung auf S. 470) ist ausgesprochen, daß auf dieser Fähigkeit des Geistes, ein Ding a mit einem Ding a' zu vergleichen, oder a auf a' zu beziehen, oder dem a ein a' entsprechen zu lassen, ohne welche überhaupt kein Denken möglich ist, auch die gesamte Wissenschaft der Zahlen beruht. Die Durchführung

Dedekind's analysis of number 57

This remark is attached to a discussion of the general notion of mapping. The evolution of that notion in Dedekind's work is one of the foci of our second paper, but the material from the manuscript *Gedanken über Zahlen* reveals already crucial aspects of this development and its significance.

The manuscript contains three distinct layers.[36] In its initial attempt to characterize natural numbers via chains, the first layer uses the notions *mappable, corresponding*, and *image*, which match (Dedekind 1871, Section I of §159 in Supplement X) as well as (1872c) in terminology and outlook. In its second attempt, calling a chain now a *group* (sic), the manuscript introduces for the first time in Dedekind's writings the term *mapping* (Abbildung). Dedekind distinguishes without any explanation between *injective* (deutliche) and *non-injective* (undeutliche) mappings. The second layer is the longest and most intricate one, and it alone discusses finite cardinals. The third layer is close to the eventual presentation of this material in (1888) and takes mappings officially as objects of study; it matches the remarks and note in (1863, 3rd edition) mentioned above.

5.2 Thoughts on numbers. Let us indicate briefly the *common* arithmetic content. In each layer Dedekind considers a system S and a (n arbitrary) mapping φ from S to S.[37] If φ is injective, the system S is called *infinite* just in case there is a proper subset U of S, such that the system $\varphi(S)$ of images is a subset of U. The other notions are defined relative to S and φ. A subset K of S is called a *chain* if and only if it is closed under φ. A subset B of S is called *dependent on* A if and only if B is a subset of any chain that contains A, and (A) is the system of all things dependent on A. Finally, Dedekind establishes as the central claim that (A) *is a chain*. As a justification for induction one can easily show that, given two subsets A and K of S,

If $A \subseteq K$ and $\varphi(K) \subseteq K$, then $(A) \subseteq K$.

Assume $A \subseteq K$ and $\varphi(K) \subseteq K$, consider an arbitrary a in (A), and distinguish two cases. In the first case a is in A, then—by the assumption $A \subseteq K$—a is in K. In the second case $\{a\}$ is dependent on A, but not in A, i.e., contained in any chain that contains A. But K is such a chain; thus $\{a\}$ is a subset of K, and a is an element of K. This sequence of steps anticipates that in (1888), except for the definition of (A) via the dependency relation.

---

dieses Gedankens ist seitdem veröffentlicht in meiner Schrift "Was sind und was sollen die Zahlen?" (Braunschweig 1888); . . . "

[36] The first layer extends in (Dugac 1976) from p. 293 to p. 297, the second from p. 297 to p. 304, and the third from p. 304 to p. 309. The order of the layers reflects, quite clearly, the temporal evolution of Dedekind's ideas, with only one exception: much of the material in the right-hand columns on pp. 293–294 must have been added later. In particular, we conjecture that the remarks quoted above from p. 293 of the manuscript (at the very beginning of this part of our paper) are from a later date; they fit systematically best with the beginning of the third layer. The material on p. 294 uses notations that are introduced and explained only on p. 308, respectively on p. 301.

[37] It should be emphasized that the (in our view, original part of the) first layer does not have the explicit notation φ for a mapping; §159 of (1871) does, but only for substitutions, i.e., isomorphisms between fields.

The second layer defines the dependency relation just for elements and calls the system (a) of all elements dependent on a the *sequence of* a. For injective mappings and infinite systems S Dedekind establishes as a theorem that (1) *is an infinite system*. Every element of (1) is called a *number*; proof by induction is justified as above, and the issue of definition by recursion is raised here, briefly. Dedekind notes on the margin:

The proof of the correctness of the method of proof from n to $n+1$ is correct; in contrast, the proof (completeness) of the definition of concepts by the method from n to $n+1$ is not yet sufficient at this point; the existence (consistent) of the concept remains in doubt. This will become possible only by *injectivity*, by the consideration of the system [n]!!!!!! Foundation.[38]

This is a pregnant remark and, together with theorems established on pp. 300–304, points ahead to central issues in (1888). To support that claim, we have to explain first of all the notation $[n]$. Informally, $[n]$ is the system of all numbers less than or equal to n, for any n in (1); systematically, $[n]$ is defined as the system of numbers not contained in (n'), and it is shown to be finite. (In (1888) the systems $[n]$ are denoted by Z_n.) Dedekind formulates as a theorem that *a system B is infinite, if every system* $[n]$ *can be mapped injectively into B*. He remarks on the margin, "To prove this is circuitous, but possible". (Umständlich, aber möglich zu beweisen.) This is, of course, the central and deep fact used to establish in §14 of (1888) that Dedekind's definition of infinite is equivalent to the standard one.[39] That proof requires definition by recursion and a form of the axiom of choice:[40] to secure generally the existence of a mapping satisfying recursion equations the systems $[n]$ are invoked and the *injectivity* of the mapping φ is needed (Remark 130 of (1888)). All of this seems to be hinted at in the remark quoted above; it is a dramatic step for gaining a proper perspective.

The third layer is a very polished version of the considerations leading up to Theorem 31 that states, (A) *is a chain*. But this time there is a most interesting and important note next to the statement of the theorem: "(A) is the 'smallest' chain that contains the system A." The layer ends with brief remarks on the "direct treatment of the system Z of natural (i.e., whole positive rational) numbers". We quote those in full and mention that Dedekind wrote next to the sentence just quoted "better N than Z":

Characteristic of the system Z. There is an injective mapping from Z — if T is a part of Z, then the image of T is denoted by T' — which has the following property.

---

[38](Dugac 1976, p. 300): "Der Beweis der Richtigkeit der Beweismethode von n auf $n+1$ ist richtig; dagegen ist der Beweis (Vollständigkeit) der Begriffserklärung durch die Methode von n auf $n+1$ an dieser Stelle noch nicht genügend; die Existenz (widerspruchsfrei) des Begriffs bleibt zweifelhaft. Dies wird erst möglich durch die *Deutlichkeit*, durch die Betrachtung des Systems [n]!!!!!! Fundament."

[39]Such a standard definition is given, for example, in Bolzano's *Paradoxien des Unendlichen*, Sections 8–9.

[40]That is now well-known and was first established in (Tarski 1924); additional details are found in (Belna 1996) on p. 41. As a matter of historical record, Zermelo remarked already on the use of the axiom of choice in Dedekind's proof on p. 188 of (Zermelo 1908a).

Dedekind's analysis of number

I Z' is a part of Z.

II There is a number (i.e., a thing contained in Z), which is not contained in Z'. This number shall be called "one" and is denoted by 1.

III A number chain (i.e., each part T of Z, whose image T' is a part of T) that contains the number 1 is identical with Z.[41]

This "characteristic" corresponds perfectly to the axiomatic conditions for a simply infinite system in (1888), i.e., we have here the very first formulation of the so-called Peano Axioms.[42]

5.3 Axioms for numbers.   In the systematic analysis of (1888) we use Dedekind's letter to Keferstein, but also his official reply (1890b) to Keferstein's review of (1888). Dedekind makes his methodological considerations much more explicit in these documents than in the essay itself. Indeed, in the letter Dedekind poses these motivating questions:

What are the mutually independent fundamental properties of the sequence N, that is, those properties that are not derivable from one another but from which all others follow? And how should we divest these properties of their specifically arithmetic character so that they are subsumed under more general notions and under activities of the understanding without which no thinking is possible at all but *with* which a foundation is provided for the reliability and completeness of proofs and for the formulation of consistent definitions of concepts?[43]

---

[41]*Charakteristik des Systems Z*. Es giebt eine deutliche Abbildung von Z — ist T ein Theil von Z, so soll das Bild von T mit T' bezeichnet werden -, welche folgende Eigenschaft besitzt.
  I. Z' ist Theil von Z.
  II. Es giebt eine Zahl (d.h. ein in Z enthaltenes Ding), welche nicht in Z' enthalten ist. Diese Zahl soll "Eins" heissen und mit 1 bezeichnet werden.
  III. Eine Zahlkette (d.h. jeder Theil T von Z, dessen Bild T' ein Theil von T ist), welche die Zahl 1 enthält, ist identisch mit Z.

[42]Peano mentions in the Introduction of his (1889): "In this paper I have used the research of others." In particular, he states later in the paragraph that begins with the sentence just quoted, "Also quite useful to me was the recent work by Dedekind, *Was sind und was sollen die Zahlen* (Braunschweig 1888), in which questions pertaining to the foundations of numbers are acutely examined." (p. 103) (Belna 1996), on p. 60, refers to a text from 1891, in which "Peano recognizes that his axioms are 'due to Dedekind' and drawn from # 71 of the latter's book". Stein remarks in his (2000b) that "Giuseppe Peano directly borrowed his axioms for arithmetic" from Dedekind's characterization of the system of natural numbers as a simply infinite system. Peirce made priority claims at a number of occasions; they are discussed very well, and accorded their proper place, in (Belna 1996) on pp. 57–59. It is quite clear from the above discussion that Dedekind gives an analysis of natural numbers in (1872/1878) that culminates in their axiomatic characterization. However, the further claim — as found in (Belna 1996) on p. 58 and (Stein 2000b) — that there is no essential difference (except by the absence of the theorem concerning the existence of infinite systems) between the (1872/1878) manuscript and (1888) is not correct; for example, none of the metamathematical results and broader conceptual reflections discussed in Section 6 are contained in (1872/1878).

[43](van Heijenoort 1967, pp. 99–100), except for a correction in the very last sentence, where "formulation of consistent definitions of concepts" replaces "construction of consistent notions and definitions". The German text, also reprinted in (Sinaceur 1974, p. 272), is as follows: "Welches sind die von einander unabhängigen Grundeigenschaften dieser Reihe N, d.h. diejenigen Eigenschaften, welche sich nicht aus einander ableiten lassen, aus denen aber alle anderen folgen? Und wie muss man diese Eigenschaften ihres spezifisch arithmetischen

When one poses the problem in this way, Dedekind continues, then one is forced to accept the following facts: the number sequence N is a system of elements or individuals, called *numbers*; the relation between these elements is given by a mapping φ from N to N; φ must be similar (ähnlich, this term replaces "deutlich" used in the earlier discussion); the image of N under φ is a proper part of N, and 1 is the only element not in the image. The central methodological problem, Dedekind emphasizes, is the precise characterization of just those individuals that are obtained by iterated application of φ to 1; this is to be achieved in general logical terms, not presupposing arithmetic notions. Before addressing this central problem, Dedekind introduces as above, relative to a system S and an arbitrary mapping φ from S to S, the general concept of a chain. Then he defines directly, using the insight gained in the third layer of (1872/1878), the *chain A_0 of a system A* as the intersection of all chains containing A. A_0 obviously contains A as a subset, is closed under the operation φ, and is minimal among the chains that contain A, i.e., if $A \subseteq K$ and $\varphi(K) \subseteq K$, then $A_0 \subseteq K$. These properties characterize A_0 uniquely. From the minimality of A_0 it is easy to prove a general induction principle in the form:

(*) if $A \subseteq \Sigma$ and $\varphi(A_0 \cap \Sigma) \subseteq \Sigma$, then $A_0 \subseteq \Sigma$

Σ denotes the extension of any property E pertaining to the elements of S.

After this preparatory step Dedekind specializes the consideration to the chain N of the system $\{1\}$ for the similar mapping φ, i.e., the *simply infinite system* $(N, \varphi, 1)$. The essence of this system is given by the axiomatic conditions α, β, γ, and δ of *Erklärung 71* in corresponding order: $\varphi(N) \subseteq N$, $N = 1_0$, $1 \notin \varphi(N)$, and φ is a similar mapping. Condition β expresses in Dedekind's notation that N is the chain $\{1\}_0$ of the system $\{1\}$; it is the basis for the usual induction principle for natural numbers formulated now as follows:

(**) If $\{1\} \subseteq \Sigma$ and $\varphi(N \cap \Sigma) \subseteq \Sigma$, then $N \subseteq \Sigma$

The considerations leading to (**) are completely parallel to those for (*) above. Indeed, reordering conditions α, β, γ, and δ, reformulating them a little, and using (**) as the induction principle yields:

$1 \in N$,

$(\forall n \in N)\, \varphi(n) \in N$,

$(\forall n, m \in N)(\varphi(n) = \varphi(m) \Rightarrow n = m)$,

$(\forall n \in N)\, \varphi(n) \neq 1$, and

$(1 \in \Sigma\ \&\ (\forall n \in N)(n \in \Sigma \Rightarrow \varphi(n) \in \Sigma)) \Rightarrow (\forall n \in N)\, n \in \Sigma$.

Characters entkleiden, der Art, dass sie sich allgemeinen Begriffen und solchen Tätigkeiten des Verstandes unterordnen, *ohne* welche überhaupt kein Denken möglich ist, *mit* welchen aber auch die Grundlage gegeben ist für die Sicherheit und Vollständigkeit der Beweise, wie für die Bildung widerspuchsfreier Begriffs-Erklärungen?"

Dedekind's analysis of number 61

These statements make explicit the principles underlying Dedekind's earlier "characteristic of the system Z" and are mere notational variants of the five axioms for the positive integers formulated in Peano's (1889).

Hilbert's axiomatization for N in his (1905a) also uses these axioms, clearly extracted from Dedekind's characterization of simply infinite systems.[44] Hilbert's syntactic consistency proof in that paper was to guarantee the existence of the "smallest infinite". Thus, his proof was to serve the dual purpose of Dedekind's argument for the existence of a simply infinite system. Already in his (*1899b) and (1899a) Hilbert intended to ensure the existence of a set, here the set of real numbers, by a "direct" proof of the consistency of an appropriate axiomatic theory. The theory was formulated in the style of Dedekind: one considers a system of objects satisfying certain axiomatic conditions, and the systematic development of the theory makes use of these conditions only. In contrast to Dedekind, Hilbert called a theory consistent if it does not allow to establish in finitely many steps a contradiction; note that this is only a quasi-syntactic specification of consistency, as the steps that are allowed in proofs were not made explicit. We turn our attention now to Dedekind's way of thinking about, and addressing, the issue of existence and consistency; this is the last component in the assembly of a Dedekindian perspective on numbers and the nature of mathematics.

6 Abstract type

The number sequence N is characterized completely as the abstract type of a simply infinite system, Dedekind writes to Keferstein; how is this to be understood? The answer to the question will evolve through a sequence of detailed metamathematical, reflective steps concerning simply infinite systems. The steps are guided by the systematic insights gained in the earlier investigations; thus, Dedekind is concerned with the "possibility of the creation of a simply infinite system" — to use the language of the early axiomatic analysis of number systems reported in Section 3; cf. in particular Section 3.3. We first discuss the existence proof for simply infinite systems and then complement the literal uniqueness of the chain of the system $\{1\}$ by the completely new sense of uniqueness "up to isomorphism". Finally, we describe Dedekind's view of the science of numbers or arithmetic that is based on the metamathematical work.

6.1 Logical existence. A simply infinite system is defined as a triple $(N, \varphi, 1)$ or, in contemporary model-theoretic terminology, as a structure that satisfies the conditions α, β, γ, and δ from Dedekind's *Erklärung 71*. We saw that these conditions correspond to the so-called Peano Axioms. Given the earlier concerns, it is perfectly natural for Dedekind to ask: "Does such

---

[44] For details, see (Sieg 2002, pp. 366–371). Hilbert does not formulate the induction principle; he just claims that it can be formulated in a way that is suitable for his investigations.

a system exist at all in our realm of thoughts?"[45] The affirmative answer to this question is given by a *logical existence proof*, and Dedekind explains to Keferstein that without such a proof "it would remain always doubtful, whether the concept of such a system does not perhaps contain internal contradictions".[46] In his official response (1890b) to Keferstein's review article Dedekind asserts more strongly, "as long as such a proof has not been given one may fear that the above definition of the system N contains an internal contradiction, whereby the certainty of arithmetic would be lost".[47] That is the reason, he emphasizes in his letter, why the proofs for theorems 66 and 72 of his essay are necessary.

The crucial considerations are presented in the proof of theorem 66. Theorem 72 just states that every infinite system contains a simply infinite one as a part, and *that* assertion can be established straightforwardly. To establish theorem 66, i.e., the claim that there is an infinite system, Dedekind formulates and proves the claim for a specific system, namely, for his *Gedankenwelt*. Dedekind's *Gedankenwelt* is defined as "the totality S of all things that can be an object of my thinking". For an arbitrary element s of S, the thought s' that "s can be an object of my thinking" is itself an element of S. The operation φ that leads from s to s' is injective, and the set of images S' is a proper part of S, as Dedekind's own self, for example, is in S but not in S'. Thus, S together with φ is indeed an infinite system. The terminology of "thing" and "system" was already introduced at the very beginning of the (1872/1878) manuscript and not only in (1888): "A thing", it says there, "is any object of our thinking; ..." and "A system ... S of things is determined, if one can judge of any thing, whether or not it belongs to the system." Dedekind notes there also that such a system of things is treated as a new thing when contrasted with the other things. These remarks can be found in (Dugac 1976, p. 293); we mention them here to make perfectly clear that the use of these notions in the argument above does not locate it close to the actual writing of (1888).

In (1890b) Dedekind reproduces the proof of theorem 66, asserts that he considers it not only as *correct*, but as *rigorously correct* (streng richtig), and explicates it in an informative way without, as he claims, adding anything new. The explication consists in expanding the specification of φ by a parenthetical remark. Instead of considering "the thought s' that ...", Dedekind considers here "the thought s' (expressible in the form of a sentence or judgment) that ...". This seems to indicate directly that Dedekind's thoughts are not to be viewed as psychological ideas. There is also indirect evidence: Frege asserts in his manuscript *Logik* that he uses the word "Gedanke" in

---

[45]The German text is: "Existiert überhaupt ein solches System in unserer Gedankenwelt?" — In (van Heijenoort 1967) "Gedankenwelt" is misleadingly translated as "realm of ideas".

[46]The German text is: "Ohne den logischen Existenz-Beweis würde es immer zweifelhaft bleiben, ob nicht der Begriff eines solchen Systems vielleicht innere Widersprüche enthält."

[47](Sinaceur 1974, p. 266). The German text is: "... so lange ein solcher Beweis nicht geliefert ist, darf man befürchten, dass die obige Definition des Systems N einen inneren Widerspruch enthält, womit dann die Gewissheit der Arithmetik hinfällig würde."

Dedekind's analysis of number 63

an unusual way and remarks that "Dedekind's usage agrees with mine".[48] Such a Fregean understanding is reinforced, when Dedekind continues his explication by claiming that the thought s' can be an object of his thinking. After all, "I may think, e.g., of this thought s', that it is obvious, that it has a subject and a predicate, etc". (Ich darf z.B. von diesem Gedanken s' denken, dass er selbstverständlich ist, dass er ein Subjekt und ein Prädikat besitzt u.s.w.). Consequently, the thought s' is an element of S.

In (1888), Dedekind writes in the footnote to theorem 66, "A similar consideration is found in §13 of Bolzano's *Paradoxien des Unendlichen* (Leipzig 1851)." The similarity of their considerations is particularly striking, when we compare Bolzano's argument with Dedekind's in the explicated form pertaining to thoughts that are expressible in the form of sentences. Bolzano establishes that "the set of sentences and truths in themselves" (die Menge der Sätze und Wahrheiten an sich) is an infinite multiplicity. This is achieved by considering first any truth T whatsoever and then using the construction principle *the proposition A is true* to step from any true proposition A to a distinct new and true proposition. Bolzano concludes that this set of all propositions constructed from T "enjoys a multiplicity surpassing every individual integer" and is therefore infinite, according to his definition.[49] (The characterization of "this" set, or of similarly constructed ones, as the chain of $\{T\}$ was for Dedekind according to his letter to Keferstein, "... one of the most difficult points of my analysis and its mastery required lengthy reflection". (p. 100))

Excursion. The need to prove the existence of an infinite system is not even discussed in (1872/1878). A proof is given in the manuscript from 1887 that precedes the final writing of (1888), and this seems to be the first appearance in Dedekind's manuscripts or published writings. Through a letter from Cantor to Dedekind dated 7 October 1882 we know that the former sent with his letter also a copy of Bolzano's booklet to Dedekind.[50] These three facts are taken as evidence, for example by Dugac, that Dedekind adapted Bolzano's considerations concerning the objective existence of the infinite. (Cf. (Dugac 1976, pp. 81 and 88), but also (Sinaceur 1974, p. 254), (Belna 1996, pp. 37, 38, and 54ff) and (Ferreirós 1999, pp. 243–246); on p. 243 Ferreirós takes it for granted that

---

[48]Quoted from p. 138 of (Frege 1969); cf. ibid. pp. 147–148, where Frege analyzes Dedekind's proof, approvingly. — McCarty asserts in his (1995) that Section 66 distinguishes itself "as the most blatantly psychologistic". To support this claim in note 5, p. 93, and also to bolster his contention of a strong connection between Kant and Dedekind on p. 71, McCarty relies on the mistranslation of "Gedankenwelt" as "realm of ideas" in (van Heijenoort 1967). McCarty writes on p. 71: "... we will find the mathematical objects of Dedekind among the pure ideas of Kant. Dedekind did, after all, write to Keferstein that he must locate the infinite system of natural numbers 'in the realm of our ideas'."

[49](Bolzano 1851, p. 258). Bolzano's definition, given on p. 254, is as follows: "... I propose the name *infinite multitude* for one so constituted that every single finite multitude represents only a part of it." Note that Bolzano uses finite multitude and whole number synonymously; see also note 57.

[50](Dugac 1976, p. 256). Cantor characterizes Bolzano's booklet as "a peculiar little work" (ein merkwürdiges Werkchen) of which he happened to have a second copy.

Dedekind knew Bolzano's proof when giving his own and "transformed it to suit his different philosophical ideas and his strict definition of infinity".)

We first describe in (almost tedious) detail, where the claim and proof for the existence of an infinite system occur first, namely, in the fourth section of (1887a) that is entitled "The finite and infinite". It starts out with a definition.

40. Definition. S is called an *infinite* system, if there is an injective mapping from S, such that the image of S is a *proper* part of S; in the opposite case S is called a *finite* system.[51]

This is followed by the remark that "all hitherto known definitions of the finite and the infinite are completely useless, to be rejected by all means".[52] Next comes a proposition, numbered 41, which states that the union S of the singleton $\{a\}$ and T is finite, if T is finite.[53] As in other manuscripts of Dedekind's, the pages of (1887a) are vertically divided in half. The main text is written on one half, whereas the other half is reserved for later additions. On this particular page a number of important additions have been made. Already its first line indicates that the manuscript is still being reorganized in significant ways: Dedekind refers to remarks on a separate page and writes that the "first two propositions of §7 belong here". This is followed by three propositions, numbered 40^x, 40^{xx}, and 40^{xxx}, the last of which claims: "There are infinite systems." Dedekind adds parenthetically, "Remarks on separate page", and mentions there that the following proposition can be added immediately to the *fundamental definition 40*:

Proposition: There are infinite systems; the system S of all those things s (this word understood in the sense given in the introduction) that *can be* objects of my thinking, is infinite (my realm of thoughts).[54]

The proposition is established by a proof of roughly the same character as that given in the sources we discussed already.[55]

In the preface to the second edition of (1888), Dedekind emphasizes that Cantor and Bolzano had also recognized the property he uses as the definition of an infinite system. However,

... neither of these authors made the attempt to use this property as the definition of the infinite and to establish upon this foundation with rigorous logic the science

---

[51](Dedekind 1887a). "40. Erklärung: S heißt ein <u>unendliches</u> System, wenn es eine derartige deutliche Abbildung von S gibt, daß das Bild von S ein <u>echter</u> Teil von S ist; im entgegengesetzten Fall heißt S ein <u>endliches</u> System." (The underlining is Dedekind's.)

[52](Dedekind 1887a). "Anmerkung: alle bisher bekannten Definitionen des Endlichen und Unendlichen sind gänzlich unbrauchbar, durchaus zu verwerfen."

[53](Dedekind 1887a). "41. *Satz*: Ist $S = M(a, T)$, wo a ein Element von S, und T ein endliches System bedeutet, so ist auch S ein endliches System." M is the union operation. This is essentially Theorem 70 in (1888).

[54](Dedekind 1887a). "Satz. Es giebt unendliche Systeme; das System S aller derjenigen Dinge s (dieses Wort in dem in der Einleitung angegebenen Sinne verstanden), welche Gegenstand meines Denkens *sein können*, ist unendlich (meine Gedankenwelt)."

[55]More will be said on the details of the reorganization in Section 5 of (1887a) at another occasion.

Dedekind's analysis of number

of numbers. But this is precisely the content of my difficult labor, which in all its essentials I had completed several years before the publication of Cantor's memoir [i.e., (Cantor 1878)] and at a time when the work of Bolzano was completely unknown to me, even by name.[56]

Whether and how Dedekind was influenced by Bolzano's work in formulating his proof of theorem 66 remains a topic of speculation. The known facts, as we recounted them, allow a different interpretation than that given by Dugac and Ferreirós: the manuscript of 1887 is so different from (1872/1878) that one might conjecture with good reason that Dedekind had other intermediate manuscripts or, at least, additional notes to bridge this remarkable conceptual and mathematical gap. The issue of providing "models" for axioms had been pressing already at that time, as we pointed out in Section 4. Given Dedekind's own remarks concerning the connection with Bolzano's and Cantor's work (indicated in the above quote), we speculate that he must have completed one such intermediate manuscript *no later than* 1878. Dedekind's own remarks, quoted above at the end of Section 4, about these early considerations do not shed decisive light on the issue at hand: in his response to Weber's inquiry concerning the status of *Was sind und was sollen die Zahlen?* from 1878 he gives a description that fits the available material in (1872/1878), and views it as a "rough draft"; in the Introduction to (1888) the earlier material is said to include also the justification of definition by recursion (that is barely hinted at in the folder of (1872/1878)); finally, in his remark concerning Cantor and Bolzano in the preface to the second edition of (1888) we just quoted, he claims to have completed the work "in all its essentials" *several years* before the appearance of Cantor's 1878 paper. *End of Excursion*.

Dedekind himself just points out a "similarity" between Bolzano's and his own considerations. We point to a central dissimilarity and, without further elaboration, to the fact that Dedekind's formulations are dramatically more rigorous.[57] Bolzano bases his considerations concerning the *objective existence* of the infinite implicitly on the existence of the species of integers and explicitly on the existence of the set of sentences and truths in themselves, whereas Dedekind uses only one universal system, his

---

[56] (Ewald 1996, p. 796). The German text — in (Dedekind 1888, IX–X) — is as follows: "... keiner der genannten Schriftsteller hat den Versuch gemacht, diese Eigenschaft zur Definition des Unendlichen zu erheben und auf dieser Grundlage die Wissenschaft von den Zahlen streng logisch aufzubauen, und gerade hierin besteht der Inhalt meiner mühsamen Arbeit, die ich in allem wesentlichen schon mehrere Jahre vor dem Erscheinen der Abhandlung von G. Cantor [i.e., (Cantor 1878)] und zu einer Zeit vollendet hatte, als mir das Werk von Bolzano selbst dem Namen nach gänzlich unbekannt war."

[57] We discuss how Dedekind obtains natural numbers below. Compare that approach to Bolzano's quick step in §8, where — after describing the formation of series that start from a particular individual of a species A and proceed by adjoining a fresh individual from that species — says: "Such multitudes I call *finite* [endlich] or *countable* [zählbar], or quite boldly: *numbers*; and more specifically: *whole* numbers — under which the first term shall also be comprised." Here and in §13, where Bolzano establishes the existence of an infinite set, the proper general ("logical") characterization of the set of objects that are obtained from an initial one via some successor operation is completely missing.

Gedankenwelt; a simply infinite system and the natural numbers are obtained from it.

6.2 Mathematical uniqueness.

How then are natural numbers obtained in Dedekind's case? Any infinite system whatsoever has as a part a simply infinite one that is unique as a minimal chain (of a chosen element 1), as we observed above. To insist on minimality has the metamathematical reason emphasized by Dedekind in both (1890a) and (1890b), namely, that it excludes "intruders"; these intruders are, in modern terminology, non-standard elements. The minimality captures the informal, motivating idea that every element of the chain is obtained by the finite iteration of the operation φ applied to 1.[58] This is also the basis for establishing that simply infinite systems are unique in a novel sense.

Given the analysis via minimal chains, it is most direct to use the general concept of a mapping and to conceive of a bijection ψ between two arbitrary simply infinite systems based on operations φ and θ, respectively. ψ would map the first element of one system to the first element of the other; in addition, the bijection would satisfy the recursion equation $\psi(\varphi(n)) = \theta(\psi(n))$. It is *one thing* to graphically draw such a connection, but *quite another thing* (i) to have the appropriate mathematical (or logical) notions to capture the essence of the situation and (ii) to prove the unique existence of such a structure-preserving mapping. The *one thing* is undoubtedly in everybody's mind, certainly Bolzano's and also Kronecker's, for example, in §1 of his (1887) entitled *Definition des Zahlbegriffs*. The *other thing* is what Dedekind does in §9 of (1888)!

Dedekind isolates the crucial feature in theorem 126, *Satz der Definition durch Induktion*: let $(N, \varphi, 1)$ be a simply infinite system, let θ be an arbitrary mapping from a system Ω to itself, and let ω be an element of Ω; then there is exactly one mapping ψ from N to Ω that satisfies the conditions

I $\psi(N) \subseteq \Omega$,
II $\psi(1) = \omega$,
III $\psi(\varphi(n)) = \theta(\psi(n))$.[59]

The justification requires subtle metamathematical considerations; i.e., a proof by induction of the existence of approximations to the intended mapping for initial segments of N. The basic idea was used later in axiomatic set theory and extended to transfinite recursion; Gödel used it within formal arithmetic.[60] In the context of his investigation, Dedekind draws two conclusions with the help of theorem 126: on the one hand, all simply infinite systems are similar (theorem 132), and on the other hand, any system that is similar to a simply infinite one is itself simply infinite (theorem 133).

---

[58] Cf. (Sinaceur 1974, p. 268).

[59] Dedekind points out on p. 27 of (1888) what is obvious, namely, that condition I is a consequence of II and III; he only includes it on account of greater clarity (*Deutlichkeit*).

[60] In contrast to induction, the recursion principle is not correct for arbitrary chains; that is discussed in *Bemerkung* 130 of (1888).

Dedekind's analysis of number

These results, together with some observations in remark 134 to which we will return below, "justify completely" the explication of the concept of number Dedekind provided already in *Erklärung* 73:

If in the consideration of a simply infinite system N ordered by a mapping φ we entirely neglect the special character of the elements, simply retaining their distinguishability and taking into account only the relations in which they are placed to one another by the ordering mapping φ, then these elements are called *natural numbers* or *ordinal numbers* or simply *numbers*, and the base-element 1 is called the *base-number* of the *number-series* N. With reference to this freeing of the elements from every other content (abstraction) we are justified in calling the numbers a free creation of the human mind.[61]

In the earlier manuscript (1887a) one finds, after an almost identical remark, a more expanded and explicit formulation concerning the result of the abstraction; Dedekind writes there:

By this abstraction, the originally given elements n of N are turned into new elements n, namely into numbers (and N itself is consequently also turned into a new abstract system $\mathcal{N}$). Thus, one is justified in saying that the numbers owe their existence to an act of free creation of the mind. For our mode of expression, however, it is more convenient to speak of the numbers as of the original elements of the system N and to disregard the transition from N to $\mathcal{N}$, which itself is an injective mapping. Thereby, as one can convince oneself using the theorems regarding definition by recursion, nothing essential is changed, nor is anything obtained surreptitiously in illegitimate ways.[62]

This is *Dedekind abstraction* in its clearest and most direct formulation (and obviously follows the spirit of the remarks concerning the free creation of the real numbers in (1872c)). Though these newly created objects are indeed the numbers, there is nevertheless no need to insist on treating them as the subject of the science of numbers. After having established as proposition 106 the similarity of all simply infinite systems, Dedekind

---

[61](Ewald 1996, p. 809). The German text is: "Wenn man bei der Betrachtung eines einfach unendlichen, durch eine Abbildung φ geordneten Systems N von der besonderen Beschaffenheit der Elemente gänzlich absieht, lediglich ihre Unterscheidbarkeit festhält und nur die Beziehungen auffaßt, in die sie durch die ordnende Abbildung φ zueinander gesetzt sind, so heißen diese Elemente *natürliche Zahlen* oder *Ordinalzahlen* oder auch schlechthin *Zahlen*, und das Grundelement 1 heißt die *Grundzahl* der *Zahlenreihe* N. In Rücksicht auf diese Befreiung der Elemente von jedem anderen Inhalt (Abstraktion) kann man die Zahlen mit Recht eine freie Schöpfung des menschlichen Geistes nennen." — The resonance with the remarks concerning the real numbers in (1872c) is not accidental, as Dedekind makes quite clear in his letter to Weber of 24 January 1888; the letter is found in (Dedekind 1932, pp. 488–490).

[62]The German text is found at the beginning of Section 5 of (1887a): "Da durch diese Abstraction die ursprünglich vorliegenden Elemente n von N (und folglich auch N selbst in ein neues abstraktes System $\mathcal{N}$) in neue Elemente n, nämlich in Zahlen umgewandelt sind, so kann man mit Recht sagen, daß die Zahlen ihr Dasein einem freien Schöpfungsacte des Geistes verdanken. Für die Ausdrucksweise ist es aber bequemer, von den Zahlen wie von den ursprünglichen Elementen des Systems N zu sprechen, und den Übergang von N zu $\mathcal{N}$, welcher selbst eine deutliche Abbildung ist, außer Acht zu lassen, wodurch, wie man sich mit Hilfe der Sätze über Definition durch Recursion ... überzeugt, nichts Wesentliches geändert, auch Nichts auf unerlaubte Weise erschlichen wird."

concludes his deliberations in (1887a) under the heading "Creation of the pure natural numbers":

> It follows from the above, that the laws regarding the relations between the numbers are completely independent from the choice of that simply infinite system N, which we called the number sequence, and that they are also <u>independent</u> from the mapping of N that orders N as a simple sequence.[63]

How does that claim follow "from the above?" — In remark 134 of (1888) Dedekind gives a rough argument that can be interpreted as showing that "categoricity" implies "elementary equivalence". The arguments for theorems 132 and 133 make use of the canonical bijection that transforms elements of one simply infinite system into corresponding elements of the other and that can even be claimed to transform, as Dedekind does, the successor operation of one system into that of the other. Thus, if one considers only propositions in which the particular character of the elements is neglected and only notions are used that arise from the successor function in one system, then these propositions have quite general validity for any other simply infinite system.[64] This gives finally the complete justification of the above remarks and allows the proper characterization of the "object of the *science of numbers* or *arithmetic*" as presented in the second half of the long note 73:

> The relations or laws which are derived entirely from the conditions α, β, γ, δ in 71, and therefore are always the same in all ordered simply infinite systems, whatever names may happen to be given to the individual elements (compare 134), form the next [in (Ewald 1996), one finds "first" here] object of the *science of numbers* or *arithmetic*.[65]

We begin to explore the meaning of this characterization next.

6.3 The science of numbers. From the very start, two broad and intimately connected issues were of paramount importance for Dedekind's foundational reflections, namely, finding fundamental concepts and principles, but also using them for the systematic development of a subject. As the reader may recall, in (1854) Dedekind views introducing a concept as formulating an hypothesis concerning the inner nature of a science. In (1888) introducing the concept "simply infinite system" is more than formulating an hypothesis concerning the essential character of number theory:

---

[63]This is the full text of section 107 in (1887a); the German is: "Aus dem Vorhergehenden ergiebt sich, daß die Gesetze über die Beziehungen zwischen den Zahlen gänzlich unabhängig von der Wahl desjenigen einfach unendlichen Systems N sind, welches wir die Zahlenreihe genannt haben, sowie auch <u>unabhängig</u> von der Abbildung von N, durch welche N als einfache Reihe geordnet ist."

[64]This is nothing but a paraphrase of Dedekind's considerations in # 134 of (1888).

[65](Ewald 1996, p. 809). The German text from (Dedekind 1888) is: "Die Beziehungen oder Gesetze, welche ganz allein aus den Bedingungen α, β, γ, δ in 71 abgeleitet werden und deshalb in allen geordneten einfach unendlichen Systemen immer dieselben sind, wie auch die den einzelnen Elementen zufällig gegebenen Namen lauten mögen (vgl. 134), bilden den nächsten Gegenstand der *Wissenschaft von den Zahlen* oder der *Arithmetik*."

it rather emerges from a deep insight into a capacity of the human mind — "without which no thinking is possible" and with which we have the essential basis for erecting "the entire science of numbers". (That is forcefully expressed in the Preface to the first edition of (1888, pp. iii–iv).) The principle of proof by induction and that of definition by recursion (or induction) can be obtained, and these principles allow the unique characterization of the number sequence — a most significant theoretical insight.

Both principles are also indispensable for a systematic development: proof by induction is the pervasive form of argumentation in number theory, and definition by recursion yields the standard operations like addition, multiplication, and exponentiation. As to the definition principle Dedekind emphasizes in his letter to Keferstein the need "to formulate the *definitions* of operations on numbers consistently for *all* numbers n". This is one aspect of Dedekind's general concern to provide the conceptual tools for the development of arithmetic and to establish their efficacy in recognizing general truths. Let us recall the parallel considerations in (1872c). There, the notion of continuity allows the characterization of the real numbers, the operations on reals are defined via operations on the rationals, and their basic properties are verified. Then Dedekind develops a fundamental part of analysis and establishes, in particular, that the principle of continuity implies the theorem that bounded increasing sequences have a limit. This theorem is actually shown to imply the principle of continuity; thus, we have here — as far as we know — the very first theorem of "reverse mathematics". This equivalence has for Dedekind significant methodological impact; in his letter to Lipschitz dated 27 July 1876 he emphasizes this equivalence, when making the point that his definition of irrational numbers has not created any number, "which was not already grasped more or less clearly in the mind of every mathematician".[66]

The reflections concerning induction and recursion have consequently two fundamental goals: to serve as the methodological frame for Dedekind's answer to the question *Was sind die Zahlen?* and to provide the systematic tools for developing number theory. We saw how they are used to justify the abstractionist move, when the natural numbers were viewed as a "free creation of the human mind"; we also discussed the basic role of induction and recursion in the development of number theory. However, we did not address the question, "What is number theory?" — For Dedekind, as is clearly stated in the second half of note 73, number theory is the "theory" of the sequence N, which is completely characterized as the abstract type of a simply infinite system. Thus, we are facing two central methodological

---

[66]The full German text in (Dedekind 1932, p. 475), is as follows: "Ebenso wenig habe ich gemeint, durch meine Definition der irrationalen Zahlen irgend eine Zahl erschaffen zu haben, die nicht vorher schon in dem Geiste eines jeden Mathematikers mehr oder weniger deutlich aufgefaßt war; dies geht aus meiner ausdrücklichen Erklärung (S. 10 und 30) hervor, daß die durch meine Definition der irrationalen Zahlen erreichte Vollständigkeit oder Stetigkeit (A) des reellen Zahlengebietes wesentlich äquivalent ist mit dem von allen Mathematikern anerkannten und benutzten Satze (B): 'Wächst eine Größe beständig, aber nicht über alle Grenzen, so nähert sie sich einem Grenzwerth'."

questions, which emerge from the general discussion in remark 134 and note 73:

(i) How are the concepts characterized that arise out of the successor operation? (Begriffe, die aus der Anordnung φ entspringen, remark 134)

(ii) How are the laws obtained that are derived exclusively from the conditions for simply infinite systems? (Gesetze, welche ganz allein aus den Bedingungen α, β, γ, δ in 71 abgeleitet werden, note 73)?

The answer to both of these questions is "By logic!", i.e., logic is to specify principles of concept formation (or of definition) and principles of proof; these "logical" principles are not explicitly formulated. A satisfactory analysis has to be sufficiently restricted in order to ensure, what Dedekind argued for, namely, that any proposition using only "those" concepts and having been inferred by only "those" proof principles is valid for any simply infinite system. Dedekind had an appropriate abstract understanding of "theory" already in the 1870s; that is clear not only from his contemporaneous mathematical work, but also from more general methodological discussions, for example, in the letters to Lipschitz dated 10 June and 27 July 1876. (We quoted from the second letter already earlier.) In the second letter he claims, for us most interestingly, that the continuity of space is by no means inseparably connected to Euclid's geometry. He proposes to establish the claim by an analysis of the whole system of Euclidean geometry making clear that the continuity principle is not being used. In a parenthetical addition he remarks on a sure, infallible method for such an analysis. This method consists in replacing all "terms of art" (Kunstausdrücke) by arbitrary newly invented and until now meaningless words; he continues:

... the system must not collapse [by such a replacement], if it has been constructed correctly, and I claim for example, that my theory of the real numbers withstands this test.[67]

These remarks precede Hilbert's pronouncement on "Tische", "Stühle", and "Bierseidel" by quite a few years! In her brief note to (1872c), Emmy Noether points to the letters Dedekind sent in 1876 to Lipschitz and asserts that they not only contain pertinent remarks on the essay itself but also express, in her view, an "axiomatic standpoint" (axiomatische Auffassung).[68] Dedekind's above remarks are the most explicit and concise expression of this standpoint.

---

[67] The German text of the whole passage, on p. 479 of (Dedekind 1932), is: "... eine untrügliche Methode einer solchen Analyse besteht für mich darin, alle Kunstausdrücke durch beliebige neu erfundene (bisher sinnlose) Worte zu ersetzen, das Gebäude darf, wenn es richtig konstruiert ist, dadurch nicht einstürzen, und ich behaupte z.B., daß meine Theorie der reellen Zahlen diese Probe aushält."

[68] In (Dedekind 1932, p. 334).

7 Concluding remarks

Let us return to the narrower historical context sketched in our Introduction and elaborated in Section 3. Kronecker and Helmholtz share, so we claim there, the "naïve" starting-point with Dedekind, but it is only Dedekind who builds a conceptual framework, in which he can express sharply the (naïve) analysis, carry out fruitful metamathematical investigations, and provide the tools for a systematic treatment of number theory. The underlying distinctive methodological themes and their evolution will be at the center of a second essay entitled *Dedekind's structuralism: mappings and models*. We will argue in particular for what was already indicated above: Dedekind's *Stetigkeit und irrationale Zahlen* is a significant stepping-stone in this development.

The essay is commonly viewed as providing the final step in a *genetic* presentation of the reals via cuts. However, from the perspective of (1888) and the (unpublished) work that contains all its central notions already before 1878, (1872c) can be seen as containing a thoroughly axiomatic characterization of the reals as a complete ordered field together with a semantic consistency proof for these axioms; that was observed already in (Sieg 1990a, pp. 264–5). Dedekind's investigation of the correspondence between the geometric line and the system of all cuts contains the crucial elements of a proof of the categoricity of the axioms. What is missing at this stage of his foundational reflections in (1872c) is the *general concept of mapping*. In our second essay we will discuss the emergence of this notion in Dedekind's mathematical and foundational work, as well as the detailed connections with (what Hilbert and Bernays later called) *existential axiomatics* and the *reductive structuralism* that arose from Hilbert's Program, properly understood.

There remains a great deal of important historico-analytical work that can and should be done, in spite of Ferreirós's marvelous book. The latter has opened a larger vista for Dedekind's work and is wonderfully informative in so many different and detailed ways. However, it seems to us to be deeply conflicted about, indeed, sometimes to misjudge, the general character of the foundational essays and manuscripts. Let us mention three important aspects. First of all, there is *no program* of a constructionist sort in (1854) that is then being pursued in Dedekind's later essays, as claimed on pp. 217–218. Secondly, there is no conflict, and consequently no choice has to be made, between a genetic and an axiomatic approach for Dedekind. That conflict is frequently emphasized. It underlies the long metadiscussion (on pp. 119–124) where the question is raised, why authors around 1870, including of course Dedekind, pursued the genetic and not the axiomatic approach. In that meta-discussion Ferreirós seeks reasons "for the limitations of thought in a period", but only reveals the limitations of our contemporary perspective. Finally, there is *no supersession* of Dedekind's "deductive method" (described on pp. 246–248) by the axiomatic method of Hilbert's, but the former is rather the very root of the latter. Hilbert's first axiomatic formulations in *Über den Zahlbegriff* and *Grundlagen der Geometrie* are patterned after Dedekind's. Indeed, Hilbert is a logicist in Dedekind's spirit at that point, and it is no accident that, as late as 1917/1918, he was attracted by attempts to provide a logicist foundation of mathematics.[69]

We mention three broad and temporally distinct directions for such historico-analytic work, namely, (i) a thorough-going exploration of the early mathematical and philosophical context of Dedekind's work, in particular, the impact of Gauss, Dirichlet, Riemann, and Herbart, (ii) a detailed examination of the deep interaction between Dedekind's foundational and mathematical work, in particular, the work on algebraic number theory in the 1870s, and (iii) a thorough investigation of Dedekind's influence on Hilbert's mathematical work, in particular, on the *Zahlbericht*.

---

[69] See (Sieg 1999) and (Mancosu 1999a).

I.2

Methods for real arithmetic

1 Immediate context

Hilbert's synthesizing perspective and the reasons for his insistence on formulating sharp problems when confronting the foundations of mathematics come to life in the long preamble to his Paris address. His account of mathematics emphasizes the subtle and ever recurring interplay of rigorous abstract thought and concrete experience. Turning his attention to the principles of analysis and geometry, he asserts that the "most suggestive and notable achievements of the last century are ... the arithmetical formulation of the concept of the continuum ... and the discovery of non-Euclidean geometry ...". Then he states Cantor's continuum problem as his first and the consistency issue for the arithmetical axioms as his second problem:

> ... I wish to designate the following as the most important among the numerous questions which can be asked with regard to the axioms: *To prove that they are not contradictory, that is, that a finite number of logical steps based upon them can never lead to contradictory results.*[1]

As to the arithmetical axioms, Hilbert points to his paper *Über den Zahlbegriff* delivered at the Munich meeting of the German Association of Mathematicians in December of 1899. Its title indicates a part of its intellectual context, as Kronecker had published only twelve years earlier a well-known paper with the very same title. In his paper Kronecker sketched a way of introducing irrational numbers, without accepting the general notion. It is precisely to this general concept Hilbert wants to give a proper foundation — using the axiomatic method and following Dedekind.

Hilbert connected the consistency problem also to Cantorian issues. From his correspondence with Cantor we know that he had been aware

---

[1](Ewald 1996, pp. 1102–03). This is part of the German text from (Hilbert 1900a, p. 55): "Vor allem möchte ich unter den zahlreichen Fragen, welche hinsichtlich der Axiome gestellt werden können, dies als das wichtigste Problem bezeichnen, *zu beweisen, daß dieselben untereinander widerspruchslos sind, d.h., daß man auf Grund derselben mittels einer endlichen Anzahl von logischen Schlüssen niemals zu Resultaten gelangen kann, die miteinander in Widerspruch stehen.*"

of Cantor's "inconsistent multiplicities" since 1897. He recognized their effect on Dedekind's logical grounding of analysis and tried to blunt it by a consistency proof for the arithmetical axioms. Such a proof, he stated, provides "the proof of the existence of the totality of real numbers, or — in the terminology of G. Cantor — the proof that the system of real numbers is a consistent (finished) set".[2] In the Paris Lecture Hilbert re-emphasized and expanded this point by saying,

> In the case before us, where we are concerned with the axioms of real numbers in arithmetic, the proof of the consistency of the axioms is at the same time the proof of the mathematical existence of the complete system ["complete system" is Ewald's translation for "Inbegriff", WS] of real numbers or of the continuum. Indeed, when the proof for the consistency of the axioms shall be fully accomplished, the doubts, which have been expressed occasionally as to the existence of the complete system of real numbers, will become totally groundless.[3]

He proposed then to extend this approach to the Cantorian Alephs. Consequently and more generally, the consistency of appropriate axiom systems was to guarantee the existence of sets or, as he put it in 1905, to provide an *objective criterion* for the consistency of multiplicities.

The background for these foundational issues is explored in Part 2 under the heading *Existential axiomatics*. Part 3 is concerned with *Direct consistency proofs* for equational calculi. Part 4 looks at *Finitist proof theory* and its problems, namely, how first to treat quantifier-free systems of arithmetic and then to deal with quantifiers via the ϵ-substitution method. Part 5 discusses *Incompleteness* and related mathematical as well as philosophical problems; the essay ends with a very brief outlook beyond 1939. [This is the overview of *Hilbert's proof theory*. Its Part 2 is given here; Part 4 is reprinted as essay (II.1) and Part 5 as (II.5). WS]

2 Existential Axiomatics

The axiom system Hilbert formulated for the real numbers in 1900 is not presented in the contemporary logical style. Rather, it is given in an algebraic way and assumes that a structure exists whose elements satisfy the axiomatic conditions. Hilbert points out that this *axiomatic* way

---

[2](Ewald 1996, p. 1095). This is part of the German paragraph in (Hilbert 1900b, p. 261): "Um die Widerspruchslosigkeit der aufgestellten Axiome zu beweisen, bedarf es nur einer geeigneten Modifikation bekannter Schlußmethoden. In diesem Nachweis erblicke ich zugleich den Beweis für die Existenz des Inbegriffs der reellen Zahlen oder — in der Ausdrucksweise G. Cantors — den Beweis dafür, daß das System der reellen Zahlen eine konsistente (fertige) Menge ist."

[3](Ewald 1996, p. 1104). Here is the German text from (Hilbert 1900a, p. 56): "In dem vorliegenden Falle, wo es sich um die Axiome der reellen Zahlen in der Arithmetik handelt, ist der Nachweis für die Widerspruchslosigkeit der Axiome zugleich der Beweis für die mathematische Existenz des Inbegriffs der reellen Zahlen oder des Kontinuums. In der That, wenn der Nachweis für die Widerspruchslosigkeit der Axiome völlig gelungen sein wird, so verlieren die Bedenken, welche bisweilen gegen die Existenz des Inbegriffs der reellen Zahlen gemacht worden sind, jede Berechtigung."

Methods for real arithmetic 75

of proceeding is quite different from the *genetic method* standardly used in arithmetic; it rather parallels the ways of geometry.[4]

Here [in geometry] one begins customarily by assuming the existence of all the elements, i.e. one postulates at the outset three systems of things (namely, the points, lines, and planes), and then — essentially after the model of Euclid — brings these elements into relationship with one another by means of certain axioms of linking, order, congruence, and continuity.[5]

These geometric ways are taken over for arithmetic or rather, one might argue, are re-introduced into arithmetic by Hilbert; after all, they do have their origin in Dedekind's work on arithmetic and algebra. In the 1920s, Hilbert and Bernays called this methodological approach *existential axiomatics*, because it assumes the existence of suitable systems of objects. The underlying ideas, however, go back in Hilbert's thinking to at least 1893.

Dedekind played a significant role in their evolution not only directly through his foundational essays and his mathematical work, but also indirectly, as his collaborator and friend Heinrich Weber was one of Hilbert's teachers in Königsberg and a marvelous interpreter of Dedekind's perspective. For example, Weber's *Lehrbuch der Algebra* and some of his expository papers are frequently referred to in Hilbert's lectures around the turn from the 19th to the 20th century. It is a slow evolution, which is impeded by the influence of Kronecker's views as to how irrational numbers are to be given. The axiomatic method is to overcome the problematic issues Kronecker saw for the general concept of irrational number. Hilbert's retrospective remarks in (*1904) make this quite clear: this general concept created the "greatest difficulties", and Kronecker represented this point of view most sharply. Hilbert seemingly realized only in (*1904) that the difficulties are overcome, in a certain way, when the concept of natural number is secured. "The further steps up to the irrational numbers", he claims there, "are then taken in a logically rigorous way and without further difficulties."[6] Let us look at the details of this remarkable evolution, as far as it can be made out from the available documents.

---

[4]In the 1905 Lecture Notes there is a detailed discussion of the genetic foundation (genetische Begründung) with references to textbooks by Lipschitz, Pasch, and Thomae. Hilbert formulates there also the deeply problematic aspects of this method; see the beginning of section 2.2.

[5](Ewald 1996, p. 1092). Here is the German text from (Hilbert 1900b, p. 257): "Hier [in der Geometrie] pflegt man mit der Annahme der Existenz von sämtlichen Elementen zu beginnen, d.h. man setzt von vornherein drei Systeme von Dingen, nämlich die Punkte, die Geraden und die Ebenen, und bringt sodann diese Elemente — wesentlich nach dem Vorbilde von Euklid — durch gewisse Axiome, nämlich die Axiome der Verknüpfung, der Anordnung, der Kongruenz und der Stetigkeit, miteinander in Beziehung."

Hilbert obviously forgot to mention, between the axioms of order and of congruence, the axiom of parallels.

[6]These matters are discussed in (Hilbert *1904, pp. 164–167). One most interesting part consists of these remarks on pp. 165–166: "Die Untersuchungen in dieser Richtung [foundations for the real numbers] nahmen lange Zeit den breitesten Raum ein. Man kann den

2.1 Axioms for continuous systems. Hilbert introduces the axioms for the real numbers in (1900b) as follows: "We think a system of things, and we call them numbers and denote them by $a, b, c \ldots$ We think these numbers in certain mutual relations, whose precise and complete description is obtained through the following axioms."[7] Then the axioms for an ordered field are formulated and rounded out by the requirement of continuity via the Archimedean axiom and the axiom of completeness (Axiom der Vollständigkeit). This formulation is not only in the spirit of the geometric ways, but actually mimics the contemporaneous and paradigmatically modern axiomatic presentation of *Grundlagen der Geometrie*.

We think three different systems of things: we call the things of the first system points and denote them by $A, B, C, \ldots$; we call the things of the second system lines and denote them by $a, b, c, \ldots$; we call the things of the third system planes and denote them by $\alpha, \beta, \gamma, \ldots$; $\ldots$ We think the points, lines, planes in certain mutual relations . . . ; the precise and complete description of these relations is obtained by the *axioms of geometry*.[8]

Five groups of geometric axioms follow and, in the first German edition, the fifth group consists of just the Archimedean axiom. In the French edition of 1900 and the second German edition of 1903, the completeness axiom is included. The latter axiom requires in both the geometric and the arithmetic case that the assumed structure is maximal, i.e., any extension satisfying all the remaining axioms must already be contained in it. This formulation is frequently criticized as being metamathematical and, to boot, of a peculiar sort.[9] However, it is just an ordinary mathematical formulation, if the algebraic character of the axiom system is kept in

---

Standpunkt, von dem dieselben ausgingen, folgendermaßen charakterisieren: Die Gesetze der ganzen Zahlen, der Anzahlen, nimmt man vorweg, begründet sie nicht mehr; die Hauptschwiergkeit wird in jenen Erweiterungen des Zahlbegriffs (irrationale und weiterhin komplexe Zahlen) gesehen. Am schärfsten wurde dieser Standpunkt von Kronecker vertreten. Dieser stellte geradezu die Forderung auf: Wir müssen in der Mathematik jede Tatsache, so verwickelt sie auch sein möge, auf Beziehungen zwischen ganzen rationalen Zahlen zurückführen; die Gesetze dieser Zahlen andrerseits müssen wir ohne weiteres hinnehmen. Kronecker sah in den Definitionen der irrationalen Zahlen Schwierigkeiten und ging soweit, dieselben garnicht anzuerkennen."

[7](Hilbert 1900b, pp. 257–258). The German text is: "Wir denken ein System von Dingen; wir nennen diese Dinge Zahlen und bezeichnen sie mit $a, b, c, \ldots$ Wir denken diese Zahlen in gewissen gegenseitigen Beziehungen, deren genaue und vollständige Beschreibung durch die folgenden Axiome geschieht."

[8](Hilbert 1899a, p. 4) in (Hallett and Majer 2004, p. 437). The German text is: "Wir denken drei verschiedene Systeme von Dingen: die Dinge des ersten Systems nennen wir *Punkte* und bezeichnen sie mit $A, B, C, \ldots$; die Dinge des zweiten Systems nennen wir *Geraden* und bezeichnen sie mit $a, b, c, \ldots$; die Dinge des dritten Systems nennen wir *Ebenen* und bezeichnen sie mit $\alpha, \beta, \gamma, \ldots$; $\ldots$ Wir denken die Punkte, Geraden, Ebenen in gewissen gegenseitigen Beziehungen . . . ; die genaue und vollständige Beschreibung dieser Beziehungen erfolgt durch die Axiome der Geometrie."

[9]In (Torretti 1984), for example, one finds on p. 234 this observation: ". . . the completeness axiom is what nowadays one would call a metamathematical statement, though one of a rather peculiar sort, since the theory with which it is concerned includes Axiom V2 [i.e., the completeness axiom, WS] itself."

Methods for real arithmetic 77

mind. In the case of arithmetic we can proceed as follows: call a system A *continuous* when its elements satisfy the axioms of an ordered field and the Archimedean axiom, and call it *fully continuous* if and only if A is continuous and for any system B, if $A \subseteq B$ and B is continuous, then $B \subseteq A$.[10] Hilbert's arithmetic axioms characterize the fully continuous systems uniquely up to isomorphism.

Hilbert thought about axiom systems in this existential way already in *Die Grundlagen der Geometrie*, his first lectures on the *foundations* of geometry. He had planned to give them in the summer term of 1893, but their presentation was actually shifted to the following summer term. Using the notions "System" and "Ding" so prominent in Dedekind's (1888), he formulated the central question as follows:

What are the necessary and sufficient and mutually independent conditions a system of things has to satisfy, so that to each property of these things a geometric fact corresponds and conversely, thereby making it possible to completely describe and order all geometric facts by means of the above system of things.[11]

At some later point, Hilbert inserted the remark that this axiomatically characterized system of things is a "complete and simple image of geometric reality".[12] The image of "image" is attributed on the next page to Hertz and extended to cover also logical consequences of the images; they are to coincide with images of the consequences: "The axioms, as Hertz would say, are images or symbols in our mind, such that consequences of the images are again images of the consequences, i.e., what we derive logically from the images is true again in nature." In the Introduction to the Notes for the 1898–99 lectures *Elemente der Euklidischen Geometrie* written by Hilbert's student Hans von Schaper, this central question is explicitly connected with Hertz's (1894):

Using an expression of Hertz (in the introduction to the *Prinzipien der Mechanik*) we can formulate our main question as follows: What are the necessary and

---

[10]This is dual to Dedekind's definition of simply infinite systems: here we are focusing on maximal extensions, and there on minimal substructures. Indeed, the formulation in the above text is to be taken in terms of substructures in the sense of contemporary model theory.

[11](Toepell 1986, pp. 58–59) and (Hallett and Majer 2004, pp. 72–73). The German text is: "Welches sind die nothwendigen und hinreichenden und unter sich unabhängigen Bedingungen, die man an ein System von Dingen stellen muss, damit jeder Eigenschaft dieser Dinge eine geometrische Thatsache entspricht und umgekehrt, so dass also mittelst obigen Systems von Dingen ein vollständiges Beschreiben und Ordnen aller geometrischen Thatsachen möglich ist."

[12](Toepell 1986, p. 59) and (Hallett and Majer 2004, pp. 73 and 74). The German text is: "... so dass also unser System ein vollständiges und einfaches Bild der geometrischen Wirklichkeit werde." — The next German text is: "Die Axiome sind, wie Hertz sagen würde, Bilder oder Symbole in unserem Geiste, so dass Folgen der Bilder wieder Bilder der Folgen sind, d.h. was wir aus den Bildern logisch ableiten, stimmt wieder in der Natur."

On the very page of the last quotation, in note 7 of (Hallett and Majer 2004), the corresponding and much clearer remark from (Hertz 1894) is quoted: "Wir machen uns innere Scheinbilder oder Symbole der äußeren Gegenstände, und zwar machen wir sie von solcher Art, daß die denknotwendigen Folgen der Bilder stets wieder die Bilder seien von naturnotwendigen Folgen der abgebildeten Gegenstände."

sufficient and mutually independent conditions a system of things has to be subjected to, so that to each property of these things a geometric fact corresponds and conversely, thereby having these things provide a complete "image" of geometric reality.[13]

One can see in these remarks the shape of a certain logical or set theoretic structuralism (that is *not* to be equated with contemporary philosophical forms of such a position) in the foundations of mathematics and of physics; geometry is viewed as the most perfect natural science.[14] How simple or how complex are these systems or structures supposed to be? What characteristics must they have in order to be adequate for both the systematic internal development of the corresponding theory and external applications? What are the things whose system is implicitly postulated? What is the mathematical connection, in particular, between the arithmetic and geometric structures?

Dedekind gives in *Stetigkeit und irrationale Zahlen* a precise notion of continuity for analysis; however, before doing so he shifts his considerations to the geometric line. He observes that every point effects a cut, but — and that he views as central for the *continuity* or *completeness* of the line — every cut is effected by a point.[15] Transferring these reflections to cuts of rational numbers and, of course, having established that there are infinitely many cuts of rationals that are not effected by rationals, he "creates" a new, an irrational number α for any cut not effected by a rational: "We will say that this number α corresponds to that cut or that it effects this cut."[16] Dedekind then introduces an appropriate ordering between cuts and establishes that the system of all cuts is indeed continuous. He shows that the continuity principle has to be accepted, if one wants to have the least upper bound principle for monotonically increasing, but bounded sequences of reals, as the two principles are indeed equivalent. He writes:

This theorem is equivalent to the principle of continuity, i.e., it loses its validity as soon as even just a single real number is viewed as not being contained in the domain R; ...[17]

---

[13](Toepell 1986, p. 204) and (Hallett and Majer 2004, p. 303). The German text is: "Mit Benutzung eines Ausdrucks von Hertz (in der Einleitung zu den *Prinzipien der Mechanik*) können wir unsere Hauptfrage so formulieren: Welches sind die nothwendigen und hinreichenden und unter sich unabhängigen Bedingungen, denen man ein System von Dingen unterwerfen muss, damit jeder Eigenschaft dieser Dinge eine geometrische Thatsache entspreche und umgekehrt, damit also diese Dinge ein vollständiges und einfaches "Bild" der geometrischen Wirklichkeit seien."

[14]There is a complementary discussion on pp. 103–104 and 119–122 of (Hallett and Majer 2004). Hilbert's perspective is echoed in the views of Ernest Nagel and Patrick Suppes; cf. also van Fraassen, *The Scientific Image*.

[15]Dedekind uses Stetigkeit and Vollständigkeit interchangeably. The reader should notice the freedom with which Dedekind reinterprets in the following considerations an axiomatically characterized concept, namely a certain order relation: it is interpreted for the rational numbers, the geometric line, and the system of all cuts of rationals.

[16](Dedekind 1872c, p. 13).

[17](Dedekind 1872c, p. 20). The German text is: "Dieser Satz ist äquivalent mit dem Prinzip der Stetigkeit, d.h. er verliert seine Gültigkeit, sobald man auch nur eine einzige reelle Zahl in dem Gebiete R als nicht vorhanden ansieht; ..."

Methods for real arithmetic 79

As it happens, Hilbert's first formulation of a geometric continuity principle in (*1894b) (p. 92 of (Hallett and Majer 2004)) is a least upper bound principle for bounded sequences of geometric points.

Dedekind emphasizes in (1872c) the necessity of rigorous arithmetic concepts for analyzing the notion of space. He points out, reflecting the view of Gauss concerning the fundamental difference between arithmetic and geometry, that space need not be continuous.[18] The required arithmetization of analysis is established in Dedekind's thinking about the matter through three steps, namely, i) a logical characterization of the natural numbers, ii) the extension of that number concept to integers and rational numbers, and iii) the further extension to all real numbers we just discussed. The first objective is accomplished in (1888) via simply infinite systems, a notion that emerged in the 1870s after Dedekind had finished his essay (1872c). [See essay (I.1) of this volume; in that paper one finds also a discussion of Dedekind's approach to the second issue based on unpublished manuscripts. WS] Dedekind gives his final characterization in article 71 of (1888) in the existential, algebraic form that is used later by Hilbert. But he avoids self-consciously the term axiom and speaks only of conditions (Bedingungen), which are easily seen to amount in this case to the well-known Peano axioms.

Dedekind sought the ultimate basis for arithmetic in logic, broadly conceived. That view is formulated in the preface to (1888):

In calling arithmetic (algebra, analysis) merely a part of logic, I already claim that I consider the number concept as entirely independent of notions or intuitions of space and time, that I consider it rather as an immediate product of the pure laws of thought.[19]

Important for our considerations is the fact that Dedekind had developed towards the end of the 1870s not only the general concept of "system", but also that of a "mapping between systems". The latter notion corresponds for him to a crucial capacity of our mind (pp. iii–iv of the preface to (1888)). He uses these concepts to formulate and prove that all simply infinite systems are isomorphic. He concludes from this result, in contemporary model theoretic terms, that simply infinite systems are all elementarily equivalent and uses that fact to characterize the science of arithmetic:

The relations or laws, which are solely derived from the conditions $\alpha, \beta, \gamma, \delta$ in 71 and are therefore in all ordered systems always the same ... form the next object of the *science of numbers* or of *arithmetic*.[20]

---

[18](Dedekind 1888, p. vii), a particular countable model of reals is described in which all of Euclid's construction can be interpreted.

[19](Dedekind 1888, p. iii). The German text is: "Indem ich die Arithmetik (Algebra, Analysis) nur einen Teil der Logik nenne, spreche ich schon aus, daß ich den Zahlbegriff für gänzlich unabhängig von den Vorstellungen oder Anschauungen des Raumes und der Zeit, daß ich ihn vielmehr für einen unmittelbaren Ausfluß der reinen Denkgesetze halte."

[20](Dedekind 1888, p. 17). The German text is: "Die Beziehungen oder Gesetze, welche ganz allein aus den Bedingungen $\alpha, \beta, \gamma, \delta$ in 71 abgeleitet werden und deshalb in allen geordneten Systemen immer dieselben sind ... bilden den nächsten Gegenstand der *Wissenschaft von den Zahlen* oder der *Arithmetik*."

Natural numbers are not viewed as objects of a particular system, but are obtained by free creation or what W.W. Tait has called "Dedekind abstraction"; that is clearly articulated in article 73 of (1888).

Similar metamathematical investigations are adumbrated for real numbers in *Stetigkeit und irrationale Zahlen*. Dedekind can be interpreted, in particular, as having characterized *fully continuous systems* as ordered fields that satisfy his principle of continuity, i.e., the principle that every cut of such a system is effected by an element. The system of all cuts of rational numbers constitutes a model; that is in modern terms the essential content of sections 5 and 6 of (1872c), where Dedekind verifies that this system satisfies the continuity principle. In contemporaneous letters to Lipschitz, Dedekind emphasizes that real numbers should not be identified with cuts, but rather viewed as newly created things corresponding to cuts.

The informal comparison of the geometric line with the system of cuts of rational numbers in sections 2 and 3 of (1872c) contains almost all the ingredients for establishing rigorously that these two structures are isomorphic; missing is the concept of mapping. As I mentioned, that concept was available to Dedekind in 1879 and with it these considerations can be extended directly to show that arbitrary fully continuous systems are isomorphic. Finally, the methodological remarks about the arithmetic of natural numbers can now be extended to that of the real numbers. As a matter of fact, Hilbert articulates Dedekind's way of thinking of the system of real numbers in his (1922), when describing the *axiomatische Begründungsmethode* for analysis:

The continuum of real numbers is a system of things that are connected to each other by certain relations, so-called axioms. In particular the definition of the real numbers by the Dedekind cut is replaced by two continuity axioms, namely, the Archimedian axiom and the so-called completeness axiom. In fact, Dedekind cuts can then serve to determine the individual real numbers, but they do not serve to define [the concept of] real number. On the contrary, conceptually a real number is just a thing of our system. . . . The standpoint just described is altogether logically completely impeccable, and it only remains thereby undecided, whether a system of the required kind can be thought, i.e., whether the axioms do not lead to a contradiction.[21]

That is fully in Dedekind's spirit: Hilbert's critical remarks about the definition of real numbers as cuts do not apply to Dedekind, as should be clear from the above discussion, and the issue of consistency had been an explicit part of Dedekind's logicist program.

---

[21](Hilbert 1922, pp. 158–159). The German text is: "Das Kontinuum der reellen Zahlen ist ein System von Dingen, die durch bestimmte Beziehungen, sogenannte Axiome, miteinander verknüpft sind. Insbesondere treten an Stelle der Definition der reellen Zahlen durch den Dedekindschen Schnitt die zwei Stetigkeitsaxiome, nämlich das Archimedische Axiom und das sogenannte Vollständigkeitsaxiom. Die Dedekindschen Schnitte können dann zwar zur Festlegung der einzelnen reellen Zahlen dienen, aber sie dienen nicht zur Definition der reellen Zahl. Vielmehr ist begrifflich eine reelle Zahl eben ein Ding unseres Systems. . . . Der geschilderte Standpunkt ist vollends logisch vollkommen einwandfrei, und es bleibt nur dabei unentschieden, ob ein System der verlangten Art denkbar ist, d.h. ob die Axiome nicht etwa auf einen Widerspruch führen."

Methods for real arithmetic   81

2.2 Consistency and logical models.

Logicians in the 19th century viewed consistency of a notion from a semantic perspective. In that tradition, Dedekind addressed the problem for his characterization of natural numbers via simply infinite systems. The methodological need for doing that is implicit in (1872c), but is formulated most clearly in a letter to Keferstein written in 1889, a year after the publication of *Was sind und was sollen die Zahlen?*

After the essential nature of the simply infinite system, whose abstract type is the number sequence N, had been recognized in my analysis (71, 73) the question arose: does such a system *exist* at all in the realm of our thoughts? Without a logical proof of existence, it would always remain doubtful whether the notion of such a system might not perhaps contain internal contradictions. Hence the need for such a proof (articles 66 and 72 of my essay).[22]

Dedekind attempted in article 66 to prove the existence of an infinite system and to provide in article 72 an example of a simply infinite system within logic. In other words, he attempted to establish the existence of a *logical* model and to guarantee in this way that the notion of a simply infinite system indeed does not contain internal contradictions.

Hilbert turned to considerations for natural numbers only around 1904. Until then he had taken for granted their proper foundation and focused on the theory of real numbers: as I reported earlier from his (*1904), Hilbert thought that the concept of irrational numbers created the greatest difficulties. These difficulties have been overcome, Hilbert claims in (*1904). When the concept of natural number is secured, he continues there, the further steps towards the real numbers can be taken without a problem. This dramatic change of viewpoint helps us to understand more clearly some aspects of Hilbert's earlier considerations related to the consistency of the arithmetic of real numbers. What did Hilbert see as the "greatest difficulties" for the general concept of irrational numbers? That question will be explored now.

In contrast to Dedekind, Hilbert had formulated a quasi-syntactic notion of consistency in *Über den Zahlbegriff* and in *Grundlagen der Geometrie*, namely, no finite number of logical steps leads from the axioms to a contradiction. This notion is "*quasi*-syntactic", as no deductive principles are explicitly provided. Hilbert did not seek to prove consistency by syntactic methods. On the contrary, the relative consistency proofs given in *Grundlagen der Geometrie* are all straightforwardly semantic. To prove the consistency of arithmetic one needs, so Hilbert in *Über den Zahlbegriff*, "a suitable modification of familiar methods of reasoning". In the Paris Lecture Hilbert suggests finding a direct proof and remarks more expansively:

---

[22]In (van Heijenoort 1967, p. 101). The German text is: "Nachdem in meiner Analyse der wesentliche Charakter des einfach unendlichen Systems, dessen abstracter Typus die Zahlenreihe N ist, erkannt war (71, 73), fragte es sich: *existirt* überhaupt ein solches System in unserer Gedankenwelt? Ohne den logischen Existenz-Beweis würde es immer zweifelhaft bleiben, ob nicht der Begriff eines solchen Systems vielleicht innere Widersprüche enthält. Daher die Nothwendigkeit solcher Beweise (66, 72 meiner Schrift)."

I am convinced that it must be possible to find a direct proof for the consistency of the arithmetical axioms [as proposed in *Über den Zahlbegriff* for the real numbers; WS], by means of a careful study and suitable modification of the known methods of reasoning in the theory of irrational numbers.[23]

Hilbert believed at this point, it seems to me, that the genetic build-up of the real numbers could *somehow* be exploited to yield the blueprint for a consistency proof in Dedekind's logicist style. There are difficulties with the genetic method that prevent it from providing directly a proper foundation for the general concept of irrational numbers. Hilbert's concerns are formulated most clearly in these retrospective remarks from (*1905b):

It [the genetic method, WS] defines things by generative processes, not by properties what really must appear to be desirable. Even if there is no objection to defining fractions as systems of two integers, the definition of irrational numbers as a system of infinitely many numbers must appear to be dubious. Must this number sequence be subject to a law, and what is to be understood by a law? Is an irrational number being defined, if one determines a number sequence by throwing dice? These are the kinds of questions with which the genetic perspective has to be confronted.[24]

Precisely this issue was to be overcome (or to be sidestepped) by the axiomatic method. In *Über den Zahlbegriff* Hilbert writes:

Under the conception described above, the doubts which have been raised against the existence of the totality of real numbers (and against the existence of infinite sets generally) lose all justification; for by the set of real numbers we do not have to imagine, say, the totality of all possible laws according to which the elements of a fundamental sequence can proceed, but rather — as just described — a system of things whose mutual relations are given by the finite and closed system of axioms I–IV, ...[25]

---

[23](Ewald 1996, p. 1104). The German text from (Hilbert 1900a, p. 55) is: "Ich bin nun überzeugt, daß es gelingen muß, einen direkten Beweis für die Widerspruchslosigkeit der arithmetischen Axiome zu finden, wenn man die bekannten Schlußmethoden in der Theorie der Irrationalzahlen im Hinblick auf das bezeichnete Ziel genau durcharbeitet und in geeigneter Weise modifiziert."

In (Bernays 1935a, pp. 198–9), this idea is re-emphasized: "Zur Durchführung des Nachweises gedachte Hilbert mit einer geeigneten Modifikation der in der Theorie der reellen Zahlen angewandten Methoden auszukommen."

[24](Hilbert *1905b, pp. 10–11). The German text is: "Sie definiert Dinge durch Erzeugungsprozesse, nicht durch Eigenschaften, was doch eigentlich wünschenswert erscheinen muß. Ist nun auch gegen die Definition der Brüche als System zweier Zahlen nichts einzuwenden, so muß doch die Definition der Irrationalzahlen als ein System von unendlich vielen Zahlen bedenklich erscheinen. Muß diese Zahlenreihe einem Gesetz unterliegen, und was hat man unter einem Gesetz zu verstehen? Wird auch eine Irrationalzahl definiert, wenn man eine Zahlenreihe durch Würfeln feststellt? Dieser Art sind die Fragen, die man der genetischen Betrachtung entgegenhalten muß."

[25](Ewald 1996, p. 1095). Here is the German text from (Hilbert 1900b, p. 261): "Die Bedenken, welche gegen die Existenz des Inbegriffs aller reellen Zahlen und unendlicher Mengen überhaupt geltend gemacht worden sind, verlieren bei der oben gekennzeichneten Auffassung jede Berechtigung: unter der Menge der reellen Zahlen haben wir uns hiernach nicht etwa die Gesamtheit aller möglichen Gesetze zu denken, nach denen die Elemente

Methods for real arithmetic

He articulates in the Paris Lecture exactly the same point by re-emphasizing "the continuum . . . is not the totality of all possible series in decimal fractions, or of all possible laws according to which the elements of a fundamental sequence may proceed". Rather, it is a system of things whose mutual relations are governed by the axioms. The completeness axiom can be plausibly interpreted as guaranteeing the continuity of the system without depending on any particular method of generating real numbers.

Going back to the 1889–90 lectures *Einführung in das Studium der Mathematik*, one sees that Hilbert then shared a Kroneckerian perspective on irrationals. He starts out with a brief summary of "Das Nothwendigste über den Zahlbegriff" and uses Dedekind cuts to define the notion of irrational number. He shifts for the practical presentation of real numbers from cuts to fundamental sequences and asserts, making the problematic aspect fully explicit, "It is absolutely necessary to have a general method for the representation of irrational numbers."[26] The need to have a method of representation is not bound to fundamental sequences. Cuts are to be given by a *rule* that allows the partition of all rationals into smaller and greater ones: the real number effecting the cut is taken to be "the carrier of that rule". (This perspective is re-iterated in the later lectures, in particular, in (*1899b, pp. 10–11).) So we have a sense of the difficulties as Hilbert saw them. In order to support the claim that Hilbert thought, nevertheless, that the genetic build-up of the real numbers could in some way be exploited to yield a logicist consistency proof, we look closely at Hilbert's treatment of arithmetic in later lectures, in particular, (*1894a), (*1897b), and (*1899b).

Let me begin with some deeply programmatic, logicist statements found in the Introduction to the notes for *Elemente der Euklidischen Geometrie* presented in 1898–99. Hilbert maintains there, "It is important to fix precisely the startingpoint of our investigations: as given we consider the laws of pure logic and in particular all of arithmetic."[27] He adds parenthetically, "On the relation between logic and arithmetic cf. Dedekind, *Was sind und was sollen die Zahlen?*" Clearly, for Dedekind arithmetic is a part of logic, and for Hilbert in 1899 that is not a novel insight. On the contrary: Hilbert opens the lectures (*1889/90) with the sentence, "The theoretical foundation for the mathematical sciences is the concept of positive integer, defined

---

einer Fundamentalreihe fortschreiten können, sondern vielmehr — wie eben dargelegt ist — ein System von Dingen, deren gegenseitige Beziehungen durch das obige endliche und abgeschlossene System von Axiomen I–IV gegeben sind, . . ."

[26](Hilbert *1889/90, p. 3). The German text is: "Es ist unumgänglich nöthig, eine allgemeine Methode zur Darstellung der irrationalen Zahlen zu haben."

[27] Clearly, arithmetic is here taken in the narrower sense of the theory of natural numbers. The German text is found in (Toepell 1986, pp. 203–4), and in (Hallett and Majer 2004, p. 303): "Es ist von Wichtigkeit, den Ausgangspunkt unserer Untersuchungen genau zu fixieren: Als gegeben betrachten wir die Gesetze der reinen Logik und speciell die ganze Arithmetik. (Ueber das Verhältnis zwischen Logik und Arithmetik vgl. Dedekind, Was sind und was sollen die Zahlen?) Unsere Frage wird dann sein: Welche Sätze müssen wir zu dem eben definierten Bereich "adjungieren", um die Euklidische Geometrie zu erhalten?"

In (Toepell 1986) one finds additional remarks of interest for the issues discussed here, namely, on pp. 195, 206, and 227–8.

as *Anzahl*." This remark is deepened in the long introduction to the lectures on *Projektive Geometrie* given in 1891, where Hilbert emphasizes that the results of pure mathematics are obtained by pure thought (reines Denken). He lists as parts of pure mathematics number theory, algebra, and complex analysis (Funktionentheorie). Mirroring Dedekind's perspective on the arithmetization of analysis from the preface to (1888, p. vi), Hilbert asserts that the results of pure mathematics *can* and *have to be reduced* to relations between integers:

> *Nowadays* a proposition is being considered as proved only, when it *ultimately expresses* a relation between integers. Thus, the *integer is the element*. We can *obtain by pure thinking* the concept of integer, for example <by> *counting thoughts themselves*. *Methods, foundations of pure mathematics* belong to the sphere of pure thought.[28]

This very paragraph ends with the striking assertion, "I don't need anything but purely logical thought, when dealing with number theory or algebra." Hilbert expresses here very clearly a thoroughgoing Dedekindian logicism with respect to natural numbers.[29]

Die Grundlagen der Geometrie from 1894 have a special section on "Die Einführung der Zahl". After all, Hilbert argues, "In all of the exact sciences one obtains precise results only after the introduction of number. It is always of deep epistemological significance to see in detail, how this measuring is done."[30] The number concept discussed now is that of real numbers; it is absolutely central for the problems surrounding the quadrature of the circle treated in the winter term 1894–95. The quadrature problem is the red thread for historically informed lectures entitled, not surprisingly, *Quadratur des Kreises*. Its solution requires the subtlest mathematical speculations with considerations from geometry as well as from analysis and number theory. In the winter term 1897–98 Hilbert presents these topics again, but this time in the *second* part of the course; its new *first* part is devoted entirely to a detailed discussion of the concept of number. That expansion is indicated by the new title of the lectures, *Zahlbegriff und Quadratur des Kreises*.

Hilbert remarks that the foundations of analysis have been examined through the work of Weierstrass, Cantor, and Dedekind. He lists three important points:

---

[28](Hallett and Majer 2004, p. 22). The German text is: "Es gilt heutzutage ein Satz erst dann als bewiesen, wenn er eine Beziehung zwischen ganzen Zahlen *in letzter Instanz zum Ausdruck bringt*. Also die *ganze Zahl ist das Element*. Zum Begriff der ganzen Zahl können wir auch *durch reines Denken gelangen*, etwa <indem ich> die *Gedanken selber zähle*. *Methoden, Grundlagen der reinen Mathematik* gehören dem reinen Denken an."

[29](Hallett and Majer 2004, p. 22). The German sentence is: "Ich brauche weiter nichts als *rein logisches* Denken, wenn ich mit Zahlentheorie oder Algebra mich beschäftige." In his extensive editorial note 6 on p. 23 of (Hallett and Majer 2004), Haubrich also points out the deep connection to Dedekind's views.

[30](Hallett and Majer 2004, p. 85). The German text is: "In allen exakten Wissenschaften gewinnt man erst dann präzise Resultate, wenn die Zahl eingeführt ist. Es ist stets von hoher erkenntnistheoretischer Bedeutung zu verfolgen, wie dies Messen geschieht."

Methods for real arithmetic 85

Today [we take as a] principle: Reduction to the laws for the whole rational numbers and purely logical operations.

1, 2, 3 ... = Anzahl = natural number

How one gets from this number [concept] to the most general concept of complex number, that is the foundation, i.e., the theory of the concept of number.[31]

In the lectures (*1897b) the genetic build-up is to be captured axiomatically; that I take to be the gist of the discussion of the third point in the above list, namely, how one gets from the natural numbers to the most general concept of complex number. The motivating focus lies on the solution of problems by an extension of the axioms (Lösung von Problemen durch Erweiterung der Axiome); the discussion ends with the remark, "We have extended the number concept in such a way, or to put it in a better way, we have enlarged the stock of axioms in such a way, that we can solve all definite equations."[32] At each step in this extending process consistency of the axioms is required, but Hilbert does not give arithmetical or logical proofs, only geometric pictures.

In the winter term 1899–1900, Hilbert again gave the lectures *Zahlbegriff und Quadratur des Kreises* and refined the discussion of the axioms for the reals significantly; it is a transition to his paper *Über den Zahlbegriff*. Pages 38–43 are added to the notes from the previous lecture, and Hilbert extends on them the discussion of "komplexe Zahlensysteme" from *Grundlagen der Geometrie*.[33] He formulates the continuity principle as "Bolzano's Axiom", i.e., every fundamental sequence has a limit; he shows also that this axiom has the Archimedian one as a consequence. Strikingly, he actually claims (on p. 42) that the full system is consistent! Before completing his (1900b), Hilbert obviously recognized i) that he could formulate the second continuity principle in such a way that it did not imply the Archimedian axiom, and ii) that the consistency of the system had *not* been established. So we are back to the issue from the very beginning of this subsection. The long digression does not provide a clear answer, but a seemingly rhetorical question: How could Hilbert think about addressing the consistency problem by "a careful study and suitable modification of the known methods of resoning in the theory of irrational numbers", if he did not have in mind a programmatically guided refinement of the steps presented here, i.e., of the foundation they constitute?

---

[31](Hilbert *1897b, p. 2). The German text is: "Heute Grundsatz: Zurückführung auf die Gesetze der ganzen rationalen Zahlen und rein logische Operationen.
1, 2, 3 ... = Anzahl = natürliche Zahl
Wie man von dieser Zahl zum allgemeinsten Begriff der komplexen Zahl kommt, das ist jenes Fundament, d.h. die Lehre vom Zahlbegriff."

[32](Hilbert *1897b, p. 23). The German text is: "Wir haben den Zahlbegriff so erweitert, oder besser gesagt, den Bestand der Axiome so vermehrt, dass wir alle definiten Gleichungen lösen können."

Hallett gives a complementary account of the completeness principle that emphasizes its role in Hilbert's work concerning geometry; that is presented in section 5 of his Introductory Note to (Hilbert 1899a), (Hallett and Majer 2004, pp. 426–435).

[33] Indeed, pages 38 and 39 must be the page proofs of the *Festschrift*, pp. 26–28.

2.3 Consistency and proofs.

Hilbert knew about the difficulties in set theory through his correpondence with Cantor, but he did not — as far as I can see — move away from this programmatic position and the associated strategy for proving consistency until 1903 or, at the latest, 1904. One reason for making this conjecture rests on the fact that Hilbert gave his lectures on *Zahlbegriff und Quadratur des Kreises* again in 1901–02; we have only the "Disposition" of those lectures, but it is very detailed and does not contain any indication whatsoever of substantive changes. In the summer term of 1904, Hilbert again lectured on *Zahlbegriff und Quadratur des Kreises*. In the notes written by Max Born one finds a dramatic change: Hilbert discusses for the first time the paradoxes in detail and sketches various foundational approaches.[34] These discussions are taken up in Hilbert's talk at the Heidelberg Congress in August of that year, where he presents a syntactic approach to the consistency problem. The goal is still to guarantee the existence of a suitable system, and the method seems inspired by one important aspect of the earlier investigations, as the crucial new step is the *simultaneous development of arithmetic and logic*!

Hilbert's view of the geometric axioms as characterizing a system of things that presents a "complete and simple image of geometric reality" is, after all, complemented by a traditional one: the axioms must allow to establish, purely logically, all geometric facts and laws. This role of the axioms, though clearly implicit in the formulation of the central question of (*1894a), is emphasized in the notes *Elemente der Euklidischen Geometrie* and described in the Introduction to the *Festschrift* in a methodologically refined way:

> The present investigation is a new attempt of formulating for geometry a *simple* and *complete* system of mutually independent axioms; it is also an attempt of deriving from them the most important geometric propositions in such a way that the significance of the different groups of axioms and the import of the consequences of the individual axioms is brought to light as clearly as possible.[35]

This second important role of axioms is reflected also in the Paris Lecture, where Hilbert states that the totality of real numbers is ". . . a system of things whose mutual relations are governed by the axioms set up and for which all propositions, and only those, are true which can be derived from the axioms by a finite number of logical inferences." As in Hilbert's opinion

---

[34] Peckhaus characterizes 1903 in his book (1990) as a "Zäsurjahr" in Göttingen; see pp. 55–58. Peckhaus also points out that Dedekind, in that very year, did not allow the republication of his (1888), because — as Dedekind put it in the preface to the third edition of 1911 — he had doubts concerning some of the important foundations of his view. Further evidence for this conjecture is presented at the beginning of section 3 below [This is section 3 of *Hilbert's proof theory*. WS], in particular Hilbert's own recollections and remarks by Bernays.

[35] (Hilbert 1899a, Einleitung, p. 3) and (Hallett and Majer 2004, p. 436). The German text is: "Die vorliegende Untersuchung ist ein neuer Versuch, für die Geometrie ein *einfaches* und *vollständiges* System von einander *unabhängiger* Axiome aufzustellen und aus denselben die wichtigsten geometrischen Sätze in der Weise abzuleiten, dass dabei die Bedeutung der verschiedenen Axiomgruppen und die Tragweite der aus den einzelnen Axiomen zu ziehenden Folgerungen möglichst klar zu Tage tritt."

Methods for real arithmetic

"the concept of the continuum is strictly logically tenable in this sense only", the axiomatic method has to confront two fundamental problems that are formulated in (1900b) at first for geometry and then also for arithmetic:

> The necessary task then arises of showing the consistency and the completeness of these axioms, i.e., it must be proved that the application of the given axioms can never lead to contradictions, and, further, that the system of axioms suffices to prove all geometric propositions.[36]

It is not clear, whether completeness of the respective axioms requires the proof of *all* true geometric or arithmetic statements, or whether provability of those that are part of the established corpora is sufficient; the latter would be a *quasi-empirical* notion of completeness. It should be mentioned that Dedekind attended in his (1888) consciously to the quasi-empirical completeness of the axiomatic conditions characterizing simply infinite systems. Through the proof principle of induction and the definition principle of recursion he provided the basic principles needed for the actual development of elementary number theory. The same observation can be made for the principle of continuity and Dedekind's treatment of analysis in his (1872c).

But what are the logical steps that are admitted in proofs? Frege criticized Dedekind on that point in the Preface of his *Grundgesetze der Arithmetik*, claiming that the brevity of Dedekind's development of arithmetic in (1888) is only possible, "because much of it is not really proved at all". Frege continues:

> ... nowhere is there a statement of the logical or other laws on which he builds, and, even if there were, we could not possibly find out whether really no others were used — for to make that possible the proof must be not merely indicated but completely carried out.[37]

Apart from making the logical principles explicit there is an additional aspect that is hinted at in Frege's critique (and detailed in other writings). Though Frege's critique applies as well to Hilbert's *Grundlagen der Geometrie*, Poincaré's 1902 review of that book brings out this additional aspect in a dramatic way, namely, the idea of formalization as machine executability. Poincaré writes: "M. Hilbert has tried, so-to-speak, putting the axioms in such a form that they could be applied by someone who doesn't understand their meaning, because he has not ever seen either a point, or a line, or a plane. It must be possible, according to him [Hilbert, WS], to reduce reasoning to purely mechanical rules." Indeed, Poincaré suggests giving the axioms to a reasoning machine, like Jevons's logical piano, and observing whether all of geometry would be obtained. Such formalization

---

[36] (Ewald 1996, p. 1092–3). The German text from (Hilbert 1900b) is: "Es entsteht dann die notwendige Aufgabe, die Widerspruchslosigkeit und Vollständigkeit dieser Axiome zu zeigen, d.h. es muß bewiesen werden, daß die Anwendung der aufgestellten Axiome nie zu Widersprüchen führen kann, und ferner, daß das System der Axiome zum Nachweis aller geometrischen Sätze ausreicht."

[37] *Grundgesetze der Arithmetik*, (Geach and Black 1977, p. 139).

might seem "artificial and childish", were it not for the important question of completeness:

> Is the list of axioms complete, or have some of them escaped us, namely those we use unconsciously? ... One has to find out whether geometry is a logical consequence of the explicitly stated axioms, or in other words, whether the axioms, when given to the reasoning machine, will make it possible to obtain the sequence of all theorems as output [of the machine, WS].[38]

At issue is, of course, what logical steps can the reasoning machine take? In Hilbert's lectures one finds increasingly references to, and investigations of, logical calculi. As to the character of these calculi, see Peckhaus's contribution to this volume [I.e., to the volume in which my *Hilbert's proof theory* appeared]. The calculi Hilbert considers remain very rudimentary until the lectures (*1917/18), where modern mathematical logic is all of a sudden being developed. Back in the Heidelberg talk, the novel syntactic approach to consistency proofs is suggested not for an axiom system concerning real numbers but rather concerning natural numbers; that will be the central topic of the next section. [That refers to Part 3 of my *Hilbert's proof theory* that is not reprinted here; however, the proof theoretic treatment is discussed in section 2 of essay (II.2).] The logical calculus, if it can be called that, is extremely restricted — it is purely equational!

In late 1904, possibly even in early 1905, Hilbert sent a letter to his friend and colleague Adolf Hurwitz. Clearly, he was not content with the proposal he had detailed in Heidelberg:

> It seems that various parties started again to investigate the foundations of arithmetic. It has been my view for a long time that exactly the most important and most interesting questions have not been settled by Dedekind and Cantor (and a fortiori not by Weierstrass and Kronecker). In order to be forced into the position to reflect on these matters systematically, I announced a seminar on the "logical foundations of mathematical thought" for next semester.[39]

The lectures from the summer term 1905, *Logische Prinzipien des mathematischen Denkens*, are as special as those of 1904, but for a different reason: one finds in them a critical examination of logical principles and a realization that a broader logical calculus is needed that captures, in particular, universal statements and inferences. Hilbert discusses in some detail (on pp. 253–266) the shortcomings of the approach he pursued in his Heidelberg address; see section 3.1 below [Again, this refers to my *Hilbert's*

---

[38](Poincaré 1902b, pp. 252–253)

[39]Hilbert mentions in this letter that (Zermelo 1904) had just been published. Zermelos's paper is an excerpt of a letter to Hilbert that was written on 24 September 1904. The German text is from (Dugac 1976, p. 271): "Die Beschäftigung mit den Grundlagen der Arithmetik wird jetzt, wie es scheint, wieder von den verschiedensten Seiten aufgenommen. Dass gerade die wichtigsten u. interessantesten Fragen von Dedekind und Cantor noch nicht (und erst recht nicht von Weierstrass und Kronecker) erledigt worden sind, ist eine Ansicht, die ich schon lange hege und, um einmal in die Notwendigkeit versetzt zu sein, darüber im Zusammenhang nachzudenken, habe ich für nächsten Sommer ein zweistündiges Colleg über die "logischen Grundlagen des math. Denkens" angezeigt."

Methods for real arithmetic 89

proof theory]. In his subsequent lectures on the foundations of mathematics, Hilbert does not really progress beyond the reflections presented in (*1905b). In a letter from Hessenberg to Nelson (dated 7 February 1906 and which I received through Peckhaus), one finds not only a discussion of Russell's early logicism, but strikingly, Dedekind and Hilbert are both described as logicists. Hessenberg remarks:

Dedekind's standpoint is also a logicist one, but altogether noble and nowhere brazenly dogmatic. One sees and feels everywhere the struggling, serious human being. But given all the superficialities I have found up to now in Russell, I can hardly assume that he is going to provide more than something witty. Zermelo believes that also Hilbert's logicism cannot be carried out, but considers Poincaré's objections as unfounded.[40]

It is only in the Zürich talk *Axiomatisches Denken* that a new perspective emerges. In that pivotal essay, Hilbert remarks that the consistency of the axioms for the real numbers can be reduced, by employing set theoretic concepts, to the very same question for integers; that result is attributed to the theory of irrational numbers developed by Weierstrass und Dedekind. Hilbert continues, insisting on a logicist approach:

In only two cases is this method of reduction to another more special domain of knowledge clearly not available, namely, when it is a matter of the axioms for the *integers* themselves, and when it is a matter of the foundation of *set theory*; for here there is no other discipline besides logic to which it were possible to appeal.

But since the examination of consistency is a task that cannot be avoided, it appears necessary to axiomatize logic itself and to prove that number theory as well as set theory are only parts of logic.[41]

Hilbert remarks that Russell and Frege provided the basis for this approach. Given Hilbert's closeness to Dedekind's perspective, it is not surprising that he was attracted by the logicist considerations of *Principia Mathematica*. In 1920, however, Hilbert and Bernays reject the logicist program of the Russellian kind. The reason is quite straightforward: for the development of mathematics the axiom of reducibility is needed, and that is in essence a return to existential axiomatics with all its associated

---

[40]The German text is: "Dedekinds Standpunkt ist ja auch logizistisch, aber durchaus vornehm and nirgends unverfroren dogmatisch. Man sieht und fühlt überall den kämpfenden ernsten Menschen. Aber bei den Oberflächlichkeiten, die ich bis jetzt in Russell gefunden habe, kann ich kaum annehmen, daß er mehr als geistreiches bringt. Zermelo ist der Ansicht, daß auch Hilberts Logizismus undurchführbar ist, hält aber Poincarés Einwendungen für unbegründet."

[41](Ewald 1996, p. 1113). Here is the German text from (Hilbert 1918, p. 153): "Nur in zwei Fällen nämlich, wenn es sich um die Axiome der *ganzen Zahlen* selbst und wenn es sich um die Begründung der *Mengenlehre* handelt, ist dieser Weg der Zurückführung auf ein anderes spezielleres Wissensgebiet offenbar nicht gangbar, weil es außer der Logik überhaupt keine Disziplin mehr gibt, auf die alsdann eine Berufung möglich wäre.

Da aber die Prüfung der Widerspruchslosigkeit eine unabweisbare Aufgabe ist, so scheint es nötig, die Logik selbst zu axiomatisieren und nachzuweisen, daß Zahlentheorie sowie Mengenlehre nur Teile der Logik sind."

problems. In particular, no reduction of epistemological interest has been achieved.

I want to end this section by pointing out that Zermelo's axiomatization of set theory in 1908, inspired by both Dedekind and Hilbert, is another instance of existential axiomatics. Points 1) through 4) in the first section of (Zermelo 1908c) are structured in the same way as the axiomatic formulations in Hilbert's (1899a) and (1900b). According to Ebbinghaus,[42] Zermelo had intended to give a consistency proof for his axiom system, but was encouraged by Hilbert to publish the axiomatization first and leave the consistency proof for another occasion. Clearly, as to the nature of Zermelo's considerations at that point I am not sure at all. Zermelo just remarks in his paper, "I have not yet even been able to prove rigorously that my axioms are consistent, though this is certainly very essential; instead I have had to confine myself to pointing out now and then that the antinomies discovered so far vanish one and all if the principles here proposed are adopted as a basis."[43] His later investigations on the foundations of set theory in (Zermelo 1930) are parallel to Dedekind's metamathematical reflections described above.

---

[42]Ebbinghaus quotes in his book (2007, p. 78), from a letter of Zermelo to Hilbert, dated 25 March 1907 in which Zermelo writes: "According to your wish I will finish my set theory as soon as possible, although actually I wanted to include a proof of consistency."

[43]All the remarks from (Zermelo 1908c) are found in (van Heijenoort 1967, pp. 200–201).

I.3

Hilbert's programs: 1917–1922

Abstract. Hilbert's finitist program was not created at the beginning of the twenties solely to counteract Brouwer's intuitionism, but rather emerged out of broad philosophical reflections on the foundations of mathematics and out of detailed logical work; that is evident from notes of lecture courses that were given by Hilbert and prepared in collaboration with Bernays during the period from 1917 to 1922. These notes reveal a dialectic progression from a critical logicism through a radical constructivism toward finitism; the progression has to be seen against the background of the stunning presentation of mathematical logic in the lectures given during the winter term 1917/18. In this paper, I sketch the connection of Hilbert's considerations to issues in the foundations of mathematics during the second half of the 19th century, describe the work that laid the basis of modern mathematical logic, and analyze the first steps in the new subject of proof theory. A revision of the standard view of Hilbert's and Bernays's contributions to the foundational discussion in our century has long been overdue. It is almost scandalous that their carefully worked out notes have not been used yet to understand more accurately the evolution of modern logic in general and of Hilbert's Program in particular. One conclusion will be obvious: the dogmatic formalist Hilbert is a figment of historical (de)construction! Indeed, the study and analysis of these lectures reveal a depth of mathematical-logical achievement and of philosophical reflection that is remarkable. In the course of my presentation many questions are raised and many more can be explored; thus, I hope this paper will stimulate interest for new historical and systematic work.

Received May 15, 1997; revised October 19, 1998.
Versions of this paper were presented at the workshop *Modern Mathematical Thought* in Pittsburgh (September 21–24, 1995), at the conference *Philosophy of Mathematics, Logic, and Wittgenstein* in Chicago (May 31–June 2, 1996), at *Gödel '96* in Brno (August 25–29, 1996), and at the Annual Meeting of the ASL in Boston (March 22–25, 1997). Critical questions and constructive suggestions from all audiences helped me to clarify some issues; I am particularly grateful to W. A. Howard, G. H. Müller, H. Stein, and A. Urquhart. I want to acknowledge the continuing, fruitful discussions with my Hilbert-Edition-colleagues W. Ewald, M. Hallett, R. Haubrich, and U. Majer, but also with my students J. Byrnes and M. Ravaglia. Finally, I profited from remarks by Jeremy Avigad, Andreas Blass, Charles Parsons, and an anonymous referee; the latter pointed me to work reported in (Abrusci 1987) and (Moore 1997).

Introduction

At the very end of a sequence of lectures he gave in 1919 under the title *Natur und mathematisches Erkennen*, Hilbert emphasized that some physical paradoxes had directed his discussion away from the methods of physics to the general philosophical problem, "whether and how it is possible to understand our thinking by thinking itself and to free it from any paradoxes".[1] Hilbert saw this problem also at the basis of his work in mathematical logic. One might ask polemically, whether there is more to Hilbert's contribution to that problem than the narrow and technical consistency program pursued in Göttingen during the twenties. A critical reader of the relevant historical and philosophical literature, and even of some of Hilbert's own writings, almost certainly would be inclined to give a negative answer.

During the last ten or fifteen years a more positive and more accurate perspective on the work of the Hilbert School has been emerging, for example, in papers by Feferman, Hallett, Sieg, and Stein. This has been achieved mainly by bringing out the rich context in which the published work is embedded: important connections have been established, on the one hand, to foundational work of the 19th century (that had been viewed as largely irrelevant) and, on the other hand, to a general reductive program (that evolved out of Hilbert's Program and underlies implicitly most modern proof theoretic investigations).[2] However, it is crucial to gain a better understanding of the development of Hilbert's thought on the foundations of arithmetic, where arithmetic is understood in a broad sense that includes elementary number theory and reaches all the way to set theory. Admittedly, this is just one aspect of Hilbert's work on the foundations of mathematics, as it disregards the complex interactions with his work on the foundations of geometry and of the natural sciences. It is, nevertheless, a most significant aspect, as it reveals a surprising internal dialectic progression (in an attempt to address broad philosophical issues) and throws a distinctive new light on the development of modern mathematical logic.

Standard wisdom partitions Hilbert's work on the foundations of arithmetic with some justification into two periods. The first period is taken to extend from 1900 to 1905, the second from 1922 to 1931. The periods are marked by dates of outstanding publications. Hilbert published in 1900 and 1905 respectively *Über den Zahlbegriff* and *Über die Grundlagen der Logik und Arithmetik*. According to the standard view, the considerations of the latter paper were taken up around 1921, were quickly expanded into the proof theoretic program, and were exposed first in 1922 through Hilbert's

---

[1] This quotation is found on page 117 of (Hilbert *1919).

[2] (Abrusci 1981) and (Toepell 1986) contain much valuable information concerning earlier roots of Hilbert's foundational work around the turn of the century, in Toepell's case for geometry. Peckhaus gives in his (1990) a detailed account of subsequent developments in Göttingen up to 1917; this includes a discussion of some of Hilbert's lectures (e.g., those of 1905), but also of Hilbert's "Personalpolitik" concerning Zermelo and Nelson.

Neubegründung der Mathematik and Bernays's *Über Hilberts Gedanken zur Grundlegung der Arithmetik*. This "continuity" is pointed out also by Hilbert and Bernays without emphasizing their early mathematical-logical work or the exploration of alternative foundational perspectives. Finally, it is argued that the pursuit of the program was halted in 1931 by Gödel's paper *Über formal unentscheidbare Sätze der Principia Mathematica und verwandter Systeme I*.

This partition of Hilbert's work does not include, or accommodate easily, the programmatic paper *Axiomatisches Denken* published in 1918. The paper had been presented already in September 1917 to the Swiss Mathematical Society in Zürich and advocates a logicist reduction of mathematics. In sharp contrast, the 1922 papers by Hilbert and Bernays seem to set out the philosophical and mathematical-logical goals of *the* Hilbert Program. This remarkable progression is not at all elucidated by publications, but it can be analyzed by reference to notes for courses Hilbert gave during that period in Göttingen. The lectures were prepared with the assistance of Bernays who wrote all the notes, except that Schönfinkel helped prepare the notes for the summer term 1920. I will discuss this development, after sketching in Part 1 connections to foundational investigations of the 19th century; Part 2 describes the strikingly novel treatment of general logical and metamathematical issues, whereas Part 3 is devoted to the emergence of specifically proof theoretic investigations. Thus, here is a first attempt to bridge the gap in the published record between Hilbert's Zürich Lecture and the proof theoretic papers from 1922; the study and analysis of these lectures reveal a depth of mathematical-logical achievement and of philosophical reflection that is remarkable.[3] In the course of my presentation many questions are raised and many more can be explored; thus, I hope this paper will stimulate interest for new historical and systematic work.

---

[3] There is some work that covers this period of Hilbert's foundational investigations. Abrusci, in his (1987), lists 25 lectures concerned with foundational matters that were given by Hilbert between 1898 and 1933. On pp. 335–8 he attempts to give a rough impression of the richness of these lectures by highlighting the contents of some. He emphasizes that the lectures "testify the remarkable Hilbert's interest [sic] in the foundations of mathematics during the years 1905–1917" and that the 1917–8 notes are the beginning of the golden period of Hilbert's logical and foundational investigations. This paper is a very brief, tentative description; it promises, but does not provide, a sustained analysis. Peckhaus describes in his (1995) Hilbert's development from "Axiomatik" to "Beweistheorie", but disregards all the lectures between 1917 and 1922 and mentions, for the step to finitist proof theory, only Hilbert's publications starting with (Hilbert 1922). Moore discusses in Section 8, pp. 113–6, of his (1988) the 1917–8 lectures as part of a general account of the "emergence of first-order logic"; however, the really novel aspects of these lectures, emphasized below in Part 2, are not brought out. As far as the emergence of proof theory is concerned, Moore's brief discussion starts with the first publication of Hilbert's investigations in 1922. The account of Hilbert's (and Bernays's) contribution to the emergence of mathematical logic is deepened in (Moore 1997). Moore focuses also here more broadly on "standard" metalogical issues, whereas I concentrate on (finitist) consistency and the emergence of proof theory. There is a significant difference in our overall analyses of the developments between 1917 and 1922; cf. note 45.

1 Before 1917: Axiomatic method and consistency

Hilbert viewed the *axiomatic method* as holding the key to a systematic organization of any sufficiently developed subject; he also saw it as providing the basis for metamathematical investigations of independence and completeness issues and for philosophical reflections. However, *consistency* was Hilbert's central concern ever since he turned his attention to the foundations of analysis in the late 1890s. For analysis, Dedekind and Kronecker had put forward two radically different kinds of arithmetizations in response to Dirichlet's demand that *any* theorem of algebra and higher analysis be formulated as a theorem about natural numbers.

1.1 Arithmetization strict and logical.

Kronecker admitted as objects of analysis only natural numbers and constructed from them, in now well known ways, integers, rationals, and even algebraic reals. The general notion of irrational number was rejected, however, because of two restrictive methodological requirements: concepts must be decidable, and existence proofs must be carried out in such a way that they present objects of the appropriate kind. For Kronecker there could be no infinite mathematical objects, and geometry was banned from analysis even as a motivating factor. (Hilbert's critical, but also appreciative discussion in his lectures during the summer term 1920 emphasizes these broad methodological points.) Clearly, this procedure is strictly arithmetic, and Kronecker believed that analysis could be re-obtained by following it. It is difficult for me to judge to what extent Kronecker pursued a program of developing parts of analysis in an elementary, constructive way. Such a program is not chimerical, as mathematical work during the last two decades has established that a good deal of analysis and algebra can be done in conservative extensions of primitive recursive arithmetic.

In contrast to Kronecker, Dedekind defined a general notion of real number, motivated cuts explicitly in geometric terms, and used infinite sets of natural numbers as respectable mathematical objects. The principles underlying the definition of cuts were for Dedekind logical ones which allowed the "creation" of new numbers, such that their system has "the same completeness or ... the same continuity as the straight line". Dedekind emphasized in a letter to Lipschitz that this *continuous completeness* is essential for a scientific foundation of the arithmetic of real numbers, as it relieves us in analysis of the necessity to assume existences without sufficient proof. Indeed, it provides the answer to Dedekind's rhetorical question:

> How shall we recognize the admissible existence assumptions and distinguish them from the countless inadmissible ones ...? Is this to depend only on the success, on the accidental discovery of an internal contradiction?[4]

Dedekind is considering here assumptions about the existence of individual real numbers. Such assumptions are not needed when a complete system

---
[4] Letter to Lipschitz of 27 July 1876; in (Dedekind 1932, p. 477).

is investigated: the question concerning the existence of particular reals is shifted to the question concerning the existence of their *complete system*.

If we interpret the essay *Stetigkeit und irrationale Zahlen* in light of Dedekind's considerations in *Was sind und was sollen die Zahlen?* and his letter to Keferstein, we can describe his procedure in an extremely schematic and yet accurate way: the essays present informal analyses that lead with compelling directness to the axioms for a *complete ordered field*, respectively to those for a *simply infinite system*. Then models for these axioms are given in logical terms; thus, the consistency of the axiomatically characterized notions seemed to be secured on logical grounds.[5] With respect to simply infinite systems Dedekind wrote to Keferstein on 27 February 1890:

After the essential nature of the simply infinite system, whose abstract type is the number sequence N, had been recognized in my analysis . . . , the question arose: does such a system exist at all in the realm of our ideas? Without a logical proof of existence it would always remain doubtful whether the notion of such a system might not perhaps contain internal contradictions. Hence the need for such a proof (articles 66 and 72 of my essay).[6]

Dedekind viewed these considerations not as specific for the foundational context of his essays, but rather as paradigmatic for a mathematical procedure to introduce axiomatically characterized notions.[7]

1.2 Consistency of sets and theories.[8] The origins of Hilbert's Program can be traced back to these foundational problems in general and to Dedekind's proposed solution in particular. Hilbert turned his attention to them, as he recognized that some observations of Cantor had an absolutely devastating effect on Dedekind's essays.[9] Cantor had remarked in letters dated 26 September and 2 October 1897 that he had been led "many years ago" to the necessity of distinguishing two kinds of totalities (multiplicities, systems), namely *absolutely infinite* and *completed* ones. In his letter to Dedekind of 28 July 1899, totalities of the first kind are called *inconsistent* and those of the second kind *consistent*. This distinction avoids, in a trivial way, the contradiction that arose from assuming, as Dedekind had done, that the totality of all things is consistent.

In 1899 Hilbert wrote *Über den Zahlbegriff*, his first paper addressing foundational issues of analysis. He intended—never too modest about aims—to rescue the set theoretic arithmetization of analysis from the

---

[5]That such a proof is intended also in *Stetigkeit und irrationale Zahlen* is most strongly supported by the discussion in (Dedekind 1888, p. 338).

[6]In (van Heijenoort 1967, p. 101). The essay Dedekind refers to is (Dedekind 1888).

[7]Cf. the discussion of ideals in (Dedekind 1877), where he draws direct parallels to the steps taken here. (Dedekind 1877, pp. 268–269), in particular the long footnote on p. 269.

[8]Note that Hilbert talked in his (Hilbert 1900b) about consistent sets; if he had followed strictly Cantor's terminology, he would have mentioned only sets—which are defined as consistent multiplicities.

[9]In particular in section 66 of *Was sind und was sollen die Zahlen?* That is clear from Cantor's response of 15 November 1899 to a letter of Hilbert's (presumably not preserved).

Cantorian difficulties. To that end he gave a categorical axiomatization of the real numbers based on Dedekind's work, claimed that its consistency can be proved by a "suitable modification of familiar methods"[10], and remarked that such a proof constitutes "the proof for the existence of the totality of real numbers or—in the terminology of G. Cantor—the proof of the fact that the system of real numbers is a consistent (completed) set". In his subsequent Paris address Hilbert went even further and claimed that the existence of Cantor's higher number classes and of the alephs can be proved in an analogous way.[11] For the real numbers he suggested more specifically that the familiar inference methods of the theory of irrational numbers have to be modified with the aim of obtaining a "direct" consistency proof; such a direct proof would show that one cannot obtain from the axioms, by means of a finite number of logical inferences, results that contradict each other.[12] Hilbert realized soon that the consistency problem, even for the theory of real numbers, could not be solved as easily as he had thought. Bernays commented later that "the considerable difficulties of this task emerged" when Hilbert actually tried to prove these consistency claims.

In his address to the International Congress of Mathematicians, Heidelberg 1904, Hilbert examined more systematically various attempts at providing foundations for analysis, including Cantor's. The critical attitude towards Cantor, that was implicit in *Über den Zahlbegriff*, was made explicit here. Hilbert accused Cantor of not giving a rigorous criterion for distinguishing consistent from inconsistent totalities, as Cantor's conception "leaves latitude for subjective judgment and therefore affords

---

[10](Hilbert 1899a, p. 261). The German original is: "Um die Widerspruchsfreiheit der aufgestellten Axiome zu beweisen, bedarf es nur einer geeigneten Modifikation bekannter Schlußmethoden." (Bernays 1935a) reports on pp. 198–199 in very similar words, but with a mysterious addition: "Zur Durchführung des Nachweises gedachte Hilbert mit einer geeigneten Modifikation der in der Theorie der reellen Zahlen angewandten Methoden auszukommen."

[11]Cantor, by contrast, insists in his letter to Dedekind of 28 August 1899 that even finite multiplicities cannot be proved to be consistent. The fact of their consistency is a simple, unprovable truth—"the axiom of arithmetic"; the fact of the consistency of multiplicities that have an aleph as their cardinal number is in exactly the same way an axiom, the "axiom of the extended transfinite arithmetic". (Cantor 1932, pp. 447–8).

[12]This is part of Hilbert's formulation of the second problem; more fully we find: "Vor allem aber möchte ich unter den zahlreichen Fragen, welche hinsichtlich der Axiome gestellt werden können, dies als das wichtigste Problem bezeichnen, *zu beweisen, daß man auf Grund derselben mittels einer endlichen Anzahl von logischen Schlüssen niemals zu Resultaten gelangen kann, die miteinander in Widerspruch stehen.*" Howard Stein pointed out to me that a syntactic view of the consistency problem is entertained here by Hilbert, though there is no indication of the logical matters that have to be faced: that is done vaguely and programmatically in 1904, but concretely and systematically only in the winter term 1917/18. Indeed, already in section 9 (pp. 19/20) of (Hilbert 1899a) a syntactic formulation of consistency is given; however, the immediately following argument for consistency is a thoroughly semantic one: "Um dies [die Widerspruchslosigkeit, WS] einzusehen, genügt es, eine Geometrie anzugeben, in der sämtliche Axiome . . . erfüllt sind."—The general problematic, indicated here only indirectly, was most carefully analyzed by Bernays in his (1950).

no objective certainty".[13] He suggested again that consistency proofs for suitable axiomatizations provide an appropriate remedy and described in greater detail how he envisioned such a proof: develop logic together with analysis in a common frame, so that proofs can be viewed as *finite mathematical* objects; then show that such formal proofs cannot lead to a contradiction. Here we have seemingly in very rough outline Hilbert's Program; but it should be noticed that the point of consistency proofs is still to guarantee the existence of sets, that the logical frame is only vaguely conceived, and that a reflection on the mathematical means admissible in consistency proofs is completely lacking. Indeed, as we will see, the path to the program is still rather circuitous.

One reason for the circuitous route is, so it seems, the critique of the enterprise by Poincaré; the latter agrees with Hilbert on the fundamental point that mathematical existence can mean only freedom from contradiction: "If therefore we have a system of postulates, and if we can demonstrate that these postulates imply no contradiction, we shall have the right to consider them as representing the definition of one of the notions entering therein."[14] But any such proof (for systems that involve an infinite number of consequences) requires the principle of complete induction; this point is re-emphasized over and over in Poincaré's remarks on Hilbert's 1905 paper. At one point he summarizes matters as follows:

So, Hilbert's reasoning not only assumes the principle of induction, but it supposes that this principle is given us not as a simple definition, but as a synthetic judgment a priori.

To sum up:

A demonstration [of consistency] is necessary.

The only demonstration possible is the proof by recurrence.

This is legitimate only if we admit the principle of induction and if we regard it not as definition but as a synthetic judgment.[15]

Only after exploring alternative foundational approaches did Hilbert "return" to proof theory and address explicitly Poincaré's objection; I will resume that discussion in 3.2 below.

1.3 Developments from 1905 to 1917. In contrast to an almost universally held opinion, Hilbert continued to be concerned with the foundations of mathematics. There is no record of publications supporting this claim, but Hilbert gave a number of lecture courses on

---

[13](Hilbert 1905a) in (van Heijenoort 1967, p. 131).

[14](Poincaré 1905, p. 1026). Having discussed Mill's view of (mathematical) existence and characterizing the latter's opinion as "inadmissible", Poincaré writes in the immediately preceding paragraph: "Mathematics is independent of the existence of material objects; in mathematics the word exist can have only one meaning, it means free from contradiction. . . . in defining a thing, we affirm that the definition implies no contradiction."

[15](Poincaré 1906, p. 1059). Howard Stein raised in discussion the question whether Poincaré's criticism had the effect of postponing the development of proof theory; it seems to me that indeed it did.

the topic between 1905 and 1917, and extensive notes of his lectures are available. The lectures on *Logische Prinzipien des mathematischen Denkens* from the summer term 1905 are preserved in two different sets of notes, one of which was prepared by Max Born; Hilbert lectured on *Zahlbegriff und Prinzipienfragen der Mathematik* (Summer 1908), on *Elemente und Prinzipienfragen der Mathematik* (Summer 1910), on *Grundlagen der Mathematik und Physik* (Summer 1913), *Prinzipien der Mathematik* (Summer 1913), *Probleme und Prinzipien der Mathematik* (Winter 1914/15), and on *Mengenlehre* (Summer 1917). Let me describe paradigmatically some crucial features of the 1910-lectures; the notes were written by Richard Courant.

The lectures start out with a *plan* dividing the course into three parts. Part I is to deal with "The Quadrature of the Circle and Related Problems" and is clearly based on lectures Hilbert had given repeatedly under that title, e.g., in 1904; Part II is called "Problems of Analysis and Mechanics"; the content of Part III is indicated by "Critique of Basic Notions. Axiomatic Method. Logic and Mathematical Thought". The final result, shaped no doubt by the exigencies of ordinary academic life, is quite different and develops only a fraction of what was announced for Part III. Quantitatively, one finds six handwritten pages of a total of 162 pages under the heading "Chapter 5: *On Logical Paradoxes and Logical Calculus*". Yet, what there is—is of genuine interest. Hilbert discusses first Richard's paradox and dismisses it as easily solvable:

One just has to look at this whole argument without prejudice to recognize that it is completely inadmissible. The ambiguous, subjective character of language does not allow us to assert the exact claim that certain words must always refer to one and the same concept; this remark is already sufficient to recognize the fallacy.[16]

Concerning the Russell-Zermelo paradox, Hilbert claims that it was removed from set theory by Zermelo, but that "it has not yet been resolved in a satisfactory way as a logical antinomy" (p. 159). With this enigmatic remark he moves on to the last point of the lectures, a sketch of basic ideas for a logical calculus that will be taken up again later. "We assume that we have the capacity to name things by signs, that we can recognize them again. With these signs we can then carry out operations that are analogous to those of arithmetic and that obey analogous laws."[17] This remark is followed by a brief algebraic description of sentential logic and the programmatic formulation of the task of a logical calculus, "to draw

---

[16] Man braucht die ganze Argumentation nur vorurteilslos anzusehen, um zu erkennen, dass sie völlig unzulässig ist. Schon der Einwand, dass der vieldeutige, subjektive Charakter der Sprache es nicht gestattet, die exakte Behauptung aufzustellen, dass bestimmte Worte stets einen und denselben Begriff bezeichnen müssen, reicht hin, um den Trugschluss zu erkennen. (p. 158)

[17] Wir gehen von der Annahme aus, dass wir die Fähigkeit haben, Dinge durch Zeichen zu benennen, dass wir sie wiederzuerkennen vermögen. Mit diesen Zeichen werden wir dann gewisse Operationen ausführen können, die denen der Arithmetik analog sind und analogen Gesetzen folgen. (p. 159)

logical inferences by means of purely formal operations with letters". Some examples of such inferences are then presented.

These lectures do not break new ground, but they do provide clarifications, broader perspectives, and a sharpening of central problems; the main issues are closely related to those I discussed above, but in none of the lectures, except those from the summer term of 1905, does Hilbert take up the proof theoretic approach of his Heidelberg paper. All of this can be seen from the lectures on set theory given in the summer term 1917, most poignantly when comparing them to the lectures given just a few months later in the winter term 1917/18. Chapter I of the set theory notes treats rational, algebraic, and transcendental numbers; under the heading "The Numbers and their Axioms", Chapter II presents a version of the axiom system for the reals formulated in *Über den Zahlbegriff* and supplements it by investigations of independence questions familiar from *Grundlagen der Geometrie*; Chapter III focuses on the concept of set, in particular, on that of an ordered and well-ordered set. Finally, in Chapter IV, Hilbert intends to deal with "Application of Set Theory to Mathematical Logic". I am not sure how to understand the last heading, as there is no discussion of mathematical logic in that chapter! But there *is* again a discussion of Richard's paradox and of the Russell-Zermelo antinomy. This time the fundamental problem is seen as related to what Hilbert calls "genetische Definitionen". The remarks warrant discussion: they point to the past as represented by Kronecker and by his own 1905 lectures, and to the future, i.e., to a fully developed finitist standpoint.

1.4 Genetic definitions. These definitions include all impredicative ones. One example is given by the set theoretic definition of inductively generated classes as the smallest sets satisfying certain closure conditions; another example is extracted in Hilbert's analysis of Dedekind's proof of the existence of an infinite system. Dedekind's proof involves the "system of all things that can be the object of my thought" and thus a system whose definition employs universal quantification. Hilbert does not emphasize in either example that the range of the quantifier must include the set that is being defined, and that is of course the characteristic feature of impredicative definitions. Instead, Hilbert simplifies matters in a quite radical way by taking a "new and unusual" standpoint that disapproves of the use of words like "all", "every", or "and so on". Hilbert views the use of these words as characteristic of genetic definitions and as pervasive in mathematics.

There is no need to consider irrational numbers; the geometric series $1 + 1/2 + 1/4 + 1/8 +$ "and so on" is already an example. Not even formulas in which finite, but only indeterminate whole numbers n occur are immune to our critique. To be able to apply them one sets $n = 1, 2, 3, 4, 5$, "and so on". Kronecker who intended to reduce all of mathematics to the whole numbers was consequently not radical enough, for 'n' does occur in his formula. He should have restricted himself to the specific numbers 7, 15, 24. Thus, one sees what kind of difficulties have to be faced when calculating with letters. Already the simple formula $a + b = b + a$ can be attacked.[18]

Finally, closing the circle to the earlier considerations, Hilbert views sets that can be given *only* through genetic definitions as *inconsistent*.

The natural numbers are given in Dedekind's set theoretic way as well as in the informal Kroneckerian way by genetic definitions; thus, Hilbert rejects the natural numbers as the fundamental system for mathematics. I take it that these reflections constitute the reasoned rejection of the "genetic method" as described in (Hilbert 1900b); the discussion of the genetic and axiomatic method is concluded there as follows: "Despite the high pedagogic and heuristic value of the genetic method, for the final presentation and the complete logical grounding of our knowledge the axiomatic method deserves to be preferred." In the present lecture notes he follows Peano in giving an axiom system for natural numbers and remarks, against Poincaré, that this is but a first step in the foundational investigation:

... if we set up the axioms of arithmetic, but forego their further reduction and take over uncritically the usual laws of logic, then we have to realize that we have not overcome the difficulties for a first philosophical-epistemological foundation; rather, we have just cut them off in this way.[19]

Hilbert answers the question "To what can we further reduce the axioms?" by "To the laws of logic!" He claims that if we try to achieve such a reduction to logic,

... we are facing one of the most difficult problems of mathematics. Poincaré has even the view that this is not at all possible. But with that view one could rest content only if it had been proved that the further reduction of the axioms for arithmetic is impossible; but that is not the case. Next term, I hope to be able to examine more closely a foundation for logic.[20]

---

[18] Wir brauchen nicht einmal Irrationalzahlen zu betrachten, schon die geometrische Reihe $1 + 1/2 + 1/4 + 1/8 +$ 'und so weiter' ist ein Beispiel dafür. Ja nicht einmal Formeln, in denen endliche, aber nur unbestimmte ganze Zahlen n vorkommen, halten unserer Kritik stand. Denn um sie anwenden zu können, setzt man $n = 1, 2, 3, 4, 5$, 'und so weiter'. Kronecker, der die ganze Mathematik auf die ganzen Zahlen zurückführen wollte, war also noch nicht radikal genug; denn in seiner Formel kommt das 'n' vor. Er hätte sich vielmehr auf spezielle Zahlen 7, 15, 24 beschränken müssen. Man sieht also, was für Schwierigkeiten der Rechnung mit Buchstaben entgegenstehen. Schon die einfache Formel $a + b = b + a$ ist anfechtbar. (p. 137)— Indeed, Hallett in his (1995) makes it clear that this is really an "old" standpoint of Hilbert's going back to 1904/5; cf. also note 49 below.

[19] ... wenn wir die Axiome der Arithmetik aufstellen, aber auf eine weitere Zurückführung derselben verzichten und die gewöhnlichen Gesetze der Logik ungeprüft übernehmen, so müssen wir uns bewusst sein, dass wir dadurch die Schwierigkeiten einer ersten philosophisch-erkenntnistheoretischen Begründung nicht überwunden, sondern nur kurz abgeschnitten haben. (p. 146)

[20] ... so stehen wir vor einem der schwierigsten Probleme der Mathematik überhaupt. Poincaré vertritt sogar den Standpunkt, dass dies garnicht möglich ist, aber damit könnte man sich erst zufrieden geben, wenn der Unmöglichkeitsbeweis für die weitere Zurückführung der Axiome der Arithmetik geführt wäre, was nicht der Fall ist. Auf eine Begründung der Logik hoffe ich im nächsten Semester näher eingehen zu können. (pp. 145–6)

One has again the sense that the exigencies of academic life and the complexity of the issues diverted Hilbert's attention to his own great dissatisfaction. That is, I assume, what motivated Hilbert's action in the spring (or fall) of 1917: he invited Paul Bernays to assist him in efforts to examine the foundations of mathematics.[21] Bernays returned to Göttingen, where he had been a student, and started to work with Hilbert on lectures that were offered in the winter term 1917/18 under the title *Prinzipien der Mathematik*.

2 From 1917 to 1920: Logic and metamathematics

As background for the 1917/18 lectures one should keep in mind that Hilbert saw himself as pursuing one of the most difficult problems of mathematics, i.e., its reduction to logic. In *Axiomatisches Denken* he had formulated matters as follows:

The examination of consistency is an unavoidable task; thus, it seems to be necessary to axiomatize logic itself and to show that number theory as well as set theory are just parts of logic. This avenue, prepared for a long time, not least by the deep investigations of Frege, has finally been taken most successfully by the penetrating mathematician and logician Russell. The completion of this broad Russellian enterprise of axiomatizing logic might be viewed quite simply as the crowning achievement of the work of axiomatization.[22]

The detailed pursuit of that goal required the presentation of a formal language (for capturing the logical form of informal statements), the use of a formal calculus (for representing the structure of logical arguments), and the formulation of "logical" principles (for defining mathematical objects). This is carried through with remarkable focus, elegance, and directness. From the very beginning, the logical and mathematical questions are mixed with, or rather driven by, philosophical reflections on the foundations of mathematics, and we find penetrating discussions of the axiom of reducibility that become increasingly critical and lead ultimately to the rejection of the logicist enterprise (in 1920; cf. 2.3).

2.1 Future and past. The collaboration of Hilbert and Bernays led to a remarkable sequence of lectures, where we can witness the creation of modern mathematical logic and the emergence of proof theory.

---

[21]According to Constance Reid, pp. 150–1, Hilbert invited Bernays in the spring of 1917; Bernays, however, writes in his biographical note: "Im Herbst 1917 wurde ich von Hilbert anlässlich seines in Zürich gehaltenen Vortrages *Axiomatisches Denken* aufgefordert, an seinen wieder aufgenommenen Untersuchungen über die Grundlagen der Arithmetik als sein Assistent mitzuwirken."

[22]Da aber die Prüfung der Widerspruchslosigkeit eine unabweisbare Aufgabe ist, so scheint es nötig, die Logik selbst zu axiomatisieren und nachzuweisen, daß Zahlentheorie sowie Mengenlehre nur Teile der Logik sind. Dieser Weg, seit langem vorbereitet—nicht zum mindesten durch die tiefgehenden Untersuchungen von Frege—ist schließlich am erfolgreichsten durch den scharfsinnigen Mathematiker und Logiker Russell eingeschlagen worden. In der Vollendung dieses großzügigen Russellschen Unternehmens der *Axiomatisierung der Logik* könnte man die Krönung des Werkes der Axiomatisierung überhaupt erblicken. (p. 153)

The relevant lectures are: *Prinzipien der Mathematik* (Winter 1917/18); *Logik-Kalkül* (Winter 1920); *Probleme der mathematischen Logik* (Summer 1920); *Grundlagen der Mathematik* (Winter 1921/22); *Logische Grundlagen* (Winter 1922/23). In the winter term 1919 Hilbert gave related lectures entitled *Natur und mathematisches Erkennen*. In presenting the lectures from 1917/18 and the further development in those from 1920, I will highlight three groups of issues, namely, logical, mathematical, and general metamathematical ones. Proof theoretic issues began to emerge only in 1920 and are the topic of Part 3. But I want to make first a few remarks about the immediate historical context of these lectures.

A polished presentation of the material developed in this sequence of lectures (leaving out the specifically proof theoretic considerations) is found in Hilbert and Ackermann's book, *Grundzüge der theoretischen Logik*, published in 1928. Indeed, the basic structure of the book is the same as that of the 1917/18 notes, large parts of the texts are identical, and there are hardly any new metamathematical results (except for important results that had been obtained in the meantime, like special cases of the decision problem, the Löwenheim Skolem theorem). In the preface to the book Hilbert wrote:

In preparing the above lectures [WS 17/18, WS 20, WS 21/22] I received support and advice in essential ways from my colleague P. Bernays; the latter also wrote the notes for these lectures most carefully.—Using and supplementing the material that had been accumulated in this way, W. Ackermann . . . provided the present organization and gave the definitive presentation of the total material.[23]

The fact that the supplements by Ackermann are minimal is historically important, as the book has been taken falsely, for example in (Goldfarb 1979), as the endproduct of a cumulative development. This one misjudgment informs others; for example, it is claimed that quantifiers were properly understood only in the book of 1928, and as evidence Goldfarb adduces that ". . . in his [Hilbert's] early presentations of axiom systems [in (Hilbert 1922) and (Hilbert 1923)] we first meet some quantifier-free number theoretic and analytic axioms; the so-called transfinite axioms which introduce quantification then follow. The direction, in short, is the reverse of that which would highlight the underlying nature of quantificational logic. . . ." (p. 359) That expresses a deep misunderstanding of the early published work on proof theory and its systematic background. As we will see shortly, the 1917/18 notes contain first-order logic in a fully developed form. Restricted calculi were introduced for programmatic reasons and not, as has been suggested, because of a "finitist prejudice" or because fuller calculi had yet to be developed.

---

[23]Bei der Vorbereitung der genannten Vorlesungen [WS 17/18, WS 20, WS 21/22] bin ich von meinem Kollegen P. Bernays wesentlich unterstützt und beraten worden; derselbe hat diese Vorlesungen auch aufs sorgfältigste ausgearbeitet.—Unter Benutzung und Ergänzung des so entstandenen Materials hat W. Ackermann . . . die vorliegende Gliederung und definitive Darstellung des Gesamtstoffes durchgeführt. ["WS" stands for "Winter Semester".]

Hilbert and Bernays's achievements during this period are overshadowed by Gödel's subsequent work—that is inspired by it and that builds on it. The book with Ackermann, though recognized as a landmark, has been severely criticized (e.g., by Goldfarb, Dreben, and van Heijenoort); the most substantial critical remarks are taken up in 2.3 below. Seeing the 1928 book as the product of a sustained development in Göttingen makes it extremely difficult to appreciate the novelty and originality of the very early published work. It makes it even more difficult, on the one hand, to understand how Hilbert and Bernays's work was influenced by contemporaneous work in logic (e.g., that of Russell and Whitehead or that in the algebraic tradition of Schröder) and, on the other hand, to appreciate in what respects it was strikingly different.

As to the influence of contemporaneous logical work, I learned through a personal communication from Alasdair Urquhart that Hilbert and Russell exchanged some postcards between 1916 and 1919; for details see Appendix B. The most relevant information for the discussion here is Hilbert's claim made on his postcard to Russell dated April 12, 1916, "that we have been discussing in the Math. Society your theory of knowledge already for a long time, and that we had intended, just before the outbreak of the war, to invite you to Göttingen, so that you could give a sequence of lectures on your solution to the problem of the paradoxes." The notes for lectures Hilbert gave before the winter term 1917/18, even for those of the immediately preceding summer term 1917, do not contain any reference to *Principia Mathematica* nor any hint of a Russellian influence. There is only one exception I discovered; in his lectures *Probleme und Prinzipien der Mathematik* given in the winter term 1914/15, Hilbert mentions Russell and remarks briefly that type theory contains something true, but that it has to be deepened significantly. Here is a real gap in our historical understanding; we also do not have a sense of Bernays's possibly pivotal role or of other influences, like Weyl's through his book *Das Kontinuum*. This gap is puzzling and very much worth closing. (The detailed analysis of Behmann's dissertation, described in Mancosu's 1998 manuscript, will undoubtedly throw light on the Russellian influence; cf. note 67.)

2.2 Languages and calculi. The lectures given during the summer term 1917 do not contain a proper logical system: what indication of logical matters one finds there is of a very restricted algebraic sort. An algebraic motivation is still present in the lectures of the following winter term, but only in a broad methodological sense. We read on page 63 for example: "The logical calculus consists in the application of the formal methods of algebra to logic." However, the general and explicit goal is to develop a symbolic language and a suitable logical calculus that allow a thoroughgoing formalization of mathematics, in particular of analysis.

The 1917/18 notes consist of 246 type-written pages and are divided into two parts. Part 1, *Axiomatische Methode*, gives on sixty-two pages Hilbert's standard account of the axiomatic method, in particular, as it applies to geometry. Part 2, *Mathematische Logik*, is a beautifully organized, almost definitive presentation of the very core of modern mathematical logic. The material is organized under the chapter headings:

1. The sentential calculus;
2. The predicate calculus and class calculus [the former is just monadic logic];
3. Transition to the function calculus [i.e., first-order logic];
4. Systematic presentation of the function calculus;
5. The extended function calculus.

Chapters 1 through 4 lead, in part, to a systematic formulation of first-order logic; every step taken in expanding the logical framework is semantically motivated and carefully argued for. This material was novel at the time; by now it is all too familiar and will not be discussed, except to note and emphasize one important difference: the languages contain *sentential* and *function* (i.e., relation) *variables*. Weyl presented in his almost contemporaneous book *Das Kontinuum* the language of first-order logic in a very similar way.[24] He did not introduce a logical calculus, but discussed very informatively the main task of logic, namely, to describe the syntactic, formal structures that would allow one to establish all the semantic, logical consequences of given assumptions; cf. the brief discussion in Remark 2 at the end of 2.3 below. This main task of logic is partially resolved for first-order logic in the 1917/18 notes, where, at the very end of Chapter 4, the suitability of the calculus for the formal-axiomatic presentation of theories is re-examined:

> The calculus is well suited for this purpose mainly for two reasons: one, because its application prevents that—without being noticed—assumptions are used that have not been introduced as axioms, and, furthermore, because the logical dependencies so crucial in axiomatic investigations are represented by the symbolism of the calculus in a particularly perspicuous way.[25]

Chapter 5 takes a noteworthy turn. After all, if only a formalization of logical reasoning were aimed for, no additional work beyond that of chapters 1 through 4 would be needed. The logical calculus is to play, however, an important role for the investigation of mathematical theories and their relation to logic.

> Not only do we want to develop individual theories from their principles in a purely formal way, but we also want to investigate the foundations of the mathematical

---

[24]The basic formulation goes back to (Weyl 1910), where the language is also built up using disjunction, negation, and existential quantification.

[25]Für diesen Zweck ist der Kalkül vor allem aus zwei Gründen sehr geeignet, einmal weil bei seiner Anwendung verhütet wird, dass man unbemerkt Voraussetzungen benutzt, die nicht als Axiome eingeführt sind, und weil ferner durch die Symbolik des Kalküls die logischen Abhängigkeits-Verhältnisse, auf die es ja bei der axiomatischen Untersuchung ankommt, in besonders prägnanter Weise zur Darstellung gelangen. (p. 187)

theories and examine what their relation to logic is and how far they can be built up from purely logical operations and concepts; and for this purpose the logical calculus is to serve as an auxiliary tool.[26]

If one wants to use the calculus for that logicist purpose, one is led to extend the rules of formally operating within the calculus in "a certain direction". Up to now, statements and functions had been sharply separated from objects; correspondingly, indeterminate statement- and function-signs (i.e., sentential and function variables) had been strictly separated from variables that can be taken as arguments, but this is being changed:

—we will allow now that statements and functions can be taken as values of logical variables in the same way as proper objects and that indeterminate statement signs and function signs can appear as arguments of symbolic expressions.[27]

A free Fregean expansion of the function calculus leads, however, to contradictions. Reflecting on the principles on which this expansion is based, a "logical circle" is discovered. The domain, associated with the original (first-order) function calculus and providing the logical meaning of quantifiers, was expanded by new kinds of objects, namely statements, predicates, and relations. Then new symbolic expressions were admitted, whose "logical meaning [as they involve quantifiers] requires a reference to the totality of statements, respectively of functions".

This way of proceeding is indeed suspicious, insofar as those expressions that gain their meaning only through reference to the totality of statements, respectively functions are counted then among the statements and functions; on the other hand, in order to be able to refer to the totality of statements and functions, we have to view the statements, respectively functions as being determined from the very beginning.[28]

This suspicious way of proceeding involves the logical circle, and there is reason to assume that "this circle is the cause for the presence of the paradoxes". The goal of avoiding any reference to dubious totalities of statements and functions leads "in the most natural way" to ramified type theory.

---

[26]Wir wollen nicht nur imstande sein, einzelne Theorien für sich von ihren Prinzipien aus rein formal zu entwickeln, sondern wollen die Grundlagen der mathematischen Theorien selbst auch zum Gegenstand der Untersuchung machen und sie darauf hin prüfen, in welcher Beziehung sie zu der Logik stehen und inwieweit sie aus rein logischen Operationen und Begriffsbildungen gewonnen werden können; und hierzu soll uns der logische Kalkül als Hilfsmittel dienen. (p. 188)

[27]... wir [werden] nunmehr zulassen, dass Aussagen und Funktionen in gleicher Weise wie eigentliche Gegenstände als Werte von logischen Variablen genommen werden und dass unbestimmte Aussagezeichen und Funktionszeichen als Argumente von symbolischen Ausdrücken auftreten. (p. 188)

[28]Dies Vorgehen ist nun in der Tat bedenklich, insofern nämlich dabei jene Ausdrücke, welche erst durch die Bezugnahme auf die Gesamtheit der Aussagen bezw. der Funktionen ihren Inhalt gewinnen, ihrerseits wieder zu den Aussagen und Funktionen hinzugerechnet werden, während wir doch andererseits, um uns auf die Gesamtheit der Aussagen und Funktionen beziehen zu können, die Aussagen bezw. die Funktionen als von vornherein bestimmt ansehen müssen. (p. 219)

The formal framework of ramified type theory is seen, however, as too narrow for mathematics, because it does not, for example, allow the proper formalization of Cantor's proof of the existence of uncountable sets; cf. pp. 229–30. To achieve greater flexibility for the calculus Russell's axiom of reducibility is adopted; this broader framework is then used for the development of the beginnings of analysis, in particular, the least upper bound principle is established. The notes end with the remark:

> Thus it is clear that the introduction of the axiom of reducibility is the appropriate means to turn the ramified calculus into a system out of which the foundations for higher mathematics can be developed.[29]

Is the outline I gave consistent with a formalist perspective on Hilbert, never mind the metamathematical novelties and logicist tendencies these developments exhibit?—Prima facie the answer may be "yes", but such a perspective is completely inadequate. Why that is so will be clear, I hope, from the further issues I will present.

2.3 Semantic interpretation. The formal frame I have been discussing is not only contentually motivated, but the semantics is properly specified and the central semantic notions are carefully formulated. Sometimes one finds that syntactic notions are interwoven with semantic concepts—amusing to a modern reader who is expecting a "formalist" presentation. But before giving an example, I have to discuss a *very* important, fundamental point that was hinted at already in 2.2. First-order theories are always viewed together with suitable non-empty domains, *Bereiche*, indicating the range of the individual variables of the theory, and interpretations of the nonlogical vocabulary (except, of course, the sentential and function variables). In modern terms, the theories are always presented together with a *structure*. Hilbert and Bernays call this the "existential aspect" of the axiomatic method. A significant philosophical motivation is revealed, when Hilbert reemphasizes the important role of domains as ranges for individual variables and notes: "This remark resolves the difficulties, discussed by Russell, in interpreting general judgments."[30] Weyl also emphasizes this broader point in *Das Kontinuum*, when he says that existential judgments presuppose that "the particulars of the categorial being under consideration should form a closed system of determinate, independently existing objects".[31]

---

[29] So zeigt sich, dass die Einführung des Axioms der Reduzierbarkeit das geeignete Mittel ist, um den Stufen-Kalkül zu einem System zu gestalten, aus welchem die Grundlagen der höheren Mathematik entwickelt werden können. (p. 246)

[30] Auf Grund dieser Bemerkung erledigen sich die von Russell erörterten Schwierigkeiten in der Interpretation des allgemeinen Urteils. (Hilbert *1920a, pp. 25–6).—Hilbert may refer here to the difficulties discussed by Russell already in sections II and III of his 1908 paper *Mathematical logic as based on the theory of types*. Cf. also the beginning of Part 3.1.

[31] l.c., p. 4.—The full German sentence is: "In diesem Sinne verstehen wir die Voraussetzung, daß *die Besonderungen des kategorialen Wesens, um das es sich handelt, ein geschlossenes System bestimmter, an sich existierender Gegenstände ausmachen sollen.*"

Finally, I can mention an example concerning the mixing of semantic and syntactic considerations or rather, how semantic considerations lead to restrictions on syntactic constructions. In the lectures from the winter term 1917/18 (pp. 112–113 and 129 ff), but also in later ones, e.g., from the winter term 1920 (p. 24), a many-sorted logic is introduced. The argument places, *Leerstellen*, of particular functions are taken to be related to particular domains: if an argument place is filled by the name of an object from an inappropriate domain, then the resulting formula is considered as meaningless (sinnlos). This is done similarly for quantification (p. 132); if the same quantified variable is used in two argument places that are related to different domains, the resulting formula is meaningless. Clearly, this can be reflected in a purely syntactic way, as it is done later on; the interesting point here is the direct semantic motivation for restrictive conditions.

How are expressions of the formal language to be understood, given the associated domain? After the discussion of the axiom system for the function calculus, including the specification of the syntax (pp. 129–135), there is the following remark clarifying where a semantic understanding is needed and where pure formality is essential:

This system of axioms provides us with a procedure to carry out logical proofs strictly formally, i.e., in such a way that we need not be concerned at all with the meaning of the judgments that are represented by formulas, rather we just have to attend to the prescriptions contained in the rules. However, we have to interpret the signs of our calculus when representing symbolically the premises from which we start and when understanding the results obtained by formal operations.

The logical signs are interpreted as before according to the prescribed linguistic reading; and the occurrence of indeterminate statement-signs and function-signs in a formula is to be understood as follows: for arbitrary replacements by determinate statements and functions . . . the claim that results from the formula is correct.[32]

This remark points to an answer to the question I raised; it is followed (pp. 136 and 137) by a careful explanation of why the application of the function calculus, given the semantic interpretation, "inhaltliche Auslegung" or "Deutung", leads always to correct results. The underlying concept of

---

[32]Dieses System von Axiomen liefert uns ein Verfahren, um logische Beweisführungen streng formal zu vollziehen, d.h. so, dass wir uns um den Sinn der durch die Formeln dargestellten Urteile gar nicht zu kümmern brauchen, sondern lediglich die in den Regeln enthaltenen Vorschriften zu beachten haben. Allerdings müssen wir bei der symbolischen Darstellung der Prämissen, von denen wir ausgehen, sowie bei der Interpretation der durch die formalen Operationen erhaltenen Ergebnisse den Zeichen unseres Kalküls eine Deutung beilegen.

Diese Deutung geschieht bei den logischen Zeichen in der bisherigen Weise, entsprechend der vorgeschriebenen sprachlichen Lesart; und das Auftreten von unbestimmten Aussage-Zeichen und Funktions-Zeichen in einer Formel ist so zu verstehen, dass bei jeder beliebigen Einsetzung von bestimmten Aussagen und Funktionen . . . die aus der Formel entstehende Behauptung richtig ist. (pp. 135–6)

This remark, almost verbatim, is found in Hilbert and Ackermann's book on page 54. Furthermore, in the notes from the winter term 1920, p. 31: ". . . ferner soll unter einer 'richtigen Formel' ein solcher Ausdruck verstanden werden, der bei beliebiger inhaltlicher Festlegung der vorkommenden unbestimmten Zeichen eine richtige Aussage darstellt."

correctness, *Richtigkeit*, with respect to a domain is to be understood as follows: (1) statements involving no sentential or function variables are "correct" if they are true in the domain, and that is understood informally in exactly the same way as in the model theoretic arguments for independence and relative consistency in Hilbert's *Grundlagen der Geometrie* or, for that matter, in Gödel's dissertation ((Gödel 1929) and (Gödel 1930a)); (2) if a statement does contain such variables, then the clause "for arbitrary replacements by determinate statements and functions the claim that results from the formula is correct" is invoked to define "correctness" for this broader class of statements.[33] This clarification will be used below.

In this semantic context I want to return to the discussion of the ramified theory of types. The standpoint that motivated ramified type theory was this: one takes for granted a domain of individuals with basic properties and basic relations between them. From this basis, all further predicates and relations are obtained, constructively, by the logical operations. Already in the lecture notes from the winter term 1917/18 it is acknowledged that the axiom of reducibility is in conflict with this constructive standpoint. It has to be assumed that "certain predicates and relations have to be viewed as having an independent existence, so that their manifold depends neither on actually given definitions nor on our possibilities of giving definitions".[34] This argument is concisely rehearsed in the notes from the summer term 1920; in the notes for the winter term 1921/22 it leads to an explicit rejection of the logicist route. After all, it is argued, if one chooses the basis in an arbitrary way, the axiom of reducibility is certainly not satisfied. Thus, one would have to expand the "system of basic properties and relations" in such a way that the demand of the axiom is met. The question, whether such an expansion can be achieved by a logical-constructive procedure, is answered negatively.

Thus, there remains only the possibility to assume that the system of predicates and relations of first-order is an independently existing totality satisfying the axiom of reducibility. In this way we return to the axiomatic standpoint and give up the goal of a logical foundation of arithmetic and analysis. Because now a reduction to logic is given only nominally.

I have only sketched this discussion; it is subtle and deserves a detailed analysis and careful comparison with its modified version, influenced by (Ramsey 1925), in *Hilbert and Ackermann*, but also with Weyl's considerations in *Das Kontinuum*. This is important, not least as it sounds explicitly and most clearly themes that will be found in the literature with equally good sense and balance only in Gödel's paper on *Russell's Mathematical*

---

[33]Bill Howard pointed out quite correctly that this notion is used in a context sensitive way; most often it is used in the way I just described it, namely as "true formula", but sometimes also in the sense of "provable formula". This foreshadows a certain ambiguity in *Hilbert and Ackermann*; cf. Remark 1 on completeness below.

[34](Hilbert *1917/18, p. 232).

Logic and, less systematically, in (Gödel 1933b). Let me return briefly to the discussion of metamathematical issues that are faced and formulated with a completely new rigor.

For the purpose of the logical calculus in the systematic investigation it is crucial that it allows one to recapture formally the ordinary forms of argumentation. This is clearly expressed in the 1917/18 lecture notes:

As for any other axiomatic system, one can raise also for this system the questions concerning consistency, logical dependencies, and completeness. The most important question is here that concerning completeness. After all, the goal of symbolic logic is to develop ordinary logic from the formalized assumptions. Thus, it is essential to show that our axiom system suffices for the development of ordinary logic.[35]

These notes contain prominently only one mathematically precise concept of completeness for logical calculi, namely "Post-completeness": "We will call the presented axiom system complete, if the addition of a formula, hitherto unprovable, to the system of basic formulas always leads to an inconsistent system."[36] That is quickly established for sentential logic, and the semantic completeness is mentioned and proved in a footnote (on p. 153). The latter notion is brought to the fore, unequivocally and beautifully, in Bernays's Habilitationsschrift[37] of 1918, where the completeness theorem receives its first "classical" formulation: "Every provable formula is a valid formula and vice versa."[38] For first-order logic the question of its Post-completeness is raised in the lecture notes (p. 156), and it is conjectured that the answer is negative. The proof of this fact, explicitly attributed to Ackermann, is then given in *Hilbert and Ackermann* (p. 66) where the considerations for sentential logic can also be found. The presentation follows that of the notes closely, but elevates the semantic completeness proof from the footnote into the main text (p. 33).

Remarks. (1) The semantic completeness for first-order logic is formulated as an open problem in *Hilbert and Ackermann*: "Whether the axiom system is complete at least in the sense that really all logical formulas that are correct for all domains of individuals can be derived from it is an unsolved question. We can only say purely empirically that this axiom

---

[35] Wie bei jeder Axiomatik lassen sich auch für dieses System die Fragen nach der Widerspruchslosigkeit, nach den logischen Abhängigkeiten und nach der Vollständigkeit aufwerfen. Am wichtigsten ist hier die Frage der Vollständigkeit. Denn das Ziel der symbolischen Logik besteht ja darin, aus den formalisierten Voraussetzungen die übliche Logik zu entwickeln. Es kommt also wesentlich darauf an, zu zeigen, dass unser Axiomensystem zum Aufbau der gewöhnlichen Logik ausreicht. (p. 67)

[36] Wir wollen das vorgelegte Axiomen-System vollständig nennen, falls durch die Hinzufügung einer bisher nicht ableitbaren Formel zu dem System der Grundformeln stets ein widerspruchsvolles System entsteht. (p. 152)—This completeness concept is clearly related to that formulated in the axiomatization of the real numbers and of geometry; this connection should be explored carefully.

[37] Only a much abbreviated version of this was published in 1926 as (Bernays 1926); the publication focuses on the independence results.

[38] Jede beweisbare Formel ist eine allgemeingültige Formel und umgekehrt. (p. 6)

system has always sufficed for any application."³⁹ Some recent commentators have viewed this formulation as *oddly obscure* (Goldfarb) or even *circular* (Dreben and van Heijenoort). Those views rest on a very particular reading of "logical formulas" that is narrowly correct, as Hilbert and Ackermann (following verbatim the 1917/18 lecture notes) define them on page 54 as those formulas that (i) do not contain "individuelle Zeichen" (i.e., symbols for determinate individuals and functions), and (ii) can be proved by appealing only to the logical axioms. Under this reading the formulation is indeed close to nonsensical. However, if one takes into account that "logische Formel" and "logischer Ausdruck" are used repeatedly⁴⁰ as indicating just those formulas satisfying (i), then their formulation together with the explication of correctness I reviewed earlier is exactly right. Indeed, the formulation of the completeness problem involves then *precisely* the definition of "allgemeingültig" given in (Gödel 1930a), notes 3 and 4. Gödel emphasizes in the first of these footnotes that "This paper's terminology and symbolism follows closely *Hilbert and Ackermann 1928*."⁴¹ An equally correct formulation of completeness is given in Hilbert's talk to the International Congress of Mathematicians in Bologna, 3 September 1928 (Hilbert *1929); validity is defined as non-refutability by an arithmetic model. It is also of interest to note that Hilbert contemplates there the incompleteness of axiomatic systems for "higher areas" (höhere Gebiete). (Obviously, both these observations implicitly use the Löwenheim Skolem theorem.)

(2) Weyl defined on pages 9 and 10 of *Das Kontinuum* a semantic notion of logical truth and consequence: "Some pertinent judgments we recognize as true purely on the basis of their logical structure—without regard either to the characteristics of the category of objects involved or to the extension of the basic underlying properties and relations or to the objects used in the operation of 'filling in'. . . . Such judgments which are true purely on account of their formal (logical) structure . . . we wish to call (*logically*) *self-evident*. A judgment whose negation is self-evident is called *absurd*. If $U \& \neg V$ is absurd, then the judgment V is a *logical consequence* of U; if U is true, then we can be certain that V is also true."⁴² (I have used the standard

---

³⁹l.c, p. 68. Ob das Axiomensystem wenigstens in dem Sinne vollständig ist, daß wirklich alle logischen Formeln, die für jeden Individuenbereich richtig sind, daraus abgeleitet werden können, ist eine noch ungelöste Frage. Es läßt sich nur rein empirisch sagen, daß bei allen Anwendungen dieses Axiomensystem immer ausgereicht hat.

⁴⁰For example, on pages 72, 73, and 80 in the discussion of the *Entscheidungsproblem*. Gödel uses in his (Gödel 1929) and also in (Gödel 1930a) "logischer Ausdruck" in exactly this sense.

⁴¹In Terminologie und Symbolik schließt sich die folgende Arbeit an *Hilbert und Ackermann 1928* an.

⁴²The English translation is from (Weyl 1918); the German text is this: "Unter den einschlägigen Urteilen gibt es solche, die wir als wahr erkennen auf Grund ihrer logischen Struktur—ganz unabhängig davon, um was für eine Gegenstandskategorie es sich handelt, was die zugrunde liegenden Ur-Eigenschaften bedeuten und welche Gegenstände . . . zur 'Ausfüllung' benutzt werden. Solche rein ihres formalen (logischen) Baus wegen wahren Urteile . . . wollen wir (*logisch*) *selbstverständlich* nennen. Ein Urteil, dessen Negation selbstverständlich ist, heiße *sinnwidrig*. Ist $U \& \neg V$ sinnwidrig, so ist das Urteil V eine '*logische Folge*' von U; ist U wahr, so können wir sicher sein, daß dann auch V wahr ist."

sentential connectives here.)—Recall that the very isolation of the language of first-order logic goes back to (Weyl 1910).

(3) The word 'Entscheidungsproblem' is used, as far as I can see, for the first time in these lectures in the winter term of 1922/23 on page 25, cf. also Kneser's "Mitschrift" on p. 12. Clearly, the general problem of mechanically deciding mathematical questions had been mentioned already earlier by Hilbert, for example, in *Axiomatisches Denken* and even in his Paris lecture of 1900.

Exploiting the standard arithmetic interpretation of the logical connectives, Hilbert addresses the consistency problem for logic in the lectures from the winter term 1917/18. He shows, by induction on derivations in sentential and first-order logic, that provable formulas are always true; consistency of the logical calculi is a direct consequence.[43] However, in a note the reader is warned not to overestimate the significance of this result, because "[i]t does not give us a guarantee that the system of provable formulas remains free of contradictions after the symbolic introduction of contentually correct assumptions".[44] That much more difficult problem has to be attacked in special ways—by a logicist reduction, perhaps, or by quite new ways of proceeding; we come to these new ways now.

3 From 1920 to 1922: Consistency and proof theory

A rigid and dogmatic formalist view is popularly attributed to Hilbert and his collaborators. This attribution is untempered by accessible works, for example, the two monumental volumes of *Grundlagen der Mathematik* published in 1934 and 1939, or Bernays's philosophical investigations starting with essays from 1922. The content of the early lecture notes should help to put Hilbert's views in proper perspective. Notice that, up to now, no specifically proof theoretic considerations concerning the consistency problem have been mentioned in these lectures. Indeed, the development

---

[43] This is done on pages 70 ff and 150 ff; the analogous considerations are contained in (Hilbert and Ackermann 1928) on pages 30 ff and 65 ff.

[44] Man darf dieses Ergebnis in seiner Bedeutung nicht überschätzen. Wir haben ja damit noch keine Gewähr, dass bei der symbolischen Einführung von inhaltlich einwandfreien Voraussetzungen das System der beweisbaren Formeln widerspruchslos bleibt. (p. 156)—In *Hilbert and Ackermann* there is a significant expansion of this remark: "Man darf das Ergebnis dieses Beweises für die Widerspruchsfreiheit unserer Axiome übrigens in seiner Bedeutung nicht überschätzen. Der angegebene Beweis der Widerspruchsfreiheit kommt nämlich darauf hinaus, daß man annimmt, der zugrunde gelegte Individuenbereich bestehe nur aus einem einzigen Element, sei also endlich. Wir haben damit durchaus keine Gewähr, daß bei der symbolischen Einführung von inhaltlich einwandfreien Voraussetzungen das System der beweisbaren Formeln widerspruchsfrei bleibt. Z.B. bleibt die Frage unbeantwortet, ob nicht bei der Hinzufügung der mathematischen Axiome in unserem Kalkül jede beliebige Formel beweisbar wird. Dieses Problem, dessen Lösung eine zentrale Bedeutung für die Mathematik besitzt, läßt sich in bezug auf Schwierigkeit mit der von uns behandelten Frage garnicht vergleichen. Die mathematischen Axiome setzen gerade einen unendlichen Individuenbereich voraus, und mit dem Begriff des Unendlichen sind die Schwierigkeiten und Paradoxien verknüpft, die bei der Diskussion über die Grundlagen der Mathematik eine Rolle spielen." (pp. 65–61)

towards the Hilbert Program as we think of it was completed only in the lectures given in the winter term 1921/22. Hilbert arrived at its formulation after abandoning the logicist route through two quite distinct steps, and only the second takes up the earlier suggestion of a theory of (formal) proofs.

3.1 Constructive number theory.
The first step is taken in the winter term 1920.[45] Hilbert reviews the logical development of his 1917/18 lectures in a polished form, frequently referring back to them for additional details. The last third of the notes is devoted, however, to a completely different topic. Hilbert argues that the set theoretic or logical developments of Dedekind and Frege did not succeed in establishing the consistency of ordinary number theory and concludes:

To solve these problems I don't see any other possibility, but to rebuild number theory from the beginning and to shape concepts and inferences in such a way that paradoxes are excluded from the outset and that proof procedures become completely surveyable.
Now I will show how I think of the beginning of such a foundation for number theory.[46]

The considerations are put back into the broader context of the earlier investigations, re-emphasizing the semantic underpinnings for axiom systems:

We have analyzed the language (of the logical calculus proper) in its function as a universal instrument of human reasoning and revealed the mechanism of logical argumentation.
However, the kind of viewpoint we have taken is incomplete in so far as the application of the logical calculus to a particular domain of knowledge requires an axiom system as its basis. I.e., a system (or several systems) of objects must be

---

[45] See the discussion in Appendix A concerning the sequencing of the lectures from winter term 1920 and the summer term 1920. Moore in his (1997) asserts, incorrectly, that the lectures of the winter term 1920 were given after those of the summer term of that year. This mistake leads not only to misunderstandings of the very lectures and their broader historical context (involving Brouwer and Weyl), but it is also partially responsible for a quite different overall assessment which is summarized in the abstract of the paper as follows: "By 1917, strongly influenced by PM, Hilbert accepted the theory of types and logicism-a surprising shift. But by 1922 he abandoned the axiom of reducibility and then drew back from logicism, returning to his 1905 approach to prove the consistency of number theory syntactically." Clearly, as documented here, logicism had been given up as a viable option in the summer of 1920 explicitly; implicitly, that recognition is already in the background for the lectures in the winter of 1920 discussed in this section. The special constructivist stance taken by Hilbert here and its connection to earlier reflections of Hilbert's are not recognized by Moore, thus also not the expanding step toward finitist mathematics. The latter is discussed in Section 3.2.

[46] Zur Lösung dieser Probleme sehe ich keine andere Möglichkeit, als dass man den Aufbau der Zahlentheorie von Anfang an durchgeht und die Begriffsbildungen und Schlüsse in eine solche Fassung bringt, bei der von vornherein Paradoxien ausgeschlossen sind und das Verfahren der Beweisführung vollständig überblickbar wird.
Ich will nun im Folgenden zeigen, wie ich mir den Ansatz zu einer solchen Begründung der Zahlentheorie denke. (p. 48)

given and between them certain relations with particular assumed basic properties are considered.[47]

This method is perfectly appropriate, Hilbert continues, when we are trying to obtain new results or present a particular science systematically. However, mathematical logic pursues also the goal of securing the foundations of mathematics.

For this purpose it seems appropriate to connect the mathematical constructions to what can be concretely exhibited and to interpret the mathematical inference methods in such a way that one stays always within the domain of what can be checked. And in fact one is going to start with arithmetic, as one finds here the simplest mathematical concepts.

In addition, it has been the endeavor in mathematics for a long time to reduce all conceptual systems (geometry, analysis) to the integers.[48]

This remark is followed by the development of what might be called *strict finitist number theory*. The considerations are delicate (and their detailed presentation has to wait for another occasion), but one thing is perfectly clear: here is a version of constructive arithmetic stricter than what will appear a little later as finitist mathematics. The basic and directly meaningful part consists only of *closed* numerical equations. This is in line with Hilbert's remark, quoted in Section 1.4 on genetic definitions, about Kronecker's not being sufficiently radical. Bernays pointed to the *evolution* towards finitist mathematics at a number of places; for example, in his (1954) he wrote: "Originally, Hilbert also intended to take the narrower standpoint that does not assume the intuitive general concept of numeral. That can be seen, for example, from his Heidelberg lecture (1904). It was already a kind of compromise that he adopted the finitist standpoint as presented in his publications."[49]

---

[47]... Wir haben die Sprache (des eigentlichen Logikkalküls) in ihrer Funktion als universales Instrument des menschlichen Denkens zergliedert und den Mechanismus der logischen Beweisführung blossgelegt.
Jedoch ist die Art der Betrachtungsweise, die wir angewandt haben, insofern unvollständig, als die Anwendung des Logikkalküls auf bestimmte Wissensgebiete ein Axiomensystem als Grundlage erfordert. D.h. es muss ein System (bzw. mehrere Systeme) von Gegenständen gegeben sein, zwischen denen gewisse Beziehungen mit bestimmten vorausgesetzten Grundeigenschaften betrachtet werden. (pp. 46–7)

[48]Zu diesem Zwecke erscheint es als der geeignete Weg, dass man die mathematischen Konstruktionen an das konkret Aufweisbare anknüpft und die mathematischen Schlussmethoden so interpretiert, dass man immer im Bereiche des Kontrollierbaren bleibt. Und zwar wird man hiermit bei der Zahlentheorie den Anfang machen, da hier die einfachsten mathematischen Begriffsbildungen vorliegen.
Auch ist es ja seit langem das Bestreben in der Mathematik, alle Begriffssysteme (Geometrie, Analysis) auf die ganzen Zahlen zurückzuführen. (pp. 47–8)

[49](Bernays 1954, p. 12). The German text: "Ursprünglich wollte auch Hilbert den engeren Standpunkt einnehmen, der nicht den anschaulichen Allgemeinbegriff der Ziffer voraussetzt. Das ist unter anderem aus seinem Heidelberger Vortrag (1904) zu ersehen. Es war schon eine Art Kompromiss, dass er sich zu dem in seinen Publikationen eingenommenen finiten Standpunkt entschloss."—In (Hallett 1995, pp. 169 and 173), there is further evidence of this early "strict finitist" view.

In the lectures from the winter term 1920 this "intuitive general concept of numeral" is not yet assumed; instead, general statements like $x + y = y + x$ are given a constructive and extremely rule-based interpretation:

Such an equation . . . is not viewed as a claim for all numbers, rather it is interpreted in such a way that its full meaning is given by a proof procedure: each step of the procedure is an action that can be completely exhibited and that follows fixed rules.[50]

This view entails that the equation $2 + 3 = 3 + 2$ is not a special case of the general equation $x + y = y + x$; on the contrary, having proved the latter, the former still has to be established, as the proof of the general equation yields only a guide to the proof of its instance. Hilbert points out, as a second consequence of this view, that the usual logical relations between general and existential statements do not obtain. After all, the truth of a general statement is usually equivalent to the non-existence of a counterexample. Under the given constructive interpretation the alternative between a general statement and the existence of a counterexample would be evident only with the additional assumption "every equation without a counterexample is provable from the assumed arithmetic principles", as the meaning of the general statement depends on the underlying system of inference rules.[51] The lecture notes conclude with this (judicious) statement in which Brouwer's name appears for the very first time:

This consideration helps us to gain an understanding for the sense of the paradoxical claim, made recently by Brouwer, that for infinite systems the law of the excluded middle (the "tertium non datur") loses its validity.[52]

---

[50]Eine solche Gleichung . . . wird nicht aufgefasst als eine Aussage über alle Zahlen, vielmehr wird sie so gedeutet, dass ihr Sinn sich in einem Beweisverfahren erschöpft, bei welchem jeder Schritt eine vollständig aufweisbare Handlung ist, die nach festgesetzten Regeln vollzogen wird. (p. 60)

[51]Hilbert mentions that this assumption would amount to the claim that all number theoretic questions are decidable; cf. p. 61. The relevant German text is: "Ein allgemeines Urteil im eigentlichen Sinne ist dann und nur dann richtig, wenn es kein Gegenbeispiel gibt. Bei einer symbolischen Gleichung [i.e., an equation with free variables] wissen wir freilich in dem Falle, wo uns ein Gegenbeispiel bekannt ist, dass sie nicht richtig sein kann. Wir können aber nicht sagen, dass eine symbolische Gleichung stets entweder richtig oder durch ein Gegenbeispiel widerlegbar sein muss. Denn die Bedeutung der richtigen Formeln hängt ja von dem System der Beweisregeln ab, und jenes Entweder-Oder wäre nur unter der Voraussetzung selbstverständlich, dass mit Hülfe der Beweisregeln jede nicht widerlegbare symbolische Gleichung bewiesen werden kann." (p. 61)

The obvious historical question here is, what did Hilbert know about Brouwer's views. In (Bernays 1935a) the following papers are listed, when the impact of Brouwer and Weyl in 1920 is discussed: (Brouwer 1918), (Brouwer 1919a), (Brouwer 1919b), and (Brouwer 1921) by Brouwer and (Weyl 1918) and (Weyl 1921) by Weyl. One should recall in this context also that in 1919 Brouwer had been offered a professorship in Göttingen. More precisely, according to a private communication of Dirk van Dalen, ". . . the decision of the Göttingen faculty to put Brouwer no. 1 on the list for the chair was made on 30.10.1919". Cf. also note 56.

[52]Wir gewinnen durch diese Überlegung ein Verständnis für den Sinn der neuerdings durch Brouwer aufgestellten paradoxen Behauptung, dass bei unendlichen Systemen der Satz vom ausgeschlossenen Dritten (das "tertium non datur") seine Gültigkeit verliere. (pp. 61–2)

It must have been a discouraging conclusion for Hilbert to see that this new approach could not secure the foundations of classical mathematics either. However, he overcame the setback by taking a second strategic step in the lectures for the summer term 1920 that joined the considerations concerning a thoroughly constructive foundation of number theory with the detailed formal logical work. Recall, that already in his Heidelberg talk of 1904 and again in his Zürich lecture of 1917, Hilbert had argued for a "Beweistheorie", but had not pursued his suggestion systematically. Here, in Section 7 of the notes from the summer term 1920, we do find initial steps, namely a consistency proof for an extremely restricted, quantifier-free part of elementary number theory that involves negations only as applied to equations.

These considerations, slightly modified, can be found in the first part of Hilbert's paper *Neubegründung der Mathematik*, a paper that is based on talks given in Copenhagen and Hamburg during the spring and summer of 1921; cf. Appendix A. The second part of the paper expands the basic set-up in new ways. Bernays pointed repeatedly to this "break" in the paper and describes its first part, for example in (Bernays 1935a), as "a remnant from that stage, at which this separation [between the formalism and metamathematical considerations] had not been made yet".[53] The new ways are pursued further in the lectures given during the winter term 1921/22; how direct the connections are can be appreciated from Bernays's outline, *Disposition*, for the lectures that is contained in Appendix A.

3.2 Finitist proof theory. The 1921/22 lectures contain for the first time the terms *finite Mathematik, transfinite Schlussweisen, Hilbertsche Beweistheorie*, and their third part is entitled: The founding of the consistency of arithmetic by the new Hilbertian proof theory (or in German, Die Begründung der Widerspruchsfreiheit der Arithmetik durch die neue Hilbertsche Beweistheorie). The clear separation of mathematical and metamathematical considerations allows Hilbert to address, finally, Poincaré's critique by distinguishing between contentual, metamathematical and formal, mathematical induction; this point is emphasized in early publications, namely, (Bernays 1922b), (Hilbert 1922), but most strongly in (Hilbert 1927). Hilbert claims in the last paper, presented as a talk in Hamburg, that Poincaré arrived at "his mistaken conviction by not distinguishing these two methods of induction, which are of entirely different kinds" and feels that "under these circumstances Poincaré had to reject my theory, which, incidentally, existed at that time only in its completely inadequate early stages".[54] Weyl, responding to Hilbert's talk, turns the argument around and justly claims that ". . . Hilbert's proof theory shows

---
[53](Bernays 1935a, p. 203).
[54](Hilbert 1927, p. 473). How important this critique was can be seen from Weyl's remarks below, but also from the writings of others, for example, Skolem; see his papers (Skolem 1922) and (Skolem 1930). In the introduction to (Weyl 1927) in *From Frege to Gödel*, pp. 480–1, one finds a very thoughtful discussion of the underlying issues.

Poincaré to have been exactly right on this point". After all, Hilbert has to be concerned not just with particular numerals, but "with an *arbitrary concretely given* numeral", and the contentual arguments of proof theory must "be carried out in *hypothetical generality*, on *any* proof, on *any* numeral". Weyl recognizes clearly the significance of the distinction and its importance for the fully articulated proof theoretic enterprise; he sees it as facing the two complementary tasks of formalizing classical mathematics without reducing its "inventory" and of proving the consistency within the limits of "contentual thought".

That there are limits to contentual thought (inhaltliches Denken) was established, according to (Weyl 1927), by Brouwer. An obviously related fundamental insight was obtained, as we saw, in Hilbert's notes for the winter term 1920, i.e., at the very beginning of 1920. It is of greatest interest to know in what ways Hilbert may have been influenced by Brouwer (or Weyl, as will be discussed below and in footnote 56); that there was some influence can be taken for granted; after all, Brouwer is mentioned at the end of the notes. Hilbert's insight was based on an interpretation of quantifiers that is bound up with a particular formal calculus. The understanding of quantifiers is explored anew in the context of an informal presentation of finitist number theory on pages 52 to 69 of the 1921/22 lectures and deepened in the long introduction to their third part. That part expands, as I described earlier, the second part of Hilbert's 1922 paper.

The interpretation is here no longer tied to a formal calculus that allows us to establish free-variable statements, but rather it assumes the "intuitive general concept of numeral" as part of the finitist standpoint.

In intuitive number theory, the general sentences have a purely hypothetical sense. A sentence like
$$a + b = b + a$$
only means: given two numerals a, b, the additive composition of a with b yields the same numeral as the additive composition of b with a. There is no mention of the totality of all numbers. Furthermore, the existential sentences have in intuitive number theory only the meaning of partial-judgments, i.e., they are substatements of more precisely determined statements, whose precise content, however, is inessential for many applications.

... thus, in general, a more detailed sentence complements in intuitive number theory an existential judgment; the sentence determines more precisely the content of that judgment. The existential claim here has sense only as a pointer to a search procedure which one possesses, but that ordinarily need not be elaborated, because it suffices generally to know that one has it.[55]

---

[55] In der anschaulichen Zahlentheorie haben die allgemeinen Sätze rein hypothetischen Sinn. Ein Satz wie
$$a + b = b + a$$
besagt nur: Wenn zwei Zahlzeichen a, b gegeben sind, so liefert die additive Zusammensetzung von a mit b dasselbe Zahlzeichen wie die additive Zusammensetzung von b mit a. Von der Gesamtheit aller Zahlen ist dabei nicht die Rede. Ferner, die existenzialen Sätze haben in der anschaulichen Zahlentheorie nur die Bedeutung von Partial-Urteilen, d.h. sie sind

Hilbert's programs: 1917–1922

This is exactly the understanding that is formulated in 1925 in *Über das Unendliche* (pp. 172–3) and, most extensively, in 1934 in the first volume of *Grundlagen der Mathematik*; it is also strikingly similar to Weyl's viewpoint in (Weyl 1921).[56] With this understanding of quantifiers the conclusion concerning the non-validity of the law of the excluded middle is again obtained. Hilbert points out:

Thus we see that, for a strict foundation of mathematics, the usual inference methods of analysis must not be taken as logically obvious. Rather, it is exactly the task for the foundational investigation to recognize why it is that the application of transfinite inference methods as used in analysis and (axiomatic) set theory leads always to correct results.[57]

As that recognition has to be obtained on the basis of finitist logic, Hilbert argues, we have to extend our considerations in a different direction to go beyond elementary number theory:

We have to extend the domain of objects to be considered; i.e., we have to apply our intuitive considerations also to figures that are not number signs. Thus we have good reason to distance ourselves from the earlier dominant principle according to which each theorem of pure mathematics is in the end a statement concerning

---

Teilaussagen von näher bestimmten Aussagen, deren genauer Inhalt jedoch für viele Anwendungen unwesentlich ist.
... so gehört allgemein in der anschaulichen Zahlentheorie zu einem existenzialen Urteil ein genauerer Satz, welcher den Inhalt jenes Urteils näher bestimmt. Die Existenzbehauptung hat hier überhaupt nur einen Sinn als ein Hinweis auf ein Verfahren der Auffindung welches man besitzt, das man aber für gewöhnlich nicht näher anzugeben braucht, weil es im allgemeinen genügt, zu wissen, dass man es besitzt. (pp. 67–8)

[56] Weyl's paper must have been known to Hilbert in 1921: in Hilbert's *Neubegründung der Mathematik* one finds the remark (on p. 160), "Wenn man von einer Krise spricht, so darf man jedenfalls nicht, wie es Weyl tut, von einer neuen Krise sprechen." This is obviously an allusion to the title of (Weyl 1921). According to (van Dalen 1995, p. 145), a draft of Weyl's paper was completed by May 1920, and a copy sent to Brouwer.—What is puzzling here is the circumstance that Weyl's views are, in some important respects (the understanding of quantifiers is one such point) close to the finitist standpoint; Weyl presents them as being different from Brouwer's, and Brouwer in turn recognizes immediately that Weyl is "in the restriction of the object of mathematics" even more radical than he himself; cf. (van Dalen 1995, p. 148 and p. 167). Why did it take the people in the Hilbert school such a long time to recognize that finitism was more restrictive than intuitionism? In a letter to Hilbert dated 25. X. 1925, Bernays mentions "a certain difference between the finitist standpoint and that of Brouwer"; but there is no elaboration of what this difference might be, and I don't know of any place where it is discussed by members of the Hilbert school before 1933. Indeed, in (Bernays 1930b), the mathematical methods of finitism and intuitionism are viewed as co-extensional; it is only in the context of the Gödel-Gentzen reduction of classical to intuitionistic arithmetic that both Gödel and Gentzen point out that finitism is more restrictive than intuitionism; cf. (Gödel 1933d, p. 294). This fact is then discussed in (Bernays 1935b, p. 77); the significance of the result is described in (Bernays 1967).

[57] Wir sehen also, dass für den Zweck einer strengen Begründung der Mathematik die üblichen Schlussweisen der Analysis in der Tat nicht als logisch selbstverständlich übernommen werden dürfen. Vielmehr ist es gerade erst die Aufgabe für die Begründung, zu erkennen, warum die Anwendung der transfiniten Schlussweisen, sowie sie in der Analysis und in der (axiomatisch begründeten) Mengenlehre geschieht, stets richtige Resultate liefert. (p. 4a)

integers. *This principle was viewed as expressing a fundamental methodological insight, but it has to be given up as a prejudice.*[58]

This is a strong statement against a tradition that started with Dirichlet and includes such distinguished mathematicians as Weierstrass and Dedekind; it is also a surprising statement in the sense that such an extension was obviously implicit in Hilbert's earlier formulations of "Beweistheorie", for example in his (1918). But what is the new extended domain of objects, and what has to be preserved from the "fundamental methodological insight"? As to the domain of objects, it is clear what has to be included, namely the formulas and proofs from formal theories. By contrast, geometric figures are definitely excluded; the reason for holding that geometric figures are "not suitable objects" for Hilbert's considerations is articulated as follows:

... the figures we take as objects must be completely surveyable and only discrete determinations are to be considered for them. It is only under these conditions that our claims and considerations have the same reliability and evidence as in intuitive number theory.[59]

From this *new standpoint*, as he calls it, Hilbert exploits the formalizability of a fragment of number theory in full first-order logic to formulate and prove its consistency. So, here we finally close the gap to the published record—with a fully developed programmatic perspective. I intend to give a proper mathematical exposition of this early work, including the elementary consistency proofs from (Hilbert 1905a), winter term 1920, summer term 1920, and the winter term 1921/22. The exposition will emphasize the inductive generation of syntactic structures and, based thereon, proofs by induction and definition by recursion; that is only natural, as soon as one has taken the methodological step Hilbert suggested. It was most strongly emphasized by von Neumann in his (1931).

If we take this expansion of the domain of objects seriously, we are dealing not just with numerals, but more generally with elements of inductively generated classes. (The generation has to be elementary and deterministic, in modern terminology.) A related point was made by Poincaré, when he emphasized after discussing the principle of induction for natural numbers:

---

[58](The emphasis is mine; WS.) Wir müssen den Bereich der betrachteten Gegenstände erweitern, d.h. wir müssen unsere anschaulichen Überlegungen auch auf andere Figuren als auf Zahlzeichen anwenden. Wir sehen uns somit veranlasst, von dem früher herrschenden Grundsatz abzugehen, wonach jeder Satz der reinen Mathematik letzten Endes in einer Aussage über ganze Zahlen bestehen sollte. Dieses Prinzip, in welchem man eine grundlegende methodische Erkenntnis erblickt hat, müssen wir jetzt als ein Vorurteil preisgeben. (p. 4a)

[59]An einer Forderung aber müssen wir festhalten, dass nämlich die Figuren, welche wir als Gegenstände nehmen, vollkommen überblickbar sind, und dass an ihnen nur diskrete Bestimmungen in Betracht kommen. Denn nur unter diesen Bedingungen können unsere Behauptungen und Überlegungen die gleiche Sicherheit und Handgreiflichkeit haben wie in der anschaulichen Zahlentheorie. (p. 5a)

I did not mean to say, as has been supposed, that all mathematical reasonings can be reduced to an application of this principle. Examining these reasonings closely, we should see applied there many other analogous principles, presenting the same essential characteristics. In this category of principles, that of complete induction is only the simplest of all and this is why I have chosen it as a type. (p. 1025)

The difficult issue is to recognize from Hilbert's standpoint induction and recursion principles. When discussing in his (1933b) the "unobjectionable methods" by means of which consistency proofs are to be carried out, Gödel formulates a first characteristic of the strictest form of constructive mathematics as follows:

The application of the notion of "all" or "any" is to be restricted to those infinite totalities for which we can give a finite procedure for generating all their elements (as we can see, e.g., for the totality of integers by the process of forming the next greater integer and as we cannot, e.g., for the totality of all properties of integers).

According to the second characteristic, existential statements are viewed as abbreviations indicating that an example has been found; and thus there is essentially only one way of establishing general propositions, namely, "complete induction applied to the generating process of our elements". Only decidable properties and calculable functions are to be introduced. As the latter, according to Gödel, can always be defined by complete induction, the system for this form of constructive mathematics (and Gödel assumes that this is really finitist mathematics) is "exclusively based on the method of complete induction in its definitions as well as in its proofs". Gödel believes, with Poincaré and Hilbert, that "the method of complete induction" has a "particularly high degree of evidence". But what is the nature of this evidence? In spite of much important work that has been done for elementary number theory, this is still a significant question and should be addressed. The suggestion that the work for number theory covers all the bases, because of a simple effective Gödel numbering, misses the opportunity of articulating in greater generality the evidential features of inductively generated objects, constructed in elementary and less elementary ways.[60]

Finally, there is ample room to improve our understanding of Hilbert's and Bernays's views on the matter. I take it, for example, that Gödel's attempt to characterize the finitist standpoint in his 1958 paper is in conflict with their views and with his own informal description of the central features of finitist mathematics sketched above. At issue is whether the insights needed to carry out proofs concerning finitist objects spring purely from the combinatorial (spatiotemporal) properties of the sign combinations that represent them, or whether an element of "reflection" is needed, reflection that takes into account the uniform generation of the objects. The latter is explicitly affirmed in (Bernays 1930b) and implicit, by my lights, in Hilbert's description of the "extra-logical concrete objects" that

---

[60]That is, as a matter of fact, the starting point of my systematic considerations concerning "accessible domains" in (Sieg 1990a) and (Sieg 1997a).

are needed to secure meaningful logical reasoning: such objects must not only be surveyable, but the fact that they *follow each other*, in particular, is immediately given intuitively together with the objects and cannot be further reduced.[61]

3.3 A concise review.
The dialectic of the developments that emerges from the lectures (given between 1917 and 1922) is described in Bernays's paper of 1922 and is also formulated very carefully on pp. 29–33 of the 1922/23 lectures. Here is Bernays's description that brings out the "Ansatzcharakter" of the proposed solution: in order to provide a rigorous foundation for arithmetic (that includes analysis and set theory) one proceeds axiomatically and starts out with the assumption of a system of objects satisfying certain structural conditions. However, in the assumption of such a system "lies something so-to-speak transcendental for mathematics, and the question arises, which principled position is to be taken [towards that assumption]". Bernays considers two "natural positions", positions that had been thoroughly explored as we saw. The first position, attributed to Frege and Russell, attempts to provide a foundation for mathematics by purely logical means; this attempt is judged to be a failure.

The second position is seen in counterpoint to the logical foundations of arithmetic: "As one does not succeed in establishing the logical necessity of the mathematical transcendental assumptions, one asks oneself, is it not possible simply to do without them." Thus one attempts a constructive foundation replacing existential assumptions by construction postulates; that is the second position and is associated with Kronecker, Poincaré, Brouwer, and Weyl. The methodological restrictions to which this position leads are viewed as unsatisfactory, as one is forced "to give up the most successful, most elegant, and most proven methods only because one does not have a foundation for them from a particular standpoint".

Hilbert takes from these foundational positions, Bernays continues in his analysis, what is "positively fruitful": from the first the strict formalization of mathematical reasoning; from the second the emphasis on constructions. Hilbert does not want to give up the constructive tendency, but emphasizes it on the contrary in the strongest possible terms. Finitist mathematics is viewed as part of an "Ansatz" to finding a principled position towards the transcendental assumptions:

Under this perspective[62] we are going to try, whether it is not possible to give a foundation to these transcendental assumptions in such a way that only primitive intuitive knowledge is used.[63]

---

[61]That description is found in (Hilbert 1922, pp. 162/3), but also later in (Hilbert 1926, p. 171), and (Hilbert 1927, p. 65).

[62]Of taking into account the tendency of the exact sciences to use as far as possible only the most primitive "Erkenntnismittel". That does not mean, as Bernays emphasizes, to deny any other, stronger form of intuitive evidence.

[63](Bernays 1922b, p. 11).

The program is taken as a tool for an alternative constructive foundation of all of classical mathematics. The great advantage of Hilbert's method is judged to be this: "The problems and difficulties that present themselves in the foundations of mathematics can be transferred from the epistemological/philosophical to the properly mathematical domain." So Bernays, without great fanfare, gives an illuminating summary of about four years of quite intense work!

Concluding remarks

I find absolutely remarkable the free and open way in which Hilbert and Bernays joined, in the end, a number of different tendencies into a sharply focused program with a special mathematical and philosophical perspective. The metamathematical core of the program amounts to this: classical mathematics is represented in a formal theory P, expressing "the whole thought content of mathematics in a uniform way"; based on this representation, it is *programatically* taken as a *formula game*. But the latter aspect should not be over-emphasized, as there are other important considerations, namely that intended mathematical structures are projected through their (assumed complete) formalizations into the properly mathematical domain, i.e., finitist mathematics.[64] In any event, the consistency of P has to be established within finitist mathematics F. P's consistency is in F equivalent to the reflection principle, expressing formally the soundness of P:

$$(\forall x)(\mathrm{Prf}(x, \text{'s'}) \Longrightarrow s).$$

Prf is the finitist proof predicate for P, s a finitist statement, and 's' the corresponding formula in the language of P. A consistency proof in F would show, because of this equivalence, that the formal, technical apparatus P can serve reliably as an instrument for the proof of finitist statements.

At first it seemed as if Hilbert's approach would yield proof theoretic results rather quickly and decisively: Ackermann's "proof" of the consistency of analysis was published in 1925, but had been submitted on March 30, 1924! However, difficulties emerged and culminated in the real obstacles presented by Gödel's Incompleteness Theorems. The program has been transformed, quite in accord with the broad strategy underlying Hilbert's proposal, to a *general reductive* one; here one tries to give consistency proofs for strong classical theories relative to "appropriate constructive" theories. Even Gödel found the mathematical reductive program with its attendant philosophical one attractive in the thirties; his illuminating reflections, partly in an examination of Gentzen's first consistency proof for arithmetic, are presented in previously unpublished papers[65] that are now available in the third volume of his *Collected Works*. Foundationally inspired work in

---

[64]Cf. section 2.1 of (Sieg 1990a).

[65]I am thinking in particular of (Gödel 1933b), (Gödel 1938), and (Gödel 1941). [See essay (II.4) below. WS]

proof theory is being continued, weaving strong set theoretic and recursion theoretic strands into the metamathematical work.

This expanding development of proof theory is but *one* effect of Hilbert's broad view on foundational problems and his sharply articulated questions. Another effect is plainly visible in the rich and varied contributions that were given to us by Hilbert, Bernays, and other members of the Hilbert School (Ackermann, von Neumann, Gentzen, Schütte); finally, we have to consider also the stimulus his approach and questions provided to contemporaries outside the school (Herbrand, Gödel, Church, Turing and, much earlier already, Zermelo). Indeed, there is no foundational enterprise with a more profound and far-reaching effect on the emergence and development of modern mathematical logic; it could, if we just cared to be open, have a similar effect on philosophical reflections concerned with mathematical experience: it can help us to gain a perspective that includes traditional philosophical concerns, but that, most importantly, allows us to ask questions transcending traditional boundaries.

Appendix A: Lectures and early papers. Here I am providing some information on (i) when the lectures of the winter term 1920 were most likely given (or the notes written) and (ii) the connection of the lectures during the early twenties with the first published accounts of Hilbert's proof theory, including their chronology as far as I can determine it presently. Quite a few specific issues remain that could be resolved with some additional archival work (and a little luck in finding appropriate documents).

As to (i), it is perfectly clear from the content of these lectures that they preceded those of the summer term 1920; the (small) puzzle is that all other winter term lectures have the indication of their year in the form 19xx/xx+1. This is the general rule, but for the years 1919 and 1920 there was an exception (as Ralf Haubrich found out in Göttingen in response to an inquiry of mine). Because of the end of World War I and soldiers having returned to the university, an extra semester was pressed into those two years: there was a "Zwischensemester" in 1919 (from September 22 to December 20); that was followed by the winter term 1920 and, then, by the regular summer term 1920 beginning on April 26.

At the moment I only know an upper bound for the completion of the notes for the 1917/18 lectures, as Bernays's Habilitationsschrift, submitted in 1918, mentions the "Ausarbeitung" in note 1 on page IV with correct page references to the relevant sections on sentential logic.—Now let me proceed systematically with (ii).

Hilbert's *Neubegründung der Mathematik* was based on lectures given in Kopenhagen (Spring 1921) and Hamburg (Summer 1921); the paper contains on pages 168–174 material that overlaps with material presented on pages 33–46 of *Probleme der mathematischen Logik* (summer term 1920) and on pages 174–177 material from the very beginning of Part III of *Grundlagen der Mathematik* (winter term 1921/22). The two different parts of the paper were distinguished by the editors of Hilbert's *Gesammelte*

Abhandlungen in note 2 on page 168: "Die hier folgenden Betrachtungen greifen auf ein früheres Stadium der Beweistheorie zurück, in welchem die Untersuchung sich zunächst auf einen ganz engen Formalismus beschränkte, der dann schrittweise verschiedene Erweiterungen erfuhr. Dieser Gedankengang wird im folgenden dargestellt und hernach—auf S. 174 ff— der Übergang von jenem provisorischen Ansatz zu dem in der vorliegenden Abhandlung intendierten Formalismus vollzogen." The intended formalism is further investigated in Part III of *Grundlagen der Mathematik*. This reflects, in a very understated way, the dramatic methodological shift that is analyzed in 3.1 and 3.2 above.

It seems, but we don't have any notes for this, that Hilbert gave also a course on foundational matters in the winter term 1920/21. There are a few written communications between Hilbert and Bernays; one of them is a postcard written on 22 October 1920 and sent to Bernays from Switzerland. Hilbert announces that he will be back in Göttingen on Monday night and asks Bernays to stop by on Tuesday morning (at 11 a.m.). The point of the meeting is described as follows:

Wir müssen vor Allem das Donnerstags-Colleg vorbereiten. Ich möchte gern als Einleitung etwas Allgemeines über reine Anschauung und reines logisches Denken sagen, die beide in der Math. eine so grosse Rolle spielen und übrigens auch in meiner gegenwärtigen Beweistheorie—die beständig ganz gute Fortschritte macht—gleichzeitig in merkwürdiger Verknüpfung stehen. Es wäre vielleicht angebracht, wenn ich auf solche allgemeinen Fragen am Schlusse des Collegs zu sprechen käme. Könnten Sie vielleicht sich für Donnerstag etwas Einleitendes überlegen?

At the very beginning of the postcard Hilbert had already mentioned that he quite agreed with the "Disposition" for the "Colleg" that had been drafted by Bernays.

It is of real interest to consider the "Disposition" for the 1921/22 lectures that was proposed by Bernays before returning to Göttingen for that term in his letter to Hilbert of 17 October 1921 (with annotations in Hilbert's handwriting which are not reproduced here).

I. Bisherige Methoden der Beweise für Widerspruchslosigkeit oder Unabhängigkeit.
   A. Methode der Aufweisung.
      Beispiel des Aussagenkalküls in der mathem[atischen] Logik.
   B. Methode der Zurückführung.
      Beispiele:
      1) Widerspruchslosigkeit der Euklidischen Geometrie
      2) Unabhängigkeit des Parallelenaxioms
      3) Widerspruchslosigkeit des Rechnens mit komplexen Zahlen
II. Versuche der Behandlung des Problems der Widerspruchslosigkeit der Arithmetik.
   A. Die Zurückführung auf die Logik bietet keinen Vorteil, weil der Standpunkt der Arithmetik schon der formal allgemeinste ist. (Frege; Russell)

B. Die konstruktive Arithmetik: Definition der Zahl als Zeichen von bestimmter Art.

III. Die weitere Fassung des konstruktiven Gedankens: Konstruktion der Beweise, wodurch die Formalisierung der höheren Schlußweisen gelingt und das Problem der Widerspruchslosigkeit in allgemeiner Weise angreifbar wird.

Hier würde sich dann die Ausführung der Beweistheorie anschließen.

This reflects much more clearly than the remark (from the editors in Hilbert's *Gesammelte Abhandlungen* I quoted above) the significance of the methodological step that had been taken, when "the earlier dominant principle according to which each theorem of pure mathematics is in the end a statement concerning integers" was viewed as a prejudice. Interestingly, in the case of these lectures, we not only have the above "Disposition" and the official lecture notes written by Bernays, but also the *Mitschrift* of Kneser; Kneser's notes show the lectures in real-time progress beginning with the first meeting on 31 October 1921. The *Mitschrift* shows, first of all, that the lectures proceeded according to Bernays's "Disposition" and, secondly, that towards the end of the lectures (on 27 February 1922) the logical τ-function was introduced. That function was to play a prominent role in Hilbert's 1923-paper (submitted for publication on September 29, 1922); it is replaced by the ϵ-symbol already in the lectures of the following winter term 1922/23. A detailed comparison of Bernays's Notes and Kneser's *Mitschrift* might be of genuine interest, in particular if one considers also the Notes and *Mitschrift* for the winter term 1922/23.

Bernays's *Über Hilberts Gedanken zur Grundlegung der Arithmetik* was presented at the September meeting of the German Mathematical Association (DMV) in Jena and was received for publication by the *Jahresberichte der DMV* on 13 October 1921. In the letter to Hilbert dated 17 October 1921, in which he proposed the "Disposition", Bernays also wrote: "Wie Sie wohl wissen, habe ich an der Tagung in Jena teilgenommen und dort über Ihre neue Theorie vorgetragen. Mit dem Interesse, welches mein Vortrag fand, konnte ich sehr zufrieden sein; und ich habe ihn übrigens zur Veröffentlichung in den *Jahresberichten d[er] Mathematiker V[ereinigun]g* (auf Veranlassung von Prof. Bieberbach) ausgearbeitet.— Man fragte mich des öfteren, wie es mit der Publikation Ihrer Hamburger Vorträge stehe. Ich wußte in dieser Hinsicht über Ihre Absichten nicht recht Bescheid. Jedenfalls würde Hecke diese Vorträge gern in der neuen Hamburger Zeitschrift drucken." Note that Bernays talks of Hilbert's "new" theory!

Bernays's *Die Bedeutung Hilberts für die Philosophie der Mathematik* appeared in *Naturwissenschaften*, Heft 4, with 27 January 1922 as its publication date; it must have been prepared during the late summer/fall of 1921, as Bernays refers explicitly to Hilbert's Hamburg lectures. Finally, a paper I did not discuss extensively, Hilbert's *Die logischen Grundlagen der Mathematik*, was based on a lecture given at the Leipzig meeting of the Deutsche Naturforscher-Gesellschaft in September 1922; the paper was received for publication by the Mathematische Annalen on 29 September 1922. It contains a summary of the consistency proof given in the lectures of the winter term 1921/22 and the first step toward a treatment of quantifiers.

Appendix B: Correspondence with Russell. Alasdair Urquhart informed me about the Russell-Hilbert connection. First of all, Urquhart mentioned that in the *Selected Letters of Bertrand Russell*, Volume 1, Nicholas Griffin (ed.), there is a letter from Russell to Lady Ottoline Morrell of 18 January 1914 that contains the following passage: "Littlewood tells me that Hilbert (the chief mathematical professor there) has grown interested in Whitehead's and my work, and that they think of asking me to lecture there next year. I hope they will." (p. 487) In this letter Russell says of Göttingen: "It makes one's mouth water to hear how many good students they have, and what advanced lectures are attended in large numbers."

Secondly, Urquhart also pointed out a brief passage in Constance Reid's Hilbert biography; one finds on p. 144 the following remarks: "The lack of contact with foreign mathematicians was extremely frustrating to Hilbert. Just before the war Bertrand Russell, with A. N. Whitehead, had published his *Principia Mathematica*. Hilbert was convinced that the combination of mathematics, philosophy and logic represented by Russell should play a greater role in science. Since he could not now bring Russell himself to Göttingen, he set about improving the position of his philosopher friend Leonard Nelson."

Thirdly (and most importantly), he pointed me to the exchange of postcards between Russell and Hilbert; some are preserved in the Russell Archives at McMaster University, Hamilton. Here are their transcriptions:

Postcard from Hilbert to Russell, dated 12 April 1916:

Hochgeehrter Herr Kollege.

Ich gestatte mir Ihnen mitzuteilen, dass wir uns in der math. Gesellschaft schon seit langem mit Ihrer Erkenntnistheorie beschäftigen und dass wir gerade vor Ausbruch des Krieges die Absicht hatten, Sie—von der Wohlfskehlstiftung[66] aus—nach Göttingen einzuladen, damit Sie uns persönlich über Ihre Lösung des Paradoxienproblem [sic] einen Zyklus von Vorträgen halten könnten. Ich hoffe, dass die Ausführung dieses Planes durch den Krieg nicht aufgehoben, sondern nur aufgeschoben worden ist.

Mit ausgezeichneter Hochachtung, Hilbert

Postcard from Hilbert to Russell, dated 24 May 1919:

Hochgeehrter Herr Professor:

Vielen Dank für Ihre Karte, die ich vor drei Jahren via Hecke—Basel erhielt. Hoffentlich bereuen Sie nicht die Zusage, die Sie damals hinsichtlich der Abhaltung

---

[66] Paul Wolfskehl (1856–1906) gave 100,000 German marks to the University of Göttingen to be awarded to the first person to give a (correct) proof of Fermat's Last Theorem. The interest from that fund was used in 1911 and 1912 to invite Poincaré, Lorentz, and Sommerfeld for lecture series in Göttingen.

von Vorträgen über Logik und Erkenntnistheorie in Göttingen gemacht haben. Vorläufig freilich sind die Zustände in Deutschland noch zu unerfreulich. Wir haben indess in unserem mathematisch-philosophischen Göttinger Kreise das Studium Ihrer Werke eifrig betrieben und versprechen uns viel von Ihrer persönlichen Anwesenheit in Göttingen.

In Erwartung besserer Zeiten und der Wiederherstellung der internationalen Gelehrten-Gemeinschaft bin ich mit bestem Gruss, Ihr ergebenster Hilbert

Postcard from Russell to Hilbert, dated 4 June 1919:

Dear Professor Hilbert

My best thanks for your postcard. I in no way repent of what I wrote before. I am very glad of what you say, and I hope correctly that better times for all will return sooner than now seems probable. When it is possible, I should like nothing better than to carry out your interrupted project, and to contribute what one man can to the restoration of international scientific cooperation.

Yours very truly, Bertrand Russell.

The crucial issue is: which of Russell's writings had actually been read in Göttingen?[67]—Finally, there is a (draft of a) letter of Russell's wife, written on 20 May 1924 and responding to an inquiry of Hilbert, whether Ackermann could study with Russell in England. Mrs. Russell relates: "My husband, Bertrand Russell, is away in America, but will be back before very long. He asks me to say that he would be very glad indeed to have Dr. Ackermann study with him in England."

Appendix C: Lectures from winter term 1922/23. They contain a very informative discussion of the (new) finitist standpoint and the formal presentation of mathematics that allow Hilbert to address the consistency problem in a novel way. Let me quote extensively from pp. 29–33:

In der Verfolgung dieses Zieles, das Gesamtgebäude der Mathematik zu sichern, wurden wir auf zwei Gesichtspunkte geführt:

Der eine betraf die *finite Einstellung*, welche im Bereiche der elementaren Zahlenlehre auch ausreichend ist, während für die Analysis die transfiniten Schlussweisen wesentlich und unentbehrlich sind.

Der andere Gesichtspunkt bestand in der Präzisierung der Sprache, soweit sie zur Darstellung der mathematischen Tatsachen und logischen Zusammenhänge in Betracht kommt. Die Präzisierung geschieht durch den Formalismus des logischen Kalküls, in welchem sich alle logischen Schlüsse, auch die transfiniten Schlussweisen, formal darstellen lassen.

Wenn wir nun diese beiden Gesichtspunkte nebeneinander halten, so kann uns dies darauf bringen, sie in einer neuen Weise zu verknüpfen.

Nämlich die Zeichen und Formeln des logischen Kalküls sind ja durchweg finite Objekte, wenngleich durch sie auch die transfiniten Schlüsse zur Darstellung

---

[67]Mancosu's manuscript (Mancosu 1999a) contains additional, important information that should be explored properly (but this came to my attention only in June of 1998, too late for this paper).

kommen. Wir haben also die Möglichkeit, diese Formeln selbst zum Gegenstande inhaltlicher finiter Überlegungen zu machen, ganz entsprechend wie es in der elementaren Zahlenlehre mit den Zahlzeichen geschieht.

Natürlich ist der Formalismus, mit dem wir es dann zu tun haben, viel mannigfaltiger und komplizierter als derjenige der Zahlzeichen: umfasst er doch alle (in Formeln ausgedrückten) mathematischen Beziehungen.

Dafür trägt er aber auch weiter, und wir können erwarten, mit Hilfe dieser weitergehenden Formalisierung die Gesamtmathematik in den Bereich der finiten Betrachtung zu ziehen.

After a further discussion of the logical calculus the following question is raised: "Was nützt uns nun diese Methode der Formalisierung und die Einsicht, dass die formalisierten Beweise finite Objekte sind, für unser Problem der Begründung der Analysis?" It is answered on p. 33:

Ebenso können wir nun bei der Begründung der Analysis von dem Wahrheitsgehalt der Axiome und Sätze absehen, wenn wir uns nur die Gewähr verschaffen können, dass alle Ergebnisse, zu denen die Prinzipien und die Schlussmethoden der Analysis führen, im Einklang mit einander stehen, sodass wir nicht, wie bisher immer, nur auf guten Glauben die Widerspruchsfreiheit annehmen und der Möglichkeit ausgesetzt sind, eines Tages durch ein Paradoxon überrascht zu werden, wie es z.B. Frege in so dramatischer Weise geschah.

Wenn wir uns nun auf diesen Standpunkt stellen und also die Aufgabe der Begründung ausschliesslich darin sehen, zu zeigen, dass die üblichen transfiniten Schlüsse und Prinzipien der Analysis (und Mengenlehre) nicht auf Widersprüche führen können, so wird für einen solchen Nachweis in der Tat durch unsere vorigen Gedanken eine grundsätzliche Möglichkeit eröffnet.

Denn das Problem der Widerspruchsfreiheit gewinnt nunmehr eine ganz bestimmte, greifbare Form: es handelt sich nicht mehr darum, ein System von unendlich vielen Dingen mit gegebenen Verknüpfungs-Eigenschaften als logisch möglich zu erweisen, sondern es kommt nur darauf an, einzusehen, dass es unmöglich ist, aus den in Formeln vorliegenden Axiomen nach den Regeln des logischen Kalküls ein Paar von Formeln wie A und $\neg A$ abzuleiten.

Hier kommt es zur Geltung, dass die Beweise, wenn sie auch inhaltlich sich im Transfiniten bewegen, doch, als Gegenstände genommen und formalisiert, von finiter Struktur sind. Aus diesem Grunde ist die Behauptung, dass aus bestimmten Axiomen nicht zwei Formeln A, $\neg A$ bewiesen werden können, methodisch gleichzustellen mit inhaltlichen Behauptungen der anschaulichen Zahlentheorie, wie z.B. der, dass man nicht zwei Zahlzeichen a, b finden kann, für welche $a^2 = 2b^2$ [gilt].

II Analyses: historical

A brief guide

Hilbert's papers (1922) and (1923) give a sense of the new subject of proof theory: a promising result had been obtained, namely, a finitist consistency proof for a quantifier-free system of number theory; the *Ansatz* with the τ-symbol indicated a way of extending that proof to first-order arithmetic and analysis. However, only the lecture notes from the winter terms 1921–22 and 1922–23, on which these papers were based, provide a full picture. The first essay, *Finitist proof theory: 1922–1934*, directly continues essay (I.3). It describes, using these notes and some additional material, the first steps in the newly established subject; it summarizes the developments in the 1920s and sketches refined expansions of results presented in *Grundlagen der Mathematik I*.

Bernays writes in the Introduction to *Grundlagen*, published in 1934, that a version of the book had been almost completed in 1931, when results of Gödel and Herbrand produced a deeply changed situation. Most significant were Gödel's incompleteness theorems. von Neumann and Herbrand had learned of the results already in late 1930 and early 1931; they understood them as restricting sharply the reach of the finitist program. In correspondence with them, Gödel gave reasons for holding a contrary position. But in (1933b) Gödel had changed his view, and in his (1938) he discussed ways in which Hilbert's program might be pursued relative to broader constructive bases. That is also a topic of the last essay in this group, *Hilbert and Bernays: 1939*, describing central aspects of *Grundlagen der Mathematik II*.

The essays (II.1) and (II.5) are parts of *Hilbert's proof theory*. The historical and systematic connections are illuminated in a dramatically different way by the essay (II.3), *In the shadow of incompleteness: Hilbert and Gentzen*; it reveals the important role Hilbert's last paper (1931b) played for Gentzen. The remaining two essays were written in the context of editing Gödel's *Collected Works*: (II.2), *After Königsberg*, introduced Gödel's correspondence with von Neumann; (II.4), *Gödel at Zilsel's*, is the Introductory Note Charles Parsons and I wrote for (Gödel 1938).

David Hilbert

II.1

Finitist proof theory: 1922–1934

In the winter term 1921–22, Hilbert and Bernays not only introduced the terms "Hilbertsche Beweistheorie" and "finite Mathematik", but they also used novel techniques to give a finitist proof of the restricted consistency result formulated in (Hilbert 1922). The proof is presented in Kneser's *Mitschrift* of these lectures. Indeed, the proof theoretic considerations start on 2 February and end on 23 February 1922; they include a treatment of definition by (primitive) recursion and proof by induction. The latter principle is formulated as a rule and restricted to quantifier-free formulas. The part of the official notes (1921/22) that present the consistency proof was written by Bernays after the term had ended and contains a different argument; that argument pertains only to the basic system and is sketched in (Hilbert *1922/23). I am going to describe a third modification found in Kneser's *Mitschrift* of the 1922–23 lectures. Apart from this important beginning of proof theoretic investigations, I will discuss Hilbert's *Ansatz* for the treatment of quantifiers, some relevant further developments in the 1920s due to Ackermann and von Neumann, and the refined and systematic presentation of this work in the first and second volume of *Grundlagen der Mathematik*.

1 Proof transformations

The work in 1921–22 gave a new direction to proof theoretic investigations and required the partial resolution of some hard technical, as well as deep methodological issues. In the first group of issues we find in particular the treatment of quantifiers; that is absolutely crucial, if consistency is to be established for the kind of formalisms that had been used in the 1917–18 lectures to represent classical mathematical practice. The second group of issues concerns the extent of metamathematical means (is contentual induction admitted?) and the character of the formal systems to be investigated (are they to be quasi-constructive?). The last issue has a direct resolution: it is addressed in (Hilbert *1921/22) by a modification of the logical calculus, i.e., by replacing the (in-) equalities in the axioms for negation by (negations of) arbitrary formulas B. The axioms are given in the form:

$$B \to (\neg B \to A)$$
$$(B \to A) \to ((\neg B \to A) \to A)$$

With this emendation we have the system that is used in Hilbert's (1923); the first axiom expresses the *principle of contradiction* and the second the *principle of tertium non datur*.

Though the technical resolution is straightforward it required a new philosophical perspective that is, perhaps first and best, articulated in Bernays's (1922b) that was written during the early fall of 1921. The paper starts out with a discussion of *existential axiomatics*, which presupposes, as we saw in 2 [i.e, essay (I.2) here], a system of objects satisfying the structural conditions formulated by the axioms. The assumption of such a system contains, Bernays writes, "something so-to-speak transcendent for mathematics". Thus, the question arises, "which principled position with respect to it should be taken". Bernays remarks that it might be perfectly coherent to appeal to an intuitive grasp of the natural number sequence or even of the manifold of real numbers. However, that could obviously not be an intuition in any primitive sense, and one should take very seriously the tendency in the exact sciences to use, as far as possible, only restricted means for acquiring knowledge. Bernays is thus led to a programmatic demand:

Under this perspective we are going to try, whether it is not possible to give a foundation to these transcendent assumptions in such a way that only *primitive intuitive knowledge is used*.[1]

Clearly, contentual mathematics is to be based on primitive intuitive knowledge and that includes for Bernays, already at this stage, induction.

Bernays's perspective is clearly reflected in the notes (*1921/22). The notes contain a substantial development of finitist arithmetic that involves from the very outset contentual induction ("anschauliche Induktion"). This step was not a small one for Hilbert. In 1904 he had described Kronecker as a dogmatist because of his view "that integers — and in fact integers as a general notion (parameter value) — are directly and immediately given". In the intervening years he scolded Kronecker for just that reason and accused him of not being radical enough. Bernays reports that Hilbert viewed this expansion of contentual arithmetic as a *compromise*, and it is indeed a compromise between a radical philosophical position and intrinsic demands arising from metamathematical investigations. But it is also a compromise in that it gives up the direct parallelism with the case of physics, when statements that can be verified by observation are viewed as corresponding to statements that can be checked by calculation.

---

[1](Bernays 1922b, p. 11). The German text is: "Unter diesem Gesichtspunkt werden wir versuchen, ob es nicht möglich ist, jene transzendenten Annahmen in einer solchen Weise zu begründen, daß nur *primitive anschauliche Erkenntnisse zur Anwendung kommen*."

For the central steps of the consistency argument we need a definition: a formula is called *numeric* if it is built up solely from $=$, $\neq$, numerals, and sentential logical connectives. The first step transforms formal proofs with a numeric endformula into proofs that contain only closed formulas, without changing the endformula. In the second step these modified proofs are turned into configurations that may no longer be proofs, but consist only of numeric formulas; that is achieved by reducing the closed terms to numerals. Finally, all formulas in these configurations are brought into disjunctive normal form and syntactically recognized as "true". Assume now that there is a formal proof of $0 \neq 0$; the last observation allows us to infer that the endformula of this proof is true. However, $0 \neq 0$ is not true and hence not provable.

This proof is given with many details in (*1921/22). It is sketched in (Hilbert 1923, p. 184) and presented in a slightly modified form, according to Kneser's *Mitschrift*, in the winter term of 1922–23. For that term the papers (Bernays 1922b), (Hilbert 1922), and (Hilbert 1923) were all available as indicated explicitly by Kneser. The system in (Kneser 1922/23) uses a calculus for sentential logic that incorporates the connectives & and ∨. The main point of this expanded language and calculus is to provide a more convenient formal framework for representing informal arguments. The axioms for the additional connectives are given as follows:

$A \& B \to A$ $\qquad\qquad\qquad A \& B \to B$

$A \to (B \to A \& B)$

$A \to A \vee B$ $\qquad\qquad\qquad B \to A \vee B$

$(A \to C) \to ((B \to C) \to (A \vee B \to C))$

Clearly, the axioms correspond directly to "natural" rules for these connectives, and one finds here the origin of Gentzen's natural deduction calculi. In (Bernays 1927a) the completeness of this calculus is asserted and its methodological advantages are discussed. Bernays writes:

The initial formulas can be chosen in quite different ways. In particular one has tried to get by with the smallest number of axioms, and in this respect the limit of what is possible has indeed been reached. One supports the purpose of logical investigations better, if one separates, as it is done in the axiomatics of geometry, different groups of axioms in such a way that each of these groups expresses the role of one particular logical operation.[2]

---

[2](Bernays 1927a, p. 10). The German text is: "Die Wahl der Ausgangsformeln kann auf sehr verschiedene Weise getroffen werden. Man hat sich besonders darum bemüht, mit einer möglichst geringen Zahl von Axiomen auszukommen, und hat hierin in der Tat die Grenze des Möglichen erreicht. Den Zwecken der logischen Untersuchung wird aber besser gedient, wenn wir, entsprechend wie in der Axiomatik der Geometrie, verschiedene *Axiomengruppen* voneinander sondern, derart, daß jede von ihnen die Rolle einer logischen Operation zum Ausdruck bringt."

It is informative to compare these remarks with Hilbert's in the Introduction to (1899a), quoted in I.2.2.

Then he lists the three groups, namely, axioms for the conditional $\rightarrow$, for conjunction &, disjunction $\vee$, as well as for negation $\neg$. This formal system is used in (Ackermann 1924), in *Über das Unendliche*, and in Hilbert's second Hamburg talk of 1927. Hilbert gives in the Hamburg talk an interestingly different formulation of the axioms for negation, calling them the *principle of contradiction* and the *principle of double negation*:

$(A \rightarrow (B \,\&\, \neg B)) \rightarrow \neg A$
$\neg\neg A \rightarrow A$

These are axiomatic formulations of Gentzen's natural deduction rules for classical negation; the earlier axioms for negation can be derived easily. This new formulation can be found in (van Heijenoort 1967, p. 382 and more fully on pp. 465–466).

The consistency argument for the system of arithmetic described above, but with this logical calculus, is given in (Kneser 1922/23, pp. 20–22); its form is used in the first volume of *Grundlagen der Mathematik*. It begins by separating linear proofs into proof threads ("Auflösung in Beweisfäden"). Then variables are eliminated in the resulting proof trees ("Ausschaltung der Variablen"), and the numerical value of functional terms is determined by calculation ("Reduktion der Funktionale"). That makes it possible to determine the truth or falsity of formulas in the very last step. Here that is not done by turning formulas into disjunctive normal form, but rather simply by truth-table computations ("Wahrheitsbewertung"). Incorporating definition by recursion and proof by induction into these considerations is fairly direct.

From a contemporary perspective this argument shows something very important. If a formal theory contains a certain class of finitist functions, then it is necessary to appeal to a wider class of functions in the consistency proof: an *evaluation function* is needed to determine uniformly the numerical value of terms, and such a function is no longer in the given class. The formal system considered in the above consistency proof includes primitive recursive arithmetic, and the consistency proof goes beyond the means available in primitive recursive arithmetic. At this early stage of proof theory, finitist mathematics is consequently stronger than primitive recursive arithmetic. This assessment of the relative strength is sustained, as we will see, throughout the development reported in this essay.

2 Hilbert's Ansatz: the ϵ-substitution method

Ackermann reviewed in section II of his (1924) the above consistency proof; the section is entitled *The consistency proof before the addition of the transfinite axioms* ("Der Widerspruchsfreiheitsbeweis vor Hinzunahme der transfiniten Axiome"). The very title reveals the restricted programmatic significance of this result, as it concerns a theory that is part of finitist mathematics and need not be secured by a consistency proof. The truly expanding step that goes beyond the finitist methods just described involves the treatment of quantifiers. Already in Hilbert's (1922) we find a brief indication of his *Ansatz*, which is elaborated in (Kneser 1921/22). Indeed, the treatment of quantifiers is just the first of three steps that have to be taken: the second concerns the general induction principle for number theory, and the third an expansion of proof theoretic considerations to analysis.

Hilbert sketched in his Leipzig talk of September 1922, which was published as (Hilbert 1923), the consistency proof from the previous section. However, the really dramatic aspect of his proof theoretic discussion is the treatment of quantifiers with the τ-function, the dual of the later ϵ-operator. The logical function τ associates with every predicate $A(a)$ a particular object $\tau_a(A(a))$ or simply τA. It satisfies the *transfinite axiom* $A(\tau A) \to A(a)$, which expresses, according to Hilbert, "if a predicate A holds for the object τA, then it holds for all objects a". The τ-operator allows the definition of the quantifiers:

$(a)A(a) \leftrightarrow A(\tau A)$
$(Ea)A(a) \leftrightarrow A(\tau(\neg A))$

Hilbert extends the consistency argument then to the "first and simplest case" that goes beyond the finitist system: this "Ansatz" will evolve into the ϵ-substitution method. It is only in the Leipzig talk that Hilbert gave proof theory its principled formulation and discussed its technical tools — for the first time outside of Göttingen. As far as analysis is concerned, Hilbert had taken as the formal frame for its development ramified type theory with the axiom of reducibility in (*1917/18). In his Leipzig talk he considers a third-order formulation; appropriate functionals (Funktionenfunktionen) allow him to prove (on pp. 189–191) i) the least upper bound principle for sequences and sets of real numbers and ii) Zermelo's choice principle for sets of sets of real numbers. He conjectures that the consistency of the additional transfinite axioms can be patterned after that for τ. He ends the paper with the remark:

Now the task remains of precisely carrying out the basic ideas I just sketched; its solution completes the founding of analysis and prepares the ground for the founding of set theory.[3]

In Supplement IV of the second volume of *Grundlagen der Mathematik*, Hilbert and Bernays give a beautiful exposition of analysis and emphasize, correctly, that the formalisms for their "deductive treatment of analysis" were used in Hilbert's early lectures on proof theory and that they were described in (Ackermann 1924).

The above concrete proof theoretic work was directly continued in Ackermann's thesis, on which (Ackermann 1924) is based. The paper was submitted to Mathematische Annalen on 30 March 1924 and published

---

[3](Hilbert *1922/23, p. 191). The German text is: "Es bleibt nun noch die Aufgabe einer genauen Ausführung der soeben skizzierten Grundgedanken; mit ihrer Lösung wird die Begründung der Analysis vollendet und die der Mengenlehre angebahnt sein."

in early 1925.[4] It starts out, as I mentioned, with a concise review of the earlier considerations for quantifier-free systems. The ϵ-symbol had been substituted for the τ-symbol in 1923 as is clear from (Kneser 1922/23), and in Ackermann's dissertation the ϵ-calculus as we know it replaced the τ-calculus from 1922. The transfinite axiom of the ϵ-calculus is obviously dual to the axiom for τ and is formulated in (Ackermann 1924) in section III entitled, "Die Widerspruchsfreiheit bei Hinzunahme der transfiniten Axiome und der höheren Funktionstypen". For number variables the crucial axiom is formulated as $A(a) \to A(\epsilon A)$ and allows of course the definition of quantifiers:

$(Ea)A(a) \leftrightarrow A(\epsilon A)$
$(a)A(a) \leftrightarrow A(\epsilon(\neg A))$

The remaining transfinite axioms are adopted from (Hilbert *1922/23). However, the ϵ-symbol is actually characterized as the least-number operator, and the recursion schema with just number variables is extended to a schema that permits also function variables.

The connection to the mathematical development in Hilbert's paper is finally established in section IV, where Ackermann explores the "Tragweite" of the axioms. At first it was believed that Ackermann had completed the task indicated by Hilbert at the end of the Leipzig talk and had established not only the consistency of full classical number theory but even of analysis. However, a note was added in proof (p. 9) that significantly restricted the result; von Neumann noted a mistake in Ackermann's arguments. In his own paper *Zur Hilbertschen Beweistheorie*, submitted on 29 July 1925, a consistency proof was given for a system that covers, von Neumann asserts (p. 46), Russell's mathematics without the axiom of reducibility or Weyl's system in *Das Kontinuum*. Ackermann corrected his arguments and obtained an equivalent result; that corrected proof is discussed in (Hilbert 1926) and (Bernays 1927a).[5] (It is also the basis for the corresponding presentation in the second volume of (Hilbert and Bernays 1939); see section 4 below.) In his Bologna talk of 1928 Hilbert stated, in line with von Neumann's observation, that the consistency of full number theory had been secured by the proofs of Ackermann and von Neumann; according to Bernays in his preface to the second volume of *Grundlagen der Mathematik* that belief was sustained until 1930.

A number of precise metamathematical problems were formulated in Hilbert's Bologna talk, among them the completeness question for first-order logic, which is presented in a new formulation suggested by the Löwenheim Skolem Theorem. The latter theorem also suggested, so it

---

[4]In his lecture *Über das Unendliche* given on 4 June 1925 in Münster, Hilbert presented an optimistic summary of the state of proof theoretic affairs together with bold new considerations. Even before this talk, Hilbert had turned his attention to issues beyond the consistency of number theory and analysis, namely, the solution of the continuum problem.

[5]See (van Heijenoort 1967, p. 477 and p. 489), but also note 3 on p. 489 and the introductions to these papers. [Cf. also the detailed discussion in (Zach 2004).]

seems, that higher mathematical theories might not be syntactically complete. In any event, Hilbert's Bologna address shows most dramatically through its broad perspective and clear formulation of open problems, how far mathematical logic had been moved in roughly a decade. This is a remarkable achievement with impact not just on proof theory; the (in-)completeness theorems, after all, are to be seen in this broader foundational enterprise.[6]

In terms of specifically proof theoretic results no striking progress was made, in spite of the fact that most talented people had been working on the consistency program. That state of affairs had intrinsic reasons. In the preface to the first volume of *Grundlagen der Mathematik*, Bernays asserts that a presentation of proof theoretic work had almost been completed in 1931. But, he continues, the publication of papers by Herbrand and Gödel produced a deeply changed situation that resulted in an extension of the scope of the work and its division into two volumes. The eight chapters of the first volume can be partitioned into three groups: chapters 6 to 8 give refined versions of results that had been obtained in the 1920s, investigating the consistency problem and other metamathematical questions for a variety of (sub) systems of number theory; chapters 3 to 5 develop systematically the logical framework of first-order logic (with identity); chapters 1 and 2 discuss the central foundational issues and the finitist standpoint in most informative and enlightening ways. Let me start with a discussion of the central issues as viewed in 1934.

3 Grundlagen, 1934

Chapter 1 begins with a general discussion of axiomatics, at the center of which is a distinction between *contentual* and *formal axiomatic theories*. Contentual axiomatic theories like Euclid's geometry, Newton's mechanics, and Clausius's thermodynamics draw on experience for the introduction of their fundamental concepts and principles, all of which are understood contentually. Formal axiomatic theories like Hilbert's axiomatization of geometry, by contrast, abstract away such intuitive content. They begin with the assumption of a fixed system of things (or several such systems), which is delimited from the outset and constitutes a "domain of individuals for all predicates from which the statements of the theory are built up". The existence of such a domain of individuals constitutes an "idealizing assumption that joins the assumptions formulated in the axioms".[7] Hilbert

---

[6]A remark indicating the depth of Dedekind's continued influence: Hilbert formulated as Problem I of his Bologna talk the consistency of the ϵ-axioms for function variables and commented, "The solution of Problem I justifies also Dedekind's ingenious considerations in his essay *Was sind und was sollen die Zahlen?*"

[7](Hilbert and Bernays 1934, pp. 1–2). The last quotation is part of this German text: "In der Voraussetzung einer solchen Totalität des "Individuen-Bereiches" liegt — abgesehen von den trivialen Fällen, in denen eine Theorie es ohnehin nur mit einer endlichen, festbegrenzten Gesamtheit von Dingen zu tun hat — eine idealisierende Annahme, die zu den durch die Axiome formulierten Annahmen hinzutritt."

and Bernays refer to this approach as *existential axiomatics*. They clearly consider formal axiomatics to be a sharpening of contentual axiomatics, but are also quite explicit that the two types of axiomatics complement each other and are both necessary.

The consistency of a formal axiomatic theory with a finite domain can be established by exhibiting a model satisfying the axioms. However, for theories F with infinite domains consistency proofs present a special problem, because "reference to non-mathematical objects cannot settle the question whether an infinite manifold exists; the question must be solved within mathematics itself".[8] One must treat the consistency problem for such a theory F as a *logical problem* — from a proof theoretic perspective. This involves formalizing the principles of logical reasoning and proving that from F one cannot derive (with the logical principles) both a formula and its negation. Proof theory need not prove the consistency of every F; it suffices to find a proof for a system F with a structure that is (i) surveyable to make a consistency proof for the system plausible and (ii) rich enough so that the existence of a model S for F guarantees the satisfiability of axiom systems for branches of physics and geometry. Arithmetic (including number theory and analysis) is considered to be such an F.

For a consistency argument to be foundationally significant it must avoid the idealizing existence assumptions made by formal axiomatic theories. But if a proof theoretic justification of arithmetic by elementary means were possible, might it not also be possible to develop arithmetic directly, free from non-elementary assumptions and thus not requiring any additional foundational justification? The answer to this question involves elementary presentations of parts of number theory and formal algebra; these presentations serve at the same time to illuminate the finitist standpoint. Finitist deliberations take here their purest form, i.e., the form of "*thought experiments* involving objects assumed to be concretely given".[9] The word finitist is intended to convey the idea that a consideration, a claim or definition respects (i) that *objects are representable*, in principle, and (ii) that *processes are executable*, in principle.[10]

A direct, elementary justification for all of mathematics is not possible, Hilbert and Bernays argue, because already in number theory and

---

[8](Hilbert and Bernays 1934, pp. 17). The German text is: "Auf Grund dieser Überlegungen kommen wir zu der Einsicht, daß die Frage nach der Existenz einer unendlichen Mannigfaltigkeit durch eine Berufung auf außermathematische Objekte nicht entschieden werden kann, sondern innerhalb der Mathematik selbst gelöst werden muß."

[9](Hilbert and Bernays 1934, pp. 20). The German text is: "Das Kennzeichnende für diesen methodischen Standpunkt ist, daß die Überlegungen in der Form von *Gedankenexperimenten* an Gegenständen angestellt werden, die als *konkret vorliegend* angenommen werden."

[10](Hilbert and Bernays 1934, pp. 32). The German text is: "Im gleichen Sinne wollen wir von finiten Begriffsbildungen und Behauptungen sprechen, indem wir allemal mit dem Wort "finit" zum Ausdruck bringen, daß die betreffende Überlegung, Behauptung oder Definition sich an die Grenzen der grundsätzlichen Vorstellbarkeit von Objekten sowie der grundsätzlichen Ausführbarkeit von Prozessen hält und sich somit im Rahmen konkreter Betrachtung vollzieht."

analysis one uses nonfinitist principles. While one might circumvent the use of such principles in number theory (as only the existence of the domain of integers is assumed), the case for analysis is different. There the objects of the domain are themselves infinite sets of integers, and the principle of the excluded middle is applied in this extended domain. Thus one is led back to using finitist proof theory as a tool to secure the consistency of mathematics. (This restriction is relaxed only at the end of the second volume when "extensions of the methodological framework of proof theory" are considered.) The first stage of this project, the proof theoretic investigation of appropriate logical formalisms, occupies chapters 3 through 5. These systems are close to contemporary ones; they can be traced back to the lectures given in 1917–18 and were presented also in (Hilbert and Ackermann 1928). The second stage is carried out in chapters 6 through 8 and involves the investigation of sub-systems of first-order number theory; there are three different groups of such systems.

The first group consists of weak fragments of number theory that contain few, if any, function symbols and extend predicate logic with equality by axioms for 0, successor and <; some of the fragments involve quantifier-free induction. Metamathematical relations between them are explored, and independence as well as consistency results are established. The main technique for giving consistency proofs is that discussed in section 1 above. However, since the formalisms contain quantifiers, an additional procedure is required, namely a procedure that assigns *reducts*, quantifier-free formulas acting as witnesses, to formulas containing quantifiers. The method underlying this procedure is due to Herbrand and Presburger. Consistency is inferred from results that involve the notion of *verifiability*, extending the notion of truth to formulas containing free variables, bound variables, and recursively defined function signs.[11] In order to establish the consistency of a formalism F, one proves that every formula not containing formula variables is verifiable, if it is derivable in F. Since $0 \neq 0$ is not verifiable, it is not derivable in F; thus, F is consistent.

The second group of subsystems of number theory contains formalisms arising from the elementary calculus with free variables (the quantifier-free fragment of the predicate calculus) through the addition of functions defined by recursion. Hilbert and Bernays discuss at the beginning of chapter 7 the formalization of the definition principle for recursion, reminding the reader how they viewed recursive definition of a function in their discussion of finitist number theory in chapter 1. There they emphasize that by "function" they understand an intuitive instruction ("anschauliche

---

[11] More precisely, letting A be a formula of the formalism F: (i) if A is a numeric formula, it is verifiable if it is true; (ii) if A contains free numeric variables (but no formula variables or bound variables), it is verifiable if one can show by finitist means that the substitution of arbitrary numerals for variables (followed by the evaluation of all function-expressions and their replacement through their numerical values) yields a true numeric formula; (iii) if A contains bound variables but no formula variables, it is verifiable if its reduct is verifiable (according to (i) and (ii)).

Anweisung") on account of which a numeral is associated with a given numeral, or a pair of numerals, or a triple ... of numerals. The proof method of complete induction is not considered as an "independent principle" but as a consequence that is gathered from the concrete build-up of numerals.[12] Similarly, definition by recursion is not viewed as an independent definition principle but rather as "an agreement concerning an abbreviated description of certain construction processes by means of which one obtains a numeral from one or several given numerals".[13] These processes or procedures can be mimicked formally by the introduction of function symbols together with recursion equations.[14] They treat in a first step ("zunächst") what they consider as the simplest schema of recursion (and "restrict *for the time being* the notion of recursion to it"). That simplest schema of recursion is the schema of what we (and they later on p. 326/331[15]) call *primitive recursion*, namely,

$$f(a,\ldots,k,0) = \mathfrak{a}(a,\ldots,k),$$
$$f(a,\ldots,k,n') = \mathfrak{b}(a,\ldots,k,n,f(a,\ldots,k,n)).$$

$\mathfrak{a}$ and $\mathfrak{b}$ denote previously defined functions and $a,\ldots,k,n$ are numerical variables. What then is the formalization of the intuitive procedure associated with a recursive definition that is formally reflected by the recursion schema for f? It is given by a derivation — via the recursion equations and the axioms for identity — of the equation $f(a,\ldots,k,z) = t(a,\ldots,k)$ for each numeral z, where t is a term not containing f. Then they argue that the value of $f(a,\ldots,k,z)$ can be determined for each sequence of numerals $a,\ldots,k$ and conclude their considerations with: "Thus we can completely mimic the recursive calculation procedure of finitist number theory in our formalism by the deductive application of the recursion equations."[16]

They then prove on pp. 298–300 a general *Consistency Theorem: Let F be a formalism extending the elementary calculus with free variables by verifiable axioms (that may contain recursively defined functions whose defining equations are taken as axioms) and the schema of quantifier-free induction,*

---

[12](Hilbert and Bernays 1934, p. 25). The German text is: "Wir haben es also hier nicht mit einem selbständigen Prinzip zu tun, sondern mit einer Folgerung, die wir aus dem konkreten Aufbau der Ziffern entnehmen."

[13](Hilbert and Bernays 1934, p. 27). The German text is: "Es handelt sich hier bei der Definition durch Rekursion wiederum nicht um ein selbständiges Definitionsprinzip, sondern die Rekursion hat im Rahmen der elementaren Zahlentheorie lediglich die Bedeutung einer Vereinbarung über eine abgekürzte Beschreibung gewisser Bildungsprozesse, durch die man aus einer oder mehreren gegebenen Ziffern wieder eine Ziffer enthält."

[14](Hilbert and Bernays 1934, p. 287). This is a translation of part of the following German text: "Dieses Verfahren [der Berechnung von Funktionswerten in der finiten Zahlentheorie] können wir auch im Formalismus nachbilden, indem wir allgemein die Einführung von Funktionszeichen in Verbindung mit Rekursionsgleichungen zulassen."

[15]Here and below, I give the references to both the first and second edition of *Grundlagen der Mathematik*, in case the page numbers are different.

[16](Hilbert and Bernays 1934, p. 292). The German text is: "Somit können wir das rekursive Berechnungsverfahren der finiten Zahlentheorie in unserem Formalismus durch die deduktive Anwendung der Rekursionsgleichungen vollkommen nachbilden."

then every derivable formula of F (without formula variables) is verifiable. This theorem is explicitly taken to establish the consistency of a number of formalisms including in particular that of recursive number theory. Having formally developed the latter theory to indicate the strength of this recursive treatment of number theory, they state:

This recursive number theory is closely related to intuitive number theory, as we have considered it in §2, because all of its formulas can be given a *finitist contentual interpretation*. This contentual interpretability follows from the verifiability of all derivable formulas of recursive number theory, as we established it already. Indeed, in this area verifiability has the character of a direct contentual interpretation, and thus the proof of consistency could be obtained here so easily.

As we have a finitist consistency proof for recursive arithmetic, finitist mathematics here goes beyond recursive arithmetic. Their notion of recursive number theory was at first, as I emphasized above, to involve only primitive recursions; so finitist mathematics is stronger than primitive recursive arithmetic. Indeed, after the above remark they continue the discussion of the relationship between intuitive and recursive number theory as follows:

Recursive number theory is differentiated from intuitive number theory by its formal restrictions; its only method for forming concepts, apart from explicit definitions, is the recursion schema, and its methods of derivation are also firmly delimited.

However, without taking away from recursive number theory what is the characteristic feature of its method, we can admit certain *extensions of the recursion schema* as well as of the induction schema. We'll discuss these [extensions] now briefly.

The extending formalisms involve the schema of nested recursion ("verschränkte Rekursion") that allows in particular the definition of the Ackermann function; they remark that their previous consistency results are extended to these proper extensions of recursive number theory as well; thus, the bounds of finitist mathematics are pushed still further.[17]

---

[17]The two long quotations above are both found on p. 325/330 of (Hilbert and Bernays 1934). The remark concerning the finitist provability of the consistency of the extended formalisms is on p. 325/330; more precisely, it is asserted there, all formulas (not containing formula variables) that are provable in the extended formalisms are verifiable. Here is the first German text:

"Diese rekursive Zahlentheorie steht insofern der anschaulichen Zahlentheorie, wie wir sie im §2 betrachtet haben, nahe, als ihre Formeln sämtlich *einer finiten inhaltlichen Deutung fähig* sind. Diese inhaltliche Deutbarkeit ergibt sich aus der bereits festgestellten Verifizierbarkeit aller ableitbaren Formeln der rekursiven Zahlentheorie. In der Tat hat in diesem Gebiet die Verifizierbarkeit den Charakter einer direkten inhaltlichen Interpretation, und der Nachweis der Widerspruchsfreiheit war daher auch hier so leicht zu erbringen."

And here is the second passage:
"Der Unterschied der rekursiven Zahlentheorie gegenüber der anschaulichen Zahlentheorie besteht in ihrer formalen Gebundenheit; sie hat als einzige Methode der Begriffsbildung, außer der expliziten Definition, das Rekursionsschema zur Verfügung, und auch die Methoden der Ableitung sind fest umgrenzt.

The third group of formalisms that are actually equivalent to full number theory is investigated towards the end of chapter 7 and in chapter 8. The first of these is the formalism of the axiom system (Z); call this formalism Z. Hilbert and Bernays comment that the techniques used in the consistency proofs for fragments of number theory cannot be generalized to Z, as any reduction procedure for Z would provide a decision procedure for Z and, thus, allow solving all number theoretic problems. They leave the possibility of such a procedure as an open problem (whose solution, if it exists, they view as a long way off) and focus on examining whether Z provides the means for formalizing informal number theory. With this end in mind, it is proved in chapter 8 that all recursive functions are representable in Z. For this proof they establish three separate claims: (1) the least number operator μ can be explicitly defined in terms of Russell and Whitehead's ι-symbol; (2) any recursive definition can be explicitly defined in Z_μ, i.e., Z extended by defining axioms for the μ-operator; (3) the addition of the ι-rule to Z is a conservative extension of Z.[18] Volume I concludes with the remark that the above results entail the consistency of Z_μ relative to that of Z, but that none of the results or methods considered so far suffice to show that Z is consistent.

4 Limited results for quantifiers

The consistency proofs in chapter 7.a) of the first volume are given for quantifier-free systems. In the second volume, these theories are embedded in the system of full predicate logic together with the ϵ-axioms in the form $A(a) \to A(\epsilon_x.A(x))$. The first crucial task is to eliminate all references to bound variables from proofs of theorems that do not contain them; axioms used in these proofs must not contain bound variables either. In the formulation of Hilbert and Bernays, the consistency of a system of proper axioms relative to the predicate calculus together with the ϵ-axioms is to be reduced to the consistency of the system relative to the elementary calculus (with free variables).[19] The consistency of the latter system is recognized on account of a suitable finitist interpretation. Thus, it is emphasized,

---

Allerdings können wir, ohne der rekursiven Zahlentheorie das Charakteristische ihrer Methode zu nehmen, gewisse *Erweiterungen des Schemas der Rekursion* sowie auch des Induktionsschemas zulassen. Auf diese wollen wir noch kurz zu sprechen kommen."

[18]The second claim is first established for primitive recursively defined functions, but then extended to the case of nested recursion; cf. (Hilbert and Bernays 1934, p. 421/431). The latter result is attributed to von Neumann and Gödel. As to the general issues compare von Neumann's letter to Gödel of 29 November 1930; the letter is reprinted and translated in (Gödel 2003b). [Cf. essays (II.2) and (II.3) below.]

[19](Hilbert and Bernays 1939, p. 33). Referring to the first ϵ-theorem they write there: "Die Bedeutung dieses Theorems besteht ja darin, daß es die Frage der Widerspruchsfreiheit eines Systems von eigentlichen Axiomen ohne gebundene Variablen bei Zugrundelegung des Prädikatenkalkuls und des ϵ-Axioms zurückführt auf die seiner Widerspruchsfreiheit bei Zugrundelegung des elementaren Kalkuls mit freien Variablen."

operating with the ε-symbol can be viewed as "merely an auxiliary calculus, which is of considerable advantage for many metamathematical considerations".[20]

In the framework of the extended calculus bound variables are associated with just ε-terms, as the quantifiers can be defined. The initial elimination result is the *First ε-Theorem: If the axioms $A_1, \ldots, A_k$ and the conclusion of a proof do not contain bound individual variables or (free) formula variables, then all bound variables can be eliminated from the proof.* The argument can be extended to cover proofs of purely existential formulas, but the proofs yield then as their conclusion suitable disjunctions of instances of the endformula. Based on this extension a *Consistency Theorem* of a quite general character is proved: *If the axioms $A_1, \ldots, A_k$ are verifiable, then (i) any provable formula containing at most free individual variables is verifiable, and (ii) for any provable, purely existential formula $(Ex_1)\ldots(Ex_n)A(x_1,\ldots,x_n)$ (with only the variables shown) there are variable-free terms $t_1,\ldots,t_n$ such that $A(t_1,\ldots,t_n)$ is true.* This theorem is applied to establish the consistency (i) of Euclidean and Non-Euclidean geometry without continuity assumptions in section 1.4, and (ii) of arithmetic with recursive definitions, but only quantifier-free induction in sections 2.1 and 2.2. In essence then, the consistency theorem from Herbrand 1931 (cf. section II.5.1) has been reestablished in a subtly more general way: Hilbert and Bernays allow the introduction of a larger class of recursive functions.[21] Putting the result in the appropriate systematic context, we see that the consistency proof of 1922 for the quantifier-free system of primitive recursive arithmetic has been extended to cover that system's expansion by full classical quantification theory. Indeed, the extensions from their (1934) are also covered.

The remainder of chapter 2 discusses the difficulty of extending the elimination procedure in the proof of the first ε-theorem to a system with full induction and examines Hilbert's original Ansatz for eliminating ε-symbols.[22] The next two chapters investigate the formalism for predicate logic and begin in chapter 3 with a proof of the *Second ε-Theorem: If the axioms and the conclusion of a proof (in predicate logic with identity)*

---

[20](Hilbert and Bernays 1939, p. 13). The quotation is part of this German text: "Es besteht ja keinesfalls die Notwendigkeit, das ε-Symbol in den endgültigen deduktiven Aufbau des logisch-mathematischen Formalismus einzubeziehen. Vielmehr kann das Operieren mit dem ε-Symbol als ein bloßer Hilfskalkul angesehen werden, der für viele metamathematische Überlegungen von erheblichem Vorteil ist."

[21](Hilbert and Bernays 1939, p. 52). That is made explicit in note 1 on page 52. Though this larger class of functions (all general recursive ones?) allows a generalized notion of verifiability, the latter would no longer guarantee a finitist consistency proof. They write there: "Diese [allgemeine Anweisung zur Einführung von Funktionszeichen in (Herbrand 1931)] ist insofern etwas enger als die hier formulierte Anweisung, als Herbrand verlangt, daß das Verfahren der Wertbestimmung sich, so wie bei den Schemata der rekursiven Definition, durch eine finite Deutung der Axiome ergeben soll."

[22]As to the character of the original and the later version of the elimination method and Ackermann's work see pp. 21, 29–30, 92ff, 121ff, the note on p. 121, as well as Bernays's preface.

do not contain ϵ-symbols, then all ϵ-symbols can be eliminated from the proof. Herbrand's theorem is obtained as well as a variety of criteria for the refutability of formulas in predicate logic; proofs of the Löwenheim–Skolem theorem and of Gödel's completeness theorem are also given. These considerations are used to establish results concerning the decision problem; solvable cases as well as reduction classes are discussed. In chapter 4 Gödel's "arithmetization of metamathematics" is presented in great detail and applied to obtain a fully formalized proof of the completeness theorem.

The completeness theorem can be taken as stating that the consistency of an axiom system relative to the calculus of predicate logic coincides with satisfiability of the system by an arithmetic model. The formalized proof is intended to establish a kind of finitist equivalent to a consequence of this formulation, namely, that the consistency relative to the predicate calculus guarantees consistency, as it is formulated on p. 205, in an open contentual sense ("im unbegrenzten inhaltlichen Sinne"). The finitist equivalent is formulated roughly as follows: if a formula is irrefutable in predicate logic, then it remains irrefutable in "every consistent number theoretic formalism", i.e., in every formalism that is consistent and remains consistent when the axioms of Z_μ and verifiable formulas are added. That result expresses a strong deductive closure of the predicate calculus, but only if Z_μ is consistent.[23] Thus, there is another reason for establishing finitistically the consistency of this number theoretic formalism.

---

[23](Hilbert and Bernays 1939, p. 253/263). Here is the German text of which the quoted remark is a part: "Das hiermit bewiesene Theorem hat allerdings seine Bedeutung als ein Vollständigkeitssatz, d.h. als Ausdruck einer Art von deduktiver Abgeschlossenheit des Prädikatenkalkuls, nur unter der Voraussetzung der Widerspruchsfreiheit des zahlentheoretischen Formalismus."

II.2

After Königsberg

The correspondence between von Neumann and Gödel opens with an extraordinary letter from von Neumann, written on 20 November 1930. In early September of that year, von Neumann had met Gödel at a congress in Königsberg and was informed about a theorem Gödel had just discovered—(a form of) the first incompleteness theorem. Von Neumann was deeply impressed; he turned his attention to logic again and gave lectures on proof theory in the winter term of 1930–1931. As can be gathered from Herbrand's letter to Claude Chevalley,[1] von Neumann was preoccupied with Gödel's result and, as he put it in his own letter, with the methods Gödel had used "so successfully in order to exhibit undecidable properties". In reflecting on this result and Gödel's methods, von Neumann arrived at a new result that seemed remarkable to him, namely, that the consistency of a formal theory is unprovable within that theory, if it is consistent. He formulated this "new result" in the letter to Gödel, claiming—less precisely—that the consistency of mathematics is unprovable; this strong interpretation of what we know as the second incompleteness theorem was to become a point of contention between Gödel and von Neumann.

In his next letter of 29 November, von Neumann acknowledges the receipt of a "Separatum" and a letter from Gödel.[2] It is most likely that the separatum was a copy of the abstract (1930b) that had been presented to the Vienna Academy of Sciences on 23 October 1930 and already contained the classical formulation of the second incompleteness theorem.[3]

Many thanks go to Sam Buss and John Dawson for providing me with information on (Buss 1995) and (Clote and Krajíček 1993), respectively (Hartmanis 1989); Dawson helped in addition with a question concerning (1930b). Solomon Feferman and Charles Parsons suggested substantive and stylistic improvements.

[1] Cf. the introductory note to the Gödel-Herbrand correspondence in (Gödel 2003b); Herbrand wrote the letter on 3 December 1930.

[2] Unfortunately, it seems that this letter and two others in this early correspondence have not been preserved: von Neumann acknowledges in his letter of 12 January 1931 that he had received two letters from Gödel. (The von Neumann Papers in the Library of Congress do not contain these letters.)

[3] Gödel had by this time completed his (1931d); indeed, the paper had been submitted for publication on 17 November 1930. Von Neumann acknowledged receipt of the galley proofs of

Von Neumann states in his response: "As you have established the theorem on the unprovability of consistency as a natural continuation and deepening of your earlier results, I clearly won't publish on this subject." Their differing views on the impact of this result for Hilbert's consistency program are discussed below. Two additional topics of scientific interest are addressed in later correspondence: (i) the relative consistency of the axiom of choice and the generalized continuum hypothesis, in letters from 1937 through 1939, and (ii) the feasibility of computations (related to the now famous P vs. NP problem), in Gödel's last letter to von Neumann in 1956.

As von Neumann's life and work are well known, only the briefest biographical sketch is presented.[4] Born on 28 December 1903 in Budapest (Hungary), von Neumann grew up in a wealthy Jewish family and attended the excellent Lutheran Gymnasium in Budapest from 1914 to 1921. He then entered the University of Berlin as a student of chemistry, but switched in 1923 to the Eidgenössische Technische Hochschule in Zürich, where he earned three years later a *Diplom* degree in that subject. He obtained, also in 1926, a doctoral degree in mathematics from the University of Budapest. Von Neumann spent the academic year 1926–1927 in Göttingen supported by a Rockefeller Fellowship. He was Privatdozent in Berlin (1927–1929) and Hamburg (1929–1930). In 1930 he was appointed visiting lecturer at Princeton University with the agreement that he would be back in Berlin for the winter term of 1930–31. In 1931 he was promoted to professor of mathematics at Princeton and became two years later one of the six mathematics professors at the newly founded Institute for Advanced Study, together with J. W. Alexander, A. Einstein, M. Morse, O. Veblen and H. Weyl; he kept that position for the remainder of his life. As to the correspondence with Gödel, it obviously started while von Neumann was staying in Berlin during the winter term of 1930–1931.

In the 1920s von Neumann contributed to the foundations of mathematics not only through a series of articles on set theory ((von Neumann 1923), (von Neumann 1925), (von Neumann 1926), (von Neumann 1928a), (von Neumann 1928b), (von Neumann 1929)) but also very specifically to Hilbert's emerging finitist consistency program through his paper *Zur Hilbertschen Beweistheorie*. Though published only in 1927, the paper had already been submitted for publication in July of 1925. In it von Neumann established the consistency of a formal system of first-order arithmetic with quantifier-free induction; he also gave a detailed critique of the consistency proof in (Ackermann 1924). What is of interest in the context of his early correspondence with Gödel is the general strategic attitude he took

---

(Gödel 1931d) in his letter of 12 January 1931. From (Mancosu 1999b) it is clear that Gödel had not sent the galleys before the end of December 1930; see the letters between Hempel and Kaufmann quoted there on pp. 35–36. Cf. also (Gödel 2003b, editorial note a to letter 2 on p. 338).

[4]Von Neumann was born Janos, used Johann when living in Germany and Switzerland, and switched to John after having moved to the United States. For accounts see the collection of essays by Garrett Birkhoff and others in the *Bulletin of the American Mathematics Society* **64** (3), part 2 (in particular the accessible description in (Ulam 1958)) and the book (Macrae 1992).

towards proof theoretic research. It is expressed in the following quote from the introduction to his (1927); where he formulates four guiding ideas of Hilbert's proof theory. (Note that "intuitionist" and "finitist" were evidently synonymous for von Neumann.) Viewing an intuitionist consistency proof for classical formal theories as the crucial aim, he articulates the final guiding idea as follows:

Here one has always to distinguish sharply between two different ways of "proving": between the formalized ("mathematical") proving within a formal system and the contentual ("metamathematical") proving about the system. While the former is an arbitrarily defined logical game (that must be, however, to a large extent analogous with classical mathematics), the latter is a chaining of immediately evident contentual insights. This "contentual proving" has consequently to be carried out completely within the intuitionist logic of Brouwer and Weyl: proof theory is to rebuild classical mathematics, so-to-speak, on an intuitionist basis and in this way reduce strict intuitionism ad absurdum.[5]

The strategic goal of proof theoretic research, as interpreted by von Neumann, also shaped his talk at the Second Conference for Epistemology of the Exact Sciences. The conference was held in Königsberg from 5 to 7 September 1930, and on the first day of the congress von Neumann talked about Hilbert's finitist standpoint in a plenary session, where Carnap and Heyting presented the logicist, respectively, intuitionist position.[6] On the next day Gödel described the results of his dissertation.[7] The plenary

---

[5](von Neumann 1927, pp. 2–3). The German text is this: "Hierbei muß stets scharf zwischen zwei verschiedenen Arten des "Beweisens" unterschieden werden: Dem formalisierten ("mathematischen") Beweisen innerhalb des formalen Systems, und dem inhaltlichen ("metamathematischen") Beweisen über das System. Während das erstere ein willkürlich definiertes logisches Spiel ist (das freilich mit der klassischen Mathematik weitgehend analog sein muß), ist das letztere eine Verkettung unmittelbar evidenter inhaltlicher Einsichten. Dieses "inhaltliche Beweisen" muß also ganz im Sinne der Brouwer-Weylschen intuitionistischen Logik verlaufen: Die Beweistheorie soll sozusagen auf intuitionistischer Basis die klassische Mathematik aufbauen und den strikten Intuitionismus so ad absurdum führen."

[6]Gödel reviewed the published versions of these presentations in (Gödel 1932a, Gödel 1932b, Gödel 1932c). Waismann had also given a paper in the plenary session, entitled *Das Wesen der Mathematik: Der Standpunkt Wittgensteins*; his talk was not published.

From the letters between von Neumann and Carnap, quoted in (Mancosu 1999b), we know that their Königsberg talks were published (in the form they were) only to reflect the situation before Gödel's results. Von Neumann writes, in his letter to Carnap of 7 June 1931:

"Ich halte daher den Königsberger Stand der Grundlagendiskussion für überholt, da Gödels fundamentale Entdeckungen die Frage auf eine ganz veränderte Plattform gebracht haben. (Ich weiss, Gödel ist in der Wertung seiner Resultate viel vorsichtiger, aber m. E. übersieht er die Verhältnisse an diesem Punkt nicht richtig.)

Ich habe mit Reichenbach mehrfach besprochen, ob es unter diesen Umständen überhaupt Sinn hat, mein Referat zu publicieren—hätte ich es 4 Wochen später gehalten, so hätte es ja wesentlich anders gelautet. Wir kamen schliesslich überein, es als eine Beschreibung eines gewissen, wenn auch überholten Standes der Dinge doch niederzuschreiben."

In a note to the last sentence von Neumann adds: "Ich möchte betonen: *Nichts* an Hilberts Ansichten ist *falsch*. Wären sie durchführbar, so würde aus ihnen durchaus das von ihm Behauptete folgen. Aber sie sind eben undurchführbar, das weiss ich erst seit Sept. 1930."

[7]The draft of Gödel's talk is presumably (1930c). Dawson's (1985) and (1997), and also Mancosu's (1999b), describe the early reception of the incompleteness theorems.

session was complemented on 7 September by a roundtable discussion concerning the foundations of mathematics. That discussion was chaired by Hans Hahn, and its participants included Carnap, Heyting and von Neumann, but also three additional scholars, namely, Arnold Scholz, Kurt Reidemeister and Gödel. A shortened and edited transcript of this discussion was published as (Hahn, Carnap, Gödel, Heyting, Reidemeister, Scholz and von Neumann 1931) in *Erkenntnis*. Gödel was invited by the editors of the journal to expand on the very brief remarks about the first incompleteness theorem he had made during the discussion; the resulting note was added as a *Nachtrag* to the transcript (see (Gödel 1931b)). According to (Dawson 1997), Gödel had already discussed the new discovery with Carnap and Waismann in Vienna, before the conference on 26 August 1930:

The main topic of conversation was the plan for their upcoming journey to the conference in Königsberg, where Carnap and Waismann were to deliver major addresses and where Gödel was to present a summary of his dissertation results. But then, Carnap tersely noted, the discussion turned to "Gödel's discovery: incompleteness of the system of *Principia Mathematica*; difficulty of the consistency proof".[8]

This provides a sketch of the background for the meeting at which von Neumann made the acquaintance of Gödel. In his (1981), Wang reports (Gödel's view) about the encounter with von Neumann:

In September 1930, Gödel attended a meeting at Königsberg (reported in the second volume of *Erkenntnis*) and announced his result [i.e., the first incompleteness theorem]. R. Carnap, A. Heyting, and J. von Neumann were at the meeting. Von Neumann was very enthusiastic about the result and had a private discussion with Gödel. In this discussion, von Neumann asked whether number theoretical undecidable propositions could also be constructed, in view of the fact that the combinatorial objects can be mapped onto the integers, and expressed the belief that it could be done. In reply, Gödel said, "Of course undecidable propositions about integers could be so constructed, but they would contain concepts quite different from those occurring in number theory like addition and multiplication." Shortly afterward Gödel, to his own astonishment, succeeded in turning the undecidable proposition into a polynomial form preceded by quantifiers (over natural numbers). At the same time but independently of this result, Gödel also discovered his second theorem to the effect that no consistency proof of a reasonably rich system can be formalized in the system itself.[9]

As to the discovery of the second incompleteness theorem, we thus clearly know that Gödel did not have it in Königsberg and that, in contrast, the abstract (1930b) contains its classical formulation. It was Hahn who presented the abstract on 23 October 1930 to the Vienna Academy of Sciences.

---

[8](Dawson 1997, p. 68).

[9](1981, pp. 654–655). The introductory note to the correspondence with Wang, in (Gödel 2003b), describes in section 3.2 the interaction between Gödel and Wang on which this paper is based.

The full text of Gödel's (1931d) was received for publication by the editors of *Monatshefte* on 17 November 1930.

There is genuine disagreement between Gödel and von Neumann on how the second incompleteness theorem affects Hilbert's finitist program. Von Neumann states his view strongly in his letters to Gödel of 29 November 1930 and 12 January 1931. (As to other views, cf. the introductory note to the correspondence with Herbrand and the exchange with Bernays, in particular, the letters of 24 December 1930, 18 January 1931, 20 April 1931 and 3 May 1931, in (Gödel 2003a,b).) In his letter of 29 November to Gödel, von Neumann writes:

> I believe that every intuitionistic consideration can be formally copied, because the "arbitrarily nested" recursions of Bernays-Hilbert are equivalent to ordinary transfinite recursions up to appropriate ordinals of the second number class. This is a process that can be formally captured, unless there is an intuitionistically definable ordinal of the second number class that could not be defined formally—which is in my view unthinkable. Intuitionism clearly has no finite axiom system, but that does not prevent its being a part of classical mathematics that does have one.

From the general fact of the unprovability of a system's consistency within the system, he concludes: "There is no rigorous justification of classical mathematics." In the third letter, after having received the galleys of Gödel's (1931d), he writes even more forcefully:

> I absolutely disagree with your view on the formalizability of intuitionism. Certainly, for every formal system there is, as you proved, another formal one that is (already in arithmetic and the lower functional calculus) stronger. But intuitionism is not affected by that at all.

Denoting first-order number theory by A, analysis by M and set theory by Z, von Neumann continues:

> Clearly, I cannot prove that every intuitionistically correct construction of *arithmetic* is formalizable in A or M or even in Z—for intuitionism is undefined and undefinable. But is it not a fact, that not a single construction of the kind mentioned is known that cannot be formalized in A, and that no living logician is in the position of naming such [a construction]? Or am I wrong, and you know an effective intuitionistic arithmetic construction whose formalization in A creates difficulties? If that, to my utmost surprise, should be the case, then the formalization should work in M or Z!

We know of Gödel's response to von Neumann's dicta not through a letter from Gödel, but rather through the minutes of the meeting of the Schlick Circle that took place on 15 January 1931. These minutes report what Gödel viewed as questionable, namely, the claim that the totality of all intuitionistically correct proofs is contained in one formal system. That, he emphasized, is the weak spot in von Neumann's argumentation.[10] However, we also know that by December of 1933 Gödel had changed his

---

[10] The minutes are found in the Carnap Archives of the University of Pittsburgh. Part of the German text is quoted in (Sieg 1988, note 11), and more fully in (Mancosu 1999b,

view as follows: Finitism, considered by Gödel as the strictest form of constructive mathematics, is *narrower* than intuitionism and (its practice) can be captured in a formal system. Thus he argues, alluding to the second incompleteness theorem, the hope of succeeding along the lines proposed by Hilbert "has vanished entirely in view of some recently discovered facts". That change is made explicit in his talk (1933b) to the Mathematical Association of America.[11]

Before moving on to the further correspondence, some additional remarks on von Neumann's letter of 12 January 1931 are warranted: it contains metamathematical observations of special interest. The letter starts out by reporting:

Incidentally, the other day I developed a method that always allows a finite decision for the effective provability question concerning propositions that are built up solely by means of the concepts "not", "or" (thus also "and", "follows", etc.), [and] "provable" (starting from the identical truth—consistency is for example such a proposition).

This observation seemingly anticipates (and announces a solution to) a problem Harvey Friedman formulated in (1975) as the 35th of his one hundred and two problems. (Boolos 1976) provided the first published solution.[12] A few paragraphs later von Neumann remarks on (Gödel 1931d) and, in particular, on the proof of the second incompleteness theorem sketched there.

Your paper is very nice; I am quite delighted, how briefly and elegantly you carried out the difficult and lengthy "enumeration" of formulas. However, I believe that the proof of the unprovability of consistency can be shortened, i.e., that the general formal repetition of all considerations, as you propose, can be avoided.

Gödel had indeed suggested there a formal repetition of all considerations that lead to the unprovability in P of his sentence G, assuming P's consistency. He states on p. 197, "All notions defined (or statements proved) in Section 2, and in Section 4 up to this point are also expressible (and provable) in P." In the 1934 Princeton Notes (on p. 18) it is similarly asserted, "The fairly simple arguments of this proof can be paralleled in the formal logic . . ." Von Neumann's sketch of a simplified argument is obviously intended as a proof in P (though, through his appeal to Gödel's Theorem V, it is not quite). It has the flavor of the penetrating proof given by Hilbert and Bernays in (1939) via their axiomatic derivability conditions. I.e., von Neumann separates the general conditions on the provability predicate

---

pp. 36–37). Interestingly, (Bernays 1933) uses "von Neumann's conjecture" to infer that the incompleteness theorems impose fundamental limits on proof theoretic investigations.

[11]Cf. (1933b), in (Gödel 1995, pp. 51–52) and also the introductory note to the correspondence with Herbrand.

[12]Around the same time, Claudio Bernardi and Franco Montagna also found a solution; their paper was submitted to *The Journal of Symbolic Logic* shortly after Boolos's paper (thus, not accepted and indeed never published). According to Montagna in private communication, their proof was based on the algebraic semantics of provability logic due to Roberto Magari. "But apart from translating from Logic to Algebra, the proof was very similar to that of Boolos."

needed for the proof of the second incompleteness theorem from their verification concerning a particular formal system.

Von Neumann's admiration for Gödel's work is expressed directly in his very first letter of 20 November 1930, when he calls the first incompleteness theorem "the greatest logical discovery in a long time". That admiration is also reflected, for example, in his decision to talk about the incompleteness theorems when he lectured at Princeton in the fall of 1931. Kleene reports in his (1987) (on p. 491) that through this lecture "Church and the rest of us first learned of Gödel's results". Von Neumann's friend Ulam states in his *Adventures of a Mathematician*:

> When it came to other scientists, the person for whom he [von Neumann] had a deep admiration was Kurt Gödel. This was mingled with a feeling of disappointment at not having himself thought of "undecidability." For years Gödel was not a professor at Princeton . . . Johnny would say to me, "How can any of us be called professor when Gödel is not?"[13]

The letters from 13 July 1937 through 17 August 1939 are mainly focused on practical issues surrounding the publication of Gödel's work on the relative consistency of the axiom of choice and the generalized continuum hypothesis. After von Neumann finally had the opportunity to study Gödel's lectures thoroughly, he wrote on 22 April 1939:

> I would like to convey to you, most of all, my admiration: You solved this enormous problem with a truly masterful simplicity. And you reduced to a minimum the unavoidable technical complications of the proof details by a presentation of impressive persistence and drive. Reading your investigations was really a first-class aesthetic pleasure.

It is quite impressive that von Neumann studied Gödel's investigations in sufficient detail to make also some "critical remarks"; perhaps even more impressive is his earlier letter of 28 February 1939 in which he directs Gödel to the paper (1938) by Kondô that contains, in his view, "quite remarkable and surprising results on higher projective sets". He asks Gödel, "Are such matters not important for your further investigations on the continuum hypothesis . . . ?" Gödel responds in his letter of 20 March 1939 by saying with reference to his (1938): "The result of Kondô is of great interest to me and will definitely allow an important simplification in the consistency proof of 3. and 4. of the attached offprint." (For an explanation of the nature of these results, see Solovay's introductory note to *1938–1940*, (Gödel 1990, in particular pp. 14–15).)

During the following 17 years, it seems, von Neumann and Gödel did not exchange letters; after all, they were colleagues at the Institute for Advanced Study. In the spring of 1955, von Neumann took a leave from the Institute and moved from Princeton to Washington, DC, in order to work as a member of the Atomic Energy Commission to which he had

---

[13](Ulam 1976, p. 80).

been appointed by President Eisenhower. In the preface to von Neumann's posthumous (1958), his widow Klara reports that von Neumann was diagnosed with bone cancer in August 1955. His health deteriorated quickly. By January 1956 he was confined to a wheelchair, though he still attended meetings and worked in his office. There was also some hope that X-ray treatment might be helpful. Klara von Neumann writes that by March of 1956, however, ". . . all false hopes were gone, and there was no longer any question of Johnny being able to travel anywhere. . . . In early April Johnny was admitted to Walter Reed Hospital; he never left the hospital grounds again until his death on February 8, 1957." Gödel wrote his last letter to von Neumann on 20 March 1956. He had heard, so he states in this letter, that von Neumann had undergone a radical treatment and was feeling better. "I hope and wish", Gödel continues, "that your condition will soon improve even further and that the latest achievements of medicine may, if possible, effect a complete cure." Then he formulates a striking mathematical problem and asks for von Neumann's view on it. It concerns the feasibility of computations and is closely connected to the problem that has caught, independently, the attention of mathematicians and computer scientists, the P versus NP problem.[14] For Gödel it is the question "how significantly *in general* for finitist combinatorial problems the number of steps can be reduced when compared to pure trial and error". The context in which he locates the general issue is noteworthy. Consider the question, whether a formula F in the language of first-order logic has a proof of length n, i.e., n is the number of symbols occurring in the proof. A suitably programmed Turing machine can answer this question. If $\psi(F,n)$ is the number of steps an "optimal" machine must take to obtain the answer and $\varphi(n) = \max_F \psi(F,n)$, then the important question is how rapidly $\varphi(n)$ grows. Gödel remarks that it is possible to prove that $\varphi(n) > Kn$, for some constant K. If there were a machine such that $\varphi(n)$ would grow essentially like Kn (or even Kn^2) Gödel suggests, "that would have consequences of the greatest significance. Namely, this would clearly mean that the thinking of a mathematician in case of yes-and-no questions could be completely replaced by machines, in spite of the unsolvability of the Entscheidungsproblem." In the next-to-last paragraph Gödel mentions Friedberg's recent solution of Post's problem and returns then to an issue that had been underlying much of the foundational discussion of the 1920s: In what formal frame-

---

[14]This is the question, whether the class P of functions computable in polynomial time is the same as the class NP of functions computable non-deterministically in polynomial time. For a very good introduction to the rich and multifaceted problems that fall into the NP category, see (Garey and Johnson 1979).

Part of the letter was already published in (Hartmanis 1989); the full German letter and its English translation are found in the preface to (Clote and Krajíček 1993). In both papers Gödel's question is related in informative ways to contemporary work in computational complexity. All the mathematical issues raised in Gödel's letter are addressed and resolved in (Buss 1995). In particular, Buss shows that indeed $\varphi(n) \geq Kn$, for some constant K and infinitely many n, and that the n-symbol provability question raised by Gödel is NP-complete for predicate logic and, surprisingly, even for sentential logic.

work can one develop classical analysis? Gödel reports that Paul Lorenzen has built up the theory of Lebesgue measure within ramified type theory.[15] "But", Gödel cautions, "I believe that in important parts of analysis there are impredicative inference methods that cannot be eliminated."

In the face of human mortality, Gödel thus chose to raise and discuss eternal mathematical questions.

---

[15] Gödel refers presumably to (Lorenzen 1955).

II.3

In the shadow of incompleteness: Hilbert and Gentzen*

Abstract. Gödel's incompleteness theorems had a dramatic impact on Hilbert's foundational program. That is common lore. For some, e.g. von Neumann and Herbrand, they undermined the finitist consistency program; for others, e.g. Gödel and Bernays, they left room for a fruitful development of proof theory. This paper aims for a nuanced and deepened understanding of how Gödel's results effected a transformation of proof theory between 1930 and 1934. The starting-point of this transformative period is Gödel's announcement of a *restricted unprovability result* in September of 1930; its end-point is the completion of Gentzen's first consistency proof for elementary number theory in late 1934.

Hilbert, surprisingly, is the initial link between starting-point and end-point. He addressed Gödelian issues in two strikingly different papers, (1931a) and (1931b), without mentioning Gödel. In (1931a) he takes on the challenge of the restricted unprovability result; in (1931b) he responds to the second incompleteness theorem concerning the unprovability of consistency for a system S in S. He does so by bringing in semantic considerations and by pursuing novel, but also highly problematic directions: he argues for the contentual correctness of a constructive (for him, finitist) theory that includes intuitionist number theory.

Gentzen followed the new directions in late 1931, addressed methodological issues and metamathematical problems in ingenious ways, while building on ideas and techniques that had been introduced in proof theory. Most distinctive is Gentzen's struggle with contentual correctness and its relation to consistency. That is reflected in a sequence of notes Gentzen wrote between late-1932 and late-1934, but it is also central in his classical paper (1936). The immediate lessons seem to be: (i) there is real continuity between Hilbert's proof theory and Gentzen's work, and (ii) there is deepened concern for interpreting intuitionist arithmetic (and thus understanding classical arithmetic) from a more strictly constructive perspective. The latter concern, ironically, influenced deeply Gödel's functional interpretation of intuitionist arithmetic.**

---

*Dedicated to Per Martin-Löf.
**In Fall 2009, I wrote an *Introduction* to late Hilbert papers for (Hilbert 2012) and presented, on 6 November 2009, a version in a talk (with the same title as this essay) to the *Workshop on Logical Methods in the Humanities* at Stanford University. The present essay expands that paper.

1 A puzzle.

During the last quarter century or so, the early history of *modern mathematical logic* has been explored in detail and we have gained a much richer perspective of its evolution from 19th century roots. Some purely historical insights have been astounding: The beginning of the subject, associated for many with Hilbert and Ackermann's book from 1928, was pushed back by a whole decade to lectures Hilbert gave in the winter term of 1917/18. These lectures also reveal the impact of Whitehead and Russell's *Principia Mathematica* on the logical framework for the investigations in the Hilbert School. Other issues of central importance are coming into sharper historical and systematic focus; among them is a deepened understanding of how (versions of) Gödel's theorems affected proof theory between 1930 and 1934.

1934 witnessed, of course, the publication of the first volume of Hilbert and Bernays's *Grundlagen der Mathematik*. Less publicly, Gerhard Gentzen finished his first consistency proof for full first-order arithmetic. In late 1931, he had already set himself the goal of proving this result as the central theorem of his dissertation. On the long path to his proof, Gentzen made two remarkable discoveries before the end of 1932: (i) he established, independently of Gödel, the consistency of classical arithmetic relative to its intuitionist version, and (ii) he proved the normalization theorem for (a fragment of) *intuitionist* first-order logic and recognized the subformula property of normal derivations. The latter property was extended in his (1934) to *classical* logic formulated in the sequent calculus. That paper was literally based on his *actual* dissertation in which he established the consistency of arithmetic with quantifier-free induction. Two questions naturally arise: why did he temporarily give up on his goal, and how had he intended to reach it?

In a draft of his *Urdissertation* from October 1932, still pursuing the goal of a consistency proof for full arithmetic, Gentzen listed the results he had obtained as items (I) through (IV) and formulated the crucial remaining task as item (V):

The consistency of arithmetic will be proved; in the process, the concept of an infinite sequence of natural numbers will be used, furthermore in one place the principle of the excluded middle. The proof is thus not intuitionist. Perhaps the *tertium non datur* can be eliminated.

However, it could not have been written without Menzler-Trott's historical work, von Plato's discovery of Gentzen's *Urdissertation*, and Thiel's efforts to establish a Gentzen Nachlass and to transcribe early manuscripts. I am grateful to all three, but in particular to Thiel for sending me, in the middle of December 2009, the important manuscript INH. Von Plato was deeply involved in its transcription and pointed out to me, why he considers it as very significant. INH and other manuscripts will be made available, I hope very soon, in a full edition of Gentzen's Nachlass.

Many thanks are also due to Sabine Friedrich, Ulrich Majer, Wilfried Nippel, Winfried Schultze, and Marion Sommer for exploring archival questions in Berlin, Göttingen, and Hamburg. For pertinent remarks, suggestions and additional information I thank Wilfried Buchholz, Martin Davis, John Dawson, Heinz-Dieter Ebbinghaus, Eckart Menzler-Trott, Grigori Mints, William Tait, and Christian Thiel.

In the shadow of incompleteness 157

How is this to be understood? A student in the Hilbert School plans to use the concept of an infinite sequence of natural numbers in a consistency proof, when only finite mathematical objects are to be appealed to in such a proof? Even more surprising, he thinks of using the principle of tertium non datur *in* a consistency proof, when that principle is viewed as distinctive for classical mathematical practice and has to be secured *through* a consistency proof? — This startling puzzle will be resolved in the end, but not without starting at the beginning.

2 Results, methods, and problems.

Finitist proof theory originated in lectures Hilbert and Bernays gave in February of 1922. Its gradual emergence can be documented with reference to notes for lectures on the principles of mathematics Hilbert presented between 1917–18 and 1923–24.[1] In the spring of 1922, a finitist consistency proof was obtained for a quantifier-free fragment of arithmetic and a year later for primitive recursive arithmetic. The proofs used quite novel means and involved, in particular, transformations of formal derivations.[2] The ultimate goal was to turn proofs of *numeric*[3] statements into proofs containing only numeric statements that can then be evaluated as correct. This approach was programmatically extended to theories involving quantifiers via Hilbert's 1923-*Ansatz*, the ϵ-substitution method. By 1925, Ackermann, expanding Hilbert's method, and von Neumann had obtained stronger results, the extent of which was not perfectly clear. Indeed, in his address *Über das Unendliche* presented in Münster on 4 June 1925, Hilbert was rather vague about the theories that had been proved to be consistent.[4]

No official lecture notes from the decade's second half illuminate this situation.[5] There are, however, four papers of Hilbert's that give insight into the developments in Göttingen. They fall into two distinct groups: his talks in Hamburg (July 1927) and Bologna (September 1928) constitute the first group, whereas the papers he presented in Hamburg (December 1930)

---

[1]See my paper *Hilbert's proof theory* [In this volume: chapters I.3 and II.4.]. All the Hilbert lectures I am referring to are contained in (Hilbert 2012).

[2]The first step in this and later consistency proofs is the transformation of linear derivations into tree-like structures. That is achieved through the "Auflösung in Beweisfäden". The structure of the argument is discussed not only in these lecture notes, but also in (Hilbert 1923, p. 1142) and (Ackermann 1924, section II); it is beautifully presented in (Hilbert and Bernays 1934, pp. 221–228).

[3]A formula is called *numeric* if it contains neither bound nor free variables.

[4]Ackermann's paper was submitted for publication on 30 March 1924 and von Neumann's on 29 July 1925.

[5]Hilbert gave lecture courses on *Grundlagen der Mathematik* in the winter term 1927/28 and on *Mengenlehre* in the summer term of 1929. There are no notes for the 1927/28 lectures, but for the set theory course Menzler-Trott describes in his (Menzler-Trott 2007, Note 8, p. 22) detailed notes that were taken by Lothar Collatz; see (Hilbert *1929). Apart from their mathematical value, these notes are of special interest in the context of this paper, as Gentzen attended these lectures with his friend Collatz; cf. section 5.

and Göttingen (July 1931) belong to the second group. The reason for this grouping will become apparent very soon.

In his Hamburg talk of July 1927, Hilbert describes the status of proof theoretic work as he had done in *Über das Unendliche* (p. 179), but also discusses the "considerable progress in the proof of consistency" made by Ackermann. That remark does *not* refer to (Ackermann 1924), but rather to work Ackermann had done in early 1925 and had communicated to Bernays in a letter of 25 June 1925. Almost a year later, Ackermann tells Bernays in a letter of 31 March 1926 that he has turned his attention to the "ε_f-proof", i.e., the consistency proof for analysis, and that he is trying to finish it with all his might. Hilbert begins his progress report by recalling the idea of his 1923-Ansatz.

In proving consistency for the ε-function the point is to show that from a given proof of $0 \neq 0$ the ε-function can be eliminated, in the sense that the arrays formed by it can be replaced by numerals in such a way that the formulae resulting from the logical axiom of choice by substitution, the "critical formulae", go over into "true" formulae by virtue of those replacements.[6]

The central idea is to transform, as in the earlier proofs, linear derivations into tree-like ones consisting only of numeric formulae, now also involving ε-terms, all of which can be recognized to be true. Bernays gives details in his *Zusatz zu Hilberts Vortrag*. Both Hilbert and Bernays assert unambiguously that Ackermann's considerations establish the consistency of elementary arithmetic.[7] Hilbert, optimistic as ever, believed that the methods of Ackermann could be extended further:

For the foundations of ordinary analysis his [Ackermann's, WS] approach has been developed so far that only the task of carrying out a purely mathematical proof of finiteness remains. (Hilbert 1927, p. 479)

The logical calculus underlying the proof theoretic investigations is described in great detail in this paper, but actually goes back to 1922, is used in Ackermann's (1924), is indicated in *Über das Unendliche*, and is investigated most carefully in (Hilbert and Bernays 1934, p. 66 ff). The "axioms for implication" allow the introduction or omission of an assumption, the interchange of assumptions, and the outright elimination of a

---

[6](Hilbert 1927, p. 477). Hilbert continues: Diese Ersetzungen werden nach erfolgter Elimination der freien Variablen durch schrittweises Probieren gefunden, und es muß gezeigt werden, daß dieser Prozeß jedenfalls zu einem Abschluß führt. — *This* is the fact that has to be established by "a purely mathematical finiteness proof". — A careful discussion is given in (Avigad and Zach 2007), and a beautiful contemporary presentation of the method is found in (Tait 2010).

[7]Bernays reemphasizes this point in the later discussion surrounding the second incompleteness theorem, when writing on 20 April 1931 to Gödel: "That proof [of Ackermann] — to which Hilbert referred in his lecture on *The Foundations of Mathematics* [i.e., in (Hilbert 1927)] with the addendum appended by me — I have repeatedly considered and viewed as correct." — Ackermann never published his second proof; it was only presented in the second volume of *Grundlagen der Mathematik* in section 2, see (Hilbert and Bernays 1939, pp. 121–130, and note 1 on p. 121).

In the shadow of incompleteness

proposition. The axioms for conjunction and disjunction receive a very special formulation:

$$(A \& B) \to A \text{ and } (A \& B) \to B$$
$$A \to (B \to (A \& B))$$

and

$$A \to (A \vee B) \text{ and } B \to (A \vee B)$$
$$((A \to C) \& (B \to C)) \to ((A \vee B) \to C).$$

Two axioms for negation are formulated in a third group:

$$((A \to (B \& \neg B)) \to \neg A) \text{ and } (\neg\neg A \to A).$$

The first axiom for negation is called the *principle of contradiction* and the second the *principle of double negation*. To round out the description of the calculus, the *transfinite ε-axiom* is stated as $A(a) \to A(\varepsilon(A))$. The ε-axiom allows the definition of universal and existential quantifiers as well as the proof of the appropriate principles for them. This formulation goes back to 1923 and is the basis for Hilbert's ε-substitution method, as indicated above. However, we are already in 1927, when that method had been extended by Ackermann and had been used, presumably, to prove the consistency of elementary arithmetic.

More than a year later, Hilbert gave a talk at the International Congress of Mathematicians in Bologna. His report on the status of proof theoretic research is essentially unchanged: Ackermann and, Hilbert adds on this occasion, von Neumann have secured the consistency of elementary number theory. He insists again that Ackermann has carried out the consistency proof for analysis with just one remaining task, namely, that of proving "a purely arithmetic elementary finiteness theorem".[8] Hilbert gives a wonderfully clear presentation of broad methodological issues and important metamathematical problems. I will focus on the problem of syntactic completeness for elementary number theory and analysis, as that will play a central role in the further developments.

The issue is formulated as Problem IV and, in the different form of Post-completeness, as Problem V. (The republications of the Bologna address list these as problems III and IV, respectively.) Hilbert asserts that completeness of the theories for arithmetic and analysis is generally claimed and thinks, I assume, that such claims are founded on their categoricity. The argument would proceed as follows: as all their models are isomorphic, a statement $\mathfrak{S}$ is either true in all models or false in all of them; thus, $\mathfrak{S}$ or $\neg\mathfrak{S}$ is a logical consequence of the axioms. If logical consequence were

---

[8](Bernays 1930b, p. 58) gives the same description of the status of proof theory: Durch die von Ackermann und v. Neumann geführten Beweise ist die Widerspruchsfreiheit für das erste Postulat der Arithmetik, d.h. die Anwendbarkeit des existentialen Schließens auf die ganzen Zahlen sichergestellt. Für das weitere Problem der Widerspruchsfreiheit des Allgemeinbegriffs der Zahlenmenge (bzw. der Zahlenfunktion) einschließlich des zugehörigen Auswahlprinzips liegt ein weitergeführter Ansatz von Ackermann vor.

captured by derivability in this second-order framework, then the claim would follow immediately.[9]

Hilbert continues, "The usual idea for showing that any two interpretations of number theory, respectively of analysis, must be isomorphic does not meet the demands of finitist rigor." He suggests, as a next step, transforming the standard categoricity proof for number theory into a finitist argument that would establish the following assertion:

> If for some statement $\mathfrak{S}$ the consistency with the axioms of number theory can be established, then it is impossible to also prove for $\neg \mathfrak{S}$ . . . the consistency with those axioms, and most directly connected with this: If a statement is consistent, then it is also provable.[10]

This problem is taken up as the central issue in Hilbert's third Hamburg talk that was presented to the local Philosophical Society in December 1930. In this talk, Hilbert describes first the philosophical and mathematical background for proof theory and again formulates the central goal of his foundational work:

> Indeed, I would like to eliminate once and for all the questions concerning the foundations of mathematics as such — by turning every mathematical statement into a formula that can be concretely exhibited and strictly derived, thus recasting mathematical concept formations and inferences in such a way that they are irrefutable and nevertheless provide an adequate image of the whole science. (Hilbert 1931a, p. 489)

Hilbert sketches the formal system for elementary number theory and reasserts emphatically that Ackermann and von Neumann have proved its consistency. Thus, they have validated as admissible all transfinite inferences, in particular, the principle of tertium non datur. Referring back to his Bologna talk, he then formulates as "our most important further task" to find the proof of two theorems:

1. If a statement can be shown to be consistent, then it is also provable; and furthermore,
2. If for some statement $\mathfrak{S}$ the consistency with the axioms of number theory can be established, then it is impossible to also prove for $\neg \mathfrak{S}$ the consistency with those axioms. (l.c., p. 491)

Hilbert asserts that he has succeeded in proving these theorems for "certain simple cases".

This success has been made possible by extending elementary arithmetic with, what Hilbert calls, a *new inference rule*. Hilbert's Rule (*HR*)

---

[9] This idea for a syntactic completeness "argument" is made explicit at the end of Gödel's Königsberg talk (Gödel 1930c, pp. 26–29). It is connected with the unprovability result he had "recently" proved (to be discussed below in section 3). Gödel obtained as a consequence of categoricity and syntactic incompleteness of PM the (semantic) incompleteness of calculi for higher-order logics. — Connections to similar considerations in the Introduction to Gödel's thesis (Gödel 1929, pp. 60–64) are detailed in (Kennedy 2010).

[10] (Hilbert *1929, p. 6). I indicate negations by prefixing with $\neg$ instead of by "overlining" as Hilbert does.

is viewed as finitist and allows the introduction of universally quantified formulae $(x)\mathfrak{A}(x)$ as initial ones, just in case the numeric instances $\mathfrak{A}(\mathfrak{z})$ have been established finitistically as correct for arbitrary numerals $\mathfrak{z}$. (*HR*) is not a standard inference rule that facilitates a step from one or more premises to a conclusion within a formal theory. It rather introduces *universal claims as axioms* when an appropriate finitist justification has been given for all their instances. For the theory thus extended, Hilbert proves claims 1 and 2, but only when the statements involved are purely universal. For purely existential formulae he proves claim 2 and warns that claim 1 is not a consequence for them. Thus, he has shown that the extended theory is indeed complete for "certain simple cases". His proof is presented as a direct extension of Ackermann's inductive argument treating only the additional case of (*HR*).

As it happened, Gödel reviewed Hilbert's paper for the *Zentralblatt* and wrote the brief, careful, and matter-of-fact report (Gödel 1931a). The talk provides, according to Gödel, "a substantial supplement to the formal steps taken thus far toward laying a foundation for number theory". This "substantial supplement" is obtained by extending the formal system through the "following rule of inference, which, structurally, is of an entirely new kind". Gödel describes (*HR*) as I did in the previous paragraph and formulates the consistency as well as the partial completeness results without further comment. In a letter to Heyting of 15 November 1932 he points out that Hilbert resolved the completeness problem (in spite of the new axioms) for only "a small sub-question" and that the extended system still has undecidable statements.[11] It should be noted that Gödel did not have any qualms about (*HR*): following Herbrand's (1931), he used it in the formulation of elementary arithmetic, when proving the consistency of its classical version relative to its intuitionist one in (Gödel 1933d).

In his letter to Gödel of 18 January 1931, Bernays remarks that the formalism Hilbert had used in his Hamburg talk of the preceding December introduces the principle of complete induction in two different ways, namely, (i) through (*HR*) for the quantifier-free statements of finitist arithmetic, and (ii) through the induction axioms of the theory of elementary arithmetic. He proposes a unified formulation through an "infinitary" rule that is not tied to finitist argumentation as (*HR*) is: "If $\mathfrak{A}(x_1, \ldots, x_n)$

---

[11] See (Gödel 2003b, p. 60). The detailed argument for the latter claim can be found in Gödel's letter of 2 April 1931; it was given in response to Bernays's letter of 18 January 1931 (ibid., p. 86ff). Bernays discussed (*HR*) and introduced an extended version, which is formulated in the next paragraph. Gödel's analysis applies also to that extension. — A clarifying discussion of (*HR*) and the ω-rule is found in (Feferman 1986) and (Feferman 2003), but also in (Tait 2002, pp. 417–418).

The translation is slightly modified from that in the *Collected Works*; here is the German formulation of the infinitary Bernays Rule: Ist $\mathfrak{A}(x_1, \ldots, x_n)$ eine (nicht notwendig rekursive) Formel, in welcher als freie Individuenvariablen nur $x_1, \ldots, x_n$ auftreten und welche bei der Einsetzung von irgendwelchen Zahlwerten anstelle von $x_1, \ldots, x_n$ in eine solche Formel übergeht, die aus den formalen Axiomen und den bereits abgeleiteten Formeln durch die logischen Regeln ableitbar ist, so darf die Formel $(x_1) \ldots (x_n)\mathfrak{A}(x_1, \ldots, x_n)$ zum Bereich der abgeleiteten Formeln hinzugenommen werden.

is a (not necessarily recursive) formula in which only $x_1, \ldots, x_n$ occur as free variables and which is transformed, through the substitution of any numerical values whatsoever in place of $x_1, \ldots, x_n$, into a formula that is derivable by the logical rules from the formal axioms and the formulae already derived, then the formula $(x_1) \ldots (x_n)\mathfrak{A}(x_1, \ldots, x_n)$ may be adjoined to the domain of derived formulae." (See Note 11 also for the German text.) In his last paper Hilbert will use a similarly expanded rule, but in a fully constructive context and tied to finitist argumentation; that significantly different move of Hilbert's is discussed in section 5.

The central question has been why Hilbert took on the issue of syntactic completeness with such prominence. — One answer may be that he was simply motivated to tackle a significant open problem. After all, in his Bologna talk he considered the issue as important and as difficult, and so did Bernays in his (1930). However, there is one fact that speaks against this understanding: both Hilbert and Bernays conjectured elementary number theory to be complete.[12] One is obviously tempted to ask, what was the reason for Hilbert not only to suspect, but to take for granted *now*, in late 1930, that there are elementary, formally undecidable sentences and, in addition, to expand elementary number theory in order to overcome that incompleteness at least partially — by an inference rule that is in Gödel's words "of an entirely new kind"?[13]

A key to answering the question may be the second assertion in the list of objections to proof theory Hilbert discusses in the last third of his talk. Hilbert views the objections naturally as unjustified. Here is his formulation of the second objection:

It has been said, in criticism of my theory, that the statements are indeed consistent, but that they are not thereby proved. But certainly they are provable, as I have shown here in simple cases. (Hilbert 1931a, p. 492)

Is the first sentence not an allusion to the first incompleteness theorem? But how could Hilbert have known about it? Did he know about the second theorem and von Neumann's related conjecture that the consistency of classical mathematics is unprovable? If he did, how could he take for granted without any hesitation that the consistency of elementary number theory had been established?

---

[12]Bernays remarks in his (1930b, p. 59): Von der Zahlentheorie, wie sie durch die Peanoschen Axiome, mit Hinzunahme der rekursiven Definition, abgegrenzt wird, glauben wir, daß sie in diesem Sinne deduktiv abgeschlossen ist [i.e., is syntactically complete]; die Aufgabe eines wirklichen Nachweises hierfür ist aber noch völlig ungelöst. Noch schwieriger wird die Frage, wenn wir, über den Bereich der Zahlentheorie hinaus, zu der Analysis und den weiteren mengentheoretischen Begriffsbildungen aufsteigen.

[13]Feferman makes exactly this point and conjectures in his Introductory Note to Gödel's correspondence with Bernays: "Since Hilbert had previously conjectured the completeness of Z, he would have had to have a reason to propose such an extension, and the only obvious one is the incompleteness of Z." (Feferman 2003, Note 1 on p. 44) — Z is of course elementary number theory.

3 Unprovability in general, first.

In order to make explicit the assumptions in these historical questions and to approach answers to them, let me examine, to begin with, what Gödel announced at the roundtable discussion in Königsberg on 7 September 1930. (This discussion was part of the Conference on Epistemology of the Exact Sciences. Hans Hahn was its moderator, and Carnap, Gödel, Heyting and von Neumann were among the participants. For details, see (Dawson 1997, pp. 68–71).) The stenographic transcript of Gödel's remarks was published in *Erkenntnis* together with a later *Postscript* in which he summarized, at the request of the journal's editors, the results of his classical paper (1931). In the transcript, the following theorem is stated:

(Assuming the consistency of classical mathematics) one can give examples of statements (and in fact statements of the type of Goldbach's or Fermat's) that are in fact contentually true, but are unprovable in the formal system of classical mathematics. Therefore, if one adjoins the negation of such a proposition to the axioms of classical mathematics, one obtains a consistent system in which a contentually false proposition is provable.[14]

Thus, these transcribed Königsberg remarks do not yet formulate the syntactic incompleteness of elementary number theory and neither do those at the end of the presentation of his thesis work on the completeness of first-order logic, which Gödel had delivered a day earlier, on 6 September; cf. Note 14.

Von Neumann sat at the Königsberg roundtable and talked with Gödel immediately after the session. Gödel's recollection of this conversation and his perspective on subsequent developments are reported in (Wang 1981):

Von Neumann was very enthusiastic about the result and had a private discussion with Gödel. In this discussion, von Neumann asked whether number-theoretical undecidable propositions could also be constructed in view of the fact that the combinatorial objects can be mapped onto the integers and expressed the belief that it could be done.

Von Neumann's question and conjecture point to an intriguing fact that is clearly formulated in Wang's paper: at the time of the Königsberg meeting, syntax had not been arithmetized. Rather, symbols were directly

---

[14](Gödel 1931b, p. 203 in (Gödel 1986)). I modified the translation; the German text is: Man kann (unter Voraussetzung der Widerspruchsfreiheit der klassischen Mathematik) sogar Beispiele für Sätze (und zwar solche von der Art des Goldbachschen oder Fermatschen) angeben, die zwar inhaltlich richtig, aber im formalen System der klassischen Mathematik unbeweisbar sind. Fügt man daher die Negation eines solchen Satzes zu den Axiomen der klassischen Mathematik hinzu, so erhält man ein widerspruchsfreies System, in dem ein inhaltlich falscher Satz beweisbar ist. — The background is described in Dawson's *Introductory Note*, ibid., pp. 196–199, and in (Dawson 1997, pp. 68–79). — (Gödel 1995) contains the report on his thesis work he gave to the Königsberg Congress (cf. note 9). He remarks after discussing the connection of categoricity and *Entscheidungsdefinitheit* that his result can be stated as follows: Das Peanosche Axiomensystem mit der Logik der *Principia Mathematica* als Überbau ist nicht entscheidungsdefinit.

represented in the formal theory by numerals, sentences by sequences of numerals, and proofs by sequences of sequences of numerals. The crucial syntactic notions and the substitution function are expressible in subsystems of type or set theory and, consequently, so is the undecidable statement. Gödel responded to von Neumann's query by saying: "Of course undecidable propositions about integers could be so constructed, but they would contain concepts quite different from those occurring in number theory like addition and multiplication." Remarks about the subsequent developments follow:

> Shortly afterward Gödel, to his own astonishment, succeeded in turning the undecidable proposition into a polynomial form preceded by quantifiers (over natural numbers). At the same time, but independently of this result, Gödel also discovered his second theorem to the effect that no consistency proof of a reasonably rich system can be formalized in the system itself.[15]

So it is clear that Gödel announced only *one* result at the Königsberg meeting; it was, as quoted above from (Gödel 1931b), a quite restricted unprovability result for "classical mathematics" as formulated in type or set theory. He had not yet obtained his second incompleteness theorem.

Now I come back to the question, what Hilbert may have known about Gödel's result(s) before giving his talk in Hamburg or before submitting his paper to *Mathematische Annalen* on 21 December 1930. As von Neumann is often mentioned as a possible conduit to Hilbert, let me recall two facts from von Neumann's correspondence with Gödel: first, von Neumann learned about the formulation of both incompleteness theorems around 25 November 1930 and, second, he got to know the details of Gödel's arguments only at the very beginning of 1931. It is equally crucial to realize that Hilbert delivered his lecture *Naturerkennen und Logik* in Königsberg on 8 September 1930, the very day after the roundtable discussion. His lecture was an invited address at the meeting of the Society of German Scientists and Physicians that took place from 7 to 11 September. We also know that Hilbert and von Neumann's stays in Königsberg overlapped.[16] Is it then

---

[15] (Wang 1981, pp. 654–5). Parsons's *Introductory Note* to the correspondence with Wang in (Gödel 1986) describes in section 3.2 the interaction between Gödel and Wang on which Wang's paper was based. — There are some seeming oddities with this and related other accounts: (i) Gödel parenthetically seems to claim in the published transcript (1931b) that the undecidable statement is of a restricted *number theoretic* form; that would be in striking conflict with his own account in (Wang 1981). However, it is only claimed that the statement is universal with a finitist matrix. Indeed, in the *Nachtrag* Gödel specifies, fully in accord with the report in (Wang 1981), the purely arithmetic character of the undecidable sentence. (ii) Carnap reports from an August 1930 meeting with Gödel that the latter had pointed out undecidability, but also "difficulty with consistency". That has been taken as an indication that Gödel had a version of his second theorem already then. Such an assumption is in direct conflict with the Gödel-Wang account described above; it receives a convincing explanation through Gödel's description in (l.c., p. 654), how he found the result, namely, when running into difficulties in his attempt of proving the consistency of analysis.

[16] Oystein Ore attended the meeting of the Society of German Scientists and Physicians and reports in (Reid 1970, p. 195), "I remember that there was a feeling of excitement and interest both in Hilbert's lecture and in the lecture of von Neumann on the foundations of set

In the shadow of incompleteness 165

implausible to think that von Neumann (or other colleagues who attended the roundtable discussion like Emmy Noether) talked with Hilbert about Gödel's result as formulated above? — I think not. In any event, it seems implausible that he was *not* informed then or later about a result that elicited von Neumann's deep and immediate response. If one assumes Hilbert knew just the result Gödel had announced in Königsberg, then the metamathematical considerations in his (1931a) take on the completeness issue for statements of the form of Gödel's unprovable sentence.[17]

What can we consider as the immediate effect of Gödel's Königsberg result on Hilbert's finitist consistency program? As that result does not concern consistency, there is no real direct effect — unless syntactic completeness is taken as a crucial ingredient of the program. Hilbert had formulated, as described above, the syntactic completeness question for arithmetic as well as analysis in his Bologna lecture and conjectured a positive answer. However, there was also a speculation at the Bologna Congress that some formal systems might be *incomplete*. To support this assertion I point to two sources.

1. In the first republication of the Bologna lecture, submitted to *Mathematische Annalen* on 25 March 1929, a remark was added on page 6:

In higher domains we might conceivably have a situation in which both $\mathfrak{S}$ and $\neg\mathfrak{S}$ are consistent: then the adoption as an axiom of one of the two statements $\mathfrak{S}$, $\neg\mathfrak{S}$ is to be justified by systematic advantages (principle of the permanence of laws, the possibility of further development, etc.).

It is not clear which "higher domains" are envisioned or why this remark was added. It would be of interest to have a sense, whether Skolem's results were taken into account. Skolem suggested in his (Skolem 1922, p. 299, note 9) the continuum problem to be undecided by the first-order formulation of Zermelo's axioms and gave as a reason the non-categoricity of those axioms.

---

theory — a feeling that one now finally was coming to grips with both the axiomatic foundation of mathematics and with the reasons for the applications of mathematics in the natural sciences." Dawson reports that Gödel left Königsberg only on 9 September, and he speculates that "it is very likely that he [Gödel] was in the audience" when Hilbert presented his lecture (l.c., p. 71).

[17] That seems to be in conflict with (Bernays 1935a, p. 215): "Noch ehe dieses Gödelsche Resultat bekannt war, hatte Hilbert die ursprüngliche Form seines Vollständigkeitsproblems bereits aufgegeben. In seinem Vortrag *Die Grundlegung der elementaren Zahlenlehre* [i.e., (Hilbert 1931a)] behandelte er dieses Problem für den Spezialfall von Formeln der Gestalt $(x)\mathfrak{A}(x)$, welche außer x keine gebundene Variable enthalten." The question is of course not, why did Hilbert give up on the *form of that problem* (which he did not), but rather, why did he give up on *the completeness conjecture*? — In Bernays's correspondence with Gödel (Gödel 2003a, p. 84) there is a peculiar and uncharacteristic lack of familiarity with Hilbert's Hamburg talk. Bernays attributes the insight — "Die Widerspruchsfreiheit der neuen Regel folgt aus der Methode des Ackermannschen (oder auch des v. Neumannschen) Nachweises für die Widerspruchsfreiheit von 3." — to an observation of A. Schmidt, when Hilbert's consistency proof for the new rule is explicitly an extension of Ackermann's proof (as discussed above in section 2)!

2. In a letter of 9 April 1947 to Heinrich Scholz, Bernays wrote:

The possibility of the underivability of the components of a derivable disjunction of the form

(**A**) $(X)(\varphi(X) \to \gamma(X)) \vee (X)(\varphi(X) \to \neg\gamma(X))$

was moreover already contemplated a considerable time before the appearance of the Gödel theorem. I discussed such matters at the time with Tarski at the Congress in Bologna.[18]

The proof of (**A**) is presumably based on the categoricity considerations motivating the "usual claim" of syntactic completeness for arithmetic and analysis. That claim was discussed by Hilbert in Bologna and was described above in the middle of section 2 as well as in note 9.

Finally, let me add a third perspective. Hilbert frequently emphasized a concept of "quasi-empirical" completeness: an axiomatic theory was to make possible, in an intelligibly structured way, proofs of all the elements of a given collection of mathematical facts. That is articulated in many places, but most directly in the set theory lectures from 1917. Hilbert suggests there that a collection of more or less secure facts should be shaped into a system following this general approach:

If we have certain statements in front of us and we are unable to assert anything certain about their correctness, we select some that seem to play a distinguished role as a preliminary axiom system — be it that we choose the simplest among them, be it that we prefer those that seem to have the most secure foundation or to be the most intuitive.[19]

On the next page of these notes Hilbert raises the completeness question in the form, "Is it really the case that all the facts of the collection are logical consequences of the selected particular statements, which as axioms are the basis for the system?" — When answering the question (on p. 48) for the axioms of complete ordered fields, Hilbert tentatively "shows" their syntactic completeness by an argument that reflects the one I sketched in section 2 and that exploits the categoricity of the axiom system.

As a consequence of these considerations, I look at matters in the following way. However surprising Gödel's Königsberg announcement was, it was not a complete shock to everyone, as syntactic completeness was not viewed as a *sine qua non* for proof theory or the axiomatic presentation of parts of mathematics. For von Neumann the unprovability result must have been nevertheless striking, and the reason seems to be straightforward. He had

---

[18]*Bernays Nachlass* at the ETH Zürich, Hs 975: 4123. See also (Mancosu 1999a, p. 33). Like Mancosu I thank Bernd Buldt for pointing me to this letter.

[19](Hilbert *1917, p. 40). The German text is: Haben wir gewisse Sätze vor uns, über deren Richtigkeit wir nichts Sicheres aussagen können, so greifen wir einige, die uns eine ausgezeichnete Rolle zu spielen scheinen, als ein vorläufiges Axiomsystem heraus — sei es dass wir die Einfachsten unter ihnen wählen, sei es dass wir diejenigen bevorzugen, die uns am sichersten fundiert oder auch am anschaulichsten erscheinen.

The German text of the following sentence is: Ist wirklich das gesamte vorliegende Tatsachenmaterial die logische Folgerung aus den herausgegriffenen besonderen Sätzen, die wir als Axiome dem System zu Grunde gelegt haben?

formulated the central tasks of proof theory in his talk on Hilbert's Program given just a few days earlier; that talk was published as his (1931). Von Neumann emphasized the quasi-empirical completeness requirement with an interestingly stronger condition, articulated as follows:

A construction procedure has to be given that allows producing successively all formulae, which correspond to the "provable" claims of classical mathematics. This procedure may thus be called "proving".[20]

This requirement for the proof theoretic program, von Neumann claims, has been secured by the work of Russell and his school and guarantees in particular that every correct finitist statement can be obtained by means of this construction procedure. It does not amount to a decision procedure for particular statements, as von Neumann points out (1931, p. 120), and it is not to be equated with syntactic completeness, as I want to emphasize.[21] Von Neumann's fundamental conviction that finitist mathematics can be formally captured was not undermined by Gödel's unprovability result; rather, that result pointed to limitations of finitist mathematics and led von Neumann quickly to a dramatic conclusion that is discussed in the next section. There I will also explore what we know about (i) when members of the wider Hilbert circle learned about Gödel's second incompleteness theorem, and (ii) when this theorem's full impact was realized. I will discuss von Neumann, Bernays, and Herbrand.

4 Unprovability of consistency, second.

Von Neumann was captivated at once by Gödel's result and, a couple of weeks later, made the remarkable discovery that, in his own words, "the consistency of mathematics is unprovable". I.e., he had arrived independently at a proof of Gödel's second incompleteness theorem. Taking for granted the co-extensionality of finitist and intuitionist mathematics, he argued for his discovery in two steps. Here is the first step, where 𝔚 stands for the formula expressing the consistency of the formal system under consideration: "If the consistency [of the system] is established intuitionistically, then it is possible, through a 'translation' of the contentual intuitionistic considerations into the formal [system], to prove 𝔚 also [in that system]." The possibility of doubting the translatability of finitist arguments because of Gödel's result is considered and rejected; von Neumann

---

[20](von Neumann 1931, p. 118). The German text is: Es ist ein Konstruktionsverfahren anzugeben, das sukzessiv alle Formeln herzustellen gestattet, welche den "beweisbaren" Behauptungen der klassischen Mathematik entsprechen. Dieses Verfahren heiße darum "Beweisen".

[21]The insight that syntactic completeness and decidability amounted to the same thing had to wait for the mathematical characterization of "formal" theories and decidability. Bernays discusses in his (1930b) syntactic completeness and adds on p. 59 in note 19 the remark: Man beachte, daß die Forderung der deduktiven Abgeschlossenheit [i.e., syntactic completeness, WS] noch nicht so weit geht wie die Forderung der *Entscheidbarkeit* einer jeden Frage der Theorie, welche besagt, daß es ein Verfahren geben soll, um von jedem beliebig vorgelegten Paar zweier der Theorie angehöriger, einander kontradiktorisch entgegengesetzter Behauptungen zu entscheiden, welche von beiden beweisbar ("richtig") ist.

believes that "in the present case it [the translatability] must obtain". In the second step he argues, "𝔚 is always unprovable in consistent systems, i.e., a putative effective proof of 𝔚 could certainly be transformed into a [proof of a] contradiction." Thus, the consistency of mathematics is unprovable by intuitionist means. Von Neumann conveyed these considerations to Gödel in a letter of 20 November 1930 and closed by calling Gödel's unprovability result as formulated in Königsberg "the greatest logical discovery in a long time". Gödel responded almost immediately and informed von Neumann of his *new* results and most likely sent him a copy of the abstract (1930b), which Hahn had presented to the Vienna Academy of Sciences on 23 October 1930. This abstract contains the classical formulation of both incompleteness theorems. The full text of Gödel's 1931-paper was submitted to *Monatshefte* on 17 November 1930, before von Neumann had formulated his letter to Gödel.

In his next letter of 29 November, von Neumann assured Gödel that he would not publish on the subject "as you have established the theorem on the unprovability of consistency as a natural continuation and deepening of your earlier results".[22] However, a disagreement emerged between him and Gödel on how this theorem affects Hilbert's finitist program. Its roots go back, on the one hand, to von Neumann's conviction of the "translatability" of finitist proofs and, on the other hand, to Gödel's view that there might be finitist proofs that are not obtainable in a particular formal theory. The dispute started after von Neumann had received the proof sheets of (Gödel 1931), at the end of which Gödel had expressed that view. After having thanked Gödel for the galleys, von Neumann writes on 12 January 1931, "I absolutely disagree with your view on the formalizability of intuitionism." The fact that for each formal system there is a proof theoretically stronger one, as Gödel's first result shows, does not "touch intuitionism" in von Neumann's view. He defends that belief again, as he had done in his first letter to Gödel, by claiming intuitionist or finitist arguments can be "translated" into formal ones in number theory and, if not there, certainly in analysis or set theory. In this letter, he also indicates a more "abstract" proof of the second incompleteness theorem using conditions on the provability predicate in order to show the equivalence of the unprovable Gödel sentence with the consistency statement 𝔚.

---

[22]Gödel's first two letters to von Neumann have not been preserved, it seems. The core content of the letters can be inferred from von Neumann's responses and from Gödel's discussion of von Neumann's perspective at the meeting of the Vienna Circle of 15 January 1931. In (Dawson 1997, p. 70) one finds Hempel's report on the course in proof theory he took with von Neumann in Berlin during the winter term 1930–31: ". . . in the middle of the course von Neumann came in one day and announced that he had just received a paper from a young mathematician in Vienna . . . who showed that the objectives which Hilbert had in mind . . . could not be achieved at all." In a letter of 3 December 1930 to his friend Chevalley, Herbrand writes about the same period: ". . . I have been here for two weeks, and every time I have seen von Neumann we talk about the work of a certain Gödel, who has produced very curious functions; and all of this has destroyed some quite solidly anchored ideas." (Details concerning this letter are found in the Appendix of (Sieg 1994b).)

Bernays wrote to Gödel on 24 December 1930 from Berlin, where he was spending the Christmas break with his family.[23] He had already earlier received a reprint of (Gödel 1930) and begins his letter with some comments on Gödel's completeness proof for first-order logic. Then Bernays asks for the galleys of Gödel's new investigations. As the reason for this request he notes that he has learned from Professors Courant and Schur,[24] "that you have recently succeeded in obtaining significant and surprising results in the area of foundational problems, and that you intend to publish them shortly". Let me remark, parenthetically, that I don't see any conflict between this request and the assumption that Bernays already knew about Gödel's restricted Königsberg result. In any event, Gödel answered on 31 December 1930, sent Bernays a reprint, presumably the abstract (1930b), and promised to send the galleys of his (1931). Bernays received the galleys on 14 January 1931, studied them and wrote a long, detailed letter to Gödel only four days later. He remarks, "For me that was very interesting and very instructive reading. What you have done is really an important step forward in the investigation of the foundational problems."

It took some time before Bernays saw clearly that Ackermann's and von Neumann's proofs established the consistency of arithmetic only when the induction principle is restricted to quantifier-free formulae. On 3 May 1931, Bernays writes again to Gödel and views Ackermann's considerations as proving the consistency of full arithmetic. He tries to find the reason why that proof cannot be formalized in arithmetic as required by the second incompleteness theorem. Incorrectly, he sees "the explanation of the matter" in the fact that nested recursions cannot be formalized in elementary arithmetic.[25] It was only in the brief note (Bernays 1933) for the International Congress of Mathematicians in Zürich that the limited nature of these consistency proofs was explicitly stated and publicly

---

[23] The letters I refer to in the following are all contained in (Gödel 2003a). The longer excerpts in this paragraph are found on p. 80, respectively on p. 82.

[24] Here and below are some important factual questions. When and where did Bernays learn this from Courant and Schur? How had *they* been informed? What did Bernays do in the winter term 1930–31? (It seems that he attended neither the Königsberg meeting nor the lecture Hilbert gave in December 1930 in Hamburg.) We know that Gentzen spent this very term in Berlin; did he attend von Neumann's lectures? Did he have contact there with Bernays and Herbrand? (Cf. Note 22.) - As to the latter question, Menzler-Trott is "convinced" that Gentzen talked with all three (Bernays, Herbrand, and von Neumann) while in Berlin. In the same note of 7 December 2010 in which he expressed this conviction about Gentzen, Menzler-Trott reports that Bernays participated in 1930/31 in the meetings of the *Berliner Gesellschaft für empirische Philosophie* with, among others, Dubislav, Grelling, Hempel, and Reichenbach. It may very well be that Bernays learned here about Gödel's Königsberg result, as Dubislav participated in that meeting and Hempel heard about it in von Neumann's lecture. (Cf. Note 22.) — The name of the Berliner Gesellschaft was changed in 1931, as suggested by Hilbert, to *Berliner Gesellschaft für wissenschaftliche Philosophie*.

[25] (Gödel 2003a, p. 104). It is mentioned in (Hilbert and Bernays 1934, p. 422), that nested recursions are formalizable in arithmetic and that Gödel and von Neumann discovered this fact.

acknowledged.²⁶ Here is Bernays's formulation concerning the "method of valuation" (Bewertungsmethode):

> This [method, WS] obtained its essential development by Hilbert's procedure of trial valuation. Using this procedure Ackermann and von Neumann demonstrated the consistency of number theory — admittedly, under the restrictive condition that the application of the inference from n to $(n + 1)$ is only allowed for formulae with just free variables. (Bernays 1933, p. 201)

Bernays remarks, at the end of the note, that the limitation of the proof theoretic results is apparently a fundamental one "because of Gödel's new theorem — and a related conjecture of von Neumann's — on the limits of decidability in formal systems". It seems noteworthy that Hilbert presided over the opening of this Congress; as in Bologna four years earlier, he received a standing ovation. That is reported in (Richardson 1932).

Finally, let me make a remark about Herbrand who was also working intensively at that time on the proof theory of elementary arithmetic. He was well aware of the incompleteness results through the contacts he had with both von Neumann and Bernays during his long stay in Berlin, from late November 1930 to the middle of May 1931. The central consistency result in Herbrand's (1931) is formulated for arithmetic with only quantifier-free induction, but including Hilbert's rule (HR) and an open characterization of finitist functions. This characterization also covers the functions that are definable by nested recursion, like the Ackermann function. In the last part of his paper, Herbrand discusses Gödel's results. He articulates in almost identical words what he had already claimed on 7 April 1931 in a letter to Gödel, namely, that — contrary to Gödel's opinion, but consonant with von Neumann's view — all finitist (or intuitionist) arguments can be formalized in analysis and possibly in elementary number theory. "If this were so," he concludes, "the consistency of ordinary arithmetic would already be unprovable."

In a letter to Hans Reichenbach that is undated, but was definitely written in the summer of 1931, von Neumann makes a remark on the publication of his contribution to the Königsberg Congress. In a way, he summarizes the broader state of affairs:

> Incidentally, I have decided not to mention Gödel, since the opinion that there still exists a certain hope for proof theory has found champions - *inter alia*, Bernays and Gödel himself. To be sure, in my opinion this view is erroneous; but a discussion of this question would stray outside the existing boundaries, and so I would rather treat it at another occasion.²⁷

---

²⁶In his letter of 15 November 1932 to Heyting, Gödel remarks somewhat indignantly: Bernays hat in seinem Züricher Vortrag (soviel mir bekannt) auch zugegeben, daß man die Widerspruchsfreiheit d[er] Zahlentheorie bisher nur mit einer von Herbrand gegebenen Einschränkung für die vollständige Induktion beweisen kann. (Gödel 2003b, p. 60).

²⁷(Mancosu 1999b, p. 49 and note 14). — The German text of von Neumann's note is: Übrigens habe ich mich entschlossen, Gödel nicht zu erwähnen, da die Ansicht, dass noch eine gewisse Hoffnung für die Beweistheorie existiert, Vertreter gefunden hat: u[nter] a[nderen] Bernays und Gödel selbst. Zwar ist m[eines] E[rachtens] diese Ansicht irrig, aber eine

In the shadow of incompleteness   171

So it is perhaps not too surprising that Hilbert himself did not give up on the proof theoretic program, but rather shifted its direction. As a matter of fact, he returned to a perspective that underlies the first part of his (Hilbert 1922), when formulating the base theory in a "quasi-constructive" way with a restricted treatment of negation; see (Sieg 2009b, p. 376, note 42) and (Sieg and Tapp 2012). But let me take one step at a time.

5 Hilbert's response.

Even in his last paper, *Beweis des Tertium non datur*, Hilbert does not mention Gödel by name, but the proof theoretic considerations react clearly to the second incompleteness theorem and — as Bernays put it in his 1977 interview — try to deal positively with Gödel's results. The approach in this paper is dramatically different from that in the third Hamburg talk. Then, in December 1930, Hilbert concentrated on the syntactic completeness of elementary arithmetic, assuming that its consistency had been secured by Ackermann's and von Neumann's work. Now, in July 1931, there is an almost exclusive focus on consistency and the principle of tertium non datur, for short: *tnd*.[28] The latter principle postulates, for all statements $A(x)$, that $A(x)$ either holds for all natural numbers or has a counterexample; it is formally expressed by $(x)A(x) \lor (Ex)\neg A(x)$.

The renewed focus on consistency is accompanied, however, by a radical strategic shift. Instead of aiming to prove finitistically the consistency of classical arithmetic by the ε-substitution method, Hilbert formulates a constructive (he thinks, finitist) theory of arithmetic and argues simultaneously for the correctness of its seemingly transfinite inference principles. Let me describe matters in greater detail. The principles for sentential logic — with conditionals, conjunctions and disjunctions as the basic forms — are those of the logical calculus described in section 2 above. However, only equations between numerals are negated, and the sole principle involving negations of this restricted sort is the *Axiom des Widerspuchs*, i.e., a contradiction implies any formula whatsoever.[29] For these basic literals Hilbert introduces the concepts "correct" (richtig) and "false" (falsch). After the formulation of the transfinite inference principles these concepts are extended to literals containing recursively defined function symbols and the special symbols connected with the elimination or, as Hilbert says, "application" of the existential quantifier. The concept of "correctness" is extended straightforwardly to conjunctions and

---

Diskussion dieser Frage würde aus dem vorliegenden Ra[h]men hinausführen, ich möchte daher bei einer anderen Gelegenheit darüber sprechen.

[28]This focus is fully in accord with the earlier considerations, e.g., in Hilbert's second Hamburg talk of 1927; see p. 471 and p. 479 for consistency, p. 470 and p. 476 for tertium non datur.

[29]The negation for general statements is later defined inductively by using classical equivalences; for example, the negation of the conditional $A \to B$ is given by $A \,\&\, \neg B$, that of $(x)A(x)$ by $(\exists x)\neg A(x)$. — In the earlier Hamburg lecture the *Axiom des Widerspuchs* is the principle $(A \to (B \,\&\, \neg B)) \to \neg A$. — Note that the law of excluded middle implies *tnd* via the definition of the negation of a universally quantified statement.

disjunctions. Much more problematically, a conditional is viewed as correct "if its antecedent is correct, so is its consequent".

The combination of syntactic and semantic considerations is particularly striking in the case of the "transfinite axioms and inference schemata" for quantifiers. Viewed from a syntactic perspective, existential quantifiers are analyzed essentially by the natural deduction rules for *introducing* and *eliminating* them. Universal quantifiers are *introduced* by the rule (*HR**), the extension of (*HR*) to formulae of arbitrary complexity all of whose instances are finitistically correct; they are *eliminated* via the axiom $(x)\mathfrak{A}(x) \to \mathfrak{A}(\mathfrak{a})$. However, the official partially semantic formulation of these principles is given as follows, extending the concept "correct" to quantified formulae:

If the statement $\mathfrak{A}(\mathfrak{z})$ is correct as soon as $\mathfrak{z}$ is a numeral, then the statement $(x)\mathfrak{A}(x)$ holds; in this case $(x)\mathfrak{A}(x)$ is called correct. The converse is provided by the axiom $(x)\mathfrak{A}(x) \to \mathfrak{A}(\mathfrak{a})$. These stipulations concern the introduction and application of the concept "all". For the introduction and application of the concept "there is" I may apply the following two schemata:

$$\frac{\mathfrak{A}(\mathfrak{a})}{(\exists x)\mathfrak{A}(x)}$$

where $\mathfrak{a}$ is a numeral; in this case, $(\exists x)\mathfrak{A}(x)$ is called correct. Conversely, an expression $(\exists x)\mathfrak{A}(x)$ may be replaced by $\mathfrak{A}(\eta)$, where η is a letter that has not been used yet. The following contentual understanding corresponds to this rule: to the formula $(\exists x)\mathfrak{A}(x)$ is associated the definition of a numeral η, according to which [i.e., according to the definition of η, WS] $A(\eta)$ is correct, whenever $(\exists x)\mathfrak{A}(x)$ is. (Hilbert 1931b, p. 121)

The ultimate goal is to show that this system can be expanded by instances of *tnd* without leading to contradictions.[30]

---

[30]The arguments leading to this goal are barely sketched, and some are deeply problematic. Gödel made a critical remark about Hilbert's paper in a letter to Heyting of 16 May 1933: Ich glaube überhaupt, Sie beurteilen Hilberts letzte Arbeiten etwas zu günstig. Z.B. ist doch in Göttinger Nachr. 1931 [(Hilbert 1931b), WS] kaum irgend etwas bewiesen. — In conversation with Olga Taussky-Todd, so it is reported in (Taussky-Todd 1987, p. 40), Gödel "lashed out against Hilbert's paper [(1931b), WS], saying something like 'how can he write such a paper after what I have done?' Hilbert in fact did not only write this paper in a style irritating Gödel, he gave lectures about it in Göttingen in 1932 and other places."

It is worth noting that (Hilbert 1931b) was neither reprinted in Hilbert's *Gesammelte Abhandlungen* nor mentioned in (Bernays 1935). In contrast, Bernays's paper discusses (Hilbert 1931a) on pp. 215–216: Das Verfahren, durch welches hier Hilbert die positive Lösung des Vollständigkeitsproblems (für den von ihm betrachteten Spezialfall) sozusagen erzwingt, bedeutet ein Abgehen von dem vorherigen Programm der Beweistheorie. In der Tat wird ja durch die Einführung der zusätzlichen Schlußregel die Forderung einer restlosen Formalisierung der Schlüsse fallen gelassen.

In his contribution to the *Encyclopedia of Philosophy*, Bernays mentions both papers and remarks summarily after a brief discussion of Gentzen's consistency proof: "The broadened methods [of Gentzen, WS] also permitted a loosening of the requirements of formalizing. One step in this direction, made by Hilbert himself, was to replace the schema of complete induction by the stronger rule later called infinite induction . . . " (p. 502) — On my reading of (Hilbert 1931b) there is no infinite rule *in* the system, but rather a finitistically justified introduction

Before following Hilbert's path of trying to achieve that goal, I want to make a brief remark about the proof theoretic strength of the system I just described. Gentzen's way of interpreting universal quantifiers applied to arbitrary finitistically meaningful statements, given in his (1936) on pp. 526 and 528, can be adapted to argue for the correctness of the standard introduction rule for universal quantifiers, but also for that of the rule of complete induction. Thus, the standard formalization of intuitionist arithmetic, Heyting arithmetic, is contained in Hilbert's constructive system.

To pursue the "proof" of *tnd*, some preparatory work is required. First, the discussion of the transfinite principles is exploited for arguing that "the whole system is consistent", where consistency requires, as usual, that no proof has $1 \neq 1$ as its endformula. Hilbert's reasons for asserting the consistency of the system are compressed into a single sentence: "All transfinite rules and inference schemata are consistent; for they amount to definitions." Second, Hilbert tries to show that consistency and correctness are "identical". Third, defining a statement as "false" in case it leads to a contradiction, he claims that *any statement is either false or correct*. Given the definition of "false" and the identity of consistency and correctness, Hilbert has to prove that any statement either does or does not lead to a contradiction. This metamathematical statement is an instance of *tnd*, and Hilbert views it as "necessary" for the founding of mathematics. He then uses *tnd* to show that correctness, falsity, and the generalized negation of statements (see Note 29) harmonize in the appropriate way. Having taken these preparatory steps, Hilbert proceeds (on p. 124) to argue that adding instances of *tnd* as axioms to the base system does not lead to contradictions. The indirect metamathematical argument for this claim is sketched in the next paragraph.

Consider a proof in the constructive system that uses, for simplicity's sake, a single instance of *tnd*

(1) $\quad (x)A(x) \lor (Ex)\neg A(x)$

as an initial formula. Hilbert lists the instances of the axiom schema (expressing for-all elimination) that have been used in the proof, say,

(2) $\quad (x)A(x) \to A(a_1), \ldots, (x)A(x) \to A(a_n).$

Now he forms the conjunction $A(a_1) \& \ldots \& A(a_n)$ and, using it, replaces all occurrences of the statement $(x)A(x)$ in the proof. Hilbert claims that all the transformed initial formulae are correct and that the syntactic configuration (resulting from this replacement) is a derivation from the initial formulae. The formulae resulting from those in (2), $A(a_1) \& \ldots \& A(a_n) \to A(a_i)$, are trivially provable, whereas the formula resulting from (1), namely,

(3) $\quad A(a_1) \& \ldots \& A(a_n) \lor (Ex)\neg A(x)$

---

of universally quantified statements using (*HR**). (Indeed, Hilbert's proof that the system expanded by *tnd* is consistent exploits the finiteness of proof figures; see the discussion below.)

is provable from correct instances of the law of excluded middle[31]

(4) $A(a_i) \lor \neg A(a_i)$.

Thus, one has $A(a_i) \lor (Ex)\neg A(x)$ for all i between 1 and n, and distributivity establishes (3). So we have obtained, Hilbert argues, a proof from correct initial formulae; as correctness is preserved by inferences, all of the formulae in the proof are correct and its endformula can't possibly be $1 \neq 1$.

Whatever problematic features there are in the overall considerations (and there are many), a central one is Hilbert's use of *tnd* when arguing that any statement is either correct or false. This appeal is openly acknowledged and is, of course, in conflict with the finitist position. In a letter sent from Berlin on 11 October 1931, Bernays reports to Hilbert on progress with the *Grundlagenbuch*: he just finished the presentation of Behmann's decision procedure for monadic predicate logic with identity and is about to begin replacing the consistency proof "for universal and existential quantifiers that is not correct" by another proof. Then he continues:

As for the relationship to your last article [i.e., (Hilbert 1931b), WS], it can be said in the Preface that the arguments in the book are carried out entirely within the framework of the finitist standpoint (i.e., other considerations are used at most in a heuristic sense), so that your last article, which is based on a different methodological standpoint, does not come within the scope of these considerations.[32]

The phrasing of this remark suggests that Hilbert and Bernays had discussed the issue and clearly agreed that the methods of *Beweis des Tertium non datur* go beyond the finitist standpoint. But why should they be excluded from further exploration? After all, Hilbert's self-conscious use of an instance of *tnd* in the metamathematical argument may be viewed as parallel to the use (and later removal) of that principle in proofs of his

---

[31]Hilbert argues simultaneously for the correctness of these instances of the law of excluded middle by an inductive argument on the complexity of the formulae in the instances of *tnd* used in the proof. It is difficult to see, why the matrix $A(a_i)$ should be of the appropriate form $(y)B(y,a_i)$ to allow the formulation of *tnd* and its use in the induction hypothesis. — Note that Hilbert does not mention a necessary modification in the upper part of the derivation in case axioms of the form (2) are actually used to infer $A(a_i)$ via modus ponens with $(x)A(x)$ as the minor premise. All the instances $A(a_i)$ must be inferred from $(x)A(x)$ and formed into a conjunction — *before* carrying out modus ponens ... to infer the $A(a_i)$. Gentzen will avoid for intuitionist logic such odd detours through the natural deduction formulation of the logical principles and his normalization proof!

[32](Cod. Ms. D. Hilbert 21, 5). The German text is: Was die Beziehung zu Ihrer letzten Note betrifft, so kann ja im Vorwort gesagt werden, dass die Ausführungen des Buches sich ausschliesslich im Rahmen des finiten Standpunktes bewegen (d.h. anderweitige Betrachtungen werden höchstens im heuristischen Sinne angestellt), dass daher Ihre letzte Note, die einen anderen methodischen Standpunkt zugrundelege, nicht in den Bereich dieser Betrachtungen falle. Bernays continues: Auch im Rahmen des finiten Standpunktes wird ja einiges noch ausserhalb bleiben, dessen Behandlung in einem "2. Teil" ja am Schluss des Textes angekündigt, bezw. in Aussicht genommen werden kann; nämlich, 1. die Formalisierung der "zweiten Stufe" (ε_f), 2. die Formalisierung der Metamathematik (Resultate von Gödel).

early mathematical career, for example, when solving Gordan's problem in invariant theory.[33]

Indeed, it seems that the methods were explored in critical detail only a few months later by a young student, who had been engaged with Bernays on another project. With the explicit goal of proving the consistency of arithmetic, he began working on his thesis in late 1931 and quickly obtained significant results. An outline of his *Urdissertation* from early October 1932 summarizes these results and formulates the remaining central task. The young student is Gerhard Gentzen. The outline of the thesis is organized in five parts.[34] Part I is worked out on eleven handwritten pages and presents the natural deduction calculi for intuitionist and classical first-order logic under the heading "Der Schlussweisenkalkül N1J". Gentzen indicates also the reductions that are the basic steps for normalizing arbitrary intuitionist derivations.

A one-page "Overview of my further results" follows this part, which ended with the remark that a "positive characterization of intuitionist inferences" has been given. In Part II the calculus N1J is shown to be equivalent to the "logistic calculi of intuitionist reasoning" given by Hilbert, Heyting, and Glivenko.[35] In Part III the consistency of classical arithmetic is shown — relative to its intuitionist version. Articulating the gist of the result, Gentzen writes: "It is thus possible to give so-to-speak an 'intuitionist interpretation' to the arithmetic statements." In Part IV Gentzen conjectures the subformula property for normal N1J-derivations of *logical statements*, i.e., statements that do not depend on open assumptions. At the original writing of the summary, the conjecture was established only for the calculus N2J with the Introduction and Elimination rules for conjunction and universal quantification, as well as the rule for negation introduction.

The crucial next task is formulated as Part V. I quoted it in Section 1 and hope that it is less puzzling now:

The consistency of arithmetic will be proved; in the process, the concept of an infinite sequence of natural numbers will be used, furthermore in one place the principle of

---

[33] This connection is never far from Hilbert's considerations as is obvious from his publications. In the manuscript SUB 603 one finds this remark made in a somewhat obscure context in which tertium non datur is discussed: Hiermit ist gezeigt, dass man so schliessen darf. Es ist in der Tat ein Unterschied, ob man hier Schluss mit tertium non datur anwendet oder nicht; v[er]gl[eiche] meine Beweise der Endlichkeit i[n] d[er] Invariantentheorie.

[34] For more details concerning Gentzen's manuscript, see (von Plato 2009b, section 5); that section is entitled *A newly discovered proof of normalization by Gentzen*. – The dating of this summary is a *conjecture* of mine that is supported by three facts: (i) all the results described in Parts I through III have been obtained already (and Part IV for a significant sub-calculus), (ii) the detailed normalization of intuitionist derivations (as indicated most clearly in the next longer quotation that begins with "Some thought") is not being pursued, and (iii) Part V articulates the consistency problem still in the manner of (Hilbert 1931b). The consistency problem as formulated here would naturally be dealt with by semantic considerations — and Gentzen began in October 1932 to write detailed, reflective notes on his approach to the problems he was facing in the manuscript INH discussed below.

[35] The reference is, I assume, to (Heyting 1930a) and to (Glivenko 1929); in the paper (Gentzen 1933) that derives from Part III of the *Urdissertation*, Heyting is mentioned, but oddly enough Glivenko is not.

the excluded middle. The proof is thus not intuitionist. Perhaps the *tertium non datur* can be eliminated.[36]

The connection to Hilbert's considerations in *Beweis des tertium non datur* seems unmistakable, as these remarks point exactly to the central features of Hilbert's argument, i.e., the metamathematical use of the rule (HR^*) and *tnd*. What did Gentzen intend to do in order to establish the consistency of arithmetic, i.e., of the part of intuitionist arithmetic that is needed for the interpretation of its classical version? In the next section, let me indicate first some of the historical circumstances and then an essential part of the systematic logical context.

6 The new student.

In a certain sense Gentzen was not at all a *new* student in Göttingen. After a year of studying mathematics in Greifswald under the tutelage of Hilbert's student Hellmuth Kneser, he went to Göttingen and spent the academic year 1929–30 there, i.e., from late April 1929 to early March 1930. Gentzen and his friend from Greifswald, Lothar Collatz, attended Hilbert's lectures on set theory in the summer term of 1929. From Collatz's careful notes we know that these lectures were divided into three parts. For the purposes here only the third part is of interest where Hilbert discusses, in an elementary way, mathematical logic and his proof theoretic program.[37] He presents the elements of the consistency proof for primitive recursive arithmetic, in particular and with many examples, the "Auflösung in Beweisfäden", i.e., the transformation of linear derivations into tree structures. That is the central step in preparing derivations for further proof theoretic analysis; see Note 2. It was taken in *every* consistency proof given in Göttingen starting in 1922, when Hilbert and Bernays first proved their result, and includes even Gentzen's proof in his (1936).[38]

---

[36]The German is: Die Widerspruchsfreiheit der Arithmetik wird bewiesen; dabei wird der Begriff der unendlichen Folge von natürlichen Zahlen benutzt, ferner an einer Stelle der Satz vom ausgeschlossenen Dritten. Der Beweis ist also nicht intuitionistisch. Vielleicht lässt sich das *tertium non datur* wegschaffen.

[37]The first two parts are adapted from the lectures on set theory Hilbert gave in the summer term of 1917, (Hilbert *1917). It would be of interest to examine Hilbert's possibly modified perspective on set theory. What is also of interest is Hilbert's discussion of ordinals, in particular ordinals less than ε_0.

[38]In (Gentzen 1936, p. 513) derivations are defined as sequences of (one-sided) sequents; the transformation into, essentially, tree form is made on p. 542 for the very same reason Hilbert and Bernays made it in 1922, namely, to insure that every sequent, except for the endsequent of course, is used at most once as a premise. Gentzen employed this technique also in the very first step of his relative consistency proof for classical arithmetic when transforming classical into intuitionist proofs, cf. (Gentzen 1933, pp. 126–127, section 4.21). — I mention these matters here, as they seem to answer convincingly, how Gentzen learned about the tree representation of proofs. Von Plato views the issue in a different way. In his (2010) he claims, already in the abstract, "the central component in Gentzen's work on logical calculi was the use of a tree form for derivations." Later, when comparing Gentzen's work with Einstein's in the latter's *annus mirabilis* 1905, he writes: "His [Gentzen's] amazing discovery of natural

After spending the summer term of 1930 in Munich (where he read on his own Hilbert and Ackermann's book *Grundzüge der theoretischen Logik*), he went to Berlin and studied there during the winter term 1930–31. Unfortunately, we do not know what courses he took. This remains an intriguing question: von Neumann lectured on proof theory and discussed, as I pointed out in sections 3 and 4, what he had learned about Gödel's theorems. As to Gentzen, Winfried Schultze (Director of the University Archive of the Humboldt-University in Berlin) told me in a letter of 23 February 2010:

Gerhard Gentzen, coming from Munich, registered on 29 October 1930 with [student] number 1335 of the 121st academic year at the Friedrich-Wilhelms-University in Berlin. He studied here mathematics until 11 March 1931. A final report, unfortunately, has not been preserved; thus, we cannot make any assertions concerning the question, which lectures given by which faculty member he actually attended during this semester.[39]

Gentzen returned to Göttingen for the summer term 1931. During that term Hilbert conducted a seminar on *Grundlagen der Mathematik* and submitted, on 17 July 1931, his (1931b) for publication. It would be of special interest to know some facts about this seminar, from topics treated to who actually attended. Given Hilbert's habit of discussing topics of papers first in lectures or seminars, it is most plausible that he presented aspects of his (1931b) during this term; given the customs of German university institutes and doctoral education, it is almost inconceivable that Gentzen did not attend the seminar. One might also conjecture (at the moment without any concrete archival support) that the letter Richard Courant wrote to his colleague Hermann Nohl in support of Gentzen's application to the *Studienstiftung* refers to that seminar. The letter was written on 31 July 1931:

As we agreed, I am reporting to you today about Mr. Gentzen on the basis of his seminar talk and a personal consultation. Mr. Gentzen discussed a particularly difficult topic in his seminar talk; he showed through the external as well as the intellectual grasp of the material a superior independence that marks him as a scientifically oriented human being. On account of his talk and after an oral interview I am confident that Mr. Gentzen can complete his doctorate relatively easily and that he can continue afterwards with scientific work. As his inclinations drive him

---

deduction and sequent calculus in 1932–33, with its full control over the structure of derivations, followed from the use of a **tree form** [von Plato's emphasis; WS] for derivations. That was the new, simple, and right idea." That idea was prefigured in earlier proof theoretic work; but that fact by no means distracts from the "amazing" character of the insight into the structure of normal proofs, in particular, their subformula property in first-order logic.

[39] Here is the German text: Gerhard Gentzen hat sich — von München kommend — am 29. Oktober 1930 unter der Nummer 1335 des 121. Rektorats in die Matrikel der Friedrich-Wilhelms-Universität zu Berlin eingetragen. Er studierte hier bis zum 11. März 1931 Mathematik. Ein Abgangszeugnis ist hier leider nicht überliefert, so dass wir keine Aussage darüber treffen können, welche Vorlesungen er bei wem in diesem Semester belegt hat.

obviously and very strongly in this direction, I can advise the *Studienstiftung* with full responsibility to grant him the doctoral scholarship.[40]

This is a relatively free translation of the full letter. Unfortunately, we don't know anything about the content of Gentzen's application for this scholarship.

Bernays recommended that Gentzen work, during the summer break of 1931, on the "sequent systems" that had been introduced by Paul Hertz; as to the latter's work see (Schröder-Heister 2002), but also (Bernays 1965). The paper (Gentzen 1932) would result from this work, and on 6 February 1932 it was submitted to *Mathematische Annalen*. In a letter of 13 December 1932 to Kneser, Gentzen mentions that he had turned, at the beginning of the winter term 1931/32, to more general problems of proof theory and that he had set himself almost a year ago, "the task of finding a proof of the consistency of logical deduction in arithmetic". When writing this letter he was hopeful to "finish soon".[41] By then, indeed somewhat earlier as argued in Note 34, Gentzen had discovered the consistency proof for classical arithmetic relative to intuitionist arithmetic.[42] He sent the resulting paper to Heyting in January or early February 1933 and submitted it to *Mathematische Annalen* on 15 March 1933, but withdrew it when he learned of Gödel's publication (1933d). Through the argument in this paper, the principle of *tertium non datur* could be added consistently to intuitionist arithmetic; in fact it had been proved in its interpreted form within the intuitionist theory. In a letter to Heyting written on 25 February 1933, Gentzen suggested investigating the consistency of intuitionist arithmetic, since a consistency proof for classical arithmetic had not been given so far by finitist means, "so that this original aim of Hilbert has not been achieved". He then continued:

---

[40]Majer has explored all the obvious archival issues implicit in this letter and my account above, but without any positive findings. — Here is the German text of Courant's letter: Verabredungsgemäss berichte ich Ihnen heute über Herrn Gentzen auf Grund seines Seminarvortrages und einer persönlichen Rücksprache mit ihm. Herr Gentzen behandelte in seinem Seminarvortrag ein besonders schwieriges Thema und bewies dabei in der äusseren und in der geistigen Durchdringung des Stoffes eine überlegene Selbständigkeit, die ihn durchaus als den Typus eines wissenschaftlichlich gerichteten Menschen kennzeichnet. Ich habe danach wie nach einer mündlichen Besprechung das Zutrauen, dass Herr Gentzen verhältnismässig leicht promovieren und auch dann weiter wissenschaftlich arbeiten kann. Da offenbar seine inneren Neigungen ihn sehr stark auf diese Bahn drängen, so kann ich mit voller Verantwortung der Studienstiftung den Rat geben, ihm die Promotion zu bewilligen.

[41]A longer excerpt from this letter to Kneser is quoted in (von Plato 2009b, p. 670); a translation of the full letter is in (Menzler-Trott 2007, pp. 30–31).

[42]In (Gentzen 1936, note 17, p. 532), this result is attributed to both Bernays and Gentzen himself: Das im Text genannte Ergebnis wurde etwas später, unabhängig von Gödel, auch von P. Bernays und mir gefunden. (In Note 2 to the publication of the German original, i.e., on p. 119, Gentzen describes very concisely in what way Bernays contributed.) It is not clear with respect to which event "etwas später" is to be understood. Gödel presented his result to Menger's Colloquium on 28 June 1932. In his letter to Heyting of 16 May 1933, Gödel explicitly claims that his result should have become known in Göttingen shortly after his presentation. — The proof of this result is given and refined in (Hilbert and Bernays 1939); it is presented there tellingly under the heading *Eliminierbarkeit des 'tertium non datur' für die Untersuchung der Widerspruchsfreiheit des Systems (Z)*.

In the shadow of incompleteness 179

If, on the other hand, one admits the intuitionistic position as a secure basis in itself, i.e., as a consistent one, the consistency of classical arithmetic is secured by my result. If one wished to satisfy Hilbert's requirements, the task would still remain of showing intuitionistic arithmetic consistent. This, however, is not possible by even the formal apparatus of classical arithmetic, on the basis of Gödel's result in combination with my proof. Even so, I am inclined to believe that a consistency proof for intuitionistic arithmetic, from an even more evident position, is possible and desirable. (Quoted in (von Plato 2009b, p. 672).)

Gentzen expressed the hope that he would investigate the consistency of intuitionist arithmetic "next year". "Thus", von Plato concludes, "the hopes of the previous December of 'finishing soon' a consistency proof, as in the letter to Kneser, had faded in a little over a month." Instead of pursuing the consistency proof, Gentzen turned his attention to writing *the* thesis, which was published as *Untersuchungen über das logische Schließen*. Gentzen defended it on 12 July 1933 and submitted it nine days later for publication in *Mathematische Zeitschrift*.

Why did the hopes fade for finishing a consistency proof? Can we get a sense of the difficulties Gentzen ran into? — Based on Gentzen's manuscript INH to be discussed in more detail below, von Plato suggests in his (von Plato 2009b, p. 677) that Gentzen recognized at the beginning of 1933 "that the planned approach [to the consistency proof for arithmetic in the *Urdissertation*, WS] through an extension of the method of section IV, i.e., normalization and the subformula property, from pure logic to derivations in arithmetic, did not work." In (von Plato 2010, p. 20) the task Gentzen had set himself in Part V is reformulated simply as, "To extend normalization and the subformula property to arithmetic." In contrast to von Plato, I view the *full* statement (V) as a crucial programmatic passage: it reveals, on the one hand, what Gentzen was trying to accomplish, and it indicates, on the other hand, how deeply his investigations were rooted in Hilbert's (1931b). INH and the discussion below will show the significant conceptual difficulties Gentzen had to overcome.

With this perspective we can ask, what had to be done in order to sharpen Hilbert's argument for the correctness of the basic constructive theory and then to show that the addition of *tnd* does not lead to a contradiction. The first step would extend Hilbert's considerations for quantifiers to the sentential logical connectives. That involves the formulation of introduction and elimination rules for those connectives, so that the correctness of logical steps can be viewed as "definitional" in the way Hilbert had done for quantificational inferences.[43] Given Hilbert and Bernays's formulation of the sentential logical axioms and their technique of "Auflösung in

---

[43]And as Gentzen explicitly did in his dissertation, (Gentzen 1933, p. 189): Die Einführungen stellen sozusagen die "Definitionen" der betreffenden Zeichen dar, und die Beseitigungen sind letzten Endes nur Konsequenzen hiervon, was sich etwa so ausdrücken läßt: Bei der Beseitigung eines Zeichens darf die betreffende Formel, um deren äußerstes Zeichen es sich handelt, nur "als das benutzt werden, was sie auf Grund der Einführung dieses Zeichens bedeutet". A few lines below Gentzen continues: Durch Präzisierung dieser Gedanken dürfte es möglich sein, die B(eseitigungs)-Schlüsse auf Grund

Beweisfäden", this first step is not difficult. In the tree representation of proofs the axioms for the connectives are used only at the very top, e.g., in the case of conjunction as follows:

$$
\begin{array}{cc}
\dfrac{\downarrow \quad A \quad A \to (B \to A \& B)}{\dfrac{B \quad B \to A \& B}{A \& B}} & \dfrac{\downarrow \quad A \& B \to A[B]}{A[B]}
\end{array}
$$

Why make the absolutely bureaucratic detour through the axioms? Is there any reason not to infer $A \& B$ from proofs of A and of B, or not to conclude $A[B]$ from a proof of $A \& B$? — One is led quite directly to Gentzen's calculus N1J as formulated in Part I of the *Urdissertation*. Then it is important to show that the formalization of logical principles via this calculus is equivalent to those given by Hilbert, Heyting and Glivenko. That is done with explicit reference to these three authors in Part II of the *Urdissertation*.

Now, the second question can be addressed, namely, whether the addition of the law of excluded middle leads to contradictions. Gentzen (on page 10 of the *Urdissertation*) defines $\neg A$ as $A \to \bot$. Given this definition, one sees readily that an instance of that law in the form $A \vee (A \to \bot)$ cannot lead to a contradiction, unless intuitionist logic is inconsistent. After all, its double negation is provable intuitionistically and very easily so in Gentzen's N1J.[44] The step from this straightforward argument to the full interpretation of classical into intuitionist arithmetic is clearly a significant one, but it is technically not difficult; in particular not, as Gentzen was presumably familiar with Glivenko's paper that is explicitly mentioned in the *Urdissertation*. The result is nevertheless striking and is contained in Part III.[45]

We come to Part IV and its main results: normalizability and the subformula property of normal derivations for the calculus N2J, a fragment of the calculus N1J for full intuitionist logic. Von Plato conjectures that the results for N1J were proved and added to the manuscript only in March 1933. (An English translation of the proof is found in (von Plato 2008).) At

---

gewisser Anforderungen als eindeutige Funktionen der zugehörigen E(infürungs)-Schlüsse nachzuweisen.

[44] Heyting proved this theorem as Theorem 4.8 in his (Heyting 1930a, p. 52) and remarks in a note that the proof was given in (Glivenko 1929); he emphasizes that it is connected to Brouwer's "Satz von der Absurdität der Absurdität des Satzes vom ausgeschlossenen Dritten".

[45] Documents in the Hilbert Nachlass contained in the folder SUB 603 indicate that Hilbert must have been informed about these matters by the fall of 1932; pages 44 through 47 on which pertinent remarks are made stem from *Druckfahnen* (page proofs) dated 15 July 1932. On p. 44, in particular, one finds this remark: Es sei eine Formel vorgelegt. Dann ersetze man $A \vee B$ durch $\neg(\neg A \& \neg B)$, ferner $(Ex)A(x)$ durch$\neg(x)\neg A(x)$, ferner $\neg A$ durch $A \to 1 \neq 1$. . . . Wenn dann jene Formel bewiesen ist (Tertium wird zugelassen) dann ist die entstandene Formel inhaltlich richtig. Z.B. $(x)\neg A(x) \vee (Ex)A(x)$, $(x)A(x) \vee ((x)A(x) \to w)$. [$w$ stands for a contradictory formula; WS.]

the time of writing Part IV, Gentzen did not view them as crucial for achieving the main goal of his investigation, a consistency proof for intuitionist arithmetic. He noted in the summary after having indicated the reduction steps for the normalization procedure:

Some thought is needed to recognize that in fact a correct proof is obtained in each case. I am not going to pursue this in detail, as I am not going to use those facts — I just present them for the purpose of illustration.[46]

The results are nevertheless the most surprising logical discovery of Gentzen's, and he must have seen them in a similar light. After all, when facing the impasse in establishing the consistency of arithmetic, he chose to establish his *Hauptsatz* (or cut-elimination theorem) for classical as well as intuitionist sequent calculi. How surprising the result was may also be seen indirectly from observations von Neumann made. Both in his (1927, pp. 11–12) and (1931, p. 120), he argued against the plausibility of a positive solution of the *Entscheidungsproblem* for provability in first-order logic by pointing out the following fact: the minor premise used in modus ponens and, thus, the antecedent of the major premise are completely unconstrained when trying to determine whether a particular statement can be inferred by this inference. Even in the 1990s, eminent logicians saw that as a decisive obstacle to using natural deduction calculi for automated proof search. For normal proofs the "fact" is, of course, false.

The above reconstruction of *one* path that may have led Gentzen up to Part (V) is most plausible if one focuses on the problem he was addressing and the steps he, in fact, took to solve it, but also those he consciously did not take (most importantly, the normalizability of N1J proofs). At this point, Gentzen had shown that classical arithmetic is consistent relative to intuitionist arithmetic and, in particular, that the addition of *tnd* does not lead to contradictions. The next issue was to establish the consistency of intuitionist arithmetic — following Hilbert's semi-semantic pattern. It is this intention that is signaled through the programmatic statement (V).

7 An impasse.

When Gentzen finally confronted the consistency issue head on, he must have recognized that Hilbert's considerations were methodologically unsatisfactory: he had difficulties, broadly, to reconcile them with Gödel's result and, more narrowly, to give a finitist interpretation of the conditional and hence of negation. These concerns shape the exposition in (Gentzen 1936) and actually go back to this point in time, mid-October 1932. Thanks to Gentzen's manuscript INH we don't have to resort to conjectures, as INH

---

[46](Gentzen 1932/33, p. 9). The German text is: Es bedarf einiger Überlegungen, um einzusehen, dass in der Tat jeweils wieder ein richtiger Beweis entsteht. Ich verzichte auf diese genaue Durchführung, da ich von diesen Tatsachen keinen Gebrauch machen werde, sie vielmehr nur zur Veranschaulichung vorführe.

directly mirrors his concerns with its detailed reflective observations.[47] Two thirds of this document was written between mid-October and mid-November 1932, a few pages in early 1933, and the rest in October 1934. The manuscript deals with the concept of *contentual correctness in pure number theory* and its relation to consistency proofs. The discussion is intricate and deserves a fuller treatment than I can provide here, as I want to focus on those issues that are directly connected to (Hilbert 1931b) and are explicitly taken up in (Gentzen 1936).

As we saw in section 5, the notion of *correctness* for the basic constructive theory was central in Hilbert's considerations. A more restricted semantic component played a role in Hilbert and Bernays's consistency proofs from the very beginning. Recall that in (Hilbert *1921/22) linear derivations were transformed into tree-like ones with Boolean combinations of numeric formulae at their nodes. As the (instantiated) axioms are correct and the inferences preserve correctness, all combinations occurring in such a derivation must be correct. Consequently, there cannot be a derivation with an incorrect endformula. In order to extend this *fundamental proof theoretic idea* to derivations involving quantificational logic, Hilbert introduced the ϵ-substitution method, removing quantifiers from derivations in favor of epsilon terms and requiring also their evaluation to determine correctness (of quantifier-free formulae). Now, in 1931, correctness is defined directly for quantified statements exploiting the understanding of quantifiers (with variables ranging over natural numbers) and "guaranteeing" consistency immediately. As described in section 5, Hilbert reasons in (1931b) for the consistency of his constructive theory by asserting, "All transfinite rules and inference schemata are consistent; for they amount to definitions."

A critical reader of Hilbert's paper could ask, how is the notion of correctness defined, and what principles are used to prove the correctness of all theorems? In fact, that is the opening of INH, written on 14 and 16 October 1932. Having defined contentual correctness for an intuitionist theory[48] and the concept "B is an intuitionist consequence of (assumptions) $V_1, \ldots, V_n$", Gentzen raises the questions, how far these notions are formal, and why the correctness proof by induction on the length of derivations is, according to Gödel, definitely not formal. In both questions Gentzen

---

[47] The full title of INH is *Die formale Erfassung des Begriffs der inhaltlichen Richtigkeit in der reinen Zahlentheorie, Beziehungen zum Widerspruchsfreiheitsbeweis* (The formal characterization of the concept of contentual correctness in pure number theory. Relations to the consistency proof). — The manuscript consists of 36 pages in shorthand; Thiel's typewritten transcription is 69 pages long.

[48] INH, p.2: Man definiert inhaltliche Richtigkeit so: die mathematischen Axiome sind richtig. $A \ \& \ B$ ist richtig, wenn A richtig ist und B richtig ist; $A \lor B$ ist richtig, wenn mindestens eines richtig; Ax, wenn bei jeder Zahleinsetzung für x dies richtig, ebenso $(x)Ax$; Aa, wenn eine Zahl angegeben werden kann, so daß Aa gilt, ebenso $(Ex)Ax$; $A \rightarrow B$, wenn aus der Richtigkeit von A die von B geschlossen werden kann; $\neg A$, wenn aus A der Widerspruch geschlossen werden kann. - This is in accord with Hilbert's intended definition of correctness. I have made some trivial notational changes from Thiel's transcription for easier comparability, and I separated the clauses for the different connectives by semicolons instead of commas.

takes formal to mean expressible in the formal theory, i.e., he takes the correctness notion to be definable and the correctness proof to be formalizable in the theory. A few days later, on 19 October, he gives a more general articulation of the issues and understands formal in a different way, contrasting *proof theoretic* with *semantic* consistency proofs:

> I seek to clarify the questions: what distinguishes a formal correctness or consistency proof from a contentual one, why is the former for certain inferences not even possible by these same inferences (according to Gödel), is a bridge inference involved then, how secure is that [bridge inference, WS], what are the connections with Gödel's proof, what role do the mathematical axioms play?[49]

He formulates a plan for investigating these questions in clear stages. He intends to treat first a theory consisting only of mathematical axioms (formulated as inferences), then to add induction, after that to consider the sentential logical connectives $\&$, $\vee$ as well as the quantifiers $\forall$, E, only later the conditional and negation, and finally to examine *tnd*.

After some experimentation with contentual correctness proofs for such restricted theories, Gentzen asks whether they involve inferences that are avoided by "purely formal proofs", seeking to explain the seemingly essential difference between the two kinds of correctness proofs.[50] That leads him to the question, "Whether Gödel's result is essentially based on the fact that consistency instead of correctness is being considered." On the next day, Gentzen returns to this question with a definite answer, "consistency is a much more formal property than *correctness*". He also formulates an insight that shapes his subsequent considerations, namely, that consistency is equivalent (under minimal conditions on the formal theory) to what, in contemporary proof theory, is called the *reflection principle* for numeric statements: if such statements are provable, then they hold. On the one hand, this insight captures and reiterates the core idea of the proof theoretic work of Hilbert, Bernays and Ackermann, and, on the other hand, it frames Gentzen's further discussion that addresses Hilbert's issues in an independent and exploratory way.

Calling a proof of a numeric statement a *Normalbeweis* if it contains only numeric statements, Gentzen can now express the difference between

---

[49]INH, p. 3: Ich suche Aufklärung über die Fragen: wie unterscheidet sich ein formaler Richtigkeits- bzw. Widerspruchsfreiheitsbeweis von einem inhaltlichen, wieso ist ersterer bei gewissen Schlußweisen nicht einmal mit Hilfe dieser selbst (nach Gödel) möglich, liegt dann ein Brückenschluß vor, wie groß ist dessen Sicherheit, wie sind die Zusammenhänge mit dem Gödelschen Beweis, welche Rolle spielen die mat[hematischen] Axiome?

[50]"Purely formal correctness proofs" are understood now in a way that will be sharpened in the next paragraph. — It seems that the difference formulated there between *semi-contentual and contentual*, respectively, *purely formal and formal* can be accounted for *mostly* by the formulation of correctness for numeric statements: if that is done syntactically as below, then the proofs would be semi-contentual and purely formal; if that is done by a semantic evaluation of terms, then they would be contentual and formal. On p. 5 of INH Gentzen calls his definition of "correct" for numeric statements *purely formal*; the definition specifies that the statement can be obtained syntactically by an immediate derivation from the axioms, what is called *Normalbeweis* below.

(purely) formal and (semi-) contentual correctness proofs by formulating carefully the claim each is to establish. The claim for a *purely formal correctness proof* is, "for every proof of a numeric statement there is a *Normalbeweis* of that statement", and the corresponding claim for the (semi-) contentual correctness proof is, "every proof has a correct result" [where result means endformula, WS]. Calling a formula with free and just universally quantified variables *correct* if every substitution instance has a *Normalbeweis*, Gentzen sketches then the proofs for a simple theory with quantifier-free axioms and rules for universal quantifiers. Comparing the proofs, Gentzen sees the essential difference as follows:

> The semi-contentual proof uses complete induction for a rather complicated statement. This contains Ri erg x [the result of proof x is correct, WS], and this predicate becomes ever more complicated in complicated cases. The formal proof uses complete induction for Ey. No y & erg x = erg y [there is a *Normalbeweis* y having the same result as the given proof x, WS]; this is also a statement containing logical signs; it is however of a simpler nature, also in more complicated cases.[51]

And yet, Gentzen is still not completely sure where the *Gödel-Punkt* lies. On the next page of INH he writes: "I think I see now clearly, why a consistency proof by a crude contentual interpretation is not formalizable. [It is not formalizable in the usual formalisms, WS] for the very reason that the interpretation itself is not formalizable." He emphasizes that it is necessary to see more clearly that *such* a consistency proof is not formalizable, but also that no consistency proof whatsoever can be carried out in the usual formalisms. In order to gain a clearer view, he extends the earlier comparative methodological experiments to a theory with quantifier-free axioms and complete induction. In this case the formal correctness proof involves a *transfinite* valuation of derivations of the sort used already in Ackermann's proof from 1924; see (Zach 2003). The experimenter Gentzen is prompted to make the following observation:

> In any event, I think there must be some connection between the non-formal element of the "correctness" definition (that is not formalizable) and that of the transfinite induction (that is not formalizable). After all, each of them apparently makes possible a consistency proof that cannot be formalized.[52]

Having gained clarity, at least in a broad sense, about the *Gödel-Punkt* and its location relative to proof theoretic and semantic considerations, he turns on 31 October to the last point of his plan from 19 October, i.e., the investigation of systems expanded by *tnd*. Central is the characterization

---

[51] INH, p. 8: Der halbinhaltliche Beweis macht eine VJ über eine ziemlich komplizierte Aussage. Diese enthält Ri erg x, und dieses Prädikat wird in komplizierten Fällen immer komplizierter. Formale[r] Beweis macht VJ über Ey. No y & erg x = erg y, eine Aussage zwar auch mit logischen Zeichen, doch von einfacherer Natur, auch bei komplizierteren Fällen. — VJ stands, of course, for *Vollständige Induktion*.

[52] INH, p. 10: Jedenfalls muß doch meines Erachtens irgendein Zusammenhang zwischen dem unformalen Element bei der nicht formalisierbaren Definition der "Richtigkeit" und bei der nicht formalisierbaren transfiniten Induktion bestehen. Da ja anscheinend jede von beiden einen nicht formalisierbaren Widerspruchsfreiheitsbeweis ermöglicht.

of "richtig" and "falsch"; thus, Gentzen addressed the very same issue as Hilbert and recognized as well, it seems, that his own attempt of doing so needed *tnd* in the metatheory.

He decides on 2 November, after quite a bit of reflection concerning a direct treatment of *tnd*, to return to the investigation of the contentual correctness of intuitionist mathematics. After all, relative to intuitionist arithmetic he had already shown that the addition of *tnd* does not lead to a contradiction. He begins by considering the essence of a purely formal consistency proof: "The main thing for this [kind of proof, WS] is to associate with a derivation of a numeric result a *Normalbeweis* of the very same result." (INH, p. 17) This association must be intuitionistically unobjectionable and should be achieved by a finite number of reduction steps similar to those used in the normalization of derivations in intuitionist first-order logic. The stepwise procedure should *simplify* the given derivation until it is a *Normalbeweis*; the initial derivation is then called *reducible*. This idea is reemphasized a little later:

The intent is: to simplify a given proof with a numeric result in steps until there is a numeric proof for the same result . . . I want to get by with a valuation that is as simple as possible.[53]

The discussion in the first part of INH ends abruptly on 8 November, when Gentzen compares a contentual proof with a proof using his reducibility concept. He discovers a difficulty for the latter proof, but emphasizes, ". . . the difficulty is seemingly not connected with complete induction, but rather with the nature of consequence, thus to the → sign." (INH, p. 20,1) The difficulties surrounding the interpretation of the intuitionist conditional, thus of negation, are also emphasized in the writings of Bernays, for example in his (1934, pp. 71–72) and later in *Grundlagen der Mathematik II*, pp. 358–360.

Gentzen takes up the thread, most likely, in February of 1933 and then again in June of that year. Important developments are indicated, and Gentzen ends with general thoughts on consistency proofs and transfinite values of derivations:

Every proof has a (transfinite) value. Consistency of a system of proofs can only be shown by a proof that has a higher value than all of these. Thus Gödel's theorem. This one has to try to prove. When doing it the values have to be determined. Furthermore, [investigate] whether there are proofs with higher values that have nevertheless greater certainty.[54]

---

[53] INH, p. 19: Die Absicht ist: einen vorliegenden Beweis mit numerischem Ergebnis schrittweise zu vereinfachen, bis ein numerischer Beweis für dasselbe Ergebnis dasteht. . . . Ich möchte mit einer möglichst einfachen Wertung auskommen.

[54] INH, p. 23: Jeder Beweis hat einen (transfiniten) Wert. Widerspruchsfreiheit für ein System von Beweisen läßt sich nur zeigen durch einen Beweis von höherem Wert als alle diese. Daher der Satz von Gödel. Dies ist zu beweisen zu versuchen. Dabei die Werte feststellen. Ferner dann, ob es Beweise höheren Wertes und trotzdem größerer Sicherheit gibt.

This remark connects and fully reconciles his proof theoretic considerations with Gödel's second theorem. The impasse for a consistency proof he had encountered in November 1932 remains, however, an impasse in June 1933.

8 Toward a solution.

Gentzen completed a consistency proof based on his reducibility concept in late 1934, but definitely not later than sometime in October of that year. The reason for this upper bound is simple. On p. 25 of INH, dated "X. 34" (October 1934), Gentzen expresses "Nagging doubts. Concerning the value of the consistency proof". He goes back to the discussion contrasting formal and semi-contentual correctness proofs and locates his proof with the reducibility notion "as a kind of intermediate thing between the other two":

For $\&\ \forall\ \vee\ \exists$ it [the proof via reducibility, WS] runs parallel to the semi-contentual proof recognizing here the contentual [understanding, WS]. For $\supset$ it [the proof, WS] jumps off and moves over to the formal proof, intent on avoiding the contentual $\supset$.[55]

That creates the impression: one begins with the semi-contentual. This becomes increasingly ominous, as one obviously assumes exactly what one wants to prove. Finally, for $\supset$ one decides to jump off and to take refuge in the formal [considerations, WS]; in that way the proof becomes more complicated and then requires, to be carried through, auxiliary tools of a special sort, for example, the super-ordering.[56]

Intent on putting the nagging doubts aside and "regaining firm ground", Gentzen interprets, (INH, p. 29), universal and existential statements, in a first step, with just finitist matrices. Before expanding the interpretation, he formulates a plan for Part III of what will be his (1936); that part is to discuss the methodological issues he has been grappling with:

Plan for Part III: first, after the finite, the An-sich-view for the infinite, which fully corresponds to that for the finite. This we reject. Then the development of the intuitionist standpoint in words like those on pp. 21–22 of INH. (Perhaps Gödel's

---

[55] Note that Gentzen uses here for the first time in INH the horseshoe as the symbol for the conditional.

On the margin, he wrote: Richtiger gesagt: bei $\&\ \forall\ \vee\ \exists$ sind die finiten Deutungen eben noch fast gleichwertig dem An-sich-sein, bei $\supset$ geht das ... weniger! — Here is the English translation: More correctly one would say: for $\&\ \forall\ \vee\ \exists$ the finitist interpretations are still almost the same as those given by the *An-sich-Auffassung*; for $\supset$ that works less.

It should be emphasized that Gentzen, in this late part of INH from October 1934, repeatedly refers back to the INH notes from 1932; for example, on pp. 25, 29, 30 he refers back to pp. 17–18 and 19–20, 21–22, 16 and 21, respectively.

[56] INH, p. 25: Bei $\&\ \forall\ \vee\ \exists$ geht er gleichlaufend mit dem halbinhaltlichen Beweis, indem er hier das inhaltliche anerkennt. Bei $\supset$ springt er ab und geht über zum formalen Beweis, indem er das inhaltliche $\supset$ vermeiden will.

Das erweckt den Eindruck: man beginnt mit Halbinhaltlich[em]. Dies wird einem fortschreitend immer unheimlicher, da man offenbar fast genau das schon voraussetzt, was man beweisen will. Bei $\supset$ endlich entschließt man sich, abzuspringen und rettet sich ins Formale, wodurch der Beweis kompliziert wird und nachher zu seiner Rettung wieder Hilfsmittel besonderer Art, sagen wir: die Superordnung, benötigt.

In the shadow of incompleteness

transfer theorem should be formulated right away.) Then probing critical analysis, as still to be developed.[57]

This will remain the blueprint for Part III. Indeed, the structure of the paper's Part III is fully in accord with this plan.[58]

When coming back to the interpretation of universal quantifiers, Gentzen expands on the marginal note from p. 25 of INH (see fn. 55) and claims that for the theory without the conditional, no reduction is achieved, just a confirmation: "No reduction to something more simple, but rather a confirmation of the constructivity [of the interpretation, WS]." That constructivity is not confirmed by proof, but is being grasped with words. Here is the reason:

Because the difference between An-sich and constructive is not formally captured. It is only known by its meaning. And that has to be that way, as it is indeed the last, extra-mathematical foundation.[59]

The interpretation is finally given in greater detail (on p. 33 of INH) under the heading "The correctness proof by finitist interpretation for the theory without $\supset$ and $\neg$". Here is, paradigmatically and exhibiting the deep parallelism with Hilbert's rule (HR^*), the interpretation of $\forall x F x$: this statement holds (and has consequently meaning) if for every numeral ν there is a constructible derivation for $F\nu$ that has been recognized already as correct. The interconnectedness of the contentual considerations for statements and for derivations is taken on the next page, which is indeed also the last of INH, as a reason for emphasizing the significance of the reducibility notion. As to the circularity for $\supset$ Gentzen writes:

Perhaps $\supset$ does not involve a circle, or more correctly, the situation is as follows: Also for the previous [logical, WS] signs a kind of "An-sich"-meaning is assumed. But surely this is somehow of a constructive kind ... also e.g. $\forall$ involves contentually an "An-sich"-for-all. The foundation on the "existence of a [correct, WS] derivation" does not seem to be a real foundation. Indeed, it cannot be. After all,

---

[57] INH, p. 29: Plan für den III. Abschnitt: erst, nach dem Endlichen, die An-sich-Auffassung im Unendlichen, welche der im Endlichen ganz entspricht. Diese lehnen wir ab. Dann die Entwicklung des intu[itionistischen] Standpunktes mit Worten wie auf INH 21–22. (Evtl. gleich Angabe des Gödelschen Übertragungssatzes.) Dann tiefergehende Kritik, wie noch zu entwickeln.

[58] Part III is entitled "Bedenkliche und unbedenkliche Schlußweisen in der reinen Zahlentheorie" and the sections have the headings, "Die Mathematik endlicher Gegenstandsbereiche" (section 7), "Entscheidbare Begriffsbildungen und Aussagen im unendlichen Gegenstandsbereich" (section 8), "Die "an-sich"-Auffassung der transfiniten Aussagen" (section 9), "Finite Deutung der Verknüpfungszeichen $\forall$, &, $\exists$, und $\vee$ in transfiniten Aussagen" (section 10), and finally, "Die Verknüpfungszeichen $\supset$ and $\neg$ in transfiniten Aussagen; die intuitionistische Grenzziehung" (section 11).

[59] INH, p. 29: Weil eben der Unterschied von An-sich und konstruktiv nicht formal erfaßt ist. Eben nur dem Sinn nach bekannt. Und das muß auch so sein, dies ist eben das letzte, außermathematische Fundament.

the "correctness" presupposes again contentual knowledge of the meaning for the ∀, and this is as well an "An-sich"-meaning, if one wishes.[60]

Gentzen re-asserts then that ∀ and ∃ are explicated by a contentual for-all and there-is. But this contentual understanding is not that of the "An-sich"-for-all or the "An-sich"-there-is. So for him the real question is:

What differentiates the "An-sich"-∀ from the contructive ∀, though the same inferences hold for both? And what about ∃? The constructive interpretation reduces to something "conceptually simple". For ⊃ one can argue, whether there is something simpler or not. Also, the "idea of reducibility" is important.[61]

These considerations are systematically taken up in the 1936 paper. In particular, the interpretation of the conditional remains (on p. 530) a main task for the consistency proof. When discussing the interpretation of statements in the language of first-order arithmetic, Gentzen formulates at the end of section 9 two explicit goals, namely, (i) to provide a finitist meaning for these statements, i.e., "to *interpret* each such statement as expressing a determinate, finitely representable fact", and (ii) to ensure that the logical inferences are in harmony with the finitist meaning of the statements involved. Here is the last paragraph of section 9 outlining the work that is related to (i) and (ii) and that is to be accomplished in the next two sections:

In the following section 10 that [program indicated by (i) and (ii), WS] is to be carried out for a considerable class of transfinite statements and the associated inferences. In section 11, I will treat the remaining statement forms and inferences; there the method will encounter difficulties, and the significance of the *intuitionist* (1.8) *boundary* between permissible and impermissible inferences will emerge; furthermore, another even more restrictive boundary can be seen to be defensible.[62]

Indeed, Gentzen discusses in section 10 the finitist interpretation for some of the connectives in transfinite statements, namely, for universal and

---

[60] INH, p. 33: Vielleicht liegt doch kein Zirkel vor beim ⊃, oder richtiger, liegt die Sache so: Auch bei den vorigen Zeichen setzt man eine Art "An-sich"-Sinn schon voraus. Freilich ist dieser irgendwie konstruktiver Art. . . . auch bei ∀ z.B. liegt doch inhaltlich ein "An-sich"-Alle vor. Die Begründung auf das "Bestehen einer [korrekten, WS] Herleitung" scheint doch auch keine wirkliche Begründung zu sein. Kann es ja gar nicht. Die "Korrektheit" setzt eben wieder inhaltliches Wissen über den Sinn des ∀ voraus, und dies ist ebenso gut ein "An-sich"-Sinn, wenn man will.

[61] INH, p. 33: Wodurch unterscheidet sich das "An-sich"-∀ von dem konstruktiven ∀, obwohl für beide die gleichen Schlußweisen gelten? Und bei ∃? Die konstruktive Deutung führt zurück auf etwas "begrifflich Einfaches". Bei ⊃ kann man darüber streiten, ob etwas Einfacheres vorliegt oder nicht. Auch die "Idee der Reduzierbarkeit" ist wichtig.

[62] (Gentzen 1936, p. 525). The German text is: Das soll im folgenden § 10 für einen beträchtlichen Teil der transfiniten Aussagen und der zugehörigen Schlußweisen durchgeführt werden. In 11 behandle ich dann die restlichen Aussageformen und Schlußweisen; dabei stößt die Methode auf Schwierigkeiten, und es zeigt sich die Bedeutung der *intuitionistischen* (1.8) *Grenzziehung* zwischen erlaubten und unerlaubten Schlußweisen innerhalb der Zahlentheorie; ferner ergibt sich eine andere, noch engere Grenzziehung als ebenfalls verfechtbar. — On p. 532 Gentzen makes this narrower boundary explicit, namely, no general use of the conditional.

In the shadow of incompleteness   189

existential quantifiers, as well as for conjunctions and disjunctions. Before turning in section 11 to the problematic connectives (conditional and negation) and to the discussion of the intuitionist boundary, he concludes section 10 by asserting what he had indicated briefly in INH:

> One could then, proceeding from these considerations, develop a purely formal consistency proof for this part of number theory.[63] But such a proof would have little value, for in the proof itself one would have to *use* transfinite statements and the accompanying modes of inference that one wants to *"ground"* by the proof. So the proof would not be a proper *reduction*, but rather a *confirmation* of the *finitist* character of the formalized rules of inference. But one must already be clear in advance *what* counts as *finitist* (in order then to be able to carry out the consistency proof itself with finitist means of proof).[64]

That outlines a consistency proof using an appropriate truth definition or the contentual concept of *Richtigkeit*. Hilbert's way of proceeding, not clearly respecting the line between syntactic and semantic considerations, has been made blindingly clear by Gentzen.

Hilbert and Bernays, in their *Grundlagen der Mathematik* (1939, p. 390), consider a consistency proof that rests ultimately or mainly on semantic considerations as unsatisfactory for proof theory. Indeed, they recall the "formalism of recursive number theory" and their proof theoretic treatment establishing its consistency: the latter was central despite the fact that, in principle, one could simply have pointed to its finitist interpretation. (See section 2 above.) In their judgment, Gentzen's consistency proof for elementary number theory does justice to this concern. They consider as a very serious possibility that the fundamental idea of Gentzen's consistency proof (using transfinite induction up to ε_0) can be extended beyond elementary number theory to more comprehensive formalisms, involving then of course larger ordinals. Their book ends with the remark:

> If this perspective should prove accurate, Gentzen's consistency proof would open a new era for proof theory.[65]

---

[63]"Purely formal" has to be understood here differently from INH, namely, as emphasizing, "informally" so-to-speak, that the proof does not have any real content and does not provide a meaningful reduction.

[64](Gentzen 1936, p. 529). The German text is: Man könnte dann, von diesen Überlegungen ausgehend, einen rein formalen Widerspruchsfreiheitsbeweis für diesen Teil der Zahlentheorie entwickeln. Ein solcher hätte aber nicht viel Wert, denn man müßte in dem Beweis selbst transfinite Aussagen und dieselben zugehörigen Schlußweisen *benutzen*, die man durch ihn *'begründen'* will. Der Beweis würde also keine eigentliche *Zurückführung* bedeuten, wohl aber eine *Bestätigung* des *finiten* Charakters der formalisierten Schlußregeln. *Was* aber *finit* ist, darüber müßte man sich zuvor im klaren sein (um dann den Widerspruchsfreiheitsbeweis selbst mit finiten Beweismitteln führen zu können).

[65](Hilbert and Bernays 1939, p. 374). The German text is: Falls diese Perspektive sich bewähren sollte, so würde mit dem Gentzenschen Widerspruchsfreiheitsbeweis ein neuer Abschnitt der Beweistheorie eröffnet.

9 New perspectives.

Gentzen gives, in his 1936-paper, not only a consistency proof, but provides a formal substitute for the contentual correctness notion, namely that of "stability of a *reduction procedure*" (1936, p. 536). This concept, together with the translation from classical into intuitionist arithmetic, Gentzen views as giving "a particular finitist interpretation[66] of the statements [of classical arithmetic, WS], which replaces their interpretation via the *an-sich-Auffassung*".[67] In a letter to Kneser written on 5 December 1934, Gentzen reports that he is in the process of polishing his paper for publication in *Mathematische Annalen*.[68] The paper was submitted on 11 August 1935. After correspondence with Bernays and, indirectly, Gödel (with whom Bernays had discussed Gentzen's proof), Gentzen inserted sections 14.1 to 16.1 1 that replaced the earlier treatment by one involving transfinite induction up to ε_0. (No changes were made in the parts of the paper to which I have been referring.) That replacement was made in February 1936. The original argument, together with an introduction by Bernays, was published in 1974. Here remains the task of joining Gentzen's methodological goals with the details of his original argument, with the mathematical analyses of Coquand and Tait, as well as with contemporary directions of proof theory.

Ironically, Gödel was deeply influenced by Gentzen's and Hilbert's considerations. That is clear from his Lecture at Zilsel's, (Gödel 1938), in which he explored various extensions of finitist mathematics and analyzed, in particular, Gentzen's first consistency proof. One of these extensions, via finite type theories for computable functionals, was pursued in lectures Gödel gave in 1941 at Princeton and at Yale. That work led ultimately to the Dialectica interpretation and was published in (Gödel 1958); its mathematical and foundational aspects are discussed in Troelstra's most informative

---

[66] In his (1941), Bernays considers the principle of transfinite induction up to ε_0 as non-finitist, but remarks that his position should not be considered "as the standpoint of the Hilbert School". In the quotation here, but also in other places of his paper, for example on p. 564, Gentzen speaks quite clearly as providing a *finitist* interpretation. May that also be in the background for Hilbert's famous remark in the Preface to the first volume of *Grundlagen der Mathematik*, that Gödel's results do not show that proof theory cannot be carried through, but rather that the finitist standpoint has to be exploited in a sharper way in order to obtain consistency proofs for more complex formalisms? — It would be of great biographical interest, but also of significance for our understanding of the systematic proof theoretic developments, if we would know more about the relation between Gentzen and Hilbert, most importantly, during the time from 1931 to 1934, but also after November 1935, when Gentzen had been appointed as Hilbert's "Special Assistant". Menzler-Trott conjectures that, during the later period, they only talked about "newspapers, poems and popular science".

[67] This connection remains to be explored; here I point to the summary in Gentzen's paper, pp. 564–565, and the related discussion in (Sieg and Parsons 1995, pp. 83–85). The most significant connection of Gödel's to Gentzen's first consistency proof, revealed in his Zilsel Lecture, has been detailed in (Tait 2005a).

[68] (Menzler-Trott 2007, p. 55). The more extended text is: "At the moment I am preparing a consistency proof for pure (i.e., no analytic means employed) number theory, which I have finished, for publication in *Mathematische Annalen*."

Introductory Notes, (Troelstra 1990, Troelstra 1995). In the Yale lecture, (Gödel 1941), Gödel viewed his project in the same general way as Gentzen saw his, namely, as giving an interpretation of intuitionist arithmetic from a more strictly constructive standpoint. He articulates three requirements such a position should satisfy and writes then:

Let me call a system strictly constructive or finitistic if it satisfies these three requirements (relations and functions decidable, respectively, calculable, no existential quantifiers at all, and no propositional operations applied to universal propositions). I don't know if the name "finitistic" is very well chosen, but there is certainly a close relationship between these systems and what Hilbert called the "finite Einstellung". (Gödel 1941, p. 191)

Coming back to Hilbert's last paper, I quote an important passage that occurs right after the discussion of the concept of *Richtigkeit*, which — as we saw — is also central for Gentzen:

Finally the important and, for our investigation, decisive fact has to be emphasized, namely, that all the axioms and inference schemata VI, which I have called "transfinite", have a strictly finitist character: the prescriptions they contain can be executed in the finite.[69]

Hilbert's considerations in (1931b) were a crucial germ for Gentzen's work on consistency, presenting — as they did — a new perspective and pointing in novel directions. The concrete problems arising in part from difficulties that had been pushed aside, Gentzen resolved in surprising and ingenious ways, but fully in the spirit of Hilbert's view that true contentual thinking consists in operations on proofs.

There are additional Hilbert manuscripts from around this time; for example, folder SUB 603 contains a copy of Hilbert's last paper with notes in his own hand (not helpful, it appears, for understanding the paper) and many seemingly unconnected pages. On one of these loose pages one finds a remarkable general statement that conveys the impact of his reflections on Gödel's second theorem: "Consistency is naturally a relative notion; that is not an objection to my theory, but rather a necessity." Hilbert drew an arrow to "relative" and wrote at the end of the arrow "New!"[70]

---

[69](Hilbert 1931b, p. 121). The German text is: Endlich werde noch die wichtige und für unsere Untersuchungen entscheidende Tatsache hervorgehoben, die darin besteht, daß die sämtlichen Axiome und Schlußschemata VI, die ich transfinit genannt habe, doch ihrerseits streng finiten Charakter haben: die in ihnen enthaltenen Vorschriften sind im Endlichen ausführbar.

[70]That is in conflict with the earlier perspective on the finitist standpoint. — Even in *Über die Grundlagen des Denkens*, SUB 604, Hilbert explicitly asserts that the concept "widerspruchsfrei" is "absolut" (on p. 6 of the original manuscript); but that particular use may not be in conflict with the new perspective, as absolute is used here in a different sense: 'Widerspruchsfrei' — ebenso wie 'Richtig' — ist ein absoluter Begriff; denn wir setzen ausdrücklich fest, dass beim Beweisen als zulässige Begriffs- und Schlussmethoden jedesmal nur diejenigen anzusehen sind, die das Verständnis der Aussage A nötig macht, so dass hinsichtlich der Bedeutung unserer Definitionen von 'widerspruchsfrei' und 'richtig' keine Unbestimmtheit eintritt.

10 Appendix

Wilfried Buchholz pointed to a paper of Kurt Schütte that appeared in an obscure journal and was based on a talk Schütte had given on 13 May 1993 to a meeting in Munich to memorialize Hilbert who had died 50 years ago. The paper's title is *Bemerkungen zur Hilbertschen Beweistheorie*. Here are two quotations that are of logical and deep human interest:

Hilbert hat selbst keine Widerspruchsfreiheitsbeweise durchgeführt, sondern nur die Anregung dazu gegeben. Er hat bald nach Erscheinen der Gödelschen Unvollständigkeitssätze und noch vor dem Gentzenschen Widerspruchsfreiheitsbeweis in einem Göttinger Kolloquiumsvortrag vorgeschlagen, das Axiomenschema der vollständigen Induktion in Erweiterung des finiten Standpunktes durch die konstruktive Verwendung einer Schlussregel mit unendlich vielen Prämissen zu ersetzen. Diese von Hilbert vorgeschlagene Methode wurde zunächst überhaupt nicht beachtet, sondern erst fast 20 Jahre später aufgegriffen und ist heute für beweistheoretische Untersuchungen von starken Teilsystemen der Analysis unentbehrlich geworden. ....

Ich selbst hatte nur mit Hilberts Mitarbeiter Paul Bernays, der bis 1933 ausserordentlicher Professor in Göttingen war, einen engen wissenschaftlichen Kontakt gewonnen. Mit Hilbert bin ich persönlich nur zweimal zusammengekommen, nämlich erstens bei meinem Besuch in seiner Wohnung vor meiner Doktorprüfung, und zweitens in der mündlichen Prüfung vor genau 60 Jahren als letzter Doktorand von Hilbert, wobei ich beide Male seine menschliche Güte erfahren habe.

II.4

Gödel at Zilsel's

(with Charles Parsons)

The text that follows [in (Gödel 1995), WS] consists of notes for a lecture Gödel delivered in Vienna on 29 January 1938 to a seminar organized by Edgar Zilsel. The lecture presents an overview of possibilities for continuing Hilbert's program in a revised form. It is an altogether remarkable document: biographically, it provides, together with (1933b) and (1941), significant information on the development of Gödel's foundational views; substantively, it presents a hierarchy of constructive theories that are suitable for giving (relative) consistency proofs of parts of classical mathematics (see §§2–4 of the present note); and, mathematically, it analyzes Gentzen's (1936) proof of the consistency of classical arithmetic in a most striking way (see §7). A surprising general conclusion from the three documents just mentioned is that Gödel in those years was intellectually much closer to the ideas and goals pursued in the Hilbert school than has been generally assumed (or than can be inferred from his own published accounts).

1 The setting

Edgar Zilsel (1891–1944) had been connected with the Vienna Circle. By 1938 his main interest was history and sociology of science.[1] The concrete stimulus for Gödel's preparing the lecture was Zilsel's question whether

---

We are grateful to Robert S. Cohen, Cheryl Dawson, John Dawson, Warren Goldfarb, Nicolas Goodman and especially Solomon Feferman and A. S. Troelstra for information, suggestions, and/or comments.

[1] Cf. (Zilsel 1976), which collects in German translation articles published in English after Zilsel's emigration. The English versions are collected in (Zilsel 2000). Zilsel had taught physics and philosophy at the *Volkshochschule* in Vienna, but as a result of the Dollfuss coup in 1934, he was dismissed from or eased out of this position (see Behrmann 1976) and (Dvořák 1981, pp. 23–25)) and thereafter worked as a *Gymnasium* teacher. He emigrated in 1938 to England and moved on to America the next year. In 1944, while teaching at Mills College in California, he committed suicide.

Testimony differs as to whether Zilsel was a member of the Vienna Circle or merely someone sympathetic to their views who attended some of the sessions; see (Dvořák 1981, pp. 30–31).

anything new had happened in the foundations of mathematics and his request that Gödel should describe the "status of the consistency question" to the seminar. This is reported in Gödel's notes[2] concerning the organizational meeting of the seminar on 2 October 1937 at Zilsel's home; it was at this organizational meeting that Zilsel made his request and that Gödel, after some reflection, agreed to speak on the consistency question.

As Zilsel did not have a university position, the seminar was probably an informal, private affair and continued to meet at Zilsel's home.[3] Gödel's notes do not give a clear view of the seminar's intended theme, except that it was rather general. The list of names of those present or mentioned as possible participants confirms this; they spread over several fields and include no one, except for Gödel, who was much involved in mathematical logic.[4] In his immediate response to Zilsel's request Gödel suggested presenting a German version of the lecture (1933b), given in Cambridge, Mass. (published in (Gödel 1995)); but on reflection he added that that talk was "zu prinzipiell", which can be translated roughly as "too general".

The Zilsel lecture gives, as we remarked, an overview of possibilities for a revised Hilbert program. The central element of that program was to prove the consistency of formalized mathematical theories by finitist means. Gödel's 1931 incompleteness theorems have been taken to imply that for theories as strong as first-order arithmetic this is impossible, and indeed, so far as Gödel ventures to interpret Hilbert's finitism, that is Gödel's view in the present text as well as earlier in (1933b) (though not in (1931d)) and later in (1941), (1958) and (1972). The crucial questions then are what extensions of finitist methods will yield consistency proofs, and what epistemological value such proofs will have.

Two developments after (Gödel 1931d) are especially relevant to these questions. The first was the consistency proof for classical first-order arithmetic relative to intuitionistic arithmetic obtained by Gödel (1933d). The proof made clear that intuitionistic methods went beyond finitist ones (cf. footnote 10 below). Some of the issues involved had been discussed in Gödel's lecture (1933b), but also in print, for example in (Bernays 1935b) and (Gentzen 1936). Most important is Bernays's emphasis on the "abstract element" in intuitionistic considerations.[5] The second development was Gentzen's consistency proof for first-order arithmetic using as the additional principle—justified from an intuitionistic standpoint—transfinite

---

[2]The notes *Zusammenkunft bei Zilsel* are in Gödel's *Nachlass* (document no. 030114) and were transcribed from Gabelsberger by Cheryl Dawson. In editing the text and preparing this note, we have also used a document (no. 040147) entitled *Konzept* (i.e., draft), evidently an earlier draft of Gödel's notes for this lecture.

[3]Karl Popper writes (1976, p. 84) of having given a paper to a gathering there several years earlier.

It is likely that the seminar did not continue long after Gödel's talk, since the Anschluß took place only six weeks later, and not long after that Zilsel emigrated (cf. fn. 1).

[4]We are indebted to Katalin Makkai for researching this matter.

[5]Significantly, Gödel refers to that earlier discussion in (1958).

induction up to ϵ_0. Already in (1933b, p. 31) Gödel had speculated about a revised version of Hilbert's program using constructive means that extend the limited finitist ones without being as wide and problematic as the intuitionistic ones:

> But there remains the hope that in future one may find other and more satisfactory methods of construction beyond the limits of the system A [[capturing finitist methods]], which may enable us to found classical arithmetic and analysis upon them. This question promises to be a fruitful field for further investigations.

The Cambridge lecture does not suggest any intermediate methods of construction; by contrast, Gödel presents in the Zilsel lecture two "more satisfactory methods" that provide bases to which not only classical arithmetic but also parts of analysis might be reducible: quantifier-free theories for higher-type functionals and transfinite induction along constructive ordinals. Before looking at these possibilities, we sketch the pertinent features of the Cambridge talk, because they give a very clear view not only of the philosophical and mathematical issues Gödel addresses, but also of the continuity of his development.[6]

2 Relative consistency

Understanding by mathematics "the totality of the methods of proof actually used by mathematicians", Gödel sees the problem of providing a foundation for these methods as falling into two distinct parts (p. 1):

> At first these methods of proof have to be reduced to a minimum number of axioms and primitive rules of inference, which have to be stated as precisely as possible, and then secondly a justification in some sense or other has to be sought for these axioms, i.e., a theoretical foundation of the fact that they lead to results agreeing with each other and with empirical facts.

The first part of the problem is solved satisfactorily through type theory and axiomatic set theory, but with respect to the second part Gödel considers the situation to be extremely unsatisfactory. "Our formalism", he contends, "works perfectly well and is perfectly unobjectionable as long as we consider it as a mere game with symbols, but as soon as we come to attach a meaning to our symbols serious difficulties arise" (p. 15). Two aspects of classical mathematical theories (the non-constructive notion of existence and impredicative definitions) are seen as problematic because of a necessary Platonist presupposition "which cannot satisfy any critical mind and which does not even produce the conviction that they are consistent" (p. 19). This analysis conforms with that given in the Hilbert school, for example in (Hilbert and Bernays 1934), (Bernays 1935b) and (Gentzen 1936).[7] Gödel expresses the belief, again as the members of the Hilbert school did, that the

---

[6] Cf. Solomon Feferman's detailed introductory note to (1933b) in (Gödel 1995, p. 36).

[7] Of these writings (Bernays 1935b) is more ready to defend Platonism, with certain qualifications.

inconsistency of the axioms is most unlikely and that it might be possible "to prove their freedom from contradiction by unobjectionable methods".

Clearly, the methods whose justification is being sought cannot be used in consistency proofs, and one is led to the consideration of parts of mathematics that are free of such methods. Intuitionistic mathematics is a candidate, but Gödel emphasizes (p. 22) that

> the domain of this intuitionistic mathematics is by no means so uniquely determined as it may seem at first sight. For it is certainly true that there are different notions of constructivity and, accordingly, different layers of intuitionistic or constructive mathematics. As we ascend in the series of these layers, we are drawing nearer to ordinary non-constructive mathematics, and at the same time the methods of proof and construction which we admit are becoming less satisfactory and less convincing.

The strictest constructivity requirements are expressed by Gödel (pp. 23–25) in a system A that is based "exclusively on the method of complete induction in its definitions as well as in its proofs". That implies that the system A satisfies three general characteristics:[8] (A1) Universal quantification is restricted to "infinite totalities for which we can give a finite procedure for generating all their elements"; (A2) Existential statements (and negations of universal ones) are used only as abbreviations, indicating that a particular (counter-)example has been found without—for brevity's sake—explicitly indicating it; (A3) Only decidable notions and calculable functions can be introduced. As the method of complete induction possesses for Gödel a particularly high degree of evidence, "it would be the most desirable thing if the freedom from contradiction of ordinary non-constructive mathematics could be proved by methods allowable in this system A" (p. 25).

Gödel infers that Hilbert's original program is unattainable from two claims: first, *all* attempts for finitist consistency proofs actually undertaken in the Hilbert school operate within system A; second, all possible finitist arguments can be carried out in analysis and even classical arithmetic. The latter claim implies jointly with the second incompleteness theorem that finitist consistency proofs cannot be given for arithmetic, let alone analysis. Gödel puts this conclusion here quite strongly: "... unfortunately the hope of succeeding along these lines [using only the methods of system A] has vanished entirely in view of some recently discovered facts" (p. 25). But he points to interesting partial results and states the most far-reaching one, due to (Herbrand 1931) in a beautiful and informative way (p. 26):

> If we take a theory which is constructive in the sense that each existence assertion made in the axioms is covered by a construction, and if we add to this theory the non-constructive notion of existence and all the logical rules concerning it, e.g., the law of excluded middle, we shall never get into any contradiction.

---

[8]The designations (A1)–(A3) are introduced by us for ease of reference.

Gödel conjectures that Herbrand's method might be generalized to treat Russell's "ramified type theory", i.e., we assume, the theory obtained from system A by adding ramified type theory instead of classical first-order logic.[9]

There are, however, more extended constructive methods than those formalized in system A; this follows from the observation that system A is too weak to prove the consistency of classical arithmetic together with the fact that the consistency of classical arithmetic can be established relative to intuitionistic arithmetic.[10] The relative consistency proof is made possible by the intuitionistic notion of absurdity, for which "exactly the same propositions hold as do for negation in ordinary mathematics—at least, this is true within the domain of arithmetic" (p. 29). This foundation for classical arithmetic is, however, "of doubtful value": the principles for absurdity and similar notions (as formulated by Heyting) employ operations over *all* possible proofs, and the totality of all intuitionistic proofs cannot be generated by a finite procedure; thus, these principles violate the constructivity requirement (A1).

Despite his critical attitude towards Hilbert and Brouwer, Gödel dismisses neither in (1933b) when trying to make sense out of Hilbert's program in a more general setting,[11] namely, as a challenge to find consistency proofs for systems of "transfinite mathematics" relative to "constructive" theories. And he expresses his belief that epistemologically significant reductions may be obtained.

3 Layers of constructivity

In his lecture at Zilsel's, Gödel explores options for such relative consistency proofs and goes beyond (1933b) by considering—in detail—three different and, as he points out, known[12] ways of extending the arithmetic version of system A: the first one uses higher-type functionals, the second

---

[9] In *Konzept*, p. 0.1, Gödel mentions Herbrand's results again and also the conjecture concerning ramified type theory. The obstacle for an extension of Herbrand's proof is the principle of induction for "transfinite" statements, i.e., formulae containing quantifiers. Interestingly, as discovered in (Parsons 1970), and independently by Mints (1971) and Takeuti (1975, p. 175), the induction axiom schema for purely existential statements leads to a conservative extension of A, or rather its arithmetic version, primitive recursive arithmetic. How Herbrand's central considerations can be extended (by techniques developed in the tradition of Gentzen) to obtain this result is shown in (Sieg 1991).

[10] In his introductory note to (1933d), Troelstra (1986, p. 284) mentions relevant work also of Kolmogorov, Gentzen and Bernays. Indeed, as reported in (Gentzen 1936, p. 532), Gentzen and Bernays discovered essentially the same relative consistency proof independently of Gödel. According to Bernays (1967, p. 502), the above considerations made the Hilbert school distinguish intuitionistic from finitist methods. Hilbert and Bernays (1934, p. 43) make the distinction without referring to the result discussed here.

[11] This is in contrast to the lecture at Zilsel's, where he makes some (uncharacteristically polemical) remarks against the members of the Hilbert school and against Gentzen in particular; cf. fn. 45 below.

[12] When introducing the three ways of extending the basic system, Gödel starts out by saying: "Drei Wege sind bisher bekannt." (I.e., three ways are known up to now.)

introduces absurdity and a concept of consequence, and the third adds the principle of transfinite induction for concretely defined ordinals of the second number class. The first way is related to the hierarchy of functionals in (Hilbert 1926) and is a precursor of the *Dialectica* interpretation; the very brief section IV of the text is devoted to it. Of course, the second way is based on intuitionistic proposals, whereas the last way is due to Gentzen. The second and third way are discussed extensively in sections V and VI.

The broad themes of the preceding discussion are taken up in sections I through III, putting the notion of reducibility first; a theory T is called *reducible to a theory S* if and only if either

S is a subsystem of T and S proves $Wid\ S \to Wid\ T$

or

S proves $Wid\ T$.

As an example for the first type of reducibility Gödel alludes to his own relative consistency proof for the axiom of choice; as an example for the second type of reducibility he mentions the reduction of analysis to logic (meaning, we assume, simple type theory). Concerning the epistemological side of the problem, Gödel emphasizes that a proof is satisfactory in the first case only if S is a proper part of T, and in the second case only if S is more evident, more reliable than T. Though he admits that the latter criterion is subjective, he points to the fact that there is general agreement that constructive theories are better than non-constructive ones, i.e., those that incorporate "transfinite" existential quantification.

Acknowledging the vagueness of the notion of constructivity, Gödel formulates in section II what he calls a *Rahmendefinition* that incorporates the requirements (A1) through (A3) which, in (1933b), had motivated the system A. Now there are four conditions:[13] (R1) restates (A3), namely, that the primitive operations must be computable and that the basic relations must be decidable; (R2) combines aspects of (A1) and (A2) to restrict appropriately the application of universal and existential quantifiers; (R3) is an open list of inference rules and axioms that includes defining equations for primitive recursive functions, axioms and rules of the classical propositional calculus, the rule of substitution, and ordinary complete induction (for quantifier-free formulae); (R4) indicates what was in (1933b) the positive motivation for restricting universal quantification, namely, the finite generation of objects: "Objects should be surveyable (that is, denumerable)."

This point of view is modified in (1941), where we find (on pp. 2 and 3) versions of the first two conditions. The list of basic axioms and rules is introduced later in a very similar way, though not as part of the *Rahmendefinition*, but rather as part of the specification of the system Σ for finite-type functionals. There is no analogue to (R4). These changes (in exposition) are most interesting, as both (R3) and (R4) are viewed as "problematic" in the lecture at Zilsel's. As to (R3), the restriction to induction

---

[13] For convenience in our further discussions, we use (Ri) to refer to these conditions, which were simply numbered by Gödel.

Gödel at Zilsel's

for just natural numbers is viewed as problematic, because induction for certain transfinite ordinals is also evident; (R4) is problematic "because of the concept of function". Here, it seems, is the germ of Gödel's analysis in (1958) of the distinction between (strictly) finitist and intuitionistic considerations.

Gödel focuses in the present text immediately on number theory. The theory that corresponds to system A is obviously *primitive recursive arithmetic*, PRA, and it is viewed as the fundamental system in the hierarchy of constructive systems described briefly in section III.

4 Higher-type functionals

The first kind of extension of PRA consists in the introduction of (defining equations for) functionals of finite type. Gödel suggests continuing, as Hilbert does in (1926), the introduction of types into the transfinite, "if it is demanded that types only [be admitted] for those ordinal numbers which have been defined in an earlier system". The mathematical details for the basic finite type theory, as well as its transfinite extension, are sketchy. Nevertheless, this extension is most interesting for at least three reasons: It is the only extension that, according to Gödel, satisfies *all* requirements of the "Rahmendefinition"; it is connected to Hilbert and Ackermann's hierarchies of recursive functionals; and, finally, it is the first known articulated, even if very rudimentary, step in the evolution of Gödel's *Dialectica* interpretation.

Gödel indicates only by an example how to extend PRA by a recursion schema for functionals; the example he gives is as follows:

$$\Phi(f, 1, k) = f(k)$$

$$\Phi(f, n+1, k) = f(\Phi(f, n, k)).$$

This definition of the iteration functional is almost identical to the ones given in (Hilbert 1926, p. 186) and (Ackermann 1928, p. 118) (van Heijenoort 1967, pp. 388 and 495, respectively), where the iteration functional is used for defining the Ackermann function.[14] The latter function was, as will be recalled, the first example of a calculable function that cannot be defined by ordinary primitive recursion. However, it can be defined by primitive recursion if higher-type objects are allowed. Hilbert (1926, p. 186; van Heijenoort 1967, p. 389) formulated the full definition schema as follows:

$$\rho(\mathfrak{g}, \mathfrak{a}, 0) = \mathfrak{a}$$

$$\rho(\mathfrak{g}, \mathfrak{a}, n+1) = \mathfrak{g}(\rho(\mathfrak{g}, \mathfrak{a}, n), n);$$

---

[14]This "iterator" suffices to define the recursor functional, which directly yields the schema of definition by primitive recursion. See (Diller and Schütte 1971) or (Troelstra 1973, theorem 1.7.11, p. 56).

a is of arbitrary type (and ρ and $\mathfrak{g}$ must satisfy the obvious type restrictions). It is not entirely clear, but certainly most plausible, that Gödel envisioned using some form of the full schema (for what he calls here "closed systems").

Gödel claims that all his requirements are satisfied. Equality of higher-type objects could pose a problem for (R1); on this point see below. The claim that (R4) is satisfied would seem to mean that Gödel thought of the higher-type variables as ranging over the primitive recursive functionals themselves, which are given by terms. But then it is puzzling that Gödel earlier says that (R4) is "problematic . . . because of the concept of function" (p. 3).

Various claims and conjectures are formulated in subsections IV, 4 and 5. We shall make some frankly speculative remarks, which, together with (the introductory note to) (1941), may help the reader to speculate further on Gödel's statements. However, before discussing the assertions in subsection 4, we try to clarify the character of the extension procedure that is claimed to "contain" certain additions. The procedure is described more directly in a remark of *Konzept* (quoted in full below); the recursion schema mentioned there is the schema depicted above for defining the iteration functional:

In general this recursion schema ⟦is used⟧ for the introduction of Φ_i ⟦with the help of⟧ earlier f_i and the functions obtained by substituting in one another. . . . This hierarchy can be continued by introducing functions whose arguments are such Φ and admitting once again recursive definitions according to a numerical parameter. And that can even be extended into the transfinite.

Already in (Ackermann 1928, pp. 118–119; van Heijenoort 1967, p. 495) there is a discussion of a hierarchy A_i of classes of functionals. The elements of A_i are, in van Heijenoort's terminology (van Heijenoort 1967, p. 494), functionals of level i, i.e., their arguments are at most of level $(i-1)$ and their values are natural numbers. Gödel does not explicitly envisage equations of terms of higher type; if he indeed does not, then his hierarchy corresponds to Ackermann's. Otherwise, one is led quite directly to a hierarchy of classes H_i, restricted to functionals of type level $\leq i$, where numbers are of type level 0, and functionals with arguments of type σ and values of type τ are of type level $\max[\text{level}(\sigma) + 1, \text{level}(\tau)]$.[15]

Gödel began section III with the statement, "The finitary systems form a hierarchy." In section VI he uses S_1 and S_ω, evidently for systems in this hierarchy. A reasonable conjecture is that they are the systems for functionals of higher type in either the hierarchy A_i or H_i. It is, however, more likely that Gödel envisaged systems with equations only of numbers, thus closer to that of (Spector 1962) than to the **T** of (Gödel 1958).[16] Otherwise

---

[15] This reading would, on the conjecture stated immediately below, make the systems S_i correspond roughly to the subsystems $\mathbf{T}_i$ of the **T** of (Gödel 1958) described in (Parsons 1972), section 4. Parsons uses the term "rank".

[16] In his introductory note to (1958) and (1972), Troelstra points out that a subsystem $\mathbf{T}_Q$ of **T** is sufficient for Gödel's interpretation of first-order arithmetic; this subsystem is

one would expect some mention of the problems involved in interpreting equations of higher type so as to be decidable. (Hilbert also did not discuss the interpretation of such equations.) Such equations do occur in (1941, pp. 15 and 17), but from the notes of Gödel's lectures on intuitionistic logic at Princeton at the time it is clear that this use was informal, and the Σ of (1941) had only equality of numbers as a primitive.[17]

Although it would be natural today to take S_ω, as the union of the S_i and thus roughly the Σ of (1941) (or perhaps the T of (1958)), this is hard to reconcile with Gödel's later statement (p. 13) that recursion on ϵ_0 may be obtainable in S_ω. We will understand S_ω as a system containing functionals of lowest transfinite type (what on the other understanding would be $S_{\omega+1}$).

The "addition of recursion on several variables", as 4.1 asserts, is contained in the procedure since the schema allows nested recursion. The claim in 4.2, "addition of the statement Wid", seems to be clear, as S_{i+1} proves the consistency of S_i; that also fits well with the discussion of the provability predicates B_i on page 10. Finally, in 4.3 the "addition of Hilbert's rule of inference" is claimed to be contained in the procedure. In (1931c) Gödel reviews Hilbert's paper in which the infinitary rule was introduced, and in (1933d) he considers Herbrand's formulation of classical arithmetic, which includes that rule in the following form (Herbrand 1931 or Herbrand 1971, pp. 5 and 291, respectively):

Let $A(x)$ be a proposition without apparent variables ⟦i.e., a quantifier-free formula⟧; if it can be proved by intuitionistic procedures that this proposition, intuitionistically considered, is true for every x, then we add $(x)A(x)$ to the hypotheses.

As, for Herbrand, intuitionistic and finitistic considerations are coextensive, that is exactly Hilbert's formulation. A mathematically precise version of Gödel's claim follows from (Rosser 1937); cf. also Feferman's introductory note to (1931c) in (1986, p. 208).

There is no indication in this section as to how the system for functionals (of finite type) is to be used in consistency proofs of classical or, respectively, intuitionistic number theory; there is not even an explicit claim that the consistency of number theory can be established relative to $\cup S_n$, (or in S_ω; see above). In subsection 5 of the present section, Gödel formulates only negative claims and conjectures. In 5.1, he says that "with finite types one cannot prove the consistency of number theory". We assume that he intended to say what he formulated at the end of the Yale lecture, namely, the system that goes up only "to a given finite type" is not sufficient, instead the system for all finite types is needed. This can be rephrased in our terminology: No system S_i, $i < \omega$, will suffice; their union is needed for relative

---

common to T and the extensional variant. Thus, to obtain the Gödel interpretation in the latter, extensionality is needed only to derive Troelstra's "replacement schemas" (3) (Gödel 1990, p. 224).

On the problems of higher-type equality, see further the informative remarks in the same note, pp. 227–229.

[17]See the introductory note to (1941) in (Gödel 1995, p. 186).

consistency. The negative conjecture in 5.2 parallels that for transfinite induction in VI, 13: Even when the extension procedure is iterated transfinitely, along ordinals satisfying the restrictive condition mentioned above, one will not be able to prove the consistency of analysis. This conjecture is in stark contrast to that for the "modal-logical" route, which, according to Gödel in section V, 11, "leads furthest" and by means of which the consistency of analysis is "probably obtainable".

5 The modal-logical route

In section V, Gödel turns to the detailed examination of the "modal-logical" route, that is, giving consistency proofs relative to intuitionistic systems of the sort that had been first formalized by Heyting. Intuitionistic mathematics, as a framework for such proofs, does not satisfy the conditions Gödel laid down at the outset, because of the free application of negation and the conditional. Thus the now well-known translation of classical into intuitionistic arithmetic (from (Gödel 1933d)) gives a simple relative consistency proof. And Gödel was quite correct in conjecturing the possibility of extending this route to stronger classical theories: Both analysis and a (carefully formulated) set theory can be shown to be consistent relative to their versions with intuitionistic logic.[18]

Gödel reformulates his proof from (1933d), to show that full intuitionistic logic is not used. He proves (p. 7) that for every formula of number theory containing only the conditional and universal quantification, $\neg\neg A \supset A$ is provable in an intuitionistic system. The details are not entirely clear, but Gödel wants to emphasize that one does not use the principle $\neg A \supset (A \supset B)$. The logical axioms he states are all principles of positive implicational logic, except for the rule of generalization (C, p. 7) and axiom B7, which states for elementary formulae, also with free variables, $p \supset q. \supset\subset .\neg p \vee q$.[19] The role of this axiom in Gödel's intended argument is not clear.

The proof of $\neg\neg A \supset A$ proceeds by induction on the construction of A, and the atomic case is simply said to be "clear". Nothing beyond minimal logic is used in the induction step. Thus it would be for the atomic case that B7 is used.[20] The likely interpretation of Gödel's intention is that he

---

[18]For references and brief discussion, see A. S. Troelstra's introductory note to (1933d), (Gödel 1986, pp. 284–285). To the work relevant to his introductory note to (1933a), (Gödel 1986, pp. 296–298), Troelstra has suggested adding (Flagg 1986), (Flagg and Friedman 1986) and (Shapiro 1985).

[19]We use Gödel's symbols for connectives, including the unusual '⊃⊂' for the biconditional.

[20]If one sets out to prove $\neg\neg(x = y) \supset x = y$ in intuitionistic arithmetic, one will normally use $\neg A \supset (A \supset B)$ or some equivalent logical principle, and this seems to be essential. Consider $\neg\neg(Sx = 0) \supset Sx = 0$. This could not be proved by minimal logic from the axioms of intuitionistic arithmetic, because minimal logic is sound if one interprets the connectives other than negation classically, and $\neg A$ as true for any truth-value of A. But on that interpretation $\neg\neg(Sx = 0) \supset Sx = 0$ is false. But note that the equivalence of $\neg A$ and $A \supset 0 = 1$ fails on this interpretation.

assumes (as part of finitist arithmetic, in line with (R3)) some of classical truth-functional logic applied to quantifier-free formulae. One might reason as follows for atomic A: $\neg A \vee A$ follows from B5 and B7. Then we can reason by dilemma: Assuming $\neg A$, we infer $\neg\neg\neg A$ and therefore $\neg\neg\neg A \vee A$; assuming A, we infer $\neg\neg\neg A \vee A$; by disjunction-elimination we have $\neg\neg\neg A \vee A$ and by B7 $\neg\neg A \supset A$. Thus B7 is the only assumption used beyond minimal logic, but this argument does use the introduction and elimination rules for disjunction not mentioned by Gödel.[21]

In fact minimal logic is sufficient, and this use of B7 redundant, given that $\neg A$ has been defined as $A \supset 0 = 1$. For $0 = 1 \supset A$ is derivable (in fact for all A) by induction on the construction of A. In the atomic case, where A is an equation $s = t$, we use primitive recursion to define a function φ such that $\varphi(0) = s$ and $\varphi(Sx) = t$, so that $\varphi(1) = t$, and then $0 = 1 \supset s = t$ follows.[22] It is doubtful that Gödel had this argument in mind, since if so there would be no reason for the presence of B7.

If A contains only negation, the conditional and universal quantification, and if every atomic formula is negated, then the provability of $\neg\neg A \supset A$ in minimal logic follows from theorem 3.5 of (Troelstra and van Dalen 1988).[23] Thus for a formula satisfying Gödel's conditions, this will be true for the formula A^g obtained by replacing each atomic subformula P by $\neg\neg P$. Since $P \supset \neg\neg P$ is provable in minimal logic, the above reasoning using axiom B7 enables us to prove $P \leftrightarrow \neg\neg P$ for atomic P, whence minimal logic suffices to prove $A^g \leftrightarrow A$. Gödel remarks further that this proof does not require any nesting of applications of the conditional to universally quantified statements. What seems to be the case is that it is not needed essentially more than is directly involved in the construction of the formula A itself.

Gödel comments that the proof goes so easily because "*Heyting's system violates all essential requirements on constructivity*" (p. 8). This is an example of a somewhat disparaging attitude toward intuitionistic methods, at least as explained by Heyting and presumably Brouwer, when applied to the task at hand. But it is clear that Gödel's requirement (R2) is violated, as he asserts at the beginning of the section. He claims that requirement (R3) is violated because "certain propositions are introduced as evident" (p. 4). Probably he has in mind the logical axioms applied to formulae with quantifiers, and possibly also induction applied to such formulae, since these are the non-finitary axioms of intuitionistic arithmetic. But he does not

---

[21]Why does Gödel assume B7 instead of directly assuming $\neg\neg A \supset A$ for atomic A? Possibly he thought it more evident when $\neg A$ is defined as $A \supset 0 = 1$.

Gödel does remark that conjunction and disjunction are definable from negation and the conditional (p. 6), but the context indicates that this would be after classical logic has been derived for the restricted language. However, if in the atomic case we define $p \vee q$ as $\neg p \supset q$, then B7 reduces to $p \supset q \supset \subset \neg\neg p \supset q$, which yields $\neg\neg p \supset p$ by putting p for q and applying axiom 5. Possibly that is what Gödel had in mind by calling the atomic case "clear".

[22]We owe this observation to A. S. Troelstra.

[23](Troelstra and van Dalen 1988, pp. 62–68), gives a general treatment of provability by minimal logic of "negative translations", based on (Leivant 1985).

elaborate. We can understand why he thought requirement (R4) violated by turning to (1933b). There he argues that intuitionistic mathematics uses methods going beyond the system A, and says (p. 30), when commenting on the logical principle "$p \supset \neg\neg p$":

> So Heyting's axioms concerning absurdity and similar notions differ from the system A only by the fact that the substrate on which the constructions are carried out are proofs instead of numbers or other enumerable sets of mathematical objects. But by this very fact they do violate the principle, which I stated before, that the word "any" can be applied only to those totalities for which we have a finite procedure for generating all their elements.

The idea that intuitionistic mathematics has proofs as basic objects is central to his later analysis in (1958) of the distinction between finitist and intuitionistic mathematics.

In the present text Gödel considers an interpretation of intuitionistic logic starting from the idea that the conditional is to be understood in terms of absolute derivability, i.e., provability by arbitrary correct means, not limited to the resources of a single formal system. It is not clear how much Gödel was influenced by Heyting's early formulations of the "BHK-interpretation" of logical constants, which in (Heyting 1930b) and (1931) are very sketchy and incomplete.[24] In his writings of the 1930's, Gödel does not comment on Heyting's conception of a mathematical proposition as expressing an "expectation" or "intention" whose fulfillment is given by the proof of the proposition.[25] Gödel remarks only that he used the notion of derivability to interpret intuitionistic logic already in (1933a); but there he does not analyze it further, and here he states that the earlier work did not put any weight on constructivity. Intuitionistic propositional logic was interpreted in the result of adding the operator B to classical propositional logic (in effect, in a version of the modal logic S4).

In order to obtain a constructive system, Gödel proposes replacing the provability predicate B with a three-place relation $z\mathrm{B}p, q$, meaning "z is a derivation of q from p", which he says can "with enough good will" be regarded as decidable. He then formulates some axioms, but in a confusing way, since he sometimes uses B as two-place. The axioms as Gödel writes them are as follows:

---

[24](Heyting 1930a) and (1930b) present his intuitionistic formal systems without discussing questions of interpretation at all. But (Heyting 1931) at least was surely known to Gödel before his work on intuitionistic logic and arithmetic. In (1933a), fn. 1, Gödel does refer to (Kolmogorov 1932), which presents the interpretation of intuitionistic logic as a calculus of "problems"; cf. Troelstra's comment, (Gödel 1986, p. 299).

[25](Heyting 1931, p. 113); cf. (1930c, pp. 958–959). In neither of these texts does Heyting directly give an explanation of the conditional, but see (Heyting 1934, p. 14), which appeared after the remarks in (1933b) but before the present text. In fact Gödel saw earlier drafts of much of (Heyting 1934), which was the result of Heyting's work on a survey of the foundations of mathematics that was to be written jointly with Gödel. Gödel, however, never finished his part, which was to include a discussion of logicism. Heyting sent him a version of his section on intuitionism with a letter of 27 August 1932.

(1) $z\mathbf{B}p,q\ \&\ u\mathbf{B}q,r \to f(z,u)\mathbf{B}p,r$

(2) $z\mathbf{B}\varphi(x,y) \to \varphi(x,y)$

(3) $u\mathbf{B}v \to u'\mathbf{B}(u\mathbf{B}v)$

He suggests further a rule of inference:

(4)   If q has been derived with proof a, infer $a\mathbf{B}q$.

It seems reasonable to conjecture that this is to be a system based on classical propositional logic, with the presupposition that formulae are decidable. In particular, the $\varphi(x,y)$ of (2) is evidently not an arbitrary formula of, say, intuitionistic arithmetic, but presumably one constructed by finitistically admissible means plus **B**. Gödel does not make clear how the possible second (or second and third) arguments of **B** are to be constructed, although it is clear that they can contain logical operators not admissible elsewhere, such as universal quantifiers. Let us suppose that the language contains the symbol $\top$ for an unanalyzed tautology or other trivial truth. Then we read the two-place "$z\mathbf{B}q$" as an abbreviation for "$z\mathbf{B}\top,q$".[26]

Gödel asserts that these axioms are sufficient to prove, for some a,

(5) $a\mathbf{B}[(u)\sim u\mathbf{B}(0=1)]$.

Clearly from (2) we have

(6) $u\mathbf{B}(0=1) \to 0=1]$.

and, assuming enough arithmetic to prove $\sim(0=1)$,

(7) $\sim u\mathbf{B}(0=1)$.

Evidently the step to (5) is to use rule (4), but Gödel gives no indication how the universal quantifier is to be introduced. Gödel may well have understood the rule, in application to a case like this where a formula with free variable has been derived, as introducing a symbol a for the general proof; in that case there would only be a notational difference between "$a\mathbf{B}[\sim u\mathbf{B}(0=1)]$" and "$a\mathbf{B}[(u)\sim u\mathbf{B}(0=1)]$" as conclusion.[27]

Gödel inquires whether the system he has sketched is constructive in the sense he has explained. His answer (p. 9) is that the violation of requirement (R2) is avoided, since to the right of **B**, where "forbidden" logical operations occur, the formula occurs "in quotes". But requirements (R3) and (R4) are still not satisfied. This defect might be removed, if one interpreted

---

[26] A notation in *Konzept*, p. 7, indicates that Gödel thought of the two-place "$z\mathbf{B}p$" as an abbreviation for "$z\mathbf{B}Ax,p$", but he gives no explanation of what axioms the expression "Ax" refers to. In that same place (3) is stated as "$u\mathbf{B}v \to g(u)\mathbf{B}(u\mathbf{B}v)$", which makes clear that a primitive function giving the proof of $u\mathbf{B}v$ in terms of the given one of v is being assumed.

[27] P. 7 of *Konzept* contains other formulae in which the universal quantifier occurs, some of them crossed out, but no suggestion as to how an introduction such as that we have discussed is to go.

B as referring to proofs of the system itself. In the usual proof of Gödel's second incompleteness theorem, it is shown that formulae corresponding to axioms (1) and (3), and a version of rule (4), are derivable in the system. But then of course the conclusion has to be drawn that (2) is *not* derivable, and just for the case $0 = 1$ that occurs in the present argument. In view of the remark, "Essentially not the underlined—that is essentially the consistency of the system",[28] it seems that Gödel saw this; so it is not clear what he thought was the value of his suggestion. By the "introduction of types" he achieves (now using **B** for "provable") that only $\mathbf{B}_{n+1} \sim \mathbf{B}_n(0 = 1)$ holds; this seems to be the natural direction in which the interpretation of **B** as referring to formal proofs would go.[29]

Gödel's remarks present in a very sketchy way an idea that was pursued in subsequent work by G. Kreisel and others (apparently without knowledge of Gödel's earlier discussion), of developing formal theories of constructions and proofs, with a basic predicate like Gödel's $z\mathbf{B}p$. Gödel seems not to attack systematically the question that arises already at the beginning of this work in (Kreisel 1962) of interpreting intuitionistic logic by giving a definition of "$z\mathbf{B}A$", where A is an arbitrary formula of first-order logic or of some intuitionistic theory. The idea that one should use the clauses of the BHK-interpretation of the intuitionistic connectives to give an inductive definition of $z\mathbf{B}A$ is a very natural one and may well have occurred to Gödel at this time. Without it, it is hard to see how a theory on the lines Gödel sketches could serve the purpose that seems to be intended for it of being a vehicle for consistency proofs. Gödel, however, seems not concerned to develop the idea very far for this purpose, but rather to exhibit where it falls short of meeting his constraints.

Kreisel in (1962) and (1965) treated the proof relation as decidable, in agreement with Gödel's suggestion. This led to complications in the inductive definition of $z\mathbf{B}A$ for formulae of intuitionistic logic. The obvious clause for $z\mathbf{B}(A \to C)$ would be $\forall u[u\mathbf{B}A \to z(u)\mathbf{B}C]$ (thinking of z as a function), and this appears not to be decidable. Kreisel thus altered the definition to:

(*) $z\mathbf{B}(A \to C)$ iff z is a pair $\langle z_1, z_2 \rangle$ and $z_2\mathbf{B}\forall u[u\mathbf{B}A \to z_1(u)\mathbf{B}C]$.[30]

The same needed to be done for universal quantification. With an axiom like Gödel's (2), (*) and $z\mathbf{B}(A \to C)$ imply $u\mathbf{B}A \to z_1(u)\mathbf{B}C$. Thus Kreisel's definiens implies the above-mentioned obvious one.

There were considerable difficulties in developing a theory along these lines. Kreisel's ideas were naturally developed in the framework of the type-free lambda-calculus, but then the resulting theory is inconsistent; see

---

[28]There were considerable difficulties in developing a theory along these lines. Kreisel's ideas were naturally developed in the framework of the type-free lambda-calculus, but then the resulting theory is inconsistent; "The underlined" seems to be the formula expressing consistency; see (Gödel 1938, editorial note y).

[29]Possibly $\mathbf{B}_n$ is intended to mean provability in S_n; see §4 above.

[30](Kreisel 1962, p. 205); (Kreisel 1965, p. 128).

(Goodman 1970, §9).[31] Development of a theory along these lines modified so as to avoid paradox was never carried much beyond the interpretation of arithmetic; see (Goodman 1970) and (Goodman 1973). A lucid treatment of the basic issues is (Weinstein 1983). An alternative approach to a theory of constructions involved abandoning the requirement of the decidability of B. This led to various typed theories of which that of Per Martin-Löf is best known; see for example (Martin-Löf 1975), (Martin-Löf 1984), and the accounts in (Beeson 1985) and (Troelstra and van Dalen 1988). (Sundholm 1983) criticizes from this point of view the motivation of Kreisel's approach, and thus indirectly Gödel's suggestion of decidability.

6 Transfinite induction and recursion

Gödel mentions repeatedly that the modal-logical route we just discussed was the heuristic viewpoint guiding Gentzen's 1936 consistency proof for classical arithmetic. However, the characteristic principle by means of which Gentzen went beyond finitist mathematics (if that is thought of as being formalized in the system S_1) is the principle of transfinite induction for both proofs and definitions of functions. And it is to the number theoretic formulation of these principles that Gödel turns immediately—a task, incidentally, that was not taken up explicitly by Gentzen. First of all it has to be clarified how to grasp (in a mathematically expressible way) specific countable ordinals α from a finitist standpoint. That can be achieved in the system S_1 with function parameters, by considering definable linear orderings $\prec$ of the natural numbers such that, for a definable functional Φ and for all functions f, S_1 proves:

$$\sim \{f(\Phi(f)+1) \prec f(\Phi(f))\};$$

i.e., $\prec$ is provably well-founded. Such an ordering is said to represent α if it is order-isomorphic to α. Two points should be noted. First, there is obviously no function quantification in S_1; universal statements concerning functions (as well as numbers) are expressed just using free-variable statements. Second, the connection between α and $\prec$ is not formulated within finitist mathematics (and is also not needed for the further systematic considerations).

Gödel observes[32] that S_1 proves the transfinite induction principles forsuch $\prec$. The principle of proof by transfinite induction is formulated as

---

[31] Although Goodman's statement (p. 109) is more cautious, his argument seems to prove the inconsistency of the "starred theory" of (Kreisel 1962) as it stands. (Kreisel 1965) envisages the typed λ-calculus, but then it is not clear what type can be assigned to z_2 in (*). Goodman's own solution involves a stratification of constructions into levels; see (1970, §§10, 13) and the criticism in (Weinstein 1983, pp. 265–266).

[32] Treatments of transfinite induction in PRA (i.e., S_1 without function parameters) are given in (Kreisel 1959) and (Rose 1984, pp. 165 ff.).

follows: If one can prove $E(a)$ from the assumption $(x)(x \prec a \to E(x))$, then one is allowed to infer $(x)E(x)$. This inference is represented by the rule

$$\frac{(x)(x \prec a \to E(x)) \to E(a)}{(x)E(x)}$$

The corresponding *principle of definition* by transfinite induction (what would nowadays be called a schema of transfinite *recursion*) for $\prec$ is formulated in this way:[33] if $g_i, 1 \leq i \leq n$, are functions with $g_i(x) \prec x$ when x is not minimal, and A is any term (in the language of a definitional extension of S_1), then there is a unique solution to the functional equation

$$\varphi(x) = A(\varphi(g_1(x)), \ldots, \varphi(g_n(x))).$$

That can be done for (orderings representing) ordinals like $\omega + \omega$, ω^2, ω^3, etc. Gödel points out that ω^ω is a precise limit for S_1, since S_1 proves the proof principle of transfinite induction for quantifier-free formulae up to any ordinal $\alpha < \omega^\omega$, but not up to ω^ω. It seems, amazingly, that this result was rediscovered and established in detail only more than twenty years later in (Church 1960) and (Guard 1961).

In current terminology ω^ω is the *proof theoretic ordinal* of S_1. The analysis of formal theories in terms of their proof theoretic ordinals has been a major topic in proof theory ever since Gentzen's consistency proof and his subsequent analysis of the (un-)provability of transfinite induction in number theory. Returning to the present text, Gödel asserts that what can be done in the system S_1, namely, prove induction inferences from other axioms, can also be done in number theory for even larger ordinals. And, as in the case of S_1, there are ordinals for which this cannot be done in number theory. One such ordinal is the first epsilon-number ϵ_0. That ordinal is definable as the limit of the sequence

$$\alpha_1 = 2^{\omega+1} \text{ and } \alpha_{n+1} = 2^{\alpha_n}$$

where exponentiation[34] is given by

$$2^1 = 2, \ 2^\beta = \Sigma_{\alpha<\beta} 2^\alpha.$$

I.e., we associate with each element the sum of the preceding ones; Gödel views this as a "very intuitive construction procedure".[35] ϵ_0 is obtained by countably iterating the transition from α to 2^α, and if this transition were given, Gödel states in section VI, 9, ϵ_0 would be given. How could one

---

[33] A systematic treatment of so-called ordinal recursive functions, with references to the literature, in particular to the work of Péter, Kreisel, and Tait, is found in (Rose 1984).

[34] Cf. (Gödel 1938), editorial note bb.

[35] He emphasized that also in *Konzept*, p. 10, where he added "... und eine Reduktion darauf erscheint mir als wertvoll" ("... and a reduction ⟦to that construction⟧ seems to me to be valuable").

obtain this transition formally in arithmetic? One would assume that α is already represented in the sense above; then one would have to *define* an ordering that represents 2^α and prove that that ordering is a well-founded relation (using a suitable functional). Clearly, one may use both the proof and definition principle of transfinite induction for α.

In section VI, 6, Gödel gives a straightforward constructive *definition* of an ordering $\prec$ representing ϵ_0 and explains in the next subsection the obstacle against carrying out the usual proof of its well-foundedness.[36] The proof proceeds by transfinite induction and uses the impredicative induction property "being an ordinal". This property can't be formulated in the language of arithmetic since it requires genuine universal quantification over functions (to be used in the induction schema). Gentzen's consistency proof together with Gödel's second incompleteness theorem is needed to show more than the failure of the usual argument, namely, that there is no proof in number theory at all. Gödel does not remark, as he did for ω^ω and S_1, that ϵ_0 is the proof theoretic ordinal of arithmetic. That for any arithmetic statement transfinite induction up to any ordinal less than ϵ_0 is indeed provable in arithmetic is shown in (Hilbert and Bernays 1939, p. 366); it is also shown in (Gentzen 1943), where the unprovability of transfinite induction up to ϵ_0 is established without appeal to the second incompleteness theorem. For a modern and very beautiful presentation of these mathematical considerations (incorporating advances due mostly to Schütte and Tait) see (Schwichtenberg 1977).

In spite of the fact that the transition from α to 2^α (and thus ϵ_0) is not given in arithmetic, Gödel emphasizes (p. 12) with respect to the epistemological side:

... one will not deny a high degree of intuitiveness to the inference by induction on ϵ_0 thus defined, as in general to the procedure of *defining an ordinal by induction on ordinals* (even though this is an impredicative procedure).

Consequently, it is natural for Gödel to consider the system obtained from S_1 by adding the principle of transfinite induction up to ϵ_0 (for quantifier-free E). He asks as the first important, philosophical question, whether this theory is still constructive.[37] Indeed, all requirements are satisfied except for (R3); that condition demands the *exclusive* use of ordinary induction. But in a sense transfinite induction is just a generalization of ordinary induction, and Gödel thinks that therefore "... the deviation from the requirement 3 [our (R3)] is perhaps not such a drastic one" (p. 13). Gödel's reason for accepting induction up to ϵ_0 is not the special combinatorial character of ϵ_0, but rather the fact that ϵ_0 falls into a broader class of ordinals *definable by recursion on already defined ordinals*. And to this procedure, though impredicative, Gödel "will not deny a high degree of

---

[36]This (usual) argument is carefully presented in Supplement V of (Hilbert and Bernays 1939, pp. 534–5).

[37]A detailed analysis of such quantifier-free theories is given in (Rose 1984, chapters 6 and 7).

intuitiveness". Such broadened views of "constructivity" underlie developments in proof theory described, e.g., in (Feferman 1981, Feferman 1988a), but also in (Feferman and Sieg 1981a), where the use of generalized inductive definitions is emphasized. As to the use of large constructive ordinals in current proof theory, we refer to (Buchholz 1986), (Pohlers 1989) and (Rathjen 1991).

The second important, mathematical question is: How far does one get with extensions of S_1 by adding the inference rule for induction on ordinals that are obtained by ordinal recursive procedures along already obtained ordinals? As to ϵ_0, Gentzen did prove the consistency of number theory and, Gödel adds, "probably also of Weyl's *Kontinuum*".[38] As the precise theory underlying the development of analysis in (Weyl 1918) is (in one interpretation) a conservative extension of number theory, Gödel is indeed right; cf. (Feferman 1988b). He even thinks that with sufficiently large ordinals one can establish the consistency of analysis and of parts of set theory, as Gentzen had hoped. But he doubts that ordinals satisfying his principle of definability will be sufficiently large. Gödel conjectures at the end of subsection 13 and in 14 that the addition of transfinite induction for such ordinals may not lead to stronger theories than the S_i, and that transfinite induction up to ϵ_0 is already provable in S_ω (see §4 above).[39]

7 Interpreting Gentzen's consistency proof

Subsections 16 through 19 are a remarkable *tour de force*: on less than two pages Gödel analyzes, with a surprising twist, the essence of Gentzen's consistency proof for classical arithmetic and indicates precisely where in the proof the ordinal exponentiation step occurs that forces the use of all ordinals below ϵ_0. As we mentioned above, Gödel repeatedly points out that the modal-logical route, as a way of assigning a finitist meaning to transfinite statements, was the heuristic viewpoint guiding Gentzen's proof. The latter point was made already in *Konzept*[40] and is in complete accord with Gentzen's intentions; in section 13 (p. 536) of his (1936) (see also (Gentzen 1969, p. 173)) Gentzen writes:

The concept of the "*statability of a reduction rule (Reduziervorschrift)*" for a sequent, to be defined below, serves as a formal substitute for the contentual *concept of correctness*; it gives a special *finitist interpretation* of statements, which replaces the *in-itself conception* of them.[41]

---

[38] Gödel presumably meant that the method of Gentzen's proof would yield a proof of the consistency of Weyl's *Kontinuum*, not that Gentzen had literally proved this.

[39] In (Tait 1968b) it is shown that the **T** of (Gödel 1958) is closed under recursion on standard orderings of type less than ϵ_0; from (Kreisel 1959) it then follows that induction on these orderings is also derivable. This would confirm Gödel's conjecture for a suitable formulation of S_ω, (as interpreted in §4 above).

[40] In *Konzept*, page iii, we read: "Die transfiniten Aussagen erhalten einen finiten Sinn." (I.e., the transfinite statements obtain a finitist meaning.)

[41] In this and in the succeeding quotation, the translation in (Gentzen 1969) is revised. The "in-itself conception" (*an-sich Auffassung*, §9.2) is what we would call realist or Platonist.

At the very end of his paper (p. 564; 1969, p. 201) Gentzen points back to the definition of *Reduziervorschrift* in §13 and claims that the *most crucial part* of his consistency proof consists in providing a finitist meaning to the theorems of classical arithmetic:

For every arbitrary statement, so long as it has been proved, a *reduction rule* according to 13.6 *can be stated*, and this fact represents the finitist sense of the statement in question, which is gained precisely through the consistency proof.

Gödel claims that the finished proof is only remotely connected to the modal-logical one and maintains that Gentzen proves of each theorem a double-negation translation "...in a different sense from the modal-logical". Gödel formulates the different sense in a mathematically and conceptually perspicuous way: it turns out to be the sense provided by the "no-counterexample interpretation" introduced by Kreisel in (1951)!

These matters are formulated paradigmatically in subsection 16 by considering a formula

(1) $\quad (x)(\exists y)(z)(\exists u) A(x, y, z, u)$,

in prenex normal form with a decidable matrix A. Proving the negation of this formula constructively means presenting a number c, a unary function f, and a proof of

(2) $\quad \sim A(c, y, f(y), u)$.

A proof of the double negation of (1) consists then in a proof that such a c and f cannot exist: for each f and c one can find functionals $y_{f,c}$ and $u_{f,c}$ such that

(3) $\quad A(c, y_{f,c}, f(y_{f,c}), u_{f,c})$,

thus, there cannot be counterexamples c and f.[42] The functionals y and u are called by Gödel a *reduction*. In subsections 17 through 19 Gödel sketches how to find reductions for theorems in number theory from their formal proofs. (And it is for the treatment of modus ponens that ordinal exponentiation comes in.) Here is not the place to show that Gödel captures the mathematical essence of Gentzen's proof, as that would require a somewhat detailed description of that proof.

Gödel's analysis and presentation are surprising indeed. What accounts partially for the dramatic difference between Gentzen's and Gödel's presentations is the latter's free use of functionals and, to be sure, neglect of all formal details. Functionals do occur in Gentzen's presentation and also in Bernays's description of Gentzen's unpublished consistency proof in (Hilbert and Bernays 1939), but only in a cautious way to express that the *Reduziervorschriften* are independent of arbitrary choices (see, e.g., pages 536 and 537 in (Gentzen 1936)). The difficult and cumbersome presentation

---

[42] The reasoning behind the necessary shift of quantifiers is made explicit in (1941), p. 9.

(and what is perceived as an unmotivated manner of associating ordinals to derivations) resulted in a quite general dismissal of Gentzen's first consistency proof; it is the second consistency proof in (Gentzen 1938b) that has been at the center of proof theoretic research. A widely shared attitude of logicians towards (the "most crucial part" of) Gentzen's first proof can be gleaned from a remark in (Kreisel 1971). With respect to that most crucial part, i.e., the finitist sense given to logically complex theorems by the *Reduziervorschriften*, Kreisel writes (p. 252):

> He [Gentzen] has reservations about his own proposal of expressing this [finitist] sense in terms of the reductions used in his proof because the proposed sense is only "loosely connected" with the form of the theorem considered (and, it might be added, the connection is so tortuous that one couldn't possibly remember it).[43]

That is particularly striking when contrasted with Gödel's uncovering of the no-counterexample interpretation in his (clearly more sympathetic) reading in late 1937 and early 1938.

8 Concluding remarks

Gödel commented to Zilsel, as reported in his notes on the organizational meeting, that Gentzen's result is of only mathematical interest, "... ist nur mathematisch interessant"; and this judgment was not uninformed: Gödel had read (at least the unpublished version of) the consistency proof carefully and had discussed it extensively with Bernays.[44] On this point he obviously modified his views when preparing the lecture for the Zilsel seminar. When discussing the finitist character of the system obtained from PRA by the principle of transfinite induction, Gödel points out, as the reader may recall, that it violates only (R3). He continues: "This [new] inference can be considered as a generalization of ordinary induction, and in this respect the deviation from the requirement 3 is perhaps not such a drastic one." In his concluding section VII Gödel evaluates the epistemological significance of consistency proofs relative to the systems he considered; with respect to Gentzen's proof he states, "one will not be able to deny of Gentzen's proof that it reduces operating with the transfinite E to something more evident (the first ϵ-number)".

Gentzen's consistency proof meets, consequently, the general condition Gödel formulated for a "satisfying" relative consistency proof, namely, that such a proof should reduce to something that is more evident. In comparison with a reduction to the basic finitist system Gödel considers the epistemological significance to be "very much diminished". But then we

---

[43] It should be mentioned briefly that Kreisel misrepresents Gentzen's remark on p. 564 (to which he alludes in this quotation): according to Gentzen, the loose connection to the form of the theorem is *not* due to the reductions, but rather to the initial (standard) double negation translation, so that, for example, an existential statement does not have its strong finitist meaning, but only the weaker one of its translation.

[44] As to this episode, see (Kreisel 1987, pp. 173–174).

Gödel at Zilsel's

have to realize that Gödel expressed in this lecture a rather high regard for Hilbert's original program; if that could have been carried out, "that would have been without any doubt of enormous epistemological value". When comparing Gödel's philosophical remarks with those of Bernays (e.g., in (1935b)) or with the reflective considerations of Gentzen (e.g., in (Gentzen 1936) and (Gentzen 1938a)) one still finds a marked affinity of their general views.[45] It is the absolutely unencumbered mathematical analysis that most distinguishes Gödel's presentation from theirs.[46]

---

[45] Despite Gödel's rather critical remarks concerning Gentzen, e.g., on p. 13, "But here again the drive of Hilbert's pupils to derive something from nothing stands out." It is difficult to justify this remark either narrowly, as applying to Gentzen's paper, or more broadly, as applying, e.g., also to Bernays.

[46] Added for this volume:
In section 2, we described the system A that Gödel takes as expressing the strictest constructivity requirements. He also thinks that it is "most desirable" to carry out consistency proofs within A. However, Gödel argues that not even the consistency of elementary number theory can be proved there. Charles Parsons and I took for granted that the purely arithmetic theory corresponding to the system A is PRA; see the very last sentence of section 3.

That is no longer obvious to me. The major reason for that mind-change is a clearer understanding of the strength of the consistency theorem proved in (Herbrand 1931) with which Gödel was thoroughly familiar. Indeed, he had used the presentation of arithmetic from Herbrand's paper in his own (Gödel 1933d). Herbrand's theorem concerns a number of different theories distinguished from each other by the class of number theoretic functions considered; see (Sieg 2005). Gödel calls the consistency theorem the most far-reaching result that has been obtained in the pursuit of Hilbert's program. As Herbrand's strongest system contains functions that are not primitive recursive, like the Ackermann function, the metatheory in which Herbrand's proof is carried out goes beyond PRA, and Gödel's system A would also be stronger than PRA.

Herbrand's and Gödel's considerations clearly present a significant data point for the question as to the extent of finitist mathematics. This question is still being debated in a lively and informative way as witnessed by the more recent papers (Feferman and Strahm 2010) and (Tait 2002), but also by sections of (Parsons 2007). An historically and mathematically most careful analysis is found in (Ravaglia 2003).

II.5

Hilbert and Bernays: 1939

The second volume [of Grundlagen der Mathematik, WS] picks up where the first left off. As we saw [above in essay (II.1)], it presents in chapters 1 and 2 Hilbert's proof theoretic "Ansatz" based on the ϵ-symbol as well as related consistency proofs; this is the first main topic. The methods used there open a simple approach to Herbrand's theorem, which is at the center of chapter 3. The discussion of the decision problem at the end of that chapter leads, after a thorough discussion of the "method of the arithmetization of metamathematics", in the next chapter to a proof theoretic sharpening of Gödel's completeness theorem. The remainder of the second volume of *Grundlagen der Mathematik* is devoted to the second main topic, the examination of the fact, "which is the basis for the necessity to expand the frame of the contentual inference methods, which are admitted for proof theory, beyond the earlier delimitation of the 'finitist standpoint'."[1] Of course, Gödel's incompleteness theorems are at the center of that discussion, and they were also at the center for the changed situation that forced a rethinking of the proof theoretic enterprise. This rethinking took two directions, (i) exploring the extent of finitist mathematics, and (ii) demarcating the appropriate methodological standpoint for proof theory.

1 Changed situation

The consistency result in (Herbrand 1931) took an additional step beyond Ackermann's, as far as the very claim is concerned; at the time, it was the strongest proof theoretic result established by finitist means. Comparing Ackermann's result with his own, Herbrand wrote to Bernays in a letter dated 7 April 1931:

In my arithmetic the axiom of complete induction is restricted, but one may use a variety of functions other than those that are defined by simple recursion: in this

---

[1] Bernays wrote in "Zur Einführung" of the second volume of *Grundlagen der Mathematik*: "Das zweite Hauptthema bildet die Auseinandersetzung des Sachverhalts, auf Grund dessen sich die Notwendigkeit ergeben hat, den Rahmen der für die Beweistheorie zugelassenen inhaltlichen Schlußweisen gegenüber der vorherigen Abgrenzung des 'finiten Standpunktes' zu erweitern."

direction, it seems to me, that my theorem goes a little farther than yours [i.e., Ackermann's].

Gödel formulated Herbrand's central result in this beautiful and penetrating way: "If we take a theory which is constructive in the sense that each existence assertion made in the axioms is covered by a construction, and if we add to this theory the non-constructive notion of existence and all the logical rules concerning it, e.g., the law of excluded middle, we shall never get into any contradiction."[2] Obviously, that formulation covers equally well the above result obtained by Hilbert and Bernays in (1939). When Herbrand wrote his last paper (1931), he knew already of Gödel's incompleteness theorems and von Neumann's related conjecture. The latter had drawn the consequences of Gödel's results most sharply and dramatically: "If there is a finitist consistency proof at all, then it can be formalized. Therefore, Gödel's proof implies the impossibility of any [such] consistency proof."[3]

Was this the end of the proof theoretic program? An answer to this question required an answer to another question: What is finitist mathematics by means of which the proof theoretic program was supposed to be carried out? There is, as I described above [in part II.1], no sharp and clear answer to be found in Hilbert and Bernays's work from the 1920s as to the upper limits. At the very start of Hilbert's new proof theoretic investigations, finitist considerations are extended beyond primitive recursive arithmetic, and it is clear that finitist arguments make use only of proof by induction and definition by recursion within a quantifier-free framework. That is clear also from the perspective of "outsiders" in the early 1930s, cf. (Herbrand 1931) and (Gödel 1933b). At issue was, and to some extent still is, what are the recursion schemata that are finitistically allowed? Again, there is no sharp and clear answer to be extracted (or to be expected) from the early work of Hilbert and Bernays. However, one fact can be stated: von Neumann, Herbrand, Gödel in (1933b) and even Hilbert and Bernays in (1934) and (1939) consider the Ackermann function as a finitist one. I.e., the lower bound on the strength of finitist methods is given by a proper extension of primitive recursive arithmetic. The issue of an upper bound was much discussed in 1931 and subsequent years, when people tried to assess the impact of the incompleteness theorems. I mentioned already von Neumann's immediate and decidedly pessimistic judgment.

Gödel was much more cautious in his (1931d) and also in his letter to Herbrand dated 25 July 1931; in the former he asserted that the second incompleteness theorem does not contradict Hilbert's formalist viewpoint:

---

[2](Gödel 1933b, p. 52)

[3]Minutes of the meeting of the Vienna Circle on 15 January 1931; the minutes are found in the Carnap Archives of the University of Pittsburgh. The German text is: "Wenn es einen finiten Widerspruchsbeweis überhaupt gibt, dann lässt er sich auch formalisieren. Also involviert der Gödelsche Beweis die Unmöglichkeit eines Widerspruchsbeweises überhaupt." The minutes are quoted more fully in (Sieg 1988, note 11) and in (Mancosu 1999b, pp. 36–37).

... this viewpoint presupposes only the existence of a consistency proof in which nothing but finitist means of proof are used, and it is conceivable that there exist finitist proofs that cannot be expressed in the formalism of P (or of M or A).[4]

Both von Neumann and Herbrand argued in correspondence with Gödel against that position and conjectured that finitist arguments may be formalizable already in elementary number theory (and if not there, then undoubtedly in analysis). That would establish the limited reach of Hilbert's finitist consistency program.[5] What is the perspective Hilbert and Bernays took on these deliberations in volume II of *Grundlagen der Mathematik*?

2 Gödel's theorems

Hilbert and Bernays's discussion of the incompleteness theorems begins with a thorough investigation of semantic paradoxes. This investigation does not try to "solve" the paradoxes in the case of natural languages, but focuses on the question under what conditions analogous situations can occur in the case of *formalized languages*. These conditions are formulated quasi-axiomatically for deductive formalisms F taking for granted that there is a bijection between the expressions of F and natural numbers, a "Gödel-numbering". The formalism F and the numbering are required to satisfy roughly two *representability conditions*: R1) primitive recursive arithmetic is "contained in" F, and R2) the syntactic properties and relations of F's expressions, as well as the processes that can be carried out on them, are given by primitive recursive predicates and functions.

For the consideration of the first incompleteness theorem the second representability condition is made more specific. It is now required that the *substitution function* (yielding the number of the expression obtained from an expression with number **k**, when every occurrence of the number variable a is replaced by a numeral **l**) is given primitive recursively by a binary function $s(k, l)$ and the *proof predicate* by a binary relation $B(m, n)$ (holding when **m** is the number of a sequence of formulas that constitutes an F-derivation of the formula with number **n**). Consider, as Gödel did, the formula $\neg B(m, s(a, a))$; according to the first representability condition this is a formula of the formalism F and has a number, say **p**. Because of

---

[4](Gödel 1931d, p. 194). The German text is: "Es sei ausdrücklich bemerkt, daß Satz XI (und die entsprechenden Resultate über M, A) in keinem Widerspruch zum Hilbertschen formalistischen Standpunkt stehen. Denn dieser setzt nur die Existenz eines mit finiten Mitteln geführten Widerspruchsfreiheitsbeweises voraus und es wäre denkbar, daß es finite Beweise gibt, die sich in P (bzw. M, A) nicht darstellen lassen." The formalism P is that of *Principia Mathematica*, M is that of von Neumann's set theory, and A that of analysis. The letter to Herbrand is found in (Gödel 2003b).

[5]That discussion in their correspondence with Gödel is described in previous chapters of this volume, II.2 and II.3. There is also an informative letter of Herbrand to his friend Claude Chevalley; the letter was written on 3 December 1930, when Herbrand was in Berlin working on proof theory and discussing Gödel's theorems with von Neumann. (The letter is presented and analyzed in the Appendix of (Sieg 1994b).)

the defining property of $s(k,l)$, the value of $s(\mathbf{p},\mathbf{p})$ is then the number $\mathbf{q}$ of the formula $\neg B(m,s(\mathbf{p},\mathbf{p}))$. The equation $s(\mathbf{p},\mathbf{p}) = \mathbf{q}$ is provable in F; thus, $\neg B(m,s(\mathbf{p},\mathbf{p}))$ is actually equivalent to $\neg B(m,\mathbf{q})$ and expresses that "the formula with number $\mathbf{q}$ is not provable in F". As $\mathbf{q}$ is the number of $\neg B(m,s(\mathbf{p},\mathbf{p}))$, this formula consequently expresses (via the equivalence) its own underivability. The argument adapted from that for the liar paradox leads, assuming that this formula is provable, directly to a contradiction in F. But instead of encountering a paradox one infers that the formula is not provable, if the formalism F is consistent.

Hilbert and Bernays discuss — following Gödel and assuming the ω-consistency of F — the unprovability of the sentence $\neg(x)\neg B(x,\mathbf{q})$. Then they establish the Rosser version of the first incompleteness theorem, i.e., the independence of a formula R from F assuming just F's consistency. Thus, a "sharpened version" of the theorem can be formulated for deductive formalisms satisfying certain conditions: *One can always determine a unary primitive recursive function f, such that the equation $f(m) = 0$ is not provable in F, while for each numeral l the equation $f(l) = 0$ is true and provable in F; neither the formula $(x)f(x) = 0$ nor its negation is provable in F.*[6] This sharpened version of the theorem asserts that every sufficiently expressive, sharply delimited, and consistent formalism is deductively incomplete.

For a formalism F that is consistent and satisfies the restrictive conditions, the proof of the first incompleteness theorem shows the formula $\neg B(m,\mathbf{q})$ to be unprovable. However, it also shows that the sentence $\neg B(\mathbf{m},\mathbf{q})$ holds and is provable in F, for each numeral $\mathbf{m}$. The second incompleteness theorem is obtained by formalizing these considerations, i.e. by proving in F the formula $\neg B(m,\mathbf{q})$ from the formal expression C of F's consistency. That is possible, however, only if F satisfies additional conditions, the so-called *derivability conditions*. The formalized argument makes use of the representability conditions R1) and R2). The second condition now requires also that there is a unary primitive recursive function e, which when applied to the number $\mathbf{n}$ of a formula yields as its value the number of the negation of that formula. The derivability conditions are formulated as follows: D1) If there is a derivation of a formula with number $\mathbf{l}$ from a formula with number $\mathbf{k}$, then the formula $(Ex)B(x,\mathbf{k}) \to (Ex)B(x,\mathbf{l})$ is provable in F; D2) The formula $(Ex)B(x,e(k)) \to (Ex)B(x,e(s(k,l)))$ is provable in F; D3) If $f(m)$ is a primitive recursive term with m as its only variable and if $\mathbf{r}$ is the number of the equation $f(a) = 0$, then the formula $f(m) = 0 \to (Ex)B(x,s(\mathbf{r},m))$ is provable in F. Consistency is formally expressed by $(Ex)B(x,n) \to \neg(Ex)B(x,e(n))$; starting with that formula C as an assumption, the formula $\neg B(m,\mathbf{q})$ is obtained in F by a rather direct argument. So, in case the formalism F is consistent, no formalized proof of consistency, i.e., no derivation of the formula C can exist in F.[7]

---

[6](Hilbert and Bernays 1939, p. 279/288). [As in II.1, the first page number refers to the first, the second to the second edition of *Grundlagen der Mathematik II.*]

[7](Hilbert and Bernays 1939, pp. 286–288/296–297).

There are two brief remarks with which I want to complement this metamathematical discussion of the incompleteness theorems. The first simply states that verifying the representability conditions and the derivability conditions is the central mathematical work that has to be done; Hilbert and Bernays accomplish this for the formalisms Z_μ and Z.[8] Thus, the second volume of *Grundlagen der Mathematik* contains the first full argument for the second incompleteness theorem; after all, Gödel's paper contains only a minimal sketch of a proof. However, it has to be added — and that is the second remark — that the considerations are not fully satisfactory for a *general* formulation of the theorems, as there is no argument given why deductive formalisms should satisfy the particular restrictive conditions on their syntax. (This added observation points to one of the general methodological issues discussed briefly at the end of section 3.)

Existential formal axiomatics emerged in the second half of the 19th century and found its remarkable expression in 1899 through Hilbert's *Grundlagen der Geometrie*; its existential assumption constituted the really pressing issue for the various Hilbert programs during the period from 1899 to 1934, the date of the publication of the first volume of *Grundlagen der Mathematik*. The finitist consistency program began to be pursued in 1922 and is the intellectual thread holding the investigations in both volumes together. The ultimate goal of proof theoretic investigations, as Hilbert formulated it in the preface to volume I, is to recognize the usual methods of mathematics, without exception, as consistent. Hilbert continued,

With respect to this goal I would like to emphasize the following: the view, which temporarily arose and maintained that certain recent results of Gödel imply the infeasibility of my program, has been shown to be erroneous.[9]

How is the program affected by those results? Is it indeed the case, as Hilbert expressed it also in that preface of (1934), that Gödel's theorems just force proof theorists to exploit the finitist standpoint in a sharper way?

3 Completeness and bounds on finitism

The second question is raised prima facie through the second incompleteness theorem. However, Hilbert and Bernays discuss also the effect of the first incompleteness theorem and ask explicitly, whether the deductive completeness of formalisms is a necessary feature for the consistency program to make sense. They touched on this very issue already in pre-Gödel publications: Hilbert in his Bologna talk of 1928, and Bernays in his penetrating article (1930b). Hilbert formulated in his talk the question of the syntactic completeness for number theory and analysis as Problem IV;

---

[8]The considerations for the former systems start on p. 293/306, for the latter on p. 324/337.

[9](Hilbert and Bernays 1934), in Hilbert's "Zur Einführung". The German text is: "Im Hinblick auf dieses Ziel möchte ich hervorheben, daß die zeitweilig aufgekommene Meinung, aus gewissen neueren Ergebnissen von Gödel folge die Undurchführbarkeit meiner Beweistheorie, als irrtümlich erwiesen ist."

he concluded the discussion by suggesting that "in höheren Gebieten" (higher than number theory) it is thinkable that a system of axioms could be consistently extended by a statement S, but also by its negation $\neg S$; the acceptance of one of the statements is then to be justified by "systematic advantages (principle of the permanence of laws, possibilities of further developments etc.)".[10]

Hilbert conjectured that number theory is deductively complete. That is reiterated in Bernays's (1930b) and followed by the remark that "the problem of a real proof for this is completely unresolved". The problem becomes even more difficult, Bernays continues, when we consider systems for analysis or set theory. However, this "Problematik" is not to be taken as an objection against the standpoint presented:

We only have to realize that the [syntactic] formalism of statements and proofs we use to represent our conceptions does not coincide with the [mathematical] formalism of the structure we intend in our thinking. The [syntactic] formalism suffices to formulate our ideas of infinite manifolds and to draw the logical consequences from them, but in general it [the syntactic formalism] cannot combinatorially generate the manifold as it were out of itself.[11]

That is also the central point in the general discussion of the first incompleteness theorem. Indeed, Hilbert and Bernays emphasize that in formulating the problems and goals of proof theory they avoided from the beginning "to introduce the idea of a total system for mathematics with a philosophically principled significance". It suffices to characterize the actual systematic structure of analysis and set theory in such a way that it provides an appropriate frame for (the reducibility of) the geometric and physical disciplines.[12] That point was already emphasized in section 3.3 [of *Hilbert's proof theory*]. From these reflective remarks it follows that the first incompleteness theorem for the central formalisms F (of number

---

[10](Hilbert *1929, p. 6). The German text is: "In höheren Gebieten wäre der Fall der Widerspruchsfreiheit von S und der von $\neg S$ denkbar; alsdann ist die Annahme einer der beiden Aussagen S, $\neg S$ als Axiom durch systematische Vorzüge (Prinzip der Permanenz von Gesetzen, weitere Aufbaumöglichkeiten u.s.w.) zu rechtfertigen." This paragraph does not appear in the original proceedings of the Bologna Congress in 1928.

[11](Bernays 1930b, p. 59). The German text is: "Wir müssen uns nur gegenwärtig halten, daß der Formalismus der Sätze und Beweise, mit denen wir unsere Ideenbildung zur Darstellung bringen, nicht zusammenfällt mit dem Formalismus derjenigen Struktur, die wir in der Gedankenbildung intendieren. Der Formalismus reicht aus, um unsere Ideen von unendlichen Mannigfaltigkeiten zu formulieren und aus diesen die logischen Konsequenzen zu ziehen, aber er vermag im allgemeinen nicht, die Mannigfaltigkeit gleichsam aus sich kombinatorisch zu erzeugen."

[12](Hilbert and Bernays 1939, p. 280/289). The extended German text is: "Wir haben in unserer Darstellung der Ausgangsproblematik und der Zielsetzung der Beweistheorie von vornherein vermieden, den Gedanken eines Totalsystems der Mathematik in einer philosophisch prinzipiellen Bedeutung einzuführen, vielmehr uns begnügt, die tatsächlich vorhandene Systematik der Analysis und Mengenlehre als eine solche zu charakterisieren, die einen geeigneten Rahmen für die Einordnung der geometrischen und physikalischen Disziplinen bildet. Diesem Zweck kann ein Formalismus auch entsprechen, ohne die Eigenschaft der vollen deduktiven Abgeschlossenheit zu besitzen."

theory, analysis, and set theory) does not directly undermine Hilbert's program. It raises nevertheless in its sharpened form a peculiar issue: any finitist consistency proof for F would yield a finitist proof of a statement in recursive number theory — that is not provable in F. Finitist methods would thus go beyond those of analysis and set theory, even for the proof of number theoretic statements. (That is of course the situation Gödel contemplated in his remarks, when claiming that his results don't contradict Hilbert's standpoint; see section 1.) This is a "paradoxical" situation, in particular, as Hilbert and Bernays quite unambiguously state in the first volume (on p. 42), "finitist methods are included in the usual arithmetic". Consequently, even the first theorem forces us to address two general tasks, namely, (i) exploring the extent of finitist methods, and (ii) demarcating appropriately the methodological standpoint for proof theory.

Tasks (i) and (ii) are usually associated with the second incompleteness theorem, which, as emphasized in section 2, allows us to infer directly and sharply that a finitist consistency proof for a formalism F (satisfying the representability and derivability conditions) cannot be carried out in F. Hilbert and Bernays explore the extent of finitist methods in chapter 5.a) by first trying to answer the question, in which formalism their various finitist investigations can actually be carried out. The immediate claim is that most considerations can be formalized, perhaps with a great deal of effort, already in primitive recursive arithmetic. But then they assert: "At various places this formalism is admittedly no longer sufficient for the desired formalization. However, in each of these cases the formalization is possible in (Z_μ)."[13] They point to the more general recursion principles from chapter 7 of the first volume as an example of "procedures of finitist mathematics" that cannot be captured in primitive recursive arithmetic, but can be formalized in Z_μ.

In the remainder of chapter 5.3.a) they discuss "certain other typical cases", in which the boundaries of primitive recursive arithmetic are too narrow to allow a formalization of their prior finitist investigations. There is, first of all, the issue of an evaluation function that is needed for the consistency proof of primitive recursive arithmetic (already in volume I), but cannot be defined by primitive recursion (p. 341/355). Secondly, there is the general concept of a calculable function (p. 341/356). That concept is used (p. 189/198) to formulate a sharpened notion of satisfiability, i.e., *effective satisfiability*, in their treatment of solvable cases of the decision problem. Thirdly, they discuss (p. 344/358) the principle of induction for universally

---

[13](Hilbert and Bernays 1939, p. 340/354). The extended German text is: "An verschiedenen Stellen ist freilich dieser Formalismus [der rekursiven Zahlentheorie mit nur primitiver Rekursion, WS] nicht mehr für die gewünschte Formalisierung ausreichend. Doch zeigt sich dann jedesmal die Möglichkeit der Formalisierung in (Z_μ). Gewisse über die rekursive Zahlentheorie (im ursprünglichen Sinne) hinausgehende Verfahren der finiten Mathematik haben wir bereits im §7 besprochen, nämlich die Einführung von Funktionen durch verschränkte Rekursionen und die allgemeineren Induktionsschemata. Dabei erwähnten wir auch die Formalisierbarkeit dieser Rekursions- und Induktionsschemata im vollen zahlentheoretischen Formalismus."

quantified formulas used in consistency proofs. The issue surrounding this principle is settled metamathematically, as we know, by later proof theoretic work: the system of elementary number theory with this induction principle is conservative over recursive arithmetic, whether in the narrow or wider sense.[14]

As to the second issue (effective satisfiability), some remarks concerning Supplement II are relevant here, because the notion of a calculable function has to be sharpened in such a way as to be formalizable. The presentation of the negative solution of the decision problem in supplement II is preceded by an analysis of the concept "reckonable function", i.e., of a function whose values can be calculated according to elementary rules. The latter rather vague notion is sharpened in a way that is methodologically very similar to their analysis of the incompleteness theorems, namely, by formulating *recursiveness conditions* for deductive formalisms that allow equational reasoning. The central condition requires the proof predicate to be primitive recursive. It is then shown that the functions calculable in formalisms satisfying these conditions are exactly the general recursive ones. Though their analysis is not fully satisfactory for the reason mentioned in section 2, it is nevertheless a major and concluding step for the analysis of effectively calculable functions as pursued in the mid-1930s by Gödel, Church, Kleene, and others. It should be emphasized that Hilbert and Bernays are perfectly clear about one fact, namely, that this class of functions goes beyond the class of functions introduced in Herbrand's 1931 by the axioms of his "Groupe C". Indeed, they write:

General instructions for the introduction of new function symbols were given by Herbrand in his paper [(1931), WS] ... ("Groupe C"). These are somewhat narrower than the ones formulated here inasmuch as Herbrand requires that the procedure for determining values [of functions] is to follow from a finitist interpretation of the axioms, as is the case for the schemata of recursive definitions.[15]

So it is also here quite clear that Hilbert and Bernays view Herbrand's procedure of introducing function symbols as a definitely finitist one; recall that Herbrand mentions explicitly that a symbol for the Ackermann function can be introduced in this way.

4 Methodological frame

The careful re-examination of their own proof theoretic practice leads Hilbert and Bernays to the conclusion that some finitist considerations go beyond primitive recursive arithmetic, but can be formally captured in Z_μ; most of this was pointed out already in volume I. It is at exactly this point

---

[14]See (Sieg 1991), section 2.1 and references there to work by Charles Parsons.

[15]Here is the German text: "Eine allgemeine Anweisung zur Einführung neuer Funktionszeichen durch Axiome wurde von Herbrand in seiner Abhandlung [(1931), WS] ... gegeben ("Groupe C"). Diese ist insofern etwas enger als die hier formulierte Anweisung, als Herbrand verlangt, daß das Verfahren der Wertbestimmung sich, so wie bei den Schematen der rekursiven Definition, durch eine finite Deutung der Axiome ergeben soll." This is found in note 2 on p. 52 (of the second volume).

that the second incompleteness theorem provides, as the title of Chapter 5 states, the "reason for extending the methodological frame for proof theory". Already in the transition from Chapter 4 to Chapter 5, Hilbert and Bernays state specifically that consequences of the theorem force us to view the domain of the contentual inference methods used for the investigations of proof theory more broadly "than it corresponds to our development of the finitist standpoint so far".[16]

The question is, whether there are any methods that can still be called properly "finitist" and yet go beyond Z_μ. Hilbert and Bernays argue that this is not a precise question, as "finitist" is not a sharply delimited notion, but rather indicates methodological guidelines that enable us to recognize some considerations as definitely finitist and others as definitely non-finitist. The limits of finitist considerations are to be "loosened"; two possibilities for such loosenings are considered and quickly seen to be conservative.[17] Which further loosening is "admissible, if we want to adhere to the fundamental tendencies of proof theory"? Against this background two results, then quite recent, are examined: the reduction of classical arithmetic Z to the system **Z** of arithmetic with just minimal logic, and Gentzen's consistency proof for a version of **Z** (and thus of Z) using a special form of transfinite induction.

The reductive result Hilbert and Bernays formulate is a slightly stronger one than the one obtained by Gödel and, independently, by Gentzen. The proof showing that Z is consistent relative to **Z** is an elementary finitist one. Thus, the obstacle for obtaining a finitist consistency proof for Z does not lie in the fact that it contains the typically non-finitist logical principles like tertium non datur! The obstacle appears already when one tries to give a finitist consistency proof for **Z**. The consistency of Z would be established on the basis of any assumptions, "which suffice to give a verifying interpretation of the restricted formalism".[18] Such a contentual verification, based on interpretations of Kolmogoroff and Heyting, is then examined with the conclusion that it involves the intuitionist understanding of negation as absurdity. In using the underlying contentual concept

---

[16](Hilbert and Bernays 1939, p. 253/263). The German text is: "Dieser Umstand [daß die bisherigen Methoden nicht zum Nachweis der Konsistenz des vollen zahlentheoretischen Formalismus ausreichen, WS] ... findet nun eine grundsätzliche Erklärung durch ein Theorem von Gödel über deduktive Formalismen, für welches der zahlentheoretische Formalismus einen ersten Anwendungsfall bildet und dessen Konsequenzen uns dazu nötigen, den Bereich der inhaltlichen Schlußweisen, die wir für die Überlegungen der Beweistheorie verwenden, weiter zu fassen, als es unserer bisherigen Durchführung des finiten Standpunktes entspricht."

[17] In the context of this discussion, on p. 348/362, a very concise explication of "Begriff der finiten Aussage" is given that reflects faithfully the informal considerations on which I have been reporting.

[18](Hilbert and Bernays 1939, p. 357/371). That point is reemphasized after the reductive argument has been completed. The German text there is: "Stellen wir uns andererseits auf einen inhaltlichen Standpunkt, von dem aus die formalen Ableitungen in (**Z**) als Darstellung richtiger inhaltlicher Überlegungen deutbar sind, so ist für diesen auf Grund der festgestellten Beziehung zwischen den Formalismen (Z) und (**Z**) die Widerspruchsfreiheit des Systems (Z) ersichtlich."

of consequence, it is claimed, "we are totally turning away from Hilbert's methodological ideas for proof theory".[19] That is consonant with the view expressed in the first volume (p. 43) that intuitionism is a proper extension of finitist mathematics (in sharp contrast to the earlier perspective that was discussed in the second half of section 3.3 [of *Hilbert's proof theory*]). Bernays expressed that view also in contemporaneous papers and in many later comments, perhaps most dramatically in his article (1967) on Hilbert, where (on p. 502) the above relative consistency proof for Z is seen as the reason for the recognition "that intuitionistic reasoning is not identical with finitist reasoning, contrary to the prevailing views at the time".

The question is raised, whether — in a proof of the consistency of Z — the use of absurdity can be avoided, as well as the appeal to an interpretation of the formalism (viewed in contrast to its direct proof theoretic examination). It is claimed that Gentzen's consistency proof addresses both these issues. After a thorough discussion of the details of the system of ordinal notation (for ordinals less than the first epsilon number) and the justification of the principle of transfinite induction, but only the briefest indication of the structure of Gentzen's proof, the main body of the first edition of the book concludes with some extremely general remarks about the significance of that proof: it provides a perspective for the proof theoretic investigation also of stronger formalisms, when one clearly has to countenance the use of larger and larger ordinals. The volume concludes with the sentence: "If this perspective should prove its value, then Gentzen's consistency proof would open a new phase of proof theory." In this way, it seems, Bernays sees Gentzen's approach as overcoming "the temporary fiasco of proof theory" he discussed in the introduction to volume II and attributed to " ... exaggerated methodological demands put on the theory". No explicit final and definitive judgment on the methodologically appropriate character of Gentzen's consistency proof is articulated in the first edition of the book. However, in the introduction to the second edition Bernays states that the transfinite induction principle used in it is "a non-finitist tool".

In the introduction of the first edition and the detailed discussion sketched here, some see an ambiguity in Hilbert and Bernays's view as to whether the extension of the finitist standpoint necessitated by the incompleteness theorems is essentially still the finitist standpoint as articulated in the first two chapters of volume I or whether it is a proper extension compatible with the broader strategic considerations underlying proof theory. I think the ambiguity, if it is there at all, should be resolved in the latter sense; after all, the considerations in Chapter 5.5 come under the heading "Transcending the former methodological standpoint of proof theory. Consistency proofs for the full number theoretic formalism." One

---

[19](Hilbert and Bernays 1939, p. 358/372). The fuller German text is: "Jedoch entfernen wir uns mit dem inhaltlichen Folgerungsbegriff total von Hilberts methodischen Gedanken der Beweistheorie, ..."

just has to distinguish very carefully, as Bernays does, between the two different tasks I described at the very beginning of this section: (i) exploring the extent of finitist mathematics, and (ii) demarcating the appropriate methodological standpoint for proof theory.

However, there is not even a broad demarcation of a new, wider methodological standpoint for proof theory; a reason for this lack is perhaps implicit in the remarks connecting the consistency proof for Z relative to intuitionist arithmetic with Gentzen's consistency proof (p. 359/372). It is claimed, first of all, that it is "unsatisfactory from the standpoint of proof theory" to have a consistency proof for Z that "rests mainly on an interpretation of a formalism". It is observed, secondly, that the only method of going beyond the formalism Z has been the formulation of truth definitions: a classical truth definition was given for Z, and the formalization of the consistency proof based on an intuitionist interpretation would amount to using a truth definition. Thirdly and finally, it is argued that a consistency proof is desirable that rests on "the direct treatment of the formalism itself"; that is seen in analogy to obtaining the consistency of (primitive) recursive arithmetic, where Hilbert and Bernays were not satisfied with the possibility of a finitist interpretation, but rather convinced themselves of the consistency by specific proof theoretic methods. Where in this discussion is even an opening for a broader demarcation?

Bernays, in the "Nachtrag" to his (1930b), reflects on these issues and indicates, in particular, that the epistemological perspective that was underlying proof theory became problematic. Referring to his own essay he writes:

... the sharp distinction between what is intuitive and what is not, as it is used in the treatment of the problem of the infinite, apparently cannot be drawn so strictly, and the reflections on the formation of mathematical ideas still need to be worked out in more detail in this respect. Various considerations for this are contained in the following essays.[20]

Some indications of a general direction for philosophical reflections are indeed contained in the essays reprinted in (Bernays 1976a), but also in an essay that is not reprinted there, namely, (Bernays 1954). There he envisions the appeal to what he calls *sharpened axiomatics* (verschärfte Axiomatik) and, opposing it to existential axiomatics, formulates as a minimal requirement that "the objects [making up the intended model of the theory] are not taken from a domain that is thought as being already given, but are rather constituted by generative processes".[21]

---

[20](Bernays 1976a, p. 61). The German text is: "... die scharfe Unterscheidung des Anschaulichen und des Nicht-Anschaulichen, wie sie bei der Behandlung des Problems des Unendlichen angewandt wird, ist anscheined nicht so strikt durchführbar, und die Betrachtung der mathematischen Ideenbildung bedarf wohl in dieser Hinsicht noch der näheren Ausarbeitung. Für eine solche sind in den folgenden Abhandlungen verschiedene Überlegungen enthalten."
The "folgenden Abhandlungen" are referring, obviously, to the essays in (Bernays 1976a).
[21](Bernays 1954, pp. 11–12). The German text is: "Die Mindest-Anforderung an eine verschärfte Axiomatik ist die, dass die Gegenstände nicht einem als vorgängig gedachten Bereich

There is no indication in that paper or in other writings what kind of generative processes should be considered, and why that particular feature of domains should play a distinctive foundational role. Contemporary proof theoretic investigations give such indications.

---

entnommen werden, sondern durch Erzeugungsprozesse konstituiert werden." Bernays continues with a methodologically important remark: "Es kann aber dabei die Meinung sein, dass durch diese Erzeugungsprozesse der Umkreis der Gegenstände determiniert ist; bei dieser Auffassung erhält das *tertium non datur* seine Motivierung. In der Tat kann Offenheit eines Bereiches in zweierlei Sinn verstanden werden, einmal nur so, dass die Konstruktionsprozesse über jeden einzelnen Gegenstand hinausführen, und andererseits in dem Sinne, dass der resultierende Bereich überhaupt nicht eine mathematisch bestimmte Mannigfaltigkeit darstellt. Je nachdem die Zahlenreihe in dem erstgenannten oder in dem zweiten Sinne aufgefasst wird, hat man die Anerkennung des tertium non datur in bezug auf die Zahlen oder den intuitionistischen Standpunkt. Bei dem finiten Standpunkt kommt noch die Anforderung hinzu, dass die Überlegungen an Hand der Betrachtung von endlichen Konfigurationen verlaufen, somit insbesondere Annahmen in der Form allgemeiner Sätze ausgeschlossen werden."

II Systematical

A brief guide

The historical analyses ended with reflections on finitist mathematics and how one might broaden the methodological perspective for proof theory, in 1939. I'll jump now across almost four decades of proof theoretic and mathematical work to investigations that culminated in the mid-1970s with two broad insights: proof theoretically, "strong" impredicative subsystems of analysis are reducible to intuitionist theories for iterated inductive definitions, indeed, theories for just constructive number classes; mathematically, (parts of) analysis can be presented in "weak" subsystems, indeed, so weak that they are conservative over (fragments of) elementary number theory.

The proof theoretic work Buchholz, Pohlers and I had done was published in (Buchholz, Feferman, Pohlers and Sieg 1981); the developments on which our investigations built are described in Feferman's introduction to that volume, *How we got from there to here*. My views on the history and the philosophical significance of the reductions are articulated in (II.6), *Foundations for analysis and proof theory*. Friedman and Simpson refined the mathematical work under the heading *reverse mathematics* and with a programmatic direction: show that certain set theoretic principles, from weak König's Lemma to comprehension for Π_1^1-formulae, are equivalent to a wide range of important mathematical statements. The proof theoretic techniques used for the foundational reductions could be refined to extract systematically computational information; that is discussed in (II.7), *Reductions of theories for analysis*.

These broad developments were the topic of a symposium in late 1985 in which Feferman, Prawitz, and Simpson participated. My introductory remarks as its chair were expanded into (II.8), *Hilbert's program sixty years later*, and appeared in 1988 with Feferman's and Simpson's contributions in the Journal of Symbolic Logic. Essay (II.9), *On reverse mathematics*, is a review of (Simpson 1985b) and analyzes mathematical and foundational aspects of reverse mathematics. The last essay in this group, *Relative consistency and accessible domains*, interprets proof theoretic reductions as structural ones, providing via accessible domains objective underpinnings for the theories to which reductions have been achieved.

Johann von Neumann

Kurt Gödel

Gerhard Gentzen

II.6

Foundations for analysis and proof theory*

> *Douter de tout ou tout croire, ce sont deux solutions également commodes qui l'une et l'autre nous dispensent de réfléchir.*
>
> H. Poincaré

Introduction

The title of my paper indicates that I plan to write about foundations for analysis and about proof theory; however, I do not intend to write about *the* foundations for analysis and thus not about analysis viewed from the vantage point of any "school" in the philosophy of mathematics. Rather, I shall report on some mathematical and proof theoretic investigations which provide material for (philosophical) reflection. These investigations concern the informal mathematical theory of the continuum, on the one hand, and formal systems in which parts of the informal theory can be developed, on the other. The proof theoretic results of greatest interest for my purposes are of the following form:

for each F in a class of sentences, F is provable in T if and only if F is provable in T^*,

where T is a classical set theoretic system for analysis and T^* a constructive theory. In that case, T is called REDUCIBLE TO T^*, as the principles of T^* are more elementary and more restricted.

I also want to emphasize from the outset that I do not think such reductions are needed to "justify" the practice of classical mathematics. Paul Bernays remarked in 1930, and I cannot but agree with him today, that

the current discussion on the foundations of mathematics does not arise from any emergency within mathematics itself. Mathematics is in a perfectly satisfactory state of certainty concerning its methods.[1]

---
*This paper was completed in June 1981; some minor changes were made in August 1982.
[1](Bernays 1930b, p. 17)

He viewed the critical issue in the discussion as follows:

> The problems, the difficulties, and the differences of opinion begin only, when one inquires not simply after the mathematical facts, but after the grounds of mathematical knowledge and after the delimitations of mathematics.[1]

These inquiries of a reflective, philosophical character continue to have profound interest. Their fruitful pursuit, however, must be informed by detailed mathematical and metamathematical work. Before either accepting or rejecting portions of mathematics on philosophical grounds, we have to analyze more carefully and understand more adequately this central part of our intellectual experience.

But what can proof theory, you may wonder, contribute to such a better understanding? Is it not most intimately connected with a particular foundational program, namely, Hilbert's? And is that not inspired by a crude formalism which, if it managed to survive Frege's criticism by some miracle, is certainly dead since Gödel proved his incompleteness theorems? These questions will be answered implicitly in this paper. I shall explicitly discuss a modification of Hilbert's program which has been explained by Bernays in numerous writings. In that modified form the program provides a coherent framework for foundational research focusing on two complementary goals: to determine (i) which principles are used and needed in a particular branch of mathematics, primarily in classical analysis, and (ii) which constructive, and possibly more evident, principles suffice for a reduction of theories arising from (i).

Recently, most interesting discoveries related to (i) and (ii) have been made. With regard to (i) it has been found that the bulk of classical analysis can be carried out in conservative extensions of elementary number theory; with regard to (ii) it has been shown that strong impredicative classical theories for analysis can be reduced to constructively acceptable ones. I will describe these results in some detail; first, however, I will sketch their historical and systematic background. Before beginning with this sketch, let me emphasize again that I want to report on work which provides, in my view, material for philosophical reflection. I will not put forward philosophical theses here, but point out objective mathematical and metamathematical relations. Perhaps that can better than any direct argument contribute to the insight, that the (exclusive) alternative between "constructivistic" and "platonistic" foundations is "a logically inadmissible application of the tertium non datur". (That is how Zermelo characterized the alternative between Brouwer's intuitionism and Hilbert's formalism.)

1 Foundations for classical analysis

In this first part of my paper I shall outline very schematically (the development towards) the arithmetization of analysis in 19th century mathematics. The main reason for presenting this historical sketch is to point beyond the set theoretic paradoxes, which occupy such a pivotal place in the

Foundations for analysis and proof theory

modern discussion on the foundations of mathematics, to the longer standing problem of providing a sound basis for analysis. A reduction or a radical restriction to the natural numbers was viewed by many mathematicians as a way of solving this problem in a mathematically and philosophically satisfactory manner. It seems to me that "the" arithmetization of analysis was the common concern of men like Dedekind, Cantor, and Weierstrass, Frege and Russell, Kronecker and Brouwer, Hilbert and Zermelo. This last statement may be more plausible when one observes that none of these mathematicians appealed in their foundational work to other than narrowly arithmetic and (or) logical, set theoretic concepts; in particular, an appeal to geometric concepts was never contemplated.[2]

The very introduction of irrational numbers into analysis, however, is geometrically motivated; the notion of an arbitrary cut of rationals in Dedekind's work is suggested by considering completely arbitrary divisions of a straight line into two segments. That notion, or that of an arbitrary subset of natural numbers, is involved in the arithmetization of analysis as given by Dedekind and Weierstrass. This is one point I want to emphasize already now. Another point has to do with the development of Hilbert's views on how to give consistency proofs for analysis: from adapting the Dedekind-Weierstrass methods to reflecting on analysis as codified in a formal theory by elementary mathematical means. He hoped to achieve in this radically new way a more thoroughgoing, stricter arithmetization, namely, an arithmetization satisfying the finitist restrictions of Kronecker.

1.1 Arithmetization of analysis. One may go back to Bishop Berkeley's "The Analyst, or a Discourse Addressed to an Infidel Mathematician" (1734) for a vigorous and biting attack on inconsistencies in the early calculus. Even a hundred years later fundamental parts of analysis were still obscure, not only to philosophers and theologians but to very gifted mathematicians. N. H. Abel vowed in a letter of March 29, 1826 to C. Hansteen

to apply all my energy to bring a little more clarity into the surprising obscurity one finds undoubtedly in analysis today. It lacks all plan and unity ... the worst is that it has not at all been treated with rigor. There are only a very few theorems in advanced analysis which are proved with complete rigor.[3]

Abel's complaint was certainly justified: basic notions of analysis (continuity for example) were vague; geometric images often took the place of strict proof. Indeed, the central concepts of derivative and integral had neither a proper definition nor a clear range of applicability. So, it is not surprising that eminent mathematicians of the 19th century attempted to secure a rigorous basis for analysis.

Euler and Lagrange had already indicated a direction for the search of such a basis: analysis was to be built exclusively on arithmetic

---
[2] With one exception: Frege contemplated in 1924/25 to base arithmetic on geometry.
[3] N. H. Abel, *Memorial*, Kristiana, 1902, p. 23.

concepts. The tendency to consider the natural numbers as the ultimate basis in mathematics was undoubtedly strengthened by the discovery of Non-Euclidean geometry. Gauss drew philosophical consequences when distinguishing (in a letter of 1817) arithmetic from geometry by observing that only the former was *a priori*. Thirteen years later he wrote to Bessel:

> Geometry has, according to my deepest convictions, a completely different relation to our knowledge a priori than pure arithmetic; our knowledge of the former is indeed lacking in *that* complete conviction of its necessity (and thus also of its absolute truth), which is characteristic of the *latter*; we have to admit in humility, that if number is *merely* the product of our mind, space has also outside of our mind a reality, to which we cannot a priori completely prescribe its laws.[4]

Dirichlet, the successor of Gauss in Göttingen, claimed repeatedly that any theorem of analysis could be formulated as a theorem concerning the natural numbers. This is related by Dedekind.[5] Dirichlet's claim must have been based on two convictions, namely, that the advanced notions of analysis can be defined arithmetically (which had already been achieved by Cauchy for continuity), and that the real numbers can be "constructed" from the natural numbers in a purely arithmetic fashion. The construction of the set $\mathbb{R}$ of reals from the set $\mathbb{N}$ of natural numbers is for my discussion most significant; in particular the step from $\mathbb{Q}$, the rational numbers, to $\mathbb{R}$. This step was taken in different, but equivalent ways by Weierstrass, Cantor, and Dedekind.

Dedekind's construction, on which I shall concentrate, was conceived in 1858 and published 14 years later in *Stetigkeit und irrationale Zahlen*. It is thoroughly geometrically motivated! The observation that each point of a straight line partitions that line into segments S_1 and S_2, such that each point of S_1 lies to the left of each point of S_2, led Dedekind to the formulation of a principle expressing the "essence of continuity". It is the converse of the above observation:

> If all points of the straight line fall into two classes, such that each point of the first class lies to the left of each point of the second class, then there exists one and only one point producing this partition of all points into two classes, this cutting (Zerschneidung) of the straight line into two segments.[6]

The field of rational numbers $\mathbb{Q}$, whose construction from $\mathbb{N}$ is assumed in that essay, has to be completed to a "continuous manifold" (stetiges Gebiet), if one wants to pursue all properties of the straight line arithmetically. But how can this be done? By carrying the geometric continuity principle over to arithmetic and letting the real numbers just be cuts of rationals, i.e., partitions of $\mathbb{Q}$ into two nonempty classes A_1 and A_2 with the characteristic

---

[4](Gauss and Bessel 1880, p. 497).

[5](Dedekind 1932/1888, p. 338). (I refer always to the third volume of Dedekind's collected papers; "1888" after the slash indicates that the quotation is taken from *Was sind und was sollen die Zahlen?* and "1872c" that the quotation is from *Stetigkeit und irrationale Zahlen*.)

[6](Dedekind 1932/1872c, p. 322).

property that each element of A_1 is smaller than each element of A_2.[7] Dedekind proves then that the system of all cuts of rationals is a continuous manifold: each cut of real numbers is already determined by a unique cut of rationals (Satz IV). $\mathbb{R}$ is thus, Dedekind writes in a letter to Lipschitz, "das denkbar vollständigste Grössen-Gebiet".[8]

The continuity principle is not only motivated from a geometric standpoint; it is indeed intrinsic to the practice of analysis. Dedekind points that out by showing that Satz IV is equivalent to two main theorems. Let me mention the first of them; it asserts that any monotonically increasing, but bounded function (from $\mathbb{R}$ to $\mathbb{R}$) approaches a limit. Dedekind says in the introduction, when describing his reasons for developing a theory of irrationals, that in proofs of this theorem he always had to appeal to geometric evidence. Another theorem which depends crucially on the continuity principle and which is obvious when we think of continuous manifolds is this: if g is a continuous function on the closed interval $[0, 1]$ with $g(0) < 0$ and $g(1) > 0$, then g vanishes at some point in the interval.[9] (Note here that the principle does what it is supposed to do; it allows us to prove in "an arithmetic manner" geometric properties characteristic of continuous manifolds.)

Dedekind viewed a cut as a purely arithmetic phenomenon (eine rein arithmetische Erscheinung).[10] This seems to be correct, however, only if parts of set theory are adjoined to arithmetic (as it is understood now), or if arithmetic is a part of logic and the needed set theoretical principles are available as logical ones (as Dedekind believed).[11] Dedekind formulated such principles explicitly in his other foundational essay *Was sind und was sollen die Zahlen?* He took, for example, as a logical law that the extension of any predicate is a "system", his term for set, manifold, or totality. A system is completely determined, if for each thing (which for Dedekind is any object of our thinking, in particular, systems are things) it is determined whether or not it is an element of the system. Dedekind remarks in a footnote that it is quite irrelevant for his development of the general laws of

---

[7]In his essay Dedekind speaks of "creating" new numbers; in a letter to Lipschitz, (Dedekind 1932, p. 471), he emphasizes that one could "identify" the real numbers with cuts.

[8](Dedekind 1932, p. 473). In another letter to Lipschitz Dedekind points out the crucial difference between Euclid's treatment of irrational magnitudes and his own. "Euclid can apply his definition of equal proportion to magnitudes, which come up in his system, i.e. whose existence is evident [ersichtlich] for good reasons, and that is quite sufficient for Euclid." However, Dedekind — in contrast to Euclid — wants to base arithmetic on the concept of magnitude, and thus it is crucial to know "from the beginning how complete (continuous) the domain of magnitudes is, because nothing is more dangerous in mathematics, than to *assume existences without sufficient proof*...." (Dedekind 1932, p. 477).

[9]This theorem is not mentioned in Dedekind, but it motivated Bolzano to search for a purely analytic proof; and such a radical modern constructivist as Bishop finds it "intuitively appealing", but unprovable in his theory. The second theorem which is discussed by Dedekind is Cauchy's convergence criterion for a function from $\mathbb{R}$ to $\mathbb{R}$.

[10](Dedekind 1932/1888, p. 339).

[11](Dedekind 1932/1888, p. 335). Frege's views are clearly parallel; he speaks of "defining the real numbers purely arithmetically or logically". (*Grundgesetze der Arithmetik*, II. Band, 1903, p. 162.) Cf. also footnote 1, ibid. p. 155.

systems, whether their complete determination can be decided by us. This remark is directed against Kronecker who had earlier on formulated decidability conditions. I shall come back to this later; but let me note that such restrictions form one important reason for Kronecker's rejecting the general notion of irrational number.

Using such logical principles, and an assumption I shall formulate at the beginning of 1.2, Dedekind showed in that essay how the natural numbers can be defined and characterized up to isomorphism. What a masterpiece of mathematical development and conceptual analysis! (For an appreciation of the latter it is enlightening to read Dedekind's letter to Keferstein (Dedekind 1890a).) The step from $\mathbb{N}$ to $\mathbb{Q}$, which is still missing for a complete "arithmetical" construction of $\mathbb{R}$, can be taken by making use of Kronecker's *Über den Zahlbegriff*.[12] In that paper, published one year before Dedekind's essay in 1887, Kronecker showed how to eliminate from arithmetic systematically all those notions which he considered to be foreign to it, namely, the negative and rational numbers, the real and imaginary algebraic numbers.[12] Kronecker's "elimination" of negative and rational numbers can easily be turned into their "creation" in terms of equivalence classes of pairs of natural numbers, and that is quite in Dedekind's spirit.

The view that a strict and satisfactory arithmetization of analysis had been achieved was widely shared among mathematicians by the end of the 19th century, in spite of Kronecker's opposition to the very introduction of the general notion of irrational number. This conviction is most vividly expressed in Poincaré's remarks to the Second International Congress of Mathematicians (Paris, 1900):

Today, in analysis there are only natural numbers or finite or infinite systems of natural numbers. . . . Mathematics, as one says, has been arithmetized. . . . In today's analysis, if one cares to be rigorous, there are only syllogisms and appeals to that intuition of pure numbers which alone cannot deceive us. One can say that absolute rigor has been achieved today.[13]

Poincaré, in contrast to Dedekind, took the natural numbers as fundamental mathematical objects; and it may be correct that intuition of pure numbers cannot deceive us. However, in the arithmetization of analysis infinite systems are used. That concept, even when restricted by "of natural numbers", turned out to be more problematic than either Dedekind or Poincaré had thought.

---

[12](Kronecker 1887) in (Kronecker 1899, p. 260). As to Kronecker's understanding of "arithmetic" see the end of section 1.2. Here it is to be noted, that the general notion of irrational number was excluded in principle from arithmetic proper. Kronecker's paper was actually part of the reason for Dedekind's decision to publish his thoughts on the matter. It is also informative to hear how Dedekind's essay was received in Berlin. "In Berlin, Dedekind's essay *Was sind und was sollen die Zahlen?*, which had just been published, was talked about in all mathematical circles by young and old — mostly in a hostile manner." Hilbert, who had traveled in 1888 to various German universities, reported this impression in (1931a, p. 487).

[13](Poincaré 1902a, p. 120 and p. 122).

Foundations for analysis and proof theory

1.2 A set (of new and old problems). The final step of Dedekind's arithmetization of analysis (or should one say "reduction to logic"?) involved an argument for the existence of infinite systems; it made crucial use of the assumption that the "totality of all things which can be an object of my thinking"[14] is a system. Cantor told Dedekind in a letter of 28 July 1899 that this assumption leads to a contradiction. In his own investigations, Cantor had discovered the necessity of distinguishing between two types of multiplicities (systems) of things:

> For a multiplicity can be such that the assumption that all its elements "are together" leads to a contradiction, so that it is impossible to conceive of the multiplicity as a unity, as "one finished thing". Such "multiplicities" I call absolutely infinite or inconsistent multiplicities.[15]

And he added:

> As we can readily see, the "totality of everything thinkable", for example, is such a multiplicity. . . .[15]

Which objective criterion can ensure the consistency of a multiplicity and "thus" its existence as a set? Hilbert who had been informed of the Cantorian difficulties proposed a solution at least for the multiplicity of real numbers.[16] In a paper which he entitled *Über den Zahlbegriff* (undoubtedly in polemical allusion to Kronecker's essay of the same title[17]) he presented a categorical axiomatization for $\mathbb{R}$. He claimed that a consistency proof for this axiom system could be given by a suitable modification of Dedekind's (or Weierstrass's) methods in the theory of irrational numbers. In Hilbert's view, the proof of the consistency would at the same time be

> the proof of the existence of the totality of real numbers or — to use G. Cantor's terminology — the proof of the fact, that the system of real numbers is a consistent (finished) set.[18]

The same point was made in his address to the Paris Congress of Mathematicians, when discussing the second of his famous problems. Actually, he went further and claimed that the existence of Cantor's higher number classes and of the alephs could be proved in a similar manner. As in the earlier paper, he mentioned that the existence of the totality of

---

[14](Dedekind 1932/1888, p. 357).

[15]Cantor's letter has been translated and can be found in (van Heijenoort 1967).

[16](Hilbert 1900b), reprinted also in *Grundlagen der Geometrie*. The paper is dated "Göttingen, den 12. Oktober 1899".

[17]At the end of the paper, Hilbert claims that if matters are viewed his way then the reservations concerning the existence of the totality of real numbers lose all their justification.

[18](Hilbert 1900b, p. 242). Note that "consistency proof" corresponds to "Beweis der Widerspruchslosigkeit", "consistent set" to "konsistente Menge". As to my claim that Hilbert thought of Dedekind's and Weierstrass's method, see (Bernays 1967, p. 500).

all number classes or all alephs cannot be established (in this way): these multiplicities are nonconsistent, nonfinished.[19]

By 1904, Hilbert had given up the view that the consistency of his axiom system for $\mathbb{R}$ could be established by means of Dedekind's methods, suitably modified. The plausible reason for this change of mind is the fact that in the meantime more difficulties had arisen in set theory; most importantly, Zermelo and Russell had pointed out the contradiction derived from the set of all sets not containing themselves as elements.[20] Hilbert gave up the particular way in which he had hoped to prove consistency, but he did not give up the goal. The thread of this development is taken up again later. At this point I just mention that Hilbert in (1905a) proposed a simultaneous development of logic and arithmetic. As logic was understood in the broad sense of Dedekind and Frege to contain set theoretical principles, his proposal included (or rather called for) an axiomatization of such principles, which could serve as the basis for mathematics. Within four years, Zermelo and Russell had formulated theories which were suitable for that task and in which the known paradoxes could not be derived, at least not by the usual arguments.[21]

Around 1900, Zermelo had begun, under the influence of Hilbert, to turn his attention to the foundations of mathematics and in particular to the basic problems of Cantorian set theory.[22] In his paper *Untersuchungen über die Grundlagen der Mengenlehre I* he undertook to isolate the set-formation principles central for the development of set theory and thus, in Zermelo's view, of mathematics. For this he analyzed Cantor's and Dedekind's work very closely. The somewhat cautious and experimental attitude in which this enterprise was pursued is best described by Zermelo himself:

In solving the problem [of establishing the foundations for set theory] we must, on the one hand, restrict these principles sufficiently to exclude all contradictions, and, on the other, take them sufficiently wide to retain all that is valuable in this theory.[23]

It is all too well-known which principles were taken by Zermelo as starting-points for his axiom system. Quite in the Hilbertian spirit, Zermelo considered a consistency proof for his axioms to be "very essential".[24] There was, however, a further problem, the more philosophical question "about the origin of these set theoretic principles and the extent to which they

---

[19] (Hilbert 1902, pp. 73–74). These views seem to me to be quite obscure; in particular, does the axiom system have to satisfy further conditions, apart from consistency? In (Hilbert 1900b, p. 242), "endlich", "abgeschlossen", and syntactic completeness (?) are mentioned; are these properties of the axiomatization thought to be important?

[20] Cf. Zermelo's footnote 9 on p. 191 of (van Heijenoort 1967) and (Bernays 1935a, p. 199).

[21] (Zermelo 1908c) and (Russell 1908).

[22] So Zermelo in a report to the Emergency Society of German Science, reprinted in (Moore 1980).

[23] (Zermelo 1908c, p. 200).

[24] Ibid., p. 201.

Foundations for analysis and proof theory 237

are valid".²⁵ A mathematically and conceptually convincing answer to this question was given more than twenty years later in Zermelo's essay *Über Grenzzahlen und Mengenbereiche*; segments of the cumulative hierarchy are recognized as the domains, in which the axioms (expanded by foundation and replacement) are valid. One may put this differently by saying that the ZF-axioms formulate the principles underlying the construction of the cumulative hierarchy.

Russell started out with a detailed analysis of the paradoxes and proposed in (1908) as a final solution his theory of ramified types. This was motivated by broad philosophical considerations.²⁶ Yet for the development of mathematics within that logical framework, he was forced to make the assumption of the axiom of reducibility.²⁷

This assumption seems to be the essence of the usual assumption of classes; at any rate, it retains as much of classes as we have any use for, and little enough to avoid the contradictions which a less grudging admission of classes is apt to entail.²⁸

The pragmatic use of the axiomatic method (in a stricter form than Zermelo's) comes to the fore even more strikingly in *Principia Mathematica*. In that work, Whitehead and Russell engaged in a quasiempirical study. They wanted to demonstrate that contemporary mathematics could be *formally* developed in the theory of types (with the axiom of reducibility). The spirit in which the work was undertaken is worthwhile recalling. In constructing a deductive system for mathematics, they say in the preface to the first edition of *Principia Mathematica*, one has to perform two concurrent tasks: to analyze which principles are actually used and to rebuild mathematics on the basis of those principles. This is a perfectly standard account of axiomatic work in mathematics and indeed any other sufficiently developed subject. It is preceded by a rather provocative and perhaps startling claim:

We have, however, avoided both controversy and general philosophy, and made our statements dogmatic in form. The justification for this is that the chief reason in favour of any theory on the principles of mathematics must always be inductive, i.e. it must lie in the fact that the theory in question enables us to deduce ordinary mathematics. In mathematics, the greatest degree of self-evidence is usually not to be found quite at the beginning, but at some later point; hence the early deductions, until they reach this point, give reasons rather for believing the premises because true consequences follow from them, than for believing the consequences because they follow from the premises.²⁹

Whether Whitehead and Russell succeeded in avoiding controversy with their logical work, I leave for the reader to judge; they definitely did not

---

²⁵ Ibid., p. 200.
²⁶ For penetrating discussions see (Gödel 1944) and (Gandy 1973). In the latter reference in particular the section "Philosophical Framework".
²⁷ (Russell 1908, p. 167).
²⁸ Ibid., p. 168.
²⁹ (Whitehead and Russell 1910, p. v).

with their statement that "the chief reason in favour of *any* theory on the principles of mathematics must *always* be inductive". There are not only convincing conceptual analyses of mathematical principles (e.g., Dedekind's for the so-called Peano-axioms or Zermelo's for the ZF-axioms), but there are also areas of "ordinary mathematics" where the "original data" are quite dubious (e.g., infinitesimals in the early calculus). To put it squarely, the creative and critical function of fundamental theories is not appreciated here.

It is almost amusing to consider Whitehead and Russell's maxim for arguing in favour of a basic theory side by side with their view that the (Dedekind, Cantor) theory of irrational numbers is undoubtedly a part of ordinary mathematics![30] (And thus, one presumes, belongs to the "original data" to be accounted for.) The controversy surrounding the irrationals and the set theoretic principles used in their definition, after all, had not been resolved. It was also connected with broader, more philosophical issues concerning the nature of mathematics. The public debate between mathematicians flared up more intensely again after Zermelo's proof of the well-ordering theorem by means of the axiom of choice (Zermelo 1904). The reluctance of mathematicians to accept the result was founded (mostly) in the strikingly nonconstructive, purely existential character of the choice principle. Though it is obviously true for Zermelo's notion of set and his broad subset notion,[31] it loses its evidence when only definable subsets of a given domain are considered. In the famous exchange of letters between Borel, Baire, Lebesgue, and Hadamard concerning Zermelo's proof, the first three considered it to be crucial for a secure foundation of analysis to admit only definable subsets of natural numbers.[32] Lebesgue associated a negative answer to the question, Can one prove the existence of a mathematical entity without defining it? explicitly with Kronecker. Baire seemed to suggest that all infinite totalities should be banished from mathematics. He writes at the end of his letter:

And, finally, despite appearances, everything must be reducible to the finite.[33]

Baire's statement is certainly in the tradition of Kronecker. The latter had rejected from the start the set theoretic treatment of the infinite (contra Cantor), the introduction of irrational numbers (contra Weierstrass), and nonconstructive methods in algebra (contra Dedekind).[34] Kronecker had been at the center of the controversy in the 1870s and 1880s.[35] Unfortunately, Kronecker published only one paper concerned with foundational matters, his *Über den Zahlbegriff*. In that paper he formulates and

---

[30] See for example Chapter XIX of Russell's *Principles of Mathematics*.
[31] In (1930, p. 31), Zermelo considers the axiom of choice as a "general logical principle".
[32] The letters are reprinted in (Borel 1914, pp. 150–160).
[33] Ibid., p. 153.
[34] See (Kronecker 1886, pp. 334–6, in particular footnote * on p. 336).
[35] One has only to think of Cantor's complaints, Weierstrass's remarks, and Hilbert's observations.

Foundations for analysis and proof theory

pursues a radically eliminative program, as described in 1.1.[36] Arithmetic is for him, as it was for Gauss, the "Queen of Mathematics" and comprises all mathematical disciplines with the exception of geometry and mechanics. Algebra and analysis in particular fall under it. Absolutely fundamental are the natural numbers;[37] he believes, that

> we shall succeed in the future to "arithmetize" the whole content of all these mathematical disciplines [which fall under arithmetic]; i.e. to base it [the whole content] on the concept of number taken in its most narrow sense, and thus to strip away the modifications and extensions of this concept, which have been brought about in most cases by applications in geometry and mechanics.[38]

(In a footnote, referring to "extension of this concept" in the above quotation, Kronecker makes clear that he has in mind "in particular the addition of the irrational and continuous magnitudes".) Immediately after this passage with its call for strict arithmetization, Kronecker points to the *difference in principle* between arithmetic (in his sense) on the one hand and geometry and mechanics on the other: only the former is *a priori*. As support for his position, he describes Gauss's view expressed in the letter to Bessel, which I quoted when discussing the arithmetization of analysis. (As a matter of fact, that text is given by Kronecker in a footnote.)

The restriction to the natural numbers as the legitimate objects of (Kronecker's) arithmetic is accompanied by a restriction of methods. The latter is hinted at in the effective treatment of roots of algebraic equations in *Über den Zahlbegriff*;[39] it is discussed explicitly and in general terms in Hensel's foreword to Kronecker's *Vorlesungen zur Zahlentheorie*. As the general points are so characteristic for a (very restricted) constructivist approach to mathematics, let me quote these remarks.

> He [Kronecker] thought that . . . any definition should be formulated in such a way, that one can find out in a finite number of trials, whether it applies to a given magnitude or not. Similarly, a proof of existence of a magnitude can only be viewed as completely rigorous, if it contains a method which allows us to find the magnitude whose existence has been claimed.[40]

It is in this critical, constructive tradition within mathematics that Brouwer and his intuitionism can be seen.[41] Brouwer, however, went beyond restricting classical mathematics: he introduced new notions (e.g., that of a choice sequence), new methods (e.g., bar induction), and a constructive treatment of infinitary objects (ordinals or well-founded trees).

---

[36] The main mathematical work, however, is devoted to an effective analysis of the real roots of algebraic equations.

[37] And that holds no matter whether they are created by God (as remarked by Kronecker in a talk at the Berliner Naturforscher-Versammlung in 1886 (Weber 1893, p. 19)) or whether the concept of number is being developed in philosophy *before* it is investigated by mathematicians (as suggested by Kronecker at the beginning of his essay).

[38] (Kronecker 1887, p. 253); cf. also the remarks on p. 274.

[39] Ibid., p. 272.

[40] (Kronecker 1901, p. vi).

[41] One should, however, be aware of the quite different philosophical visions.

This led actually to a development of analysis which is in conflict with classical logic.[42] Brouwer attempted to rebuild on an intuitionist basis parts of classical mathematics, in particular, parts of set theory and analysis;[43] but he had no qualms about discarding what did not measure up to intuitionist principles. And that seemed to be large and substantial parts of the classical theory.

1.3 Hilbert's program. Others were not prepared to follow Brouwer or, earlier on, Kronecker in restricting mathematical methods and set theory (more than necessary to avoid contradictions). Most outspoken among those was Hilbert. Bernays is quoted in C. Reid's Hilbert biography as saying:

Under the influence of the discovery of the antinomies in set theory, Hilbert temporarily thought that Kronecker had probably been right there [i.e., right in insisting on restricted methods]. But soon he changed his mind. Now it became his goal, one might say, to do battle with Kronecker with his own weapons of finiteness by means of a modified conception of mathematics....[44]

In his Heidelberg talk of 1904, Hilbert sketched what he considered to be a refutation (sachliche Widerlegung)[45] of Kronecker's viewpoint on foundations. The ultimate goal of his proposal was the same as the one formulated in *Über den Zahlbegriff* and his Paris address, namely, to establish the existence of the set of natural and real numbers and of the Cantorian alephs;[46] and that was to be achieved, as before, by consistency proofs. In this talk, however, Hilbert indicates how such proofs might be given without presupposing basic logical, set theoretic notions. Such presuppositions seem to him to be problematic now for the foundation of arithmetic, as they in turn use basic arithmetic concepts. So he proposes a "simultaneous development of the laws of logic [understood in Dedekind's broad sense] and arithmetic".[47] The development is to be given in a stricter, more formal way so that proofs can be viewed as finite mathematical structures (endliche mathematische Gebilde). The new task is to show by elementary mathematical means that such proofs cannot lead to a contradiction.[48] However vague, provisional, and confused these suggestions may have been, they foreshadowed aspects of Hilbert's proof theoretic program of the twenties.

It is to be noted here that Hilbert was not opposed to constructive tendencies in mathematics; indeed, he later claimed that only his program does

---

[42] Due to the *intuitionist* continuity principle.
[43] Choice sequences were introduced to obtain a more adequate theory of the continuum.
[44] (Reid 1970, p. 173).
[45] (Hilbert 1905a, p. 258).
[46] (Hilbert 1905a, pp. 252–3, 257–8).
[47] In this proposal two different tasks are involved: (i) an axiomatization of set theoretic principles, and (ii) a strictly formal development of the subject.
[48] (Hilbert 1905a, pp. 251–2 and V on p. 257).

Foundations for analysis and proof theory 241

justice to them "as far as they are natural".[49] Furthermore, he admitted that "Kronecker's criticism of the usual way of dealing with the infinite was partly justified".[50] From papers written by Bernays in 1921 and 1922 we can plausibly infer that Hilbert viewed his program then as a way of mediating between two conflicting doctrines, namely, the classical set theoretic and the (intuitionist) constructive one.[51] How could that be achieved? — By an epistemological reduction! That is most clearly formulated in (Bernays 1922b, pp. 10–11) (and in a more refined form in (Bernays 1930b, pp. 54–55) and in (Hilbert and Bernays 1934, Bd. I, pp. 42–44). The axiomatic theories, which had been given for number theory, analysis, and set theory, and their implicit existential assumptions had to be justified by appealing only to the most primitive intuitive cognitions (primitiv anschauliche Erkenntnisse) in (Bernays 1922b, p. 11). This justification was not to guarantee the truth of those assumptions, but the consistent development of the axiomatic theories and thus the truth of elementary statements.[52] (This will be explained in greater detail below.) As the consistency problem could be formulated in an elementary mathematical way, there was at least a chance of solving it by constructive means. Those means were included in the most elementary part of number theory. Hilbert called this part of mathematics *finitist* and believed that it coincided essentially with what Kronecker and Brouwer (had) accepted.[53] Let me describe more precisely how such a justification was to be given. (The particular way of giving this description is due to Kreisel; see for example (Kreisel 1968).)

For Russell, the work in *Principia Mathematica* established the reduction of mathematics to logic; for Hilbert, it simply showed that (parts of) mathematics can be developed within a particular formal system S. Basic assumptions are formulated as axioms, and logical inferences are specified as rules; thus, mathematical reasoning, as far as its results are concerned, can be replaced by manipulating symbols according to fixed rules.[54] Hilbert made the crucial observation that such a formal system can be described from a metamathematical point of view in a finitist way. In particular one

---

[49](Hilbert 1935, p. 160). This is Hilbert's "Neubegründung der Mathematik. Erste Mitteilung".

[50](Bernays 1967, p. 500).

[51]With regard to my interpretative claim, see (Bernays 1922b, p. 15). The conflicting doctrines may be said to "correspond" to opposing tendencies in Hilbert's own thinking about mathematics. "On one side, he [Hilbert] was convinced of the soundness of existing mathematics; on the other hand, he had — philosophically — a strong scepticism." See Bernays in (Reid 1970, p. 173).

[52]The existence of mathematical structures is no longer guaranteed by a consistency proof — not even programmatically. They play, however, a role in mathematical thinking. This is a part of Hilbert's and, especially, Bernays's view which I find extremely fascinating — AND puzzling. See for example the pre-Gödel paper (Bernays 1930b, pp. 54–5) and (Bernays 1950) which is devoted to precisely this problem.

[53]As to Kronecker, see (Hilbert 1931a, p. 487); as to Brouwer, see (Bernays 1930b, pp. 41–2 and fn. 9 on p. 42).

[54](Bernays 1922b, p. 12).

can express finitistically that a certain configuration b constitutes a proof of a formula F in the system S; briefly,

$\text{Pr}_S(b, F)$.

And this notion satisfies Kronecker's methodological desideratum, as one can decide in a finite number of steps whether a given configuration is or is not an S-proof of F.

Using the proof predicate Pr_S one can formulate what Kreisel calls ADEQUACY CRITERIA for S with respect to finitist mathematics $\mathcal{F}$. The first criterion requires that finitist mathematics can be developed in S. For the formulation of the criterion let $\pi(\tau)$ be a finitist mapping associating derivations (formulas) in S with statements $\mathcal{A}$ of $\mathcal{F}$:[55]

(Adeq 1) $\mathcal{A} \to \text{Pr}_S(\pi(\mathcal{A}), \tau(\mathcal{A}))$.

This condition seemed to be satisfied by those theories in which Hilbert was most interested; in set theory, for example, all of elementary number theory could be developed. The second adequacy criterion is crucial. It expresses in finitist terms that S proves only correct $\mathcal{F}$-statements;[56] more concisely,

(Adeq 2) $\text{Pr}_S(b, \tau(\mathcal{A})) \to \mathcal{A}$.

Adeq 2 is a proof theoretic REFLECTION PRINCIPLE, and it is equivalent to the consistency statement for S.[57] The central question for work on Hilbert's program was consequently: *can one establish the reflection principle (or the consistency statement) for S by finitist means?* — Notice, as Hilbert did[58], that a finitist consistency proof for S would yield a method to transform any S-proof of $\tau(\mathcal{A})$ into a finitist proof of $\mathcal{A}$. In Hilbert's own language, a finitist consistency proof would allow us to eliminate ideal statements from the proof of real statements. (We shall see in 2.2 below how "literally" this can be achieved by Gentzen's method.)

Remark. Hilbert thought that finitist mathematics was a philosophically unproblematic and absolutely basic part of mathematics; by his program he hoped to have separated the foundational questions in mathematics from general philosophy, and to have formulated them in such a way that they were amenable to a *final*, mathematical solution. Proof theory was to settle these questions — once and for all! "Ich möchte nämlich die Grundlagenfragen der Mathematik als solche endgültig aus der Welt schaffen...."[59]

The central question of the program initially allowed for some successful answers. Ackermann and von Neumann gave consistency proofs for

---

[55] Cf. with (Kreisel 1968, pp. 322–3).
[56] (Hilbert and Bernays 1934, p. 42 and p. 44).
[57] This equivalence is easily established under very general conditions on S; for details see (Smoryński 1977).
[58] E.g., in Hilbert's *Grundlagen der Mathematik*, translated in (van Heijenoort 1967, p. 474).
[59] Ibid., p. 465 in translation.

Foundations for analysis and proof theory 243

elementary number theory, or so they thought. A consistency proof for analysis seemed to be within reach.[60] Elementary number theory was believed to be complete.[61] These speculations were limited severely by Gödel's incompleteness theorems. The second theorem showed immediately that the results of Ackermann and von Neumann were of restricted scope and applied only to a part of elementary number theory.[62] It implied furthermore a general restriction on consistency proofs for sufficiently strong formal theories S: consistency proofs for S cannot be given by means formalizable in S. Gödel noticed, however, at the end of his paper that this result does not contradict Hilbert's formalistic standpoint:

> For the latter presupposes only the existence of a consistency proof carried out by finitist methods, and it is conceivable that there might be finitist proofs which cannot be represented in P. . . . [63] [P is the system for which Gödel proved his theorems.]

The situation is nevertheless *prima facie* a dilemma: one has to find an elementary, constructive proof (of a finitist statement) which cannot be carried out in S, but S is supposed to be adequate for elementary number theory. Thus, the metamathematical methods cannot be fixed once and for all, but must be suitably chosen for the specific theory. Further, the consistency of a part of mathematics cannot be settled with even "relative" finality, as a simple arithmetic statement, to wit, the consistency statement, can be formulated in S but not decided by S. This serious difficulty is already implied by the first incompleteness theorem. The very choice of the formal system S is problematic, as the completeness criterion is in principle not available in the essential cases.[64]

The conclusive and final mathematical solution to foundational problems Hilbert had hoped for cannot be obtained, the separation of mathematics from philosophy not sustained; these are the general consequences of Gödel's results. Yet if one gives up the dogmatic (metamathematical) restrictions and the radical (philosophical) aims, proof theory can be fruitfully developed. As early as 1932, Gödel and Gentzen independently gave a consistency proof for classical number theory Z. Their arguments

---

[60](Hilbert *1929, p. 8) and (Bernays 1930b, p. 58); Hilbert's paper was delivered at the International Congress of Mathematicians in Bologna.

[61](Bernays 1930b, p. 59). It is important to notice here that Hilbert and Bernays did not think that any formal theory would be a final, nonextendible framework for mathematics; it would be extendible by new concepts! Nevertheless, Bernays argues that a formal theory may be such that an extension by new concepts does not lead to new results in the original (language of the) theory. This condition, he continues, is certainly satisfied if the theory is deductively closed, i.e., each sentence, which can be formulated in the theory, is either provable or refutable. It is precisely syntactic completeness in this sense which is excluded by Gödel's first incompleteness theorem.

[62](Bernays 1935a, p. 211).

[63](Gödel 1931d, p. 37).

[64]Gödel emphasized this point in his 1931 correspondence with Zermelo, published in Grattan-Guinness, 'In memoriam K. Gödel', *Hist. Math.*, **6** (1979), 294–304. Cf. also note 61.

showed that Z is reducible to intuitionist arithmetic HA (in the sense of the introduction). As HA was seen to be correct from an intuitionist viewpoint, a consistency proof had been given not by finitist but by intuitionist means.[65] Bernays remarked, when looking back at this result in 1967:

It thus became apparent that the "Finite Standpunkt" is not the only alternative to classical ways of reasoning and is not necessarily implied by the idea of proof theory. An enlarging of the methods of proof theory was therefore suggested: instead of a restriction to finitist methods of reasoning it was required only that the arguments be of a constructive character, allowing us to deal with more general forms of inference.[66]

Thus the question, Can we prove *the* system S for this part of mathematics to be consistent by finitist means? must be replaced by, Can this T, which is significant for this part of mathematics, be shown to be consistent by appropriate constructive means? Or, to paraphrase the new question, Is T reducible to a constructively justified theory? Hilbert and Bernays considered "ordinary analysis"[67] as the most significant T for which this modified goal should be pursued.

2 Proof theory of subsystems of analysis

Two groups of (meta) mathematical results will be discussed now. The results give (partial) answers to the questions I asked in my introductory remarks. Not surprisingly then, this part of my paper consists of two sections. In the first section I will describe formal theories for "ordinary" full classical analysis, some important subsystems, and a logical calculus in sequential form. In the second section I will outline some proof theoretic investigations concerning elementary number theory and impredicative subsystems of analysis. Here, I want to present matters in such a way that also the methods of proof are clearly indicated. As a matter of fact, I will restrict attention to one particularly lucid method due to G. Gentzen. It is my impression that it yields in the most direct and intelligible way partial solutions to Hilbert's reduction problem. For more information concerning this quite active field of research I have to refer to the literature; I mention in particular the survey papers (Kreisel 1968) and (Feferman 1977), and the Lecture Notes in Mathematics by Buchholz, Feferman, Pohlers, and Sieg.

2.1 Formalisms for (parts of) analysis.
Hilbert gave a formalism for analysis in lectures during the early twenties. This formalism along with two equivalent formulations is presented in Supplement IV of *Grundlagen der Mathematik*, Band II. It is indicated there in rough sketches how to

---

[65]This was incidentally quite revealing, as it had been assumed in the Hilbert school that intuitionist and finitist reasoning were identical. Cf. note 53.

[66](Bernays 1967, p. 502).

[67]Their formalism of (ordinary) analysis is described at the beginning of section 2.1.

Foundations for analysis and proof theory 245

formally develop analysis and the theory of Cantor's second number class.[68] The main formalism is equivalent to second-order number theory with the full comprehension principle

CA $(\exists Y)(\forall x)(x \in Y \leftrightarrow Fx)$

and the axiom of choice in the form

AC $(\forall x)(\exists Y) FxY \to (\exists Y)(\forall x) Fx(Y)_x,$

where $y \in (Y)_x \leftrightarrow \langle y, x \rangle \in Y$, and $\langle , \rangle$ is a standard pairing function for natural numbers. In both CA and AC, F is an arbitrary second-order formula and may contain number- and set-parameters.[69]

Part of analysis can obviously be developed in subsystems of this theory. This is a trivial observation, as one uses in each specific proof only finitely many instances of the comprehension and choice principles. The interesting question is, whether significant portions can be captured in fixed, bounded parts. To discuss this point clearly we need a more detailed description of the formalisms.[70] All theories are formulated in the language $\mathcal{L}^2$ of second-order arithmetic; $\mathcal{L}^2$ is a two-sorted language with variables (and parameters) $x, y, z, \ldots (a, b, c, \ldots)$ ranging over natural numbers and $X, Y, Z, \ldots (A, B, C, \ldots)$ ranging over subsets of $\mathbb{N}$. As nonlogical symbols $\mathcal{L}^2$ contains $0, ', f_j$, for each $j \in \mathbb{N}, =, \in$, and $\bot$. $(f_j)_{j \in \mathbb{N}}$ is interpreted as an enumeration of the primitive recursive functions; they are assumed to be unary, except for the pairing function $\langle , \rangle$. $()_0$ and $()_1$ are the related projection functions. $\bot$ abbreviates $0 = 0'$. The numerical terms of $\mathcal{L}^2$ are built up as usual; $s, t, \ldots$ are syntactic variables for them. Formulas are obtained inductively from atomic formulas $s = t$ and $t \in A$ by closing under the propositional connectives $\to, \wedge$ and quantification over both sorts (and of both kinds). These logical symbols are chosen as the language is to serve for both classical and intuitionist theories.[71] $F, G, \ldots$ are syntactic variables over formulas. The axioms for classical analysis (CA) are first of all

---

[68] For someone familiar with the development of classical analysis in ZFC it is easy to see that in such a presentation one appeals only to the first three levels of the cumulative hierarchy, considering the natural numbers as urelements. These levels are prima facie needed to accommodate reals, functions from $\mathbb{R}$ to $\mathbb{R}$, and specific functionals like the Riemann integral. Considering, however, appropriate third-order functions and coding continuous functions by sets of natural numbers, one can carry out this development in Hilbert and Bernays's main formalism which is described next.

[69] The formalism is not only of interest as a framework for mathematical practice in classical analysis, but it has also important metamathematical stability properties. First of all, there are proof theoretic equivalences to various versions of set theory WITHOUT the power-set axiom; secondly, the formalism with AC is conservative over the formalism with just CA for Π^1_4-sentences. This point has been emphasized by Kreisel. For technical details see (Apt and Marek 1974).

[70] It is given as in Feferman's and my first chapter of the *Lecture Notes* volume already mentioned.

[71] Even in the restriction to the purely number theoretic part of the language and to intuitionist logic one can define $\neg, \vee, \leftrightarrow$ from $\to, \wedge, \bot$.

those for zero and successor, pairing and projections; they include furthermore the defining equations for all one-place primitive recursive functions, the comprehension schema and the induction schema

$$F0 \wedge (\forall y)(Fy \to Fy') \to (\forall y)Fy$$

for all formulas F of $\mathcal{L}^2$. In the presence of CA, the induction schema can be replaced by the corresponding second-order axiom

$$(\forall X)[0 \in X \wedge (\forall y)(y \in X \to y' \in X) \to (\forall y)y \in X]$$

to yield an equivalent formal theory. However, as soon as the comprehension principle is restricted to a subclass of $\mathcal{L}^2$-formulas, the theory with the induction axiom is in general proof theoretically weaker than the theory with the schema. The base theory for our consideration is (PR-CA)↾; i.e. second-order arithmetic with CA restricted to PR-formulas[72] and the induction axiom. That theory is conservative over primitive recursive arithmetic. The further second-order theories will extend this basic theory. They are denoted by the additional (set-existence) principle. (Π^0_∞-CA)↾ and (Π^1_1-CA)↾ are (PR-CA)↾ together with the comprehension principle for all arithmetic and Π^1_1-formulas, respectively; (Σ^1_i-AC)↾ for $i = 1, 2$ is (PR-CA)↾ extended by the instances of AC for all Σ^1_i-formulas.[73] Dropping "↾" from the name of a theory indicates that the full induction schema is available; for example, (PR-CA) is like (PR-CA)↾ except that the induction axiom is replaced by the induction schema for all formulas of $\mathcal{L}^2$.

Hermann Weyl was the first to systematically develop analysis in a restricted framework. $\mathbb{N}$ is assumed, but subsets of the natural numbers have to be defined arithmetically; and that means now defined by a formula of $\mathcal{L}^2$ which contains only number-quantifiers. (Thus, impredicative definitions are avoided altogether.) Weyl's analysis, as presented in *Das Kontinuum*, can be formally carried out in (Π^0_∞-CA). Feferman describes matters as follows in his lecture on *Systems of Predicative Analysis*:

> The surprising result found by Weyl was that essentially the whole of analysis of continuous functions as contained, say, in the standard undergraduate course could be developed in this system [i.e., (Π^0_∞-CA)]. We have, for example, for a continuous function on a closed interval of real numbers, existence of maximum and minimum, mean-value theorem, uniform continuity, existence of Riemann integral, and the Fundamental theorem of Calculus.[74]

The crucial mathematical point is that Dedekind's continuity principle can be proved in this theory when it is formulated only for sequences of arithmetically definable reals. The least-upper-bound principle, similarly

---

[72] Quantifier-free formulas and formulas with bounded number-quantifiers. Π^0_∞-formula contain only number-quantifiers; Π^1_1-(Σ^1_1-, Σ^1_2-)formulas are of the form $(\forall X)F$ (($\exists X)F$, $(\exists X)(\forall Y)F$), where F is arithmetic.

[73] Cf. note 72.

[74] (Feferman 1964, p. 102).

restricted, can also be established. These principles suffice to prove appropriate versions of the Bolzano-Weierstrass and Heine-Borel theorems. (For this compare (Weyl 1918, Kapitel II, §§ 4–5).)

Weyl's analysis can actually be carried out in $(\Pi^0_\infty\text{-CA})\!\upharpoonright$; this fact is of substantial interest, as $(\Pi^0_\infty\text{-CA})\!\upharpoonright$ is conservative over elementary number theory Z and remains so even if the choice principle for Σ^1_1-formulas is added.[75] Feferman (1977), Friedman (1976), and Takeuti (1978) have presented finite type theories which are also conservative over Z and which allow a more far-reaching and more convenient development. Essential here is that functions and sets are not interdefinable as in ordinary set theory. For answers to the obvious question, What cannot be done in such weak systems? I have to refer to the literature, e.g., (Feferman 1977).

During the last decades, significant, stronger subsystems have been isolated; they are significant, because they allow the formalization of further substantial parts of mathematical practice, or admit natural nonmaximal models, or are based on foundational viewpoints more restrictive than, or alternative to, the set theoretic one. A paradigm of such foundational research has been carried out under the heading of predicative analysis; see (Feferman 1978). In terms of Feferman and Schütte's precise characterization of "predicative analysis", the subsystem $(\Sigma^1_1\text{-AC})$, for example, is predicative. The theories $(\Pi^1_1\text{-CA})\!\upharpoonright$ and $(\Sigma^1_2\text{-AC})$, in contrast, are impredicative. A constructive foundation for these theories will be given in section 2.2 below.

Quasi-empirical studies as those described above are, in my view, an indispensable component of work on the modified Hilbert program. The primary concern of proof theorists, however, has been the investigation of formal theories. In presenting a small part of such work, I will focus (as mentioned earlier) on Gentzen's method for analyzing the structure on derivations in formal and semiformal theories. It is based on special logical calculi, the so-called sequent calculi.[76] It is characteristic for them that they do not prove individual formulas (from assumptions), but rather pairs of finite sequences of formulas in the form

$$\Gamma \supset \Delta.$$

Γ and Δ stand for sequences $\langle F_1, \ldots, F_m \rangle$ and $\langle G_1, \ldots, G_n \rangle$, respectively. The *sequent* $\Gamma \supset \Delta$ may be interpreted as

$$(F_1 \wedge \ldots \wedge F_m) \rightarrow (G_1 \vee \ldots \vee G_n)$$

---

[75] $(\Pi^0_\infty\text{-CA})$ proves the consistency of Z, and is thus stronger than $(\Pi^0_\infty\text{-CA})\!\upharpoonright$. For a proof theoretic argument concerning $(\Sigma^1_1\text{-AC})\!\upharpoonright$ and Z, compare (Feferman and Sieg 1981a) in the LN-volume. This result is originally due to Barwise and Schlipf.

[76] Such calculi had been introduced by Hertz. In (Gentzen 1934/5) they were investigated and applied to solve some proof theoretic problems, e.g., consistency of Z^-. Gentzen's method was developed further by, among others, Lorenzen, Schütte, Takeuti, Kreisel, Feferman, and Tait.

Let me describe a calculus L^2, appropriate for the language $\mathcal{L}^2$ of analysis. The calculus has as axioms the sequents

$$F \supset F,$$

where F is an atomic formula. There are two types of rules: STRUCTURAL and LOGICAL.

Structural Rules.

$$\frac{\Gamma \supset \Delta}{F, \Gamma \supset \Delta} \qquad \frac{\Gamma \supset \Delta}{\Gamma \supset \Delta, F}$$

$$\frac{F, F, \Gamma \supset \Delta}{F, \Gamma \supset \Delta} \qquad \frac{\Gamma \supset \Delta, F, F}{\Gamma \supset \Delta, F}$$

$$\frac{\Gamma \supset \Delta}{\Gamma^p \supset \Delta} \qquad \frac{\Gamma \supset \Delta}{\Gamma \supset \Delta^p}$$

Γ, F and F, Γ simply indicate the sequences obtained from Γ by pre-(suf-)fixing F. Γ^p is a permutation of the elements of Γ.

Logical Rules.

$$\frac{F_i, \Gamma \supset \Delta}{(F_1 \wedge F_2), \Gamma \supset \Delta} \quad i = 1 \text{ or } 2 \qquad \frac{\Gamma \supset \Delta, F \quad \Gamma \supset \Delta, G}{\Gamma \supset \Delta, (F \wedge G)}$$

$$\frac{\Gamma \supset \Delta, F \quad G, \Gamma \supset \Delta}{(F \to G), \Gamma \supset \Delta} \qquad \frac{\Gamma, F \supset G, \Delta}{\Gamma \supset \Delta, (F \to G)}$$

$$\frac{Ft, \Gamma \supset \Delta}{(\forall x) Fx, \Gamma \supset \Delta} \qquad * \quad \frac{\Gamma \supset \Delta, Fa}{\Gamma \supset \Delta, (\forall x) Fx}$$

In the starred rule for the universal quantifier the parameter a must not occur in the lower sequent. There are clearly further rules, namely, dual ones for the existential number-quantifier and analogous ones for the set-quantifiers. Gentzen employed yet another rule to show that the sequent calculus is equivalent to a standard logical calculus of the Frege-Hilbert type, his SCHNITTREGEL or cut-rule:

$$\frac{\Gamma \supset \Delta, F \quad F, \Gamma_1 \supset \Delta_1}{\Gamma, \Gamma_1 \supset \Delta, \Delta_1}$$

Remark. For the formalization of intuitionist logic one requires that at most one formula occurs on the right-hand side (of $\supset$) in any sequent.

Gentzen proved a remarkable fact concerning the sequent calculus which he formulated as his HAUPTSATZ. It asserts that any derivation can be transformed into a cut-free derivation of the same endsequent.

Foundations for analysis and proof theory

Theorem (Hauptsatz). If $\mathcal{D}$ is a derivation of $\Gamma \supset \Delta$, then we can find a cut-free derivation $\mathfrak{E}$ of $\Gamma \supset \Delta$.[77]

The proof provides a method to obtain finitistically the cut-free derivation $\mathfrak{E}$ from $\mathcal{D}$. The cut-elimination theorem is remarkable, as cut-free derivations exhibit a very special feature, namely, the subformula property: if $\mathcal{D}$ is a cut-free derivation of $\Gamma \supset \Delta$, then every formula occurring in $\mathcal{D}$ is a subformula of a formula in $\Gamma \supset \Delta$. Furthermore, some of the rules can be "inverted" in the cut-free part of the calculus, for example, the quantifier-rules. Let me state one such result.

Inversion Lemma.[78] Let $\mathcal{D}$ be a cut-free derivation of $\Gamma[(\forall x)Fx] \supset \Delta$, let $\Gamma(\Delta)$ contain only Π_1^0-(Σ_1^0) formulas, and let Fa be quantifier-free; then we can find a sequence of terms $t_1, \ldots, t_n$ and a cut-free derivation $\mathfrak{E}$ of $\Gamma[Ft_1, \ldots, Ft_n] \supset \Delta$.

Let us quickly see how these results can be used to eliminate ideal statements from proofs of real ones, i.e., here quantifier-free statements without set-parameters. Z^- denotes the axioms for $0, '$, f_j for all $j \in \mathbb{N}$, $\langle,\rangle$, $()_0, ()_1$, i.e., Z^- is elementary number theory without induction. Assume F is a "real" statement in the above sense and can be proved from Z^-, i.e., there are finitely many axioms $A_1, \ldots, A_m$, and a derivation in L^2 of $\langle A, \ldots, A_m \rangle \supset F$. Then there is cut-free derivation $\mathcal{D}$ of that sequent, and indeed, by the inversion lemma, a cut-free derivation $\mathfrak{E}$ of

$$\langle A_1^*, \ldots, A_k^* \rangle \supset F,$$

where the A_j^* are instances of the Z^--axioms used in $\mathcal{D}$. Now we can observe that the endsequent is quantifier-free and (by the subformula property) *all formulas in $\mathfrak{E}$ are sentential and purely arithmetic, i.e., real.*[79]

2.2 From classical to constructive theories.

That is the direction work on the modified Hilbert program described in this section will take. Recall that in general the proof theoretic equivalence of a classical theory T and a theory T^* of constructive character has to be shown, i.e., for each F in a class $\mathfrak{C}$ of formulas

(*) T proves F iff T^* proves F

In the cases I will discuss, T^* is actually a subtheory of T and, consequently, one direction of (*) is trivial. Focusing on the other direction, we can first remark that it must be established constructively to be significant

---

[77] Indeed, given the length and cut-rank of $\mathcal{D}$ one can determine a bound on the length of $\mathfrak{E}$. This more detailed information can also be given in the case of the cut-elimination theorem for PL below; there it is crucial.

[78] $\Gamma[G]$ denotes a sequence which consists of the elements of Γ and is interspersed with occurrences of G. Π_1^0-(Σ_1^0-)formulas are of the form $(\forall x)F$ $((\exists x)F)$, where F is quantifier-free.

[79] Thus neither the addition of extra sorts nor that of quantificational logic will allow the proof of new real statements.

for Hilbert's program. Combining the constructivity requirements for the metamathematical argument and the theory T^*, the problem can be formulated more concisely as follows: we want to recognize the validity of proofs in T (for sentences F in $\mathfrak{C}$) from the standpoint of T^*. The technical problem is then to *establish in T^** all instances of the *partial reflection principle for T and $\mathfrak{C}$*

$$\mathbb{P}r_T(\ulcorner \mathcal{D} \urcorner, \ulcorner F \urcorner) \to F;$$

$\mathbb{P}r_T$ is the canonical representation of the proof predicate for T in T^*, $\ulcorner \mathcal{D} \urcorner$ and $\ulcorner F \urcorner$ are codes of a T-derivation $\mathcal{D}$ and a formula F, respectively.[80]

Let me look briefly at the paradigm of such a reduction, namely, Gödel and Gentzen's result concerning number theory. The proof of that result establishes finitistically that every negative arithmetic sentence provable in classical number theory can already be proved intuitionistically.

Remark. This proof can be extended to theories of iterated inductive definitions $\text{ID}_{<\nu}$, and full classical analysis (CA); they, too, are conservative over their formally intuitionist versions with respect to all negative arithmetic sentences.[81] In the rest of this section I will explain, how the $\text{ID}_{<\nu}$ can be reduced to a constructive theory. That solves the reduction problem for the impredicative theories $(\Pi_1^1\text{-CA})\!\upharpoonright$ and $(\Sigma_2^1\text{-AC})$, as they are proof theoretically equivalent to $\text{ID}_{<\omega}$ and $\text{ID}_{<\epsilon_0}$, respectively, by work of (Feferman 1970) and (Friedman 1970). However, let me take one step at a time!

How can it be seen that HA proves all instances of the partial reflection principle for Z and all negative arithmetic sentences? The first additional step is a direct formalization of the metamathematical argument *in* HA; that is easily done and gives in HA

$$\mathbb{P}r_Z(\ulcorner \mathcal{D} \urcorner, \ulcorner F \urcorner) \to \mathbb{P}r_{\text{HA}}(\sigma(\ulcorner \mathcal{D} \urcorner), \ulcorner F \urcorner);$$

σ is a primitive recursive function transforming classical into intuitionist proofs. In the second step one makes use of the following facts: (i) $\sigma(\ulcorner \mathcal{D} \urcorner)$ contains only a finite number of formulas and their complexity is consequently bounded by a fixed n; (ii) for formulas of bounded complexity one can give in HA a partial truth-definition T_n. The adequacy of T_n, i.e.,

$$T_n(\ulcorner F \urcorner) \leftrightarrow F,$$

can be established in HA for formulas of complexity $\leq n$, and so can the soundness of derivations containing only such formulas, i.e.,

---

[80] Notice that this is "almost" the reflection principle as formulated in (Adeq 2) above; but the partial reflection cannot be improved to $\mathbb{P}r_T(a, \ulcorner F \urcorner) \to F$ as long as $\mathfrak{C}$ includes Π_1^0-sentences and T and T^* are equiconsistent.

[81] $\text{ID}_{<\nu}$ is described below. The extension of Gödel and Gentzen's result to these theories is easily proved; using a trick of Friedman's it can be established also for Π_2^0-sentences. The work of Feferman and Friedman yields proof theoretical equivalences for classes of sentences, which include all arithmetic ones. Detailed statements of theorems and proofs are in (Feferman and Sieg 1981a).

Foundations for analysis and proof theory

$$\mathbb{Pr}_{HA}(a, \ulcorner F \urcorner) \to T_n(\ulcorner F \urcorner).$$

Detaching in the obvious way we obtain the partial reflection principle.[82]

The argument I just sketched can be carried out for any equivalent formalization of Z and HA. Its first step can be taken in primitive recursive arithmetic PRA, the quantifier-free part of Z or, equivalently, HA. In the second step HA is required for recognizing the truth of axioms (in particular, instances of induction), and simply to formulate the partial truth-definition for quantified formulas. This is in general so, even if the end-formula of a derivation $\mathcal{D}$ is quantifier-free: $\mathcal{D}$ may contain quantified formulas. Let me show now in a simple example, how the sequent calculus and the Hauptsatz are used to eliminate such "extraneous elements" from proofs. For this purpose I return to and continue the investigation of Z^- at the end of section 2.1 above. The considerations presented there, including the proof of the Hauptsatz, can be formalized in PRA to yield

$$\mathbb{Pr}_{Z^-}(\ulcorner \mathcal{D} \urcorner, \ulcorner F \urcorner) \to \mathbb{Pr}(\ulcorner \mathcal{E} \urcorner, \ulcorner \langle A_1^*, \ldots A_k^* \rangle \supset F \urcorner).$$

Due to the subformula property of cut-free derivations, the (purely sentential) complexity of formulas in $\mathcal{E}$ is bounded by that of the formulas in the endsequent. A partial truth-definition for quantifier-free formulas of bounded complexity can be defined and shown to be adequate in PRA. Defining

$$T_n(\ulcorner \Delta \supset G \urcorner) \leftrightarrow (T_n(\ulcorner F_1 \urcorner) \wedge \ldots \wedge T_n(\ulcorner F_m \urcorner) \to T_n(\ulcorner G \urcorner))$$

where Δ is $\langle F_1, \ldots, F_m \rangle$ and all formulas in $\Delta \supset G$ are of complexity $\leq n$, one shows

$$\mathbb{Pr}(a, \ulcorner \Delta \supset G \urcorner) \to T_n(\ulcorner \Delta \supset G \urcorner).$$

So we can immediately infer

$$T_n(\ulcorner A_1^* \urcorner) \wedge \ldots \wedge T_n(\ulcorner A_k^* \urcorner) \to T_n(\ulcorner F \urcorner)$$

and then F, making use of T_n's adequacy and the provability in PRA of the A_j^*. In summary, we have obtained in PRA the partial reflection principle for Z^- and quantifier-free formulas F

$$\mathbb{Pr}_{Z^-}(\ulcorner \mathcal{D} \urcorner, \ulcorner F \urcorner) \to F.$$

The considerations concerning (and involving) the partial truth definition T_n are so obvious for the simple theory Z^-, that Gentzen remarked after having described the transformation of $\mathcal{D}$ into a cut-free derivation $\mathcal{E}$:

---

[82] Partial truth-definitions and reflection principles are treated carefully in (Troelstra 1973, pp. 33–37).

That a contradiction cannot be inferred from such statements [i.e., quantifier-free ones of Z^-] by means of sentential logic is almost self-evident (selbstverständlich); and proof of that would hardly be more than a formal description (Umschreibung) of an intuitively clear fact (Sachverhalt).[83]

Nevertheless, by making this very last step explicit one has the schema of an argument which can be adapted to more complicated situations. Indeed, it can be extended easily to the theory Z_0, i.e., Z^- together with the induction schema for quantifier-free formulas. The general induction schema, however, is a genuine block to a straightforward extension. Gentzen overcame the problem in his (1936) and (1938b) consistency proofs for Z. He used in those proofs in addition to finitist principles formulated in PRA the transfinite induction and recursion principles up to the first ϵ-number ϵ_0.[84]

The role of transfinite ordinals in Gentzen's proofs was at first difficult to understand and to assess. Today, after Lorenzen's, Schütte's, and Tait's work, their role (in a version of the 1938 proof) is perfectly clear: they serve as a natural measure of the length of formulas and derivations in an infinitary propositional calculus! Let me explain this claim by outlining the argument as it is now standardly presented. The infinitary calculus PL is obtained as an extension of the propositional and purely number theoretic part of L^2 by admitting conjunctions and disjunctions of infinite sequences of formulas: if $(A_n)_{n\in\mathbb{N}}$ is a sequence of formulas, then $\Pi_n A_n$, and $\Sigma_n A_n$ are also formulas. The rules for Π are the following:

$$\frac{\Gamma \supset \Delta, A_n}{\Gamma \supset \Delta, \Pi_n A_n} \quad \text{for each } n \in \mathbb{N}$$

$$\frac{A_n, \Gamma \supset \Delta}{\Pi_n A_n, \Gamma \supset \Delta} \quad \text{for some } n \in \mathbb{N}$$

The rules for Σ are dual. The axioms of PL are the sequents $\Delta \supset A$ and $A \supset \Delta$, where Δ is the empty sequence and A is a true and a false atomic sentence, respectively.

The sentences of the language of Z can be translated into formulas of PL, the infinitary operators replacing quantifiers. For example, if Fa is quantifier-free, then $(\forall x)Fx$ is translated as $\Pi_n Fn$.[85] Z-derivations (of sentences) can be transformed into "purely logical" PL-derivations, as all

---

[83](Gentzen 1934/5, p. 416).

[84]Gentzen gave a natural system of notations for the segment of the second number class determined by ϵ_0. One can use it to define (via an appropriate effective coding) a primitive recursive well-ordering of the natural numbers. A precise formulation of transfinite induction and recursion principles is given in (Tait 1965). I denote that theory by PRA + TI(β), if those principles concern a primitive recursive well-ordering of ordinal β.

[85]n is used here as a variable ranging over natural numbers and as a syntactic variable ranging over numerals.

Foundations for analysis and proof theory

Z-axioms are provable in PL. That holds in particular for the instances of the induction principle and is seen most easily when the induction rule

$$\frac{F0 \quad (\forall x)(Fx \to Fx')}{(\forall x)Fx}$$

is used.[86] For assume that $\mathcal{D}_1$ and $\mathcal{D}_2$ are Z-derivations of $F0$ and $(\forall x)(Fx \to Fx')$, respectively; then one has (in the inductive argument on the length of derivations) PL-derivations $\mathcal{E}_1$ and $\mathcal{E}_2$ for $F0$ and $\Pi_n(Fn \to Fn')$.[87] $\mathcal{E}_1$ and $\mathcal{E}_2$ permit us to construct an infinite PL-proof of $\Pi_n Fn$ which is roughly indicated by

$$\mathcal{E}_1 \begin{cases} \downarrow \\ F0 \end{cases} \quad \cfrac{\cfrac{\downarrow}{F0} \quad \cfrac{\downarrow \;\; \Pi_n(Fn \to Fn')}{F0 \to F1}}{F1} \Bigg\} \mathcal{E}_2 \quad \cfrac{\cfrac{\downarrow \;\; \Pi_n(Fn \to Fn')}{F0 \quad F0 \to F1}}{F1} \quad \cfrac{\downarrow \;\; \Pi_n(Fn \to Fn')}{F1 \to F2} \quad \cdots}{\Pi_n Fn}$$

Measuring the length of formulas and derivations of PL with ordinals of the second number class, it is straightforward to prove the cut-elimination theorem for PL by transfinite induction. Consequently, Z-derivations can be transformed into cut-free PL-derivations. If the endformula F of a Z-derivation is Π_1^0, then the associated cut-free PL-derivation contains only subformulas of (the translation of) F. For such formulas one can give an adequate partial truth-definition even in PRA, as we have seen; and the truth of F can be established by induction on the length of its cut-free PL-derivation. The argument schema is the same as for the treatment of Z^-; but are the considerations here not highly non-constructive? Due to the fact that Z is a formal theory, it is sufficient to take into account an effectively described part of PL, all of whose derivations are less than ϵ_0. The argument sketched above can actually be formalized in PRA + TI(β), $\beta < \epsilon_0$.[88] In summary then, each instance of the partial reflection principle for Z and quantifier-free formulas F

$$\mathrm{Pr}_Z(\ulcorner \mathcal{D} \urcorner, \ulcorner F \urcorner) \to F$$

can be proved in PRA + TI(β) for some $\beta < \epsilon_0$.[89]

The investigation of formal theories via infinitary sequent calculi proved extremely fruitful for predicative analysis. It turned out that the approach

---

[86] The resulting theory is easily seen to be equivalent to Z.
[87] For simplicity I assume again that F is quantifier-free.
[88] For a beautiful presentation of a quite similar argument see (Schwichtenberg 1977).
[89] This is best possible in the sense that (i) PRA + TI(ϵ_0) allows the proof of the consistency of Z, and (ii) PRA + TI(β) is contained in Z for each $\beta < \epsilon_0$. (The latter fact was shown by Gentzen.)

can be extended to impredicative subsystems of analysis, seemingly in the most understandable and informative way, through an investigation of proof theoretically equivalent theories of transfinitely iterated inductive definitions.[90] The idea of an inductive definition is familiar from mathematics and mathematical logic; for example, the formulas of a formal language and the theorems of a formal theory are determined inductively. The one feature which gives inductively defined classes (i.d. classes) their special character and accounts in part for their foundational interest is captured by saying that they are generated according to rules or that they are obtained by iterated application of a definable operator.[91] Let me mention one i.d. class which is not given by an "elementary" inductive definition (as those above), namely, the class $\mathcal{O}_0$ of constructive ordinals. That class is given by two inductive clauses: (i) 0 is in $\mathcal{O}_0$; (ii) if a is (the Gödel-number of) a recursive function enumerating elements of $\mathcal{O}_0$, then a is in $\mathcal{O}_0$. The elements of $\mathcal{O}_0$ are thus generated by joining recursively given sequences of previously generated elements of $\mathcal{O}_0$ and can be pictured as infinite, well-founded trees. As this construction-tree is uniquely determined for each constructive ordinal a, it can be viewed as a canonical proof (built by using only rules (i) and (ii)) showing that a is in $\mathcal{O}_0$. Locally, the structure of a tree is as follows:

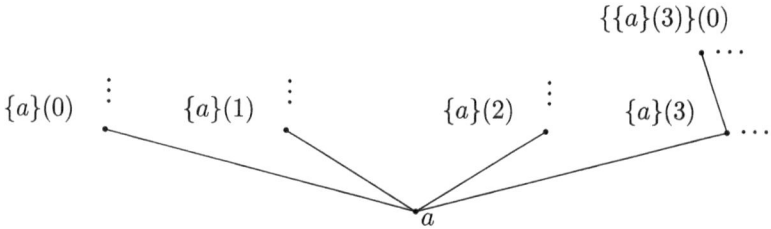

By iterating the definition of $\mathcal{O}_0$ along a given recursive well-ordering $<$ of the natural numbers one obtains the higher constructive number classes $\mathcal{O}_\nu$.[92] Without stating the inductive clauses let me just say that the elements of $\mathcal{O}_\nu$ can be pictured as infinite, well-founded trees, which branch not only over $\mathbb{N}$ but also over number classes $\mathcal{O}_\mu$, $\mu < \nu$.

The theory for $\mathcal{O}_0$, denoted by $\mathrm{ID}_1(\mathcal{O})$, is an extension of elementary number theory. Its language is that of Z expanded by the unary-predicate symbol $\mathcal{O}_0$, its axioms are those of Z (with induction for all formulas in the expanded language) together with two principles for $\mathcal{O}_0$. For their formulation let me note first, that the clauses (i) and (ii) above can be given by

---

[90] The proof theoretic work on predicative analysis is described in (Feferman 1964) and in (Tait 1968b). The first steps in analyzing impredicative subsystems of analysis were taken by Takeuti.

[91] For an attempt to describe the special features of i.d. classes I refer to (Feferman and Sieg 1981a).

[92] $\mu, \nu, \ldots$ are used here as syntactic variables over numerals to indicate that the numerals are related via the well-ordering $<$ to an ordinal of the classical second number class.

Foundations for analysis and proof theory

a formula $\mathfrak{A}$ which is arithmetic and positive in $\mathcal{O}_0$.[93] The first principle is a closure principle and expresses that $\mathfrak{A}$ leads from elements of $\mathcal{O}_0$ to elements in $\mathcal{O}_0$:

($\mathcal{O}$.1) $\mathfrak{A}(\mathcal{O}_0) \subseteq \mathcal{O}_0$;

the second principle is a minimality principle and expresses that one can give arguments by induction on $\mathcal{O}_0$:

($\mathcal{O}$.2) $\mathfrak{A}(F) \subseteq F \to \mathcal{O}_0 \subseteq F$.

(In other words, if the class given by a formula F is closed under $\mathfrak{A}$, then $\mathcal{O}_0$ is contained in that class.) The formal principles for the higher number classes $\mathcal{O}_\mu$, $\mu < \nu$, are formulated in $\mathrm{ID}_\nu(\mathcal{O})$ in analogy to those for $\mathcal{O}_0$.

($\mathcal{O}$.1)$_\nu$ $(\forall y < \nu)\, \mathfrak{A}_y(\mathcal{O}_y) \subseteq \mathcal{O}_y$;

($\mathcal{O}$.2)$_\nu$ $(\forall y < \nu)\, (\mathfrak{A}_y(F) \subseteq F \to \mathcal{O}_y \subseteq F)$.

The theory $\mathrm{ID}_<(\mathcal{O})$ is just the union of the $\mathrm{ID}_\nu(\mathcal{O})$; it is also denoted by $\mathrm{ID}_{<\alpha}$, if the classical ordinal associated with $<$ is α. In the general theories of iterated inductive definitions ID_ν and $\mathrm{ID}_<$, one has these principles for all i.d. classes, which are determined by arbitrary positive defining clauses iterated along $<$. The intuitionist versions of these theories are denoted by $\mathrm{ID}^i_\nu(\mathcal{O})$, $\mathrm{ID}^i_<(\mathcal{O})$, ID^i_ν, and $\mathrm{ID}^i_<$.

Remark. These theories have recently been analyzed from a proof theoretic point of view by Buchholz, Pohlers, and myself. The various results we obtained will be published in the LN-volume already mentioned. In the following I focus on a reductive result in my dissertation (Sieg 1977).[94]

The reductive result to be described now is this: assuming that α is a recursive ordinal of limit characteristic, the classical theory of arbitrary inductive definitions $\mathrm{ID}_{<\alpha}$ is conservative over the intuitionist theory of higher constructive number classes $\mathrm{ID}^i_{<\alpha}(\mathcal{O})$ for all negative arithmetic and Π^0_2-sentences. The crucial step in the reduction is a Gentzen-type analysis of ID_ν, $\nu < \alpha$, (which can also be carried out for ID^i_ν) showing that $\mathrm{ID}_{<\alpha}$ and $\mathrm{ID}_{<\alpha}(\mathcal{O})$ prove exactly the same arithmetic sentences. The schema of the argument is that of the proof theoretic analysis of Z^- and Z.

For the investigation of number theory the infinitary propositional calculus PL was used. Even stronger infinitary calculi PL_ν are considered here; conjunctions and disjunctions are taken over $\mathbb{N}$, as well as over all number classes $\mathcal{O}_\mu$, $\mu < \nu$. The language of PL_ν allows us to define explicitly all i.d. classes given by a positive inductive definition iterated at most ν times. This is the basis for a translation of the language of ID_ν into that of PL_ν, associating with each formula F in the former a formula F^+ in the

---

[93] Positivity is a purely syntactic concept and taken here in a very broad sense; see (Feferman and Sieg 1981a).

[94] The starting point of my work was a paper of Tait's, namely, his (1970).

latter. A derivation $\mathcal{D}$ of F in ID_ν can be turned into a PL_ν-derivation of $\mathrm{CP} \supset F^+$, where CP is the translation of the closure principle for the i.d. class mentioned in $\mathcal{D}$.[95] The natural metatheory for these considerations is $\mathrm{ID}_{\nu+1}(\mathcal{O})$, as the syntactic objects of PL_ν just are trees in $\mathcal{O}_\nu$. In this theory one can prove without using a special system of ordinal notations the cut-elimination theorem for PL_ν. Thus, the PL_ν-derivation of $\mathrm{CP} \supset F^+$ can be assumed to be cut-free. From such derivations one can eliminate CP, if F is arithmetic. (The proof of this elimination lemma is a little delicate, but can also be carried out in $\mathrm{ID}_{\nu+1}(\mathcal{O})$.) Using again a partial truth-definition and the soundness of cut-free proofs, the truth of F^+ can be established. So we obtain for ID_ν and any arithmetic F the partial reflection principle in $\mathrm{ID}_{\nu+1}(\mathcal{O})$, analogously that for ID_ν^i in $\mathrm{ID}_{\nu+1}^i(\mathcal{O})$. We know now in particular that the partial reflection principle for $\mathrm{ID}_{<\alpha}^i$, and all arithmetic formulas

(*) $\quad \mathbb{P}r_{\mathrm{ID}_{<\alpha}^i}(\ulcorner\mathcal{D}\urcorner,\ulcorner F\urcorner) \to F$

can be proved in $\mathrm{ID}_{<\alpha}^i(\mathcal{O})$.

The argument for the reduction of $(\Pi_1^1\text{-CA})\!\upharpoonright$ and $(\Sigma_2^1\text{-AC})$ to constructive theories can be quickly concluded now. It was mentioned in the *Remark* at the beginning of this subsection that $(\Pi_1^1\text{-CA})\!\upharpoonright$ and $(\Sigma_2^1\text{-AC})$ are proof theoretically equivalent to $\mathrm{ID}_{<\omega}$ and $\mathrm{ID}_{<\epsilon_0}$ respectively; these latter theories can be reduced to their intuitionist versions by an extension of the double-negation translation. As the steps involved are finitist and can be taken in PRA, one certainly has in $\mathrm{ID}_{<\epsilon_0}^i(\mathcal{O})$ for negative arithmetic and Π_2^0-sentences F

(**) $\quad \mathbb{P}r_{\Sigma_2^1\text{-AC}}(\ulcorner\mathcal{D}\urcorner,\ulcorner F\urcorner) \to \mathbb{P}r_{\mathrm{ID}_{<\epsilon_0}^i}(\rho(\ulcorner\mathcal{D}\urcorner),\ulcorner F\urcorner);$ [96]

combining (*) with (**) yields the desired partial reflection principle for $(\Sigma_2^1\text{-AC})$ in $\mathrm{ID}_{<\epsilon_0}^i(\mathcal{O})$ for the above class of sentences;

$$\mathbb{P}r_{\Sigma_2^1\text{-AC}}(\ulcorner\mathcal{D}\urcorner,\ulcorner F\urcorner) \to F$$

Similarly, one can prove the partial reflection principle for $(\Pi_1^1\text{-CA})\!\upharpoonright$ in $\mathrm{ID}_{<\omega}^i(\mathcal{O})$.[97]

Detailed axiomatic investigations of the type reported in section 2.1 above are of obvious mathematical and philosophical interest. But what has been achieved by the reductions of impredicative theories for analysis to theories for ordinals? Any answer which focuses on their foundational significance will certainly have to argue for the constructive character of the theories for ordinals, and that can be described in part as follows. The theories are based on intuitionist logic; the objects in their intended models are obtained by construction; the definition- and proof-principles which

---

[95] One can assume without loss of generality that a given derivation in ID_ν contains references to only one i.d. class.

[97] $\mathrm{ID}_{<\omega}^i(\mathcal{O})$ and $\mathrm{ID}_{<\epsilon_0}^i(\mathcal{O})$ can be interpreted in $(\Pi_1^1\text{-CA})\!\upharpoonright$, respectively $(\Sigma_2^1\text{-AC})$, such that at least all arithmetic sentences are preserved. Consequently, these theories are equiconsistent.

Foundations for analysis and proof theory

are admitted in the theories follow that construction. The objects, i.e., the constructive ordinals, are furthermore of a very special character. They reflect their buildup according to the generating clauses of their definition in a direct and locally effective way. Viewing the clauses as inference rules, the constructive ordinals are infinitary derivations and show that they fall under their definition. All of this indicates that the theories for ordinals are constructively justified and thus provide a constructive foundation for the classical theories which are reducible to them.

This answer is complemented and its claim (perhaps) supported by technical results of the following form: there is a recursive well-ordering $<$ and a theory T^* for constructive analysis, such that $\text{ID}^i_<(\mathcal{O})$ and T^* are proof theoretically equivalent. Some results of this form have been obtained by Troelstra and by Feferman and Sieg.[98] The conceptually important point, which is technically exploited for their proof, is that sufficiently strong theories for constructive analysis (should) contain principles allowing the construction of ordinals. (It is the parenthetical "should" which accounts for the "perhaps" at the beginning of this paragraph.) The reductive results make it possible now to compare the proof theoretic strength of theories for classical and constructive analysis. That can lead to highly important insights into the proof theoretic structure of parts of mathematics.[99]

So much for the more immediate foundational significance of theories for ordinals (in connection with the modified Hilbert program) and the almost direct application of the reductive results. From a broader perspective, I see these investigations as part of an attempt to take the concept of iteration or inductive definition as basic for analyzing that section of mathematics which lends itself to an arithmetic, constructive treatment.[100] It would be a step forward if a more general notion of i.d. class could be found, permitting distinctions which are mathematically intelligible and significant.[101] Work in two areas might provide complementary experience helpful for finding such a notion and suitable distinctions: (1) the proof theory of subsystems of ZF, and (2) the detailed study of the second constructive number class. Jäger and Pohlers have started work on (1); for (2) I have a concrete problem in mind. When discussing Spector's consistency proof for full classical analysis and the possibility of obtaining satisfactory consistency proofs for parts of analysis, Gödel noted:

---

[98](Troelstra 1980) and (Feferman and Sieg 1981b).

[99]An example of a striking result (but not established with reductions described here) is that obtained by Kreisel and Troelstra, showing the proof theoretic equivalence of ID_1, their system CS for intuitionist analysis with principles for choice sequences, and Kleene's theory for intuitionist analysis in (Kreisel and Troelstra 1970), and *Annals of Mathematical Logic* **3** (1971), 437–9.

[100]This suggestion clearly leaves untouched the treatment of the continuum in Dedekind's classical way using the notion of the powerset of $\mathbb{N}$. For a detailed "defense" of this geometrically motivated approach see (Bernays 1978, pp. 6–9).

[101]Indeed, uses of the notion within mathematics should be looked for.

Perhaps the most promising extension of the system T for the Dialectica-interpretation of HA is that obtained by introducing higher type computable functions for constructive ordinals.[102]

It seems to me that for such an extension one must find new constructive principles for ordinals of the *second number class*, which can be used to give Dialectica-interpretations of intuitionist theories of ordinals $ID^i_<(\mathcal{O})$ for various well-orderings $<$.

Concluding Remarks

Classical mathematical analysis can be developed in a conservative extension of elementary number theory; some impredicative subsystems of analysis do have a constructive foundation. These are, very informally stated, the main facts which are established by the work described in Part 2. They throw an ironic light on the dispute between Brouwer and Hilbert during the twenties. Keep in mind that both men intended to provide a constructive basis for mathematics, with the crucial difference that Brouwer wanted to develop mathematics exclusively by intuitionist principles "in unerschütterlicher Sicherheit", whereas Hilbert strove for a justification of (the use of) classical mathematics through finitist means.[103] They agreed, however, on one fundamental point, namely, that some assumptions of classical mathematics transcend what is given by elementary intuition. The rejection of such assumptions would lead, they both expected, to severe restrictions of classical mathematics and, indeed, of analysis. That conclusion is certainly not supported by section 2.1 above.[104]

The second fact mentioned above is not so much related to (the effect of) the rejection of classical assumptions as to the strength of the admitted constructive principles. The incompleteness theorems showed that finitist mathematics cannot support Hilbert's program; Gödel and Gentzen's consistency proof for classical number theory, on the other hand, made it clear that intuitionist principles go beyond finitist ones. It is most doubtful that they are all justified by what Brouwer called the "fundamental phenomenon of mathematical thinking, the intuition of the bare two-oneness".[105] Indeed, by section 2.2 we know that principles which are intuitionistically

---

[102] K. Gödel, *On an extension of finitary mathematics which has not yet been used*; mimeographed translation of (Gödel 1958), with additional notes. The quotation is from p. 16.

[103] Brouwer's phrase is from *Collected Works I*, A. Heyting (editor), Amsterdam, 1975, p. 412; recall, that for the Hilbert school finitist and intuitionist mathematics coincided.

[104] The use of classical logic can be seen to be inessential (at least for proofs of negative arithmetic and Π^0_2-sentences) by metamathematical means. But it should be mentioned here that E. Bishop in his *Foundations for Constructive Analysis* developed analysis in an almost Kroneckerian spirit. Work of Friedman and Feferman shows that Bishop's constructive mathematics can be carried out in conservative extensions of intuitionist number theory. For references to this area of foundational investigations see (Feferman 1979, pp. 159–4).

[105] Brouwer's *Collected Works I*, p. 85.

Foundations for analysis and proof theory

acceptable permit the reduction of impredicative classical theories — theories so problematic from a constructive point of view.[106]

Bernays pointed out that the intuitionist understanding of conditional statements is not based on elementary evidence when their antecedents are universal statements or conditionals. The reason, Bernays argues, is that intuitionist methods of proof are not fixed;[107] they are most decidedly not captured in a formal theory. As a matter of fact, they included infinitary proofs which were admitted and investigated by Brouwer. Such proofs are mathematically treated as constructive, well-founded trees or ordinals. With reference to them Brouwer wrote:

> These *mental* [gedankliche] mathematical proofs that in general contain infinitely many terms must not be confused with their linguistic accompaniments, which are finite and necessarily inadequate, hence do not belong to mathematics.[108]

He continued that this remark contains his "main argument against the claims of Hilbert's metamathematics".[109]

Zermelo, from a completely different foundational viewpoint, also asserted that finite linguistic means are inadequate for capturing the essence of mathematics and of mathematical proofs:

> ... "combinations of symbols" are *not* ... the true subject matter of mathematics, but *conceptual-ideal relations* between the elements of a conceptually posited *infinite manifold*; and our systems of symbols are in this always just *imperfect* ... auxiliary means of our *finite* mind, to conquer at least in stepwise approximation the infinite, which we cannot *directly* and *intuitively* "survey" or grasp.[110]

To overcome finitist restrictions he proposed an infinitary logic and mathematical theories which ensure that, for a given infinite domain of elements and relations between them, truth and provability coincide. Proofs are infinite well-founded trees and represent, very roughly speaking, classical truth-conditions. The epistemologically problematic aspect of such infinitary proofs is quite clearly seen by Zermelo:

> Such a "proof" contains most often *infinitely many* intermediate sentences, and it is not clear yet, in how far and by which auxilary means it [such a proof] can be made intelligible also to our *finite* mind. Basically any mathematical proof, e.g. the inference principle [Schlußverfahren] of "complete induction", is thoroughly "infinitistic", and yet we are able to comprehend it. There seem to be no fixed limits of intelligibility here.[111]

---

[106] See (Gödel 1944, p. 219) in the reprinted version published in (Benacerraf and Putnam 1983).

[107] And thus the condition that something has been proved is not intuitively determined (anschaulich bestimmt). These considerations are found in (Bernays 1935b).

[108] (Brouwer 1927, p. 460, fn. 8).

[109] Ibid., p. 460, fn. 8.

[110] (Zermelo 1931, p. 85); cf. also the newly published documents from Zermelo's *Nachlass* in (Moore 1980).

[111] (Zermelo 1935, pp. 144–5). To the contemporary work on infinitary logics I can only allude; an introductory survey is given by (Barwise 1981).

The effect of infinitary derivations can be partially obtained in a "finitist" manner. After all, some infinitistic principles (e.g., induction principles, choice principles) can be formulated, even if only inadequately, in formal languages and can be employed in finite derivations for the recognition of truths.[112] For precisely this purpose Gödel suggested the use of stronger and stronger axioms of infinity, as he saw the true reason for the incompleteness of formal systems in the fact that "the formation of ever higher types can be continued into the transfinite . . . , while in any formal system at most denumerably many of them are available".[113] When investigating subsystems of analysis in section 2.2 above, we proceeded in exactly the opposite direction: finite derivations, in which infinitistic principles are used, were expanded into infinitary proofs. Such proofs were transformed into cut-free ones and then recognized as correct (intuitionist) truth-conditions for their endformulas. These considerations were carried out by constructive, but certainly non-elementary principles.[114]

Gödel remarked that giving a constructive consistency proof for classical mathematics means

to replace its axioms [those of classical mathematics] about abstract entities of an objective Platonic realm by insights about the given operations of our mind.[115]

This poignant and schematic formulation assigns to constructive consistency proofs the task of relating two essential aspects of our mathematical experience:

– the impression that mathematics deals with abstract objects arranged in structures which are independent of us;

– the conviction that principles for some structures are a priori and more immediately evident, because the buildup of their elements corresponds so intimately to operations of our mind that we even say the objects are created by us.[116]

If constructive consistency proofs are viewed in this light, they can still be said to provide epistemological reductions, though certainly not in the strong justificatory sense dear to Hilbert. What seems to me to be essential for gaining a general perspective is the fact that they are (usually the most difficult) part of establishing equivalences between theories which are based on prima facie radically opposed, irreconcilable foundational

---

[112]The truth of initial statements and the correctness of rules are certainly assumed.

[113](Gödel 1931d, pp. 28–9, fn. 48a).

[114]Gödel discussed in his 1958 paper, referred to in note 102, a particular extension of finitist mathematics and some general points concerning such extensions. Gentzen suggested to use progressively longer segments of the second number class for consistency proofs, in the form of appropriate systems of ordinal notations.

[115](Reid 1970, p. 218); cf. also Gödel's remarks quoted in (Wang 1974, pp. 325–6).

[116]By that one may simply mean that the operation is constructive and can be grasped "directly", for example, successor operation and the joining of an effectively given sequence of trees in the definition of $\mathcal{O}$.

Foundations for analysis and proof theory 261

viewpoints. The very fact that such equivalences can be given for substantive theories, I believe, undermines traditional positions and encourages an open view for mathematical facts and an unforced attitude in foundational work.[117]

The investigations on the foundations of mathematics are being pursued vigorously. Several basic questions remain open, and we do not know what more we are destined to discover in this area. In any event, these investigations arouse in their changing aspects our curiosity — a sentiment which is brought forth only to a lesser degree by the classical areas of mathematics, which have already achieved a greater perfection.[118]

---

[117] Bernays suggests to view mathematics as "die Wissenschaft von den idealisierten Strukturen"; such idealized structures are in a complex way integrated in our broad intellectual experience. "Wenn wir die zuvor dargelegte Auffassung zugrunde legen, wonach die Mathematik die Wissenschaft von den idealisierten Strukturen ist, so haben wir damit für die Grundlagenforschung der Mathematik eine Haltung, welche uns vor übersteigerten Aporien und vor forcierten Konstruktionen bewahrt und welche auch nicht angefochten wird, wenn die Grundlagenforschung vieles Erstaunliche zutage bringt." (Bernays 1976a, p. 188). This notion invites philosophical reflection on two general questions: (1) what is the status of idealized structures? (2) what are the grounds on which infinitistic principles concerning them are assumed?

[118] (Bernays 1976a, p. 78).

II.7

Reductions of theories for analysis

Introduction

The formal theories considered here are all subsystems of second-order arithmetic with the full comprehension principle, briefly (CA). They are theories for classical analysis: Hilbert used a theory equivalent to (CA) as a formal framework for mathematical analysis in lectures during the early twenties; an extensive portion of analysis had already earlier been developed by Weyl in a weak subsystem of (CA). Weyl's work aimed at *rebuilding parts of analysis on a "sound" basis*, and that meant, above all, avoiding impredicative principles. Hilbert, in contrast, set himself the task of securing the instrumental usefulness of all of classical mathematics. He hoped to achieve that aim by *reducing analysis* and even set theory *to a fixed, absolutely fundamental part of arithmetic*, so-called finitist mathematics. The specific proposal of how to achieve such a reduction is the mathematical centerpiece of Hilbert's program; it was refuted by Gödel's Incompleteness Theorems.

A generalized reductive program has been pursued for analysis, and significant progress has been made. Indeed, progress has been made in two complementary directions. On the one hand, the work of Weyl and of other predicatively or constructively inclined mathematicians has been extended to show that all of classical analysis can be carried out in theories that are reducible to elementary arithmetic. On the other hand, strong impredicative subsystems of (CA) have been reduced to constructively acceptable theories of inductive definitions. These results lead naturally to a distinction between two types of reductions. The two types are distinguished from each other not by the techniques for obtaining them, but rather by programmatic aims. FOUNDATIONAL REDUCTIONS are to provide a constructive basis for strong and (from certain foundational perspectives) problematic parts of (CA); COMPUTATIONAL REDUCTIONS are to yield algorithmic information from proofs in weak and (even from a finitist standpoint) unproblematic parts of (CA).

It is via foundational reductions that the generalized program is pushed forward in an attempt to answer the question "WHAT MORE THAN

FINITIST MATHEMATICS DO WE HAVE TO KNOW IN ORDER TO RECOGNIZE THE (PARTIAL) SOUNDNESS OF A STRONG THEORY?". Computational reductions answer in a specific way Kreisel's question "WHAT MORE THAN ITS TRUTH DO WE KNOW, IF WE HAVE PROVED A THEOREM BY RESTRICTED MEANS (HERE: IN A WEAK SUBSYSTEM)?". Answers to the latter question are of mainly mathematical interest, whereas answers to the former question are of more philosophical significance: in any event, they provide detailed material for reflections on the epistemology of mathematics.

A concise formulation of the generalized reductive program is given below, and the aims of proof theoretic investigations are discussed in greater detail. Some of the results which have been obtained more recently are described, and there is even a new theorem or two. However, the paper is expository and reflective. I hope it will be informative and accesible to non-proof theorists, as I believe that the broad themes are, or should be, of general interest to philosophers concerned with the foundations of mathematics and the sciences. In part 1, I start out with a few historical remarks on the context in which the Hilbert program arose, because it is still widely and deeply misunderstood as an ad-hoc weapon against the growing influence of Brouwer's intuitionism. (A fuller account of this context is given in (Sieg 1984). See also (Giaquinto 1983).) At the end of part 1 the crucial features of one proof theoretic tool, sequent calculi, are presented. Part 2 gives examples of the two types of reductions. Foundational reductions of impredicative parts of (CA) to intuitionist theories of constructive, well-founded trees are given in 2.1. Part 2.2 contains computational reductions; in particular, I will show that an interesting, weak subsystem can be reduced to a proper part of primitive recursive arithmetic (PRA).

1 Reductive aims of proof theory

The problems that motivated Hilbert's program can be traced back to the central foundational issue in 19th century mathematics, namely securing a basis for analysis. A possible resolution is indicated by the slogan "Arithmetize analysis". That direction was given already by Gauss, and its meaning can be fathomed from Dirichlet's claim (reported by Dedekind) that *any* theorem of analysis can be formulated as a theorem concerning the natural numbers. For some the arithmetization of analysis was accomplished by the work of Cantor, Dedekind, and Weierstrass; for others, e.g., Kronecker, a stricter arithmetization was required which would base the whole content of all mathematical disciplines (with the exception of geometry and mechanics) "on the concept of number taken in its most narrow sense, and thus to strip away the modifications and extensions of this concept, which have been brought about in most cases by applications in geometry and mechanics" (Kronecker 1887, p. 253). (In a footnote, Kronecker makes clear that he has in mind "in particular the addition of the irrational and continuous magnitudes".) Kronecker strongly opposed Cantor's and Dedekind's free use of set theoretic notions, as it violated methodological restrictions on ("legitimate") mathematical concepts and arguments. Cantor suggested a way to turn such a restrictive position to positive work and formulated "certain advantages".

If, as it is assumed here [i.e., from such a restrictive position], only the natural numbers are real and all others just relational forms, then it can be required that the proofs of theorems in analysis are checked as to their "number-theoretic content" and that every gap, which is discovered, is filled according to the principles of arithmetic. The feasibility of such a supplementation is viewed as the true touchstone for the genuineness and complete rigor of those proofs. It is not to be denied, that in this way the foundation of many theorems can be perfected and that also other methodological improvements in various branches of analysis can be effected. Adherence to the principles justified from this viewpoint, it is also believed, secures against any kind of absurdities or mistakes. (Cantor 1932, p. 173)

Hilbert may have been influenced by such considerations when he gave up — in his Heidelberg address of 1904 — his first attempt of circumventing the Cantorian problems in set theory as far as they affected analysis. The then recently discovered elementary contradictions of Zermelo and Russell had changed his outlook on these problems. Bernays is quoted in Reid's biography of Hilbert as saying:

Under the influence of the discovery of the antinomies in set theory, Hilbert temporarily thought that Kronecker had probably been right there [i.e., right in insisting on restricted methods]. But soon he changed his mind. Now it became his goal, one might say, to do battle with Kronecker with his own weapons of finiteness by means of a modified conception of mathematics. (Reid 1970, p. 173)

The key question was, how might that be done?

1.1 Reflection. The radicalization of the axiomatic method, highlighted in Hilbert's own *Grundlagen der Geometrie*, and the fresh developments in logic due to Frege and Peano provided the basic background for Hilbert's way of answering the question. The ultimate goal of his proposal in 1904 was the same as the one he had formulated in *Über den Zahlbegriff* (1900b) and his famous Paris lecture, namely to establish by means of a consistency proof the existence of the set of natural and real numbers and of the Cantorian alephs. Hilbert indicated now a possibility of giving such a proof without presupposing basic set theoretic notions as he had done earlier. He proposed a simultaneous *formal* development of logic and arithmetic, so that proofs could be viewed as *finite* mathematical structures. The new task was to show by elementary mathematical means that such formal proofs cannot lead to a contradiction.[1]

---

[1]From a much later perspective Bernays describes the fundamental conception as follows: "... in taking the deductive structure of a formalized theory as an object, the theory is projected so-to-speak into number theory. The structure obtained in this way is in general essentially different from the structure intended by the theory. However, it [the deductive structure] can serve the purpose of recognizing the consistency of the theory from a standpoint

This formulation foreshadowed aspects of the proof theoretic program Hilbert pursued in the twenties together with, e.g., Bernays, Ackermann, von Neumann, Herbrand. There was, however, a crucial and sophisticated shift in what a consistency proof was to establish. (Recall, above it was the existence of infinite mathematical systems.) To bring this out clearly, let P be a formal theory in which mathematical practice can be represented and let F be a theory formulating principles of finitist mathematics. Under weak assumptions on P (satisfied by the usual formal theories) the consistency statement for P is equivalent to the *reflection principle*

$$\Pr(a, `\psi') \to \psi$$

Pr is the canonical proof-predicate for P, 'ψ' the translation of the F-statement ψ into the language of P. Proving the reflection principle in F amounts to recognizing — from the restricted standpoint of F — the truth of F-statements whose translations have been derived in P. As a matter of fact, the proof would yield a method turning any P-proof of 'ψ' into an F-proof of ψ. In this way a consistency proof would achieve what Hilbert used to call the elimination of ideal elements from proofs of real statements.[2] Finitist mathematics was viewed as a fixed part of elementary arithmetic, as I mentioned already; its philosophical justification seemed to be unproblematic. Thus Hilbert thought that the consistency proof for P would solve the foundational problems "once and for all" and would achieve this by mathematical considerations. "This is precisely the great advantage of Hilbert's proposal, that the problems and difficulties arising in the foundations of mathematics are transferred from the epistemological-philosophical to the genuinely mathematical domain." (Bernays 1922b, p. 19)

1.2 Incompleteness. The radical foundational aims of Hilbert's program had to be abandoned on account of Gödel's Incompleteness Theorems. A generalization of the program was developed in response to Gödel's results, and it has been pursued with great vigor and mathematical success — for parts of analysis. Bernays and later Kreisel were highly influential in this development. (For relatively recent and polished formulations, see (Bernays 1970, pp. 186–187) and (Kreisel 1968, pp. 321–323).) The basic task of the generalized reductive program can be seen as follows: find for a significant part of classical mathematical practice, formalized in a theory P^*, an appropriate constructive theory F^*, such that F^* proves the partial reflection principle for P^*. I.e., F^* proves for any P^*-derivation D

$$\Pr^*(D, `\psi') \to \psi.$$

---

which is more elementary than the assumption of the intended structure." (Bernays 1970, p. 186)

[2]Compare this with Cantor's remarks quoted above. But note that this goal is quite in accord with the positivist-instrumentalist ZEITGEIST of the early decades of this century. Vide (Bernays 1922b, pp. 18–19).

ψ is in a class Λ of F*-statements. It follows immediately that P* is conservative over F* with respect to the statements in Λ; consequently, P* is consistent relative to F*. (I made the assumption, satisfied by the theories discussed below, that F* is easily seen to be contained in P*. If that is not the case, reductions in both directions have to be established.)

The Gödel-Gentzen-reduction of classical elementary arithmetic (Z) to its intuitionist version (HA) is the early paradigm of a successful contribution to the generalized program. Clearly, (Z) is taken as P*, (HA) as F*, and Λ consists of all negative arithmetic and Π_2^0-sentences.³ It was incidentally this result which showed to the Hilbert-school that intuitionist and finitist reasoning did not coincide, "contrary to the prevailing views at the time" as Bernays puts it. In addition, it gave an important positive impetus to(wards) the generalized program.

It thus became apparent that the "finite Standpunkt" is not the only alternative to classical ways of reasoning and is not necessarily implied by the idea of proof theory. An enlarging of the methods of proof theory was therefore suggested: instead of a restriction to finitist methods of reasoning, it was required only that the arguments be of a constructive character, allowing us to deal with more general forms of inferences. (Bernays 1967)

The questions which had sweeping, general answers in the original Hilbert program had to be addressed anew, indeed in a more subtle way. Which parts of classical mathematical practice can be represented in a certain theory P*? What are (the grounds for) the principles of a corresponding constructive F*? Briefly put, if a metamathematical conservation result has been obtained, it has to be complemented by additional mathematical and philosophical work establishing its foundational interest by answering these questions. — Classical analysis was viewed as decisive for the generalized program. Its basic notions and results were carefully and in detail presented by Hilbert and Bernays (Supplement IV of their *Grundlagen der Mathematik, II*).

Remark. The language of analysis is chosen here to be that of second-order arithmetic with function variables (and quantifiers) and names for all primitive recursive functions. The axioms of the base theory (BT) include (a) the axioms for zero, successor, and the defining equations for all primitive recursive functions, (b) the induction principle for quantifier-free formulas, and (c) quantifier-free comprehension. Full classical analysis (CA) is obtained from (BT) by adding the comprehension principle for all formulas of the language. Subsystems of analysis are principally distinguished by their restricted function (or set) existence principles. They include, for example, the comprehension principle or forms of the axiom of choice only for certain classes of formulas (like Π_n^0, Π_∞^0, Π_n^1). There is a second

---

³The negative arithmetic are those sentences in the language of arithmetic which are either atomic or use in their build-up only the logical connectives $\to$, &, $\forall$. — Gödel and Gentzen established the result for negative arithmetic sentences, Kreisel (1958a) extended it to Π_2^0-sentences, and Friedman (1978) gave a very simple syntactic argument for this extension.

important distinguishing feature. The induction principle can be formulated either as a schema for all formulas of the language or as a second-order axiom. In the latter case, the principle is available only for functions (sets) which can be proved to exist in the theory. For example, $(\Pi^0_\infty\text{-CA})$ denotes the theory obtained from (BT) by adding the arithmetic comprehension principle and the full induction schema; $(\Pi^0_\infty\text{-CA})\restriction$, "restricted-$(\Pi^0_\infty\text{-CA})$", is the corresponding theory with the induction axiom. Clearly, $(\Pi^0_\infty\text{-CA})\restriction$ is equivalent to the theory obtained from (BT) by just the arithmetic comprehension principle. This is Weyl's theory. (BT) is conservative over (PRA). $(\Pi^0_\infty\text{-CA})\restriction$ is conservative over elementary arithmetic, whereas $(\Pi^0_\infty\text{-CA})$ proves the consistency of (Z).

In the case of analysis the first question can be more sharply focused: which set or function-existence principles are sufficient (and necessary) to prove central facts? The surprising answer is that arithmetic comprehension suffices. This is the outcome of work by Takeuti (1978), Feferman, e.g., (1977), and Friedman, e.g., (1980). Obviously, their investigations lie in a rather long tradition of persistent efforts to pursue analysis by restricted means. The work of constructivists like Kronecker, Brouwer, and Bishop is part of that tradition. Predicatively inclined mathematicians contributed also significantly; indeed, Weyl's *Das Kontinuum* is an early landmark of that kind of research.

From results of Friedman and Simpson one can infer that weaker theories will not do for all of analysis as some crucial theorems are actually equivalent to the arithmetic comprehenion principle. Examples of such theorems are the Bolzano-Weierstrass-Theorem and the fact that every Cauchy-sequence is convergent. Nevertheless, in a theory isolated by Friedman a good deal of analysis can be developed, and that theory is conservative over primitive recursive arithmetic (PRA). It is the first theory I consider to be of interest for computational reductions. Theories like $(\Pi^1_1\text{-CA})\restriction$ and $(\Sigma^1_2\text{-AC})$ for which foundational reductions are available are thus for the actual practice of analysis far too strong. The F^*'s to which they can be reduced are, ironically, justified from an intuitionist standpoint. They are theories for constructive trees, and we have a(n intuitionistically) convincing answer to a sharpening of the second question: what are (the principles for) constructive mathematical objects? — Before turning to the detailed discussion of these reductive results I want to outline one way of establishing conservative extension results by proof theoretic means.

1.3 Sequent calculi. A variety of technical tools have been employed in proof theory; for example the ϵ-calculus, the no-counterexample-interpretation, the Dialectica-interpretation. The tools most directly useful for the reductive aims are in my view finitary and infinitary sequent calculi for which the cut-elimination-theorem can be established. (Clearly, in the case of infinitary calculi, the proof theoretic treatment has to be kept constructive.) The subformula-property is the crucial feature of cut-free derivations. This feature makes it possible to use truth-definitions for formulas

Reductions of theories for analysis

of restricted syntactic complexity and to establish formally the truth of statements. Let me explain this for the usual finitary calculi.

The classical sequent calculi are assumed to be in the style of (Tait 1968b); i.e., finite sets of formulas are proved, negation is directly available only for atomic formulas. (For complex formulas it is defined in the obvious way.) Thus, the basic logical symbols are just $\&, \vee, \forall, \exists$. The rules of the calculi are included among the following ones, where Γ is used as a syntactic variable ranging over finite sets of formulas.

$$\underline{\text{LA}} \qquad \Gamma, \psi, \neg \psi \qquad \psi \text{ is atomic}$$

$$\underline{\&} \qquad \frac{\Gamma, \psi_1 \qquad \Gamma, \psi_2}{\Gamma, \psi_1 \& \psi_2}$$

$$\underline{\vee} \qquad \frac{\Gamma, \psi_i}{\Gamma, \psi_1 \vee \psi_2} \qquad i = 1, 2$$

$$\underline{C}(\text{ut}) \qquad \frac{\Gamma, \psi \qquad \Gamma, \neg\psi}{\Gamma}$$

When appropriate, rules for quantification are available. In case of (extensions of) number theory we have the rules

$$\underline{\forall} \qquad \frac{\Gamma, \psi a}{\Gamma, (\forall x)\psi x} \qquad (a \text{ must not occur in } \Gamma)$$

$$\underline{\exists} \qquad \frac{\Gamma, \psi t}{\Gamma, (\exists x)\psi x} \qquad ;$$

the rules for function-quantification are analogous. Identity rules can be taken in the form

$$\frac{\Gamma, \psi t}{s \neq t, \Gamma, \psi s}$$

$$\frac{\Gamma, \psi t}{t \neq s, \Gamma, \psi s} \quad ;$$

the axiom for identity is

$$\underline{\text{IA}} \qquad \Gamma, a = a \ .$$

(An intuitionist version of this calculus can be formulated in a straightforward, though more complicated way.) Derivations are built up in tree-form as usual; let me use D, E, ... as syntactic variables ranging over derivations. A derivation is called cut-free or normal, if it does not contain an occurrence of the cut-rule $\underline{C}$. Two facts concerning the invertibility of quantifier-rules will be extremely useful. I formulate them only for number-quantifiers, but analogous results hold for function-quantifiers as well.

∀-*inversion*. If D is a normal derivation of $\Delta, (\forall x)\psi x$, then there is a normal derivation E of $\Delta, \psi c$; c is a parameter not used in D and E is not longer than D.

The second fact concerns the invertibility of ∃; this is possible only in restricted contexts and is a form of Herbrand's Theorem. I refer to it also as an Herbrand-type lemma.

∃-*inversion*. Let Δ contain only purely existential formulas and let ψa be quantifier-free; if D is a normal derivation of $\Delta, (\exists x)\psi x$, then there is a finite sequence of terms $t_1, \ldots, t_n$ and a normal derivation E of $\Delta, \psi t_1, \ldots, \psi t_n$. Again, E is not longer than D.

Due to Gentzen's Hauptsatz, the fundamental fact for these finitary calculi, the restriction to normal derivations in the above lemmata is not restrictive. After all, any derivation can be effectively transformed into a normal one with the same endsequent. By inspection of the rules it can be seen immediately that all formulas occurring in a normal derivation of Δ are subformulas of elements of Δ.

Let me explain through an example how the subformula-property is exploited for proofs of (partial) reflection principles. The underlying idea is simple, pervasive, and quite elegant. Consider a fragment of arithmetic, say (HAN); it has the usual axioms for zero and successor, defining equations for finitely many primitive recursive functions, and the induction schema for just quantifier-free formulas. Consequently, all of the axioms can be taken to be in quantifier-free form. Now assume that (HAN) proves (a Π_1^0-statement and thus by ∀-inversion) a quantifier-free statement ψ. A normal derivation of Δ, ψ can then be obtained, where Δ contains only negations of HAN-axioms. These considerations can actually be carried out in (PRA), i.e.,

$$(\text{PRA}) \vdash \text{Pf}_{\text{HAN}}(\ulcorner\psi\urcorner) \to \text{Pf}^n(\ulcorner\Delta, \psi\urcorner);$$

Pf_{HAN} and Pf^n express that there is a derivation in (HAN), respectively a normal derivation in the sequent calculus. A normal derivation of Δ, ψ contains only subformulas of elements in its endsequent. So one can use an adequate, quantifier-free truth-definition Tr (for quantifier-free formulas) to show that

$$(\text{PRA}) \vdash \text{Pf}^n(\ulcorner\Delta, \psi\urcorner) \to \text{Tr}(\ulcorner\vee\Delta, \psi\urcorner);$$

$\vee\Delta, \psi$ is the disjunction of the formulas in Δ, ψ. This is possible, as the language of (HAN) contains only finitely many symbols for primitive recursive functions: we can easily define a primitive recursive valuation function for all terms built up from them. More generally, but for the same reason, (HAN) could contain all functions of a fixed segment of the Grzegorczyk-hierarchy. (For details concerning the standard part of truth-definitions see (Schwichtenberg 1977, pp. 893–894).) As Tr is provably adequate we have

$$(\text{PRA}) \vdash \text{Tr}(\ulcorner \vee \Delta, \psi \urcorner) \to \vee \Delta, \psi$$

and thus

$$(\text{PRA}) \vdash \text{Pf}_{\text{HAN}}(\ulcorner \psi \urcorner) \to \psi.$$

This step can be taken, as the axioms of (HAN) are axioms of (PRA). Note that the above is essentially Gentzen's way of establishing the consistency result due to Ackermann, von Neumann, and Herbrand.

For the last step it was obviously crucial that the axioms of (HAN) are available in (PRA). The basic structure of the argument can be modified, however, for theories P* whose principles are not recognizable (i.e., provable) in F*. The strategy is to eliminate those principles from derivations (of certain classes of formulas). Eliminations can be effected by (i) the embedding of finitary into infinitary calculi, and (ii) the extraction of information from a given finitary or infinitary normal derivation via an Herbrand-type lemma. An example of (i) is the embedding of (Z) into a calculus with the ω-rule replacing the induction principle. (See (Schwichtenberg 1977) for a lucid presentation.) Lorenzen and Schütte introduced this technique to recast Gentzen's consistency proof for (Z), and extensions have been used widely. For the foundational reductions below one uses infinitary calculi with inferences involving constructive tree classes O_α. These inferences are of the form

$$\frac{\Delta, \psi e}{\Delta, (\forall x)(O_\alpha x \to \psi x)} \quad \text{for all } e \text{ in } O_\alpha;$$

in case $\alpha = 0$, this is just a form of the ω-rule. For the reductive program it is crucial that the infinitary derivations needed for the embedding and subsequent proof theoretic investigation can be treated constructively, and that means first of all that they are constructive mathematical objects.[4] Examples of (ii) can be found in (Tait 1968b), in (Feferman and Sieg 1981b) and in (Sieg 1985a); the extracted information is used to eliminate various forms of the axiom of (dependent) choice in favor of the comprehension principle. One achieves by this technique a constructive understanding of the quantifier combination $\forall \exists$ in a classical, albeit restricted context. — This brings us directly to the eliminative considerations for foundational and computational reductions.

2 Reductive results for parts of analysis

Before describing such results, I want to make some brief remarks on one facet of proof theory which is sometimes taken as the most enlightening

---

[4] Compare this with the (modified) version of the Hilbert program as described by (Bernays 1970, p. 186) and in 1.2. Recall also that the restriction to finite derivation figures in Hilbert's metamathematics aroused very early on the criticism of people with completely different foundational views; e.g., Brouwer (1927) and Zermelo (1931, 1935). See also the last section of (Sieg 1984).

one. I am alluding to (the use of) systems of ordinal notations. Indeed, from Gentzen's (second) consistency proof for (Z) emerged a program which is more specific than the generalized reductive program described earlier. Gentzen's argument, recall, can be carried out in (PRA) extended by the principles of transfinite recursion and transfinite induction for the first ϵ-number ϵ_0. ϵ_0 is the smallest ordinal with this property. Gentzen wanted to characterize the strength of stronger and stronger theories by their uniquely determined "proof theoretic ordinals". This is a fascinating, relatively precise technical problem, and a great deal of sophisticated work has been devoted to it. The foundational interest of such investigations rests, however, on the constructive character of the ordinals used. Gentzen saw quite clearly that "one has to carry out investigations [in parallel to establishing consistency via transfinite induction] with the aim of making the validity of transfinite induction for larger and larger limit numbers constructively intelligible". (In a letter to Bernays in 1936.) The point I want to make is quite simply that the determination of the proof theoretic ordinal of a theory T is not the ultimate aim of investigations in proof theory and presumably not even the central one (unless further research *shows* that ordinals are conceptually central in constructive mathematics and thus for the general reductive aims). In any event, the foundational reductions for some impredicative parts of (CA) discussed in Part 2.1 do not use special systems of ordinal notations. The strongest of the theories treated there is (Δ_2^1-CA). Part 2.2 presents computational reductions of two weak subsystems which are conservative over (a fragment of) (PRA). Details of the argument for the latter result are indicated in Part 2.3.

2.1 Foundational reductions. Generalized inductive definitions[5] play a central role here, both technically and conceptually. Classes given by inductive definitions, i.d. classes for short, have been used in constructive mathematics ever since Brouwer. Two familiar examples of such classes are well-founded trees of finite sequences of natural numbers ("unsecured sequences") and Borel sets. The former were employed in Brouwer's justification of bar-induction; the latter in Bishop's original development of measure theory. In spite of the fact that i.d. classes can be avoided in the current practice of constructive analysis, particular ones are of intrinsic mathematical and foundational interest.

The constructive (well-founded) trees form such a distinguished class, called O. It is given by two inductive clauses, namely (i) if e is 0, then e is in O, and (ii) if e is (the Gödel-number of) a recursive function enumerating elements of O, then e is in O.[6] — The elements of O are thus generated by joining recursively given sequences of previously generated elements of O

---

[5]In contrast to elementary inductive definitions by means of which exactly the recursively enumerable sets of natural numbers are definable.

[6]The antecedents of these generating clauses can be expressed by a formula which is arithmetic in O. Their disjunction is abbreviated by $A(O, e)$.

Reductions of theories for analysis

and can be pictured as infinite, well-founded trees. Locally, the structure of such a tree is as follows:

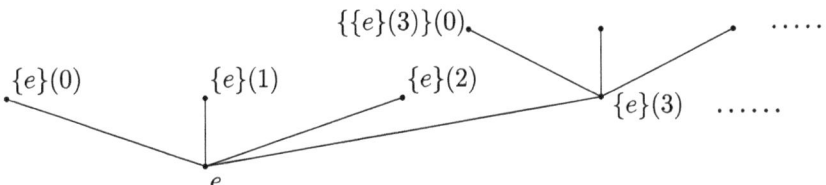

Higher tree classes are obtained by a suitable iteration of this definition along a given recursive well-ordering of the natural numbers. Suitable means here that branchings in trees are not only taken over the natural numbers but also over already given lower tree classes.

These special i.d. classes have been investigated by Brouwer, Church, Kleene, and others; they have also played a significant role in metamathematical studies of theories for classical and intuitionist analysis. Their constructive appeal consists partly in this: the trees reflect their build-up according to the generating clauses of their definition directly and locally in an effective way. If one views the clauses as inference rules, the constructive trees are infinitary derivations and show that they fall under their definition. Constructive theories for O have been formulated as extensions of intuitionist arithmetic with the following principles for O (compare note 6):

$O1.$ $(\forall x)(A(O, x) \to Ox)$

$O2.$ $(\forall x)(A(\psi, x) \to \psi x) \to (\forall x)(Ox \to \psi x)$

$A(\psi, x)$ is obtained from $A(O, x)$ by replacing all occurrences of Oz by ψz. — $O1.$ may be called a definition principle making explicit that applications of the defining clauses to elements of O yield elements of O. $O2.$ is a schematic proof principle expressing that one can give arguments by induction on O (employing any formula ψz of the language). Proofs by this principle follow or parallel the construction of the elements in O. The resulting theory is called $ID_1(O)$. For the higher tree classes the definition and proof principles can be formulated in a similar, though more complicated manner. The theory is denoted by $ID_{<\lambda}(O)$, when the iteration proceeds along arbitrary initial segments of the given well-ordering of type λ. (For the detailed formulation the reader is referred to (Feferman and Sieg 1981a).)

These theories (for higher tree classes) are also meaningful from a classical point of view, even when more general defining clauses are considered. Let P be a unary predicate variable and call a formula $A(P, x)$ positive in P if P does not occur in the scope of a negation sign or in the antecedent of a conditional. Each such formula determines an i.d. class P^A which is definable in second-order logic by

$$P^A a \leftrightarrow (\forall P)((\forall x)(A(P,x) \to Px) \to Pa);$$

for convenience we make use of a second-order language with predicate variables. P^A exists as a set by an impredicative instance of the comprehension principle, as positive formulas satisfy the following monotonicity condition

$$(\forall x)(Px \to Qx) \to (\forall x)(A(P,x) \to A(Q,x)).$$

The theories obtained from classical number theory by adding the definition and proof principles for all i.d. classes P^A given by positive A is denoted by ID_1^c. As above, one can consider iterations of such definitions and obtain theories $ID_{<\lambda}^c$. Feferman (1970) and Friedman (1970) established that for example $(\Pi_1^1\text{-CA})\restriction$, $(\Delta_2^1\text{-CR})$, and $(\Delta_2^1\text{-CA})$ are equivalent to $ID_{<\omega}^c$, $ID_{<\omega^\omega}^c$, respectively $ID_{<\epsilon_0}^c$. Thus, it is sufficient — for reducing the above impredicative subsystem — to reduce the theories of iterated inductive definitions. That was achieved among other things by Buchholz, Pohlers, and myself in 1977. Our work is reported in (Buchholz et al. 1981). (Buchholz and Pohlers determined in addition the proof theoretic ordinals of the theories.) The reduction I achieved is formulated in the next theorem.

Theorem. For any primitive recursive well-ordering of limit characteristic λ, $ID_{<\lambda}^c$ is conservative over $ID_{<\lambda}(O)$ for all negative arithmetic and Π_2^0-formulas.

The proof, inspired by (Tait 1970), employs the strong infinitary calculi described at the end of 1.3 and exploits the close interrelation between constructive trees, derivations in those calculi, and positive inductive definitions. A detailed informal description of the argument is given in (Sieg 1984), a full account in (Sieg 1981).

The reductions to theories of tree classes are most satisfactory from an intuitionist point of view. The question clearly is whether still stronger parts of (CA) can be reduced in the same way. Unfortunately, the answer is "NO". It is a well-known result of Addison and Kleene (1957) that the iteration of the hyperjump (and thus of inductive definitions) along recursive ordinals leads only to Δ_2^1-sets. Consequently, theories for i.d. classes cannot be used for reductions of $(\Pi_n^1\text{-CA})$ with $n \geq 2$. Here is a major conceptual problem, it seems to me, namely to find a broad notion of "constructive mathematical objects" and suitable principles for them which can serve as a starting point for foundational reductions of parts of analysis beyond $(\Delta_2^1\text{-CA})$.[7] There are also some precise and important open questions concerning the theories for i.d. classes (compare (Sieg 1981) and (Feferman 1982)). Problem 1: Is ID_α^c conservative over $ID_\alpha(O)$ for negative arithmetic and Π_2^0-sentences for all (primitive) recursive α? Problem 2: Is ID_α^{mon} conservative over $ID_\alpha(O)$ for all arithmetic sentences? (ID_α^{mon} is the intuitionist theory for i.d. classes given by provably monotone clauses.)[8]

---

[7] There are results for $((\Delta_2^1\text{-CA}) + BI)$: Jäger and Pohlers determined the proof theoretic ordinal, and Jäger reduced it to Feferman's constructive theory T_0. (This establishes with earlier work of Feferman's the equivalence of these theories.)

[8] Added for this volume:

2.2 Computational Reductions.

It was already pointed out in 1.2 that all of classical analysis can be formally developed in various conservative extensions of elementary arithmetic, and that one cannot do better as some central theorems are over (BT) equivalent to the arithmetic comprehension principle. If one concentrates, however, on proper parts one can get by with weaker systems. Friedman (1975) isolated a particularly interesting system. I present it in an equivalent formulation that is more suitable for the subsequent proof theoretic treatment. The theory, call it (F), extends (BT) by the axiom of choice for Σ_1^0-formulas, Σ_1^0-AC_0,

$$(\forall x)(\exists y)\psi xy \to (\exists f)(\forall x)\psi x f(x),$$

the induction principle for Σ_1^0-formulas, Σ_1^0-IA, and König's Lemma for binary trees, WKL,

$$(\forall f)[T(f) \,\&\, (\forall x)(\exists y)(lh(y) = x \,\&\, f(y) = 1) \to (\exists g)(\forall x)f(\overline{g}(x)) = 1].$$

$T(f)$ expresses that f is (the characteristic function of) a tree of 0-1-sequences, lh is the length-function for sequence numbers. $T(f)$ is the purely universal formula

$$(\forall x)(\forall y)((f(x * y) = 1 \to f(x) = 1) \,\&\, (f(x * \langle y \rangle) = 1 \to y \leq 1)).$$

(F) is surprisingly strong for mathematical work and quite weak from a metamathematical point of view, namely conservative over (PRA) for Π_2^0-statements.

Remarks. (i) Let me indicate what kind of facts can be established in (F) and point to the phenomenon of "reverse mathematics". In (F$^-$), i.e., (F) without WKL, one can establish the equivalence of the following statements: (1) WKL, (2) the Heine-Borel Theorem (every covering of the unit interval by a countable sequence of open intervals has a finite subcovering), (3) every continuous function on the unit interval is uniformly continuous, (4) every continubus function on the unit interval is bounded, (5) every continuous function on the unit interval has a supremum and attains it, (6) the Cauchy-Peano Theorem on the existence of solutions for ordinary differential equations. — The equivalence between (1) and (2) was established by Friedman; the other equivalences are due to Simpson (1982, 1985b).

(ii) Mints (1976) introduced a subsystem, proved it to be conservative over (PRA) for Π_2^0-sentences, and established Gödel's Completeness Theorem in it. The system, call it (M$^-$), is essentially (BT) together with

---

Important issues concerning theories for monotone inductive definitions have been settled by subsequent work, in particular, Problem 2 has a positive answer; see section 2 of (Feferman 2010). Problem 1 has been addressed by Avigad and Towsner in their (2009): among other results, they establish — by adapting my technique of formalizing cut-elimination for infinitary calculi and by using finite type functionals over tree ordinals — that the classical theory ID_1 can be reduced to the intuitionist $ID_1(O)$. The general case of theories for iterated inductive definitions, they claim, can be treated similarly.

the comprehension principle for Π_1^0-formulas *without* function parameters and the induction rule for Π_2^0-formulas *without* function parameters. The crucial step in the argument for the completeness theorem is a proof of WKL for primitive recursive trees. Indeed, the extension of (M$^-$) by WKL yields a theory (M) which is conservative over (PRA) for Π_2^0-sentences. (Sieg 1985a)

Friedman established the conservation result mentioned just before the remarks by model-theoretic methods. In (1985a) I gave a purely proof theoretic argument.[9] (F) is thus directly justified on finitist grounds. The result has also a computationally informative consequence.

Corollary. There is a primitive recursive function f and a proof in (PRA) of $\psi a f(a)$ in case (F) proves $(\forall x)(\exists y)\psi xy$; ψ is quantifier-free.

The corollary implies that the provably recursive functions of (F) are exactly the primitive recursive ones. It is through a characterization of the provably recursive functions of a theory T that one answer to Kreisel's question can be given. The chances for obtaining an interesting answer are obviously greater when T is extremely restricted, yet mathematically still powerful. It seems to me that the work of Friedman and Simpson has improved our chances dramatically; so much so that further systematic studies are justified, not just ad-hoc constructions.

Remark. Kreisel has urged to explore "the mathematical significance of consistency proofs", in particular in his 1958 paper of that very title. He emphasized two uses of a characterization of T's provably recursive functions: (1) to obtain more explicit solutions for Π_2^0-theorems in the sense of the above corollary;[10] (2) to give independence arguments concerning Π_2^0-sentences. To see the latter assume that $(\forall x)(\exists y)\psi xy$ is true, but that for no f in the class of provably resursive functions of T $(\forall x)\psi xf(x)$ is true; then $(\forall x)(\exists y)\psi xy$ is independent of T. A beautiful example of that strategy is given in (Graham, Rothschild and Spencer 1980), to establish the independence of the Paris-Harrington-Ramsey statement. (They make use of the fact that the provably recursive functions of (Z) are exactly the $<\epsilon_0$-recursive ones.)

From a mathematical-computational vantage point one is really looking for refinements of the corollary, i.e., for subsystems whose provably recursive functions are in a small subclass of the primitive recursive functions. I have one relevant result whose proof I will describe in the next section. Let (ET) be like (BT), but admit only names for the Kalmar-elementary

---

[9] A proof theoretic argument for this and some other results was given by A. Cantini in a very recent, unpublished paper "On the proof theory of Weak König's Lemma".

[10] In addition, they should be extracted efficiently from given proofs. That is most important, but requires work in a different direction. One has to pay close attention to the computational complexity of (presentations of) elementary syntax and of proof theoretic operations on (restricted classes of) derivations.

functions into the language. Naturally, the axioms include just the defining equations for them. (KEA) is (PRA) restricted in the same way.

Theorem. Let (K) be (ET + Σ_1^0-AC$_0$ + WKL); (K) is conservative over (KEA) for Π_2^0-sentences.

The theorem has as an immediate consequence a refinement of the earlier corollary.

Corollary. There is a Kalmar-elementary function f and a proof in (KEA) of $\psi a f(a)$ in case (K) proves $(\forall x)(\exists y)\psi xy$; ψ is quantifier-free.

Thus, the provably recursive functions of (K) are exactly the Kalmar-elementary ones. — A first candidate for testing the usefulness of (K) is Artin and Schreier's Theorem, that every positive definite rational function with rational coefficients is a sum of squares. This is a solution to Hilbert's 17th problem, and Kreisel showed that the number and degrees of the squares are bounded by a (specific) primitive recursive function of the degree and the number of variables of the given positive definite function. That there is such a primitive recursive function follows immediately from the fact that Artin and Schreier's Theorem can be established in (F), according to (Friedman, Simpson and Smith 1983). A proof in (K) would yield a Kalmar-elementary bounding function. (That an elementary function can be found was shown recently by Friedman.)

There are some obvious questions to be asked. (1) Can one push the proof theoretic considerations further and treat theories whose provably recursive functions are computationally feasible? (2) Which parts of mathematics can be developed somewhat systematically in such weak systems? And, finally, leaving aside a systematic development, (3) Which interesting Π_2^0-statements can be proved in them? — Further refined computational reductions, if accompanied by informative applications, would be of real mathematical interest.[11]

2.3 Proof theoretic facts used for establishing the theorem concerning (K).
If we want to follow the strategy suggested in 1.3, the work is cut out for us: we have to eliminate Weak König's Lemma and the Σ_1^0-axiom of choice (indeed, only the quantifier-free one, as these two forms are equivalent over (ET)). Let me formulate the two crucial elimination lemmata. In their formulation I assume that Δ consists only of purely existential formulas. $\Delta[\neg\text{QF-AC}_0]$ denotes the sequent obtained from Δ by adding instances of the negation of the quantifier-free axiom of choice; $\Delta[\neg\text{WKL}]$ and $\Delta[\neg\text{QF-CA}]$ are defined similarly.

---

[11] Added for this volume:
Fernando Ferreira started such investigations in his (Ferreira 1994). His student Gilda Ferreira made significant further headway in her dissertation (Ferreira 2006); see also their joint paper (Ferreira and Ferreira 2008).

QF-AC$_0$-*elimination.* If D is a normal derivation of $\Delta[\neg\text{QF-AC}_0]$, then there is a normal derivation E of $\Delta[\neg\text{QF-CA}]$.

I.e., the quantifier-free axiom of choice is eliminated in favor of quantifier-free comprehension; the same fact holds for Weak König's Lemma.

WKL-*elimination.* If D is a normal derivation of $\Delta[\neg\text{WKL}]$, then there is a normal derivation E of $\Delta[\neg\text{QF-CA}]$.

Assuming these two lemmata I sketch the proof of the theorem concerning (K). Notice, first of all, that a normal derivation in (K) of the Π_2^0-statement $(\forall x)(\exists y)\psi xy$ is a normal derivation with an endsequent of the form

$$\Delta[\neg\text{QF-CA}, \neg\text{QF-AC}_0, \neg\text{WKL}], (\forall x)(\exists y)\psi xy,$$

where Δ contains only negations of other, purely universal axioms of (K). By $\forall$-inversion the endsequent can be assumed to be of the form

$$\Delta[\neg\text{QF-CA}, \neg\text{QF-AC}_0, \neg\text{WKL}], (\exists y)\psi ay.$$

The main argument proceeds now by induction on (the length of) normal derivations D with endsequents of this form and establishes that such D can be transformed into normal derivations E of

$$\Delta[\neg\text{QF-CA}], (\exists y)\psi ay.$$

The induction step is trivial in case the last rule in D affects an element of Δ, an instance of $\neg\text{QF-CA}$ or $(\exists y)\psi ay$. If the last rule introduces an instance of $\neg\text{QF-AC}_0$ or $\neg\text{WKL}$ one appeals (first to the induction hypothesis and then) to the appropriate elimination-lemma. For this to be possible, the comprehension principle has to be formulated in purely universal form, e.g., by using λ-terms as $(\forall x)\lambda y.t[y, \ldots](x) = t[x, \ldots]$. This establishes the conservativeness of (K) over (ET) for Π_2^0-statements; the latter theory is easily seen to be conservative over (KEA).

Now let me turn to the proofs of the elimination lemmata. The QF-AC$_0$-elimination is obtained in a way which is completely analogous to establishing the Main-Lemma in (Feferman and Sieg 1981a, 2.1.3, p. 105). The $\exists$-inversion lemma is used crucially to define a choice-function when, in the course of an inductive argument on the length of D, one has to treat the case that the last rule in D introduces an instance of $\neg\text{QF-AC}_0$. The WKL-elimination lemma is also proved by induction on the length of normal derivations. I concentrate on the central case when the last rule in D introduces an instance of $\neg\text{WKL}$, i.e.,

$$T(f) \,\&\, (\forall x)(\exists y)(lh(y) = x \,\&\, f(y) = 1) \,\&\, \neg(\exists g)(\forall x)f(\overline{g}(x)) = 1.$$

Then there are derivations D_1, D_2, D_3 all shorter than D, of

$$\Delta[\neg\text{WKL}], T(f),$$
$$\Delta[\neg\text{WKL}], (\forall x)(\exists y)(lh(y) = x \,\&\, f(y) = 1), \text{ and}$$
$$\Delta[\neg\text{WKL}], (\forall g)(\exists x)f(\overline{g}(x)) \neq 1.$$

Reductions of theories for analysis

Using $\forall$-inversion and the induction-hypothesis one obtains derivations E_1 E_2, E_3 of
$$\Delta[\neg\text{QF-CA}], \text{T}(f),$$
$$\Delta[\neg\text{QF-CA}], (\exists y)(lh(y) = c \,\&\, f(y) = 1), \text{ and}$$
$$\Delta[\neg\text{QF-CA}], (\exists x) f(\overline{u}(x)) \neq 1.$$

$\exists$-inversion and definition by cases (available for elementary functions) provide terms t and s and derivations F_2 and F_3 of
$$\Delta[\neg\text{QF-CA}], (lh(t[c]) = c \,\&\, f(t[c]) = 1), \text{ and}$$
$$\Delta[\neg\text{QF-CA}], f(\overline{u}(s)) \neq 1.$$

Now observe (i) that t yields sequences of arbitrary length in f which do not necessarily form a branch, and (ii) that the right-most formula in the endsequent of F_3 expresses the well-foundedness of f. In short, we have a binary tree (according to E_1), that is well-founded and contains sequences of arbitrary length. To exploit this situation let me first make the simplifying assumption that s is a closed term with numerical value k. Let $t[k]$ be the 0-1-sequence $t_0, \ldots, t_{k-1}$ and define with QF-CA the function u^* by
$$u^*(n) = \begin{cases} t_n & \text{if } n < k \\ 0 & \text{otherwise} \end{cases}$$

Obviously, $\overline{u^*}(s) = t[k]$. Replacing u in F_3 by u^* yields a derivation of
$$\Delta[\neg\text{QF-CA}], f(\overline{u^*}(s)) \neq 1$$

and indeed a G_3 of
$$\Delta[\neg\text{QF-CA}], f(t[k]) \neq 1,$$

when taking into account the equation $\overline{u^*}(s) = t[k]$. A derivation G_2 is obtained from F_2 by &-inversion and replacing c by k; G_2 has the endsequent
$$\Delta[\neg\text{QF-CA}], f(t[k]) = 1,$$

Cutting G_3 with G_2 and subsequent normalizing yields a normal derivation of
$$\Delta[\neg\text{QF-CA}].$$

The simplifying assumption concerning s can be eliminated. s contains in general occurrences of u and f; both u and f can be assumed to be majorized by the function which is identically 1. Let $2_m(x)$ be defined by $2_0(x) = x$ and $2_{n+1} = 2^{2_n(x)}$. The usual argument showing that for any elementary function $g(x)$ there is a k, such that for all x : $g(x) \leq 2_k(x)$, can be specialized and carried out in (ET) to obtain a closed term s^* majorizing s. As f is a tree,
$$\Delta[\neg\text{QF-CA}], f(\overline{u}(s^*)) \neq 1$$
can be proved, and one can complete the argument as above.

3 Concluding remarks*

Many fascinating mathematical and logical problems are left open by the work reported here; I indicated some of them. These problems are partly precise (meta) mathematical, partly broad conceptual ones. The accomplished work on the open questions are both integrated to the generalized reductive program. That is not seen, in sharp contrast to the original Hilbert program, as an instrument to solve the foundational problem of mathematics once and for all; it is rather viewed as a coherent scheme guiding foundational research. This research has yielded rich (meta-) mathematical results, and it continues to provide us with detailed material for philosophical reflections on the nature of our mathematical experience.

Through this work we have gained a much clearer understanding of the mathematical strength of (weak) classical theories, the relation between classical and constructive theories, and the nature of constructive principles and their relative strength. Here is a body of definite results which should inform the philosophical discussion of foundational issues today. But the reductive results do not only provide information; they present also a deep philosophical challenge. The reductions do secure, after all, classical theories on the basis of theories which can justly be called more elementary. In this sense they are epistemological reductions, whose precise character awaits philosophical analysis. It seems to me that such an analysis would throw considerable light on issues discussed under the heading of platonistic and constructivistic tendencies in mathematics.

---

*I left out a very brief section that was entitled "Two results and three conjectures": the results turned out to be incorrect; so did the conjectures that generalised them.

II.8

Hilbert's program sixty years later

On 4 June 1925, Hilbert delivered an address to the Westphalian Mathematical Society in Münster; that was, as a quick calculation will convince you, almost exactly sixty years ago [This remark was made, correctly, in late December of 1985, WS]. The address was published in 1926 under the title *Über das Unendliche* and is perhaps Hilbert's most comprehensive presentation of his ideas concerning the finitist justification of classical mathematics and the role his proof theory was to play in it. But what has become of the ambitious program for securing all of mathematics, once and for all? What of proof theory, the very subject Hilbert invented to carry out his program? The Hilbertian ambition reached out too far: in its original form, the program was refuted by Gödel's Incompleteness Theorems. And even allowing more than finitist means in metamathematics, the Hilbertian expectations for proof theory have not been realized: a constructive consistency proof for second-order arithmetic is still out of reach. (And since that theory provides a formal framework for analysis, it was considered by Hilbert and Bernays as decisive for proof theory.) Nevertheless, remarkable progress has been made. Two separate, but complementary directions of research have led to surprising insights: classical analysis can be formally developed in conservative extensions of elementary number theory; relative consistency proofs can be given by constructive means for impredicative parts of second order arithmetic. The mathematical and metamathematical developments have been accompanied by sustained philosophical reflections on the foundations of mathematics.[1] This indicates briefly the main themes of the contributions to the symposium; in my introductory remarks I want to give a very schematic perspective, that is partly historical and partly systematic.

Received September 23, 1986; revised March 13, 1987.
[1]Most penetratingly in the writings of Paul Bernays. See his collected essays (Bernays 1976a).

1 Sets and consistent theories

We have to go back to the 19th century to gain a proper sense of the "Problematik" addressed by Hilbert's Program.² It is directly connected to the arithmetization of analysis that was attempted with strikingly different understanding by Dedekind (Cantor, and Weierstrass) on the one hand and Kronecker on the other. Dedekind showed in his booklet *Stetigkeit und irrationale Zahlen* how one can define, in a geometrically motivated way, a general notion of real number via cuts. He viewed cuts as purely arithmetical phenomena. He also believed that arithmetic is a part of logic and that the needed general (for us, set theoretic) principles were available as logical ones. Such principles were formulated in his second foundational essay *Was sind und was sollen die Zahlen?* For example, he took as a logical law that the extension of any predicate is a "system", his term for set. Furthermore, a system is "completely determined" if for each thing (i.e. any object of our thinking, including in particular systems) it is determined, whether or not it is an element of the system. Dedekind remarked in a footnote that it is quite irrelevant for the development of the general laws of systems, whether their complete determination can be decided by us.

This last remark was explicitly directed against Kronecker, who had opposed the introduction of the general notion of irrational number in lectures and in his essay *Über den Zahlbegriff*. There he expressed his belief that, and I quote,

> In the future we shall succeed in "arithmetizing" the whole content of all these mathematical disciplines [that fall under arithmetic]; i.e. to base it [the whole content] on the concept of number taken in its most narrow sense, and thus to strip away the modifications and extensions of this concept, which have been brought about in most cases by applications in geometry and mechanics.

(Note, that arithmetic comprised for Kronecker all mathematical disciplines with the exception of geometry and mechanics. He explained in a footnote to the above passage that the extension of the number concept had been effected by "the addition of the irrational and continuous magnitudes".) He showed in his paper how to "eliminate from arithmetic"—we would say, construct set theoretically—some of those notions that he considered foreign to it; namely, the negative and rational numbers, and the real and imaginary algebraic numbers. The restriction to the natural numbers as the legitimate objects of arithmetic was accompanied by a restriction of methods that was most decidedly rejected by Dedekind as I indicated above: mathematical notions should be decidable, and existence proofs should contain a method for finding the magnitude whose existence has been claimed.³

---

²In (Sieg 1984) I traced the background of Hilbert's Program in greater detail. The historical remarks here are largely based on that paper.

³These views are explicitly attributed to Kronecker in Hensel's foreword to (Kronecker 1901, p. vi).

In spite of Kronecker's protestations, the view that a strict and satisfactory arithmetization had been achieved by the work of Dedekind, Cantor, and Weierstrass was widely shared among mathematicians at the end of the 19th century. This conviction was vividly expressed by Poincaré at the Second International Congress of Mathematicians held in Paris in 1900.

Today, in analysis there are only natural numbers or finite or infinite systems of natural numbers. . . . Mathematics, as one says, has been arithmetized. . . . In today's analysis, if one cares to be rigorous, there are only syllogisms and appeals to that intuition of pure numbers that alone cannot deceive us. One can say that absolute rigor has been achieved today.[4]

Is it not ironic that Hilbert, at this very Congress, formulated the second of his famous problems dealing with the consistency of arithmetic? The background for Hilbert's concern was this. Dedekind had been informed by Cantor in a letter of 28 July 1899 that one assumption in his 1888 essay leads to a contradiction.[5] It was the assumption that, as Dedekind had put it, the totality of all things that can be the object of my thinking is a system. Cantor explained that he had discovered in his own investigations the necessity of distinguishing between two types of multiplicities (systems) of things.

For a multiplicity can be such that the assumption that ALL of its elements "are together" leads to a contradiction, so that it is impossible to conceive of the multiplicity as a unity, as "one finished thing". Such "multiplicities" I call ABSOLUTELY INFINITE or inconsistent multiplicities.

And he added: *As we can readily see, the "totality of everything thinkable", for example, is such a multiplicity* . . . —Which objective criterion can ensure the consistency of a multiplicity and "thus" its existence as a set? Hilbert, informed of the Cantorian difficulties, proposed a solution for at least the multiplicity of real numbers. In *Über den Zahlbegriff* (written in 1899 and undoubtedly entitled so as to allude polemically to Kronecker's essay), he presented a categorical axiomatization for the reals.[6] A consistency proof, he claimed, could be given by a suitable modification of Dedekind's methods and would be "the proof of the existence of the totality of real numbers or—to use G. Cantor's terminology—the proof of the fact, that the system of real numbers is a consistent (finished) set". The same point was made in the discussion of the Second Problem. Actually, there Hilbert went even further and asserted that the existence of Cantor's higher number classes and of the alephs could be proved in a similiar manner.

---

[4](Poincaré 1902a, pp. 115–130).

[5]Translation in (van Heijenoort 1967, pp. 113–117). According to Bernstein, in (Dedekind 1932, pp. 448/449), Cantor had informed Dedekind of these difficulties already in 1897. Cantor had also talked and corresponded with Hilbert on these problems at that time. See his letters to Hilbert of September 26 and October 2, 1897, published in (Purkert and Ilgauds 1987, pp. 224–227).

[6]Published in 1900.

Hilbert changed his mind under the impression of the elementary contradiction discovered by Zermelo and Russell. About this time Bernays reports in Reid's Hilbert biography:

> ... Hilbert temporarily thought that Kronecker had probably been right there [i.e. in insisting on restricted methods]. But soon he changed his mind. Now it became his goal, one might say, to do battle with Kronecker with his own weapons of finiteness by means of a modified conception of mathematics.[7]

In his Heidelberg talk of 1904 Hilbert sketched what he considered to be a refutation, "sachliche Widerlegung", of Kronecker's view on foundations.[8] The ultimate goal of his proposal was still to establish the existence of sets, e.g. the natural and real numbers, the Cantorian alephs. And as before he intended to achieve that by means of consistency proofs. But Hilbert indicated here how such proofs might be given without presupposing basic logical and set theoretic notions. He suggested a "simultaneous development of the laws of logic and arithmetic" in a stricter, formal manner. Proofs could then be viewed as finite mathematical structures, and the new task was to show that such (formal) proofs could not lead to a contradiction. However vague, provisional, and confused these suggestions may have been, they foreshadowed aspects of Hilbert's proof theoretic program of the twenties.

2 From absolute to relative consistency

In *Über das Unendliche* Hilbert formulated and supported, not altogether successfully, two bold claims:

(i) proof theory can secure the foundations of classical mathematics "once and for all", and

(ii) proof theory can answer "pre-existent questions that the theory was not specifically created to answer".

As to the second claim, Hilbert sketched a solution to Cantor's continuum problem; the proof of the "continuum theorem" turned out to be inconclusive at best. Concerning the first claim, Hilbert mentioned the consistency proof of the arithmetic axioms in finitist mathematics. By "arithmetic axioms" Hilbert meant a formal system for the real numbers, by then, full second-order arithmetic; by "finitist mathematics" he meant that elementary part of mathematics whose subject matter consists of finite strings of symbols (from some finite alphabet) and that satisfies Kronecker's methodological restrictions. A consistency proof was no longer taken to establish the existence of infinite sets, but rather thought to guarantee the finitist truth of

---

[7](Reid 1970, p. 173). Note also the revealing discussion in *Über das Unendliche*, p. 375 in (van Heijenoort 1967): "... contradictions appeared, sporadically at first, then ever more severely and ominously. They were the paradoxes of set theory, as they are called. In particular, a contradiction discovered by Zermelo and Russell had, when it became known, a downright catastrophic effect in the world of mathematics."

[8](Hilbert 1905a), translated in (van Heijenoort 1967, pp. 129–138).

those finitist statements that are provable from the arithmetic axioms. This is a direct consequence of the general fact that the consistency statement for standard formal theories is provably equivalent to the reflection principle for such theories. In our case, the principle is formulated finitistically by the schema

$$\Pr(a, `S') \Rightarrow S,$$

where $\Pr$ is the canonical proof predicate for second-order arithmetic and 'S' the second-order translation of the finitist statement S. The principle expresses, in Hilbert's terminology, that ideal elements can be eliminated from proofs of real statements; its finitist proof would actually provide a method for obtaining a finitist proof of S from a classical proof of 'S'. Note, that this is a sophisticated way of justifying the instrumental use of classical theories on the basis of finitist mathematics.[9] As this basis was considered to be absolutely fundamental, Hilbert's claim may be reformulated as saying that an absolute consistency proof for the arithmetic axioms had been found. But it was recognized before 1928 that the claim was incorrect. It was still believed, however, that a finitist consistency proof for elementary number theory had been achieved (by Ackermann and von Neumann).[10]

That claim turned out to be incorrect, too, on account of results in Gödel's fundamental paper *Über formal unentscheidbare Sätze der Principia Mathematica und verwandter Systeme I*. Indeed, the very core of Hilbert's Program seemed to be threatened. In a meeting of the Vienna Circle on 15 January 1931, Gödel reported on an argument of von Neumann's; namely, "If there is a finitist consistency proof at all, then it can be formalized. Therefore, Gödel's proof implies the impossibility of any [such] consistency proof." May we trust the judgment of this brilliant collaborator of Hilbert's and assume that Hilbert's Program was dead already in 1931? Certainly yes . . . and undoubtedly no! The familiar yes-part of this unambiguous answer is all too well-known. Assuming that finitist mathematics is a proper part of classical mathematics and (thus?) formalizable in a comprehensive theory for classical mathematical practice, like ZF, Gödel's Second Incompleteness Theorem denies the possibility of a finitist consistency proof for such a theory. As to the no-part let me report Gödel's further remark, that is incidentally in complete accord with the one he made on "Hilbert's formalistic viewpoint" in his classical paper.[11] At

---

[9]This instrumentalism is described vividly in *Über das Unendliche*, on pp. 376–377 of (van Heijenoort 1967). For details concerning the metamathematical claims made above, see (Smoryński 1977).

[10]Compare the remarks in (Bernays 1967, p. 501). In Hilbert's talk in Hamburg (1927) and that in Bologna (International Congress, 1928) the consistency of analysis is formulated as an open problem. In the latter talk Hilbert remarked: "Der Ansatz eines Beweises liegt vor. Diesen hat Ackermann schon so weit durchgeführt, dass die verbleibende Aufgabe nur noch in dem Beweise eines rein arithmetischen elementaren Endlichkeitssatzes besteht." The most detailed discussion is in (Bernays 1935a, pp. 210–211).

[11]The German text of the quotations from the 'Protokoll des Schlick Kreises' for the meeting of January 15, 1931. *Vermutung von Herrn von Neumann:* "Wenn es einen finiten

the Vienna Circle meeting Gödel argued that it is doubtful, "whether all intuitionistically correct proofs can be captured in a *single* formal system. That is the weak spot in Neumann's argumentation."

It was no weak spot in Gödel's remarks that he considered "intuitionistically correct proofs" as answering Hilbert's demand for finitist consistency proofs: *finitist and intuitionist methods were thought to be co-extensive*. The incorrectness of this identification was recognized only through Gödel and Gentzen's theorem that classical elementary number theory is consistent relative to its intuitionist version. This result was a crucial factor for the formulation of a relativized Hilbert program. Paul Bernays, to whom we owe its clear and early formulation, made the following remark on the impact of the Gödel-Gentzen result:

It thus became apparent that the "finite Standpunkt" is not the only alternative to classical ways of reasoning and is not necessarily implied by the idea of proof theory. An enlarging of the methods of proof theory was therefore suggested: instead of a restriction to finitist methods of reasoning, it was required only that the arguments be of a constructive character, allowing us to deal with more general forms of inferences.[12]

The aim of this modified program can be succinctly formulated: establish by appropriate constructive means the relative consistency of formal theories in which parts of classical mathematics can be carried out. That amounts to proving constructively for each d the partial reflection principle

$$\Pr(d, \ulcorner S \urcorner) \Rightarrow S,$$

where $\Pr$ is the canonical proof predicate for the classical theory, d is a given derivation in that theory, and S is in a class D of sentences. If, as in the cases considered here, the constructive theory is contained in the classical theory, then it follows directly that the classical theory is conservative over the constructive one for all sentences in D. This version of Hilbert's program is a general *reductive program*, not based exclusively on finitist mathematics and not fixed on establishing the consistency of all-encompassing mathematical theories. It does not claim to solve

---

Widerspruchsbeweis überhaupt gibt, dann lässt er sich auch formalisieren. Also involviert der Gödelsche Beweis die Unmöglichkeit eines Widerspruchsbeweises überhaupt." Gödel bemerkt dazu, dass es fraglich sei *"ob die Gesamtheit aller intuitionistisch einwandfreien Beweise in einem formalen System Platz findet. Das sei die schwache Stelle in der Neumannschen Argumentation."* The minutes are found in the Carnap Archives at the University of Pittsburgh and were brought to my attention by Neil Tennant. In his 1931 paper Gödel wrote: "I wish to note expressly that Theorem XI (and the corresponding results for M and A) do not contradict Hilbert's formalistic viewpoint. For this viewpoint presupposes only the existence of a consistency proof in which nothing but finitary means of proof is used, and it is conceivable that there exist finitary proofs that cannot be expressed in the formalism of P (or of M or A)." This translation is taken from (van Heijenoort 1967, p. 615).

[12]The quotation is from (Bernays 1967, p. 502). Gödel's paper is (Gödel 1933d). Gentzen's paper *Über das Verhältnis zwischen intuitionistischer und klassischer Arithmetik* was submitted and accepted by *Mathematische Annalen* in 1933, but withdrawn on account of Gödel's publication. An English translation is found in (Gentzen 1969).

3 In pursuit of the reductive program

The theory of the continuum had been and continued to be at the center of foundational concern: Weyl, for example, developed in his book *Das Kontinuum* extensive parts of analysis in a weak, predicative subsystem of second order arithmetic (CA); Hilbert used a theory equivalent to full (CA) as a formal framework for analysis in lectures during the early twenties;[14] Brouwer proposed an intuitionist theory of the continuum and established intuitionist versions of many classical theorems. Refined and detailed investigations in these traditions were carried out in the 50's and 60's. Contributors were, among others, Lorenzen, Kreisel, and most significantly, as far as the systematic mathematical work was concerned, Bishop through his book *Foundations of constructive analysis*. This quasi-empirical work of redeveloping analysis was focused, by logicians, more and more on the question: which set existence principles are sufficient (and necessary) to prove central facts? It found a most satisfactory conclusion for all of classical analysis through work of Takeuti, Feferman, and Friedman;[15] they established, in different ways and employing different formal theories, that analysis can be carried out in conservative extensions of elementary number theory. The crucial principle postulates the existence of sets given by arithmetic formulas; and it is only for sets that the induction principle is available. (This is a most significant deviation from earlier formulations of subsystems, including always the induction schema for all formulas of the language.) From results of Friedman's one can infer that weaker theories will not do for all of analysis, as the arithmetic comprehension principle is actually equivalent to crucial theorems. Examples of such theorems are the Bolzano-Weierstrass theorem, the fact that every Cauchy sequence is convergent, and the least upper bound principle for bounded sequences of reals. Nevertheless, in a theory isolated by Friedman a good deal of analysis and also of countable algebra can be developed;[16] that theory turned out to be conservative over primitive recursive arithmetic. All of this constitutes a rather remarkable and fascinating development.

---

[13] Bernays emphasized in the Nachtrag to his (1930b) the need to revise the epistemological outlook basic to the Hilbert Program. He wrote: "Im Ganzen ist die Situation nun so, dass die Hilbertsche Beweistheorie ... ein reiches Feld der Forschung geschaffen hat, dass jedoch die erkenntnistheoretischen Gesichtspunkte, von denen ihre Aufstellung ausging, problematisch geworden sind." (Bernays 1976a, p. 61).

[14] This is presented in Supplement IV of (Hilbert and Bernays 1939).

[15] Relevant sources are (Feferman 1977), (Friedman 1977), (Takeuti 1978), (Friedman 1980), and (Feferman 1985).

[16] See for example (Friedman, Simpson and Smith 1983). Friedman's and Simpson's research on "reverse mathematics" has been crucial in pursuing the aim of establishing the necessity of set existence principles for the proof of mathematical theorems. This is a venerable subject; the first instance known to me is in Dedekind's *Stetigkeit und irrationale Zahlen*: its last section establishes the equivalence of the principle of continuity to "certain fundamental theorems of infinitesimal analysis".

The mathematical work was partially motivated by the desire to secure the significance of reductions: how much mathematics can actually be done in a subsystem for which the relative consistency has been established? A systematic step in exploring reductive possibilities was taken in the marvelously informative, but never fully published *Stanford report on the foundations of analysis*. It was prompted by Spector's paper[17] and its principal purpose was, as Kreisel put it in the *Introduction*, to explore the possibility that the schema of bar recursion adjoined to the schemata of Gödel's T would permit a constructive consistency proof for classical analysis. "This involved", Kreisel wrote, "not only a careful presentation of the general nature of the problem, but also a full report of detailed results available in (a) the axiomatic system of classical analysis, (b) the constructive systems which codify the principles of proof that might be used in the consistency proof."[18] As to (b) Kreisel proposed, in Chapter 3 of the report, intuitionist theories for generalized inductive definitions; he investigated the relationship of their classical versions to subsystems of analysis (with no clear and convincing results). In his *A survey of proof theory* he posed the problem of reducing the subsystem with the Δ_2^1-comprehension principle to a suitable constructive theory of inductive definitions. That problem was solved during the next ten years also for a range of other subsystems. There was a great deal of technical development (e.g. concerning systems of ordinal notations), and a sharp push for conceptual clarification; but the details are too involved for this decidedly introductory presentation. I only want to indicate steps towards the most perspicuous and convincing solutions and state the main result very roughly. The first step was taken by Feferman and Friedman in their contributions to the Buffalo conference in 1968; they established the proof theoretic equivalence of some subsystems to classical theories of (transfinitely iterated) inductive definitions. But there seemed to be — largely on account of Zucker's work[19] — definite obstacles to a further reduction to intuitionist theories. Such further reductions to intuitionist theories of quite restricted kinds of inductive definitions were achieved independently and in a variety of ways by Buchholz, Pohlers, and Sieg in 1977. For the detailed discussion of these results, the character of the inductive definitions involved, and also the development alluded to above I have to refer to the literature mentioned in footnote 19. In any event, the outcome is this: strong impredicative subsystems of analysis have been reduced to constructively meaningful theories.

---

[17] (Spector 1962).

[18] Introduction to the *Stanford report*, p. 0.1.

[19] The papers are (Feferman 1970) and (Friedman 1970). Zucker's results were presented in his dissertation (1971); that work was published as (Zucker 1973). The work of Buchholz, Pohlers, and Sieg was published in volume 897 of the Springer Lecture Notes in Mathematics, entitled *Iterated inductive defnitions and subsystems of analysis: recent proof theoretical studies*. In the latter volume there are two papers of more introductory character: the first (by Feferman) sketches the development from 1963 to 1977 and a little bit beyond; the second (by Feferman and Sieg) gives a detailed account of the substantive results obtained prior to 1977 and a thorough discussion of the special characteristics of inductively defined classes.

Let us look at these results from the philosophical perspective underlying the reductive program. On the one hand, we have a justification for the practice of classical analysis based on intuitionist number theory; indeed, we have a justification based on finitist mathematics for significant parts of that practice.[20] On the other hand, we have reductions of strong impredicative subsystems of (CA) to constructive theories of inductive definitions. These results lead naturally to a distinction between two types of reductions. The two types are not distinguished from each other by the techniques for obtaining them, but rather by programmatic aims. *Foundational* reductions are to provide a constructive basis for strong and (from certain foundational perspectives) problematic parts of (CA); *computational reductions* are to yield algorithmic information from proofs in weak and (even from a finitist standpoint) unproblematic parts of (CA), e.g. through the characterization of the class of functions that are provably recursive in the classical theory.[21] It is via foundational reductions that the reductive program is pushed forward in an attempt to answer the question, *What more than finitist mathematics do we have to know to recognize the (partial) correctness of a strong theory?* Computational reductions answer in a specific way Kreisel's question, *What more than its truth do we know, if we have proved a theorem by restricted means (here: in a weak subsystem)?* Answers to the latter question are of mainly mathematical interest for obtaining bounds from proofs of Π_2^0-sentences and for establishing independence results;[22] answers to the former question are of more philosophical significance for assessing, in constructive terms, the epistemological demands of classical theories.

4 A guiding scheme

"Sometimes it happens that a man's circle of horizon becomes smaller and smaller; and as the radius approaches zero it concentrates on one point. And then that point becomes his standpoint." So Hilbert, according to (Reid 1970, p. 174). The finitist standpoint was not arrived at in this way; its general horizon was always comprehensively broad. It was just the metamathematical horizon that was restricted to include only finitist mathematics, because of Hilbert's epistemological radicalism. The reductive program is freed from such philosophically motivated restrictions; in consequence, it is not an instrument to solve the foundational problems of mathematics once and for all. It can and should be viewed as a coherent scheme guiding foundational research in exploring relationships between

---

[20] We have only to assume, certainly consonant with Hilbert and Bernays, that primitive recursive arithmetic is a part of finitist mathematics. The substantive thesis that primitive recursive arithmetic is essentially finitist mathematics (modulo a finitistically unproblematic arithmetization of syntax) is defended by W. W. Tait in (Tait 1981).
[21] These distinctions were made in (Sieg 1985b).
[22] Kreisel has urged the significance of such investigations for a long time, already e.g. in his (1958b, pp. 155–182).

two pervasive tendencies in mathematics, constructivist and platonist ones. (The latter in Bernays's restricted sense.[23])

The research in the pursuit of the reductive program has given us rich mathematical and metamathematical results; and there is a host of significant open problems, both logical and mathematical ones. I see two main problems; namely, (i) to reduce the subsystem of (CA) with Π_2^1- comprehension to a constructively justified theory, and (ii) to find a mathematically significant subsystem of analysis whose class of provably recursive functions consists only of computationally "feasible" ones (e.g. polynomially bounded ones). Other problems will certainly be mentioned later. Let me close by formulating in very general terms what I see as the three central, interdependent issues that have been addressed and will be at the center of the following contributions. The first concerns mathematics:

Which formal theories are adequate (and necessary) for which parts of classical mathematical practice?

The second concerns metamathematics:

To which constructive theories can classical theories be reduced?

The final one concerns philosophy. The mathematical and metamathematical research has been stimulated by foundational issues and provides detailed new material for philosophical reflection, material that deserves careful analysis; and we may ask:

What contribution to our understanding of (the nature of) mathematics do reductions make?

We share and readdress the foundational concerns of the pioneers with sharper conceptual tools, more sophisticated techniques, more subtle results; we have clear and challenging problems; and, perhaps, we have a more open and balanced philosophical perspective. I find this very exciting!

---

[23]The characteristic feature of Bernays's restricted platonism is the assumption of more and more inclusive totalities. Such an assumption goes hand in hand with a claim concerning properties P for elements of that totality, namely, it is objectively determined whether or not there is an element that has P. Bernays calls this tendency "platonist" because it moves towards considering mathematical objects as independent from the thinking subject and because it comes to the fore in Plato's philosophy. To emphasize, the restricted platonism of Bernays does not pretend to be anything but an "idealizing extrapolation of a domain of thought". See his paper (1935b). An English translation of the paper, originally published in 1935 in French, is found in (Benacerraf and Putnam 1983).

II.9

On reverse mathematics

Subsystems of second-order arithmetic or, synonymously, subsystems of analysis have been used for some time as frameworks for the formal development of classical mathematical analysis, i.e., the theory of the continuum set theoretically described by Dedekind and Cantor. Weyl, in his booklet *Das Kontinuum* (1918), developed substantial parts of analysis in a subsystem with the comprehension axiom for just arithmetic formulas; Hilbert, in lectures during the early twenties, recast analysis in full second-order arithmetic allowing third-order parameters. (The fourth supplement of Hilbert and Bernays, *Grundlagen der Mathematik*, Volume 2, is based on those lectures.) Because of this mathematical adequacy, second-order arithmetic and a variety of subsystems have been thoroughly investigated from a metamathematical perspective. Such investigations were started in the pursuit of Hilbert's program. Even after Gödel's incompleteness theorems refuted the program's original aims, giving a constructive consistency proof for analysis was viewed as "decisive for proof theory", according to Bernays in his (1935b).

The article reviewed here [i.e., (Simpson 1985a), WS] gives a lucid and informative survey of Friedman's work in two central areas of research on subsystems of analysis: to the actual development of parts of mathematics in subsystems, Friedman has contributed (programmatically) influential theorems of so-called reverse mathematics; to the metamathematical investigation of subsystems, he has contributed important conservation results and thus, in particular, relative consistency theorems. To be able to describe these results and to argue for my critical remarks (on Simpson's presentation), it is necessary to sketch a more detailed logical background. The essential set theoretic principles for these systems are the comprehension axiom

CA $\quad (\exists X)(\forall y)(y \in X \Leftrightarrow S(y))$,

the axiom of choice in the form

AC $\quad (\forall x)(\exists Y)S(x, Y) \Rightarrow (\exists Z)(\forall x)S(x, (Z)_x)$, or the axiom of dependent choice

DC
$$(\forall x)(\forall X)(\exists Y)S(x, X, Y) \Rightarrow (\forall X)(\exists Z)[X = (Z)_0 \,\&\, (\forall x)S(x, (Z)_x, (Z)_{x+1})],$$

where S is in each case an arbitrary formula of the language and thus may contain set quantifiers. Without restrictions these principles are impredicative: the sets X and Z whose existence is postulated are in general characterized by reference to all sets of natural numbers. All theories have the standard axioms for zero, successor, and—simply for convenience—all primitive recursive functions. The induction principle is formulated either as a schema or as a second-order axiom

IND $\;(\forall X)[0 \in X \,\&\, (\forall y)(y \in X \Rightarrow y' \in X) \Rightarrow (\forall x)x \in X]$.

(Comprehension and induction are always available for at least bounded arithmetic formulas.) The theories with the induction schema and with IND are equivalent in the presence of full CA; but they can be of strikingly different strength when only restricted set-existence principles are available. Theories are denoted by the name of their set-existence principle enclosed in parentheses; thus (CA) names full analysis. If ↾ follows its name, a theory uses the second-order axiom IND to formalize induction. I should remind the reader of two general results: (**1**) (CA) is proof theoretically equivalent to Zermelo-Fraenkel set theory without the power set axiom; (**2**) (AC) is conservative over (CA) for Π_4^1-formulas and properly stronger, as there is a Π_2^1-instance of AC not provable in (CA). (For a detailed discussion see technical note VI of Kreisel's (1968).) These results show that second-order arithmetic has a certain robustness; in addition, (**2**) implies that its refinement to subsystems described below as (**4**) is the best possible.

Subsystems of analysis are defined mainly by restricting S in the set-existence schemata to particular classes of formulas. In the early sixties, partly in the systematic study of predicativity, significant subsystems with S restricted to small classes of analytic formulas were isolated, namely, (Σ_1^1-DC), (Σ_1^1-AC), and (Δ_1^1-CA). Kreisel pointed out that (Σ_1^1-DC) $\supseteq$ (Σ_1^1-AC) $\supseteq$ (Δ_1^1-CA). These theories were investigated by Friedman in Chapter II of his 1967 MIT doctoral dissertation, *Subsystems of set theory and analysis*. He showed that the sub-systems (Σ_1^1-DC) and (Σ_1^1-AC) are conservative over (Σ_1^1-CA) for Π_2^1-formulas and that the first inclusion is proper; the second is also proper, as Steel later established. Most surprising was, however: (**3**) the above systems are conservative for Π_2^1-formulas over (Π_0^1-CA)$_{<\epsilon_0}$, i.e., a theory based on the transfinite iteration of the jump operator and equivalent to the theory of ramified analysis of level less than ϵ_0. This result allowed the determination of the proof theoretic ordinal of the systems, but it showed also—and this was quite unexpected—that these prima facie impredicative theories were clearly predicative in the sense of Feferman and Schütte. (See (Feferman 1964) and (Kreisel 1968, technical note V, in particular, p. 373).)

Theorem (3) turned out to be a special case of a general result: (4) the theory $(\Sigma^1_{n+1}\text{-AC})$ is conservative over $(\Pi^1_n\text{-CA})_{<\epsilon_0}$ for formulas in F_n, where F_0 is Π^1_2, F_1 is Π^1_3, and F_n is Π^1_4 for $n > 1$. This theorem of Friedman's was published in *Iterated inductive definitions and Σ^1_2-AC*. As just explained, the case $n = 0$ was of special interest for the study of predicativity; the case $n = 1$ was placed—not least by the choice of the paper's title—in the context of related foundational investigations. The most immediate context was provided by Feferman's paper *Formal theories for transfinite iterations of generalized inductive definitions and some subsystems of analysis*. Together, the papers achieved a reduction of the subsystems $(\Sigma^1_2\text{-AC})$ and, equivalently, $(\Delta^1_2\text{-CA})$ to the classical theory for less than ϵ_0-times iterated i.d. classes, i.e., classes defined by generalized inductive definitions. Well-known examples are the classes O of constructive ordinals and W of recursive well-founded trees. Indeed, $(\Sigma^1_2\text{-AC})$ was reduced to the classical theory of the tree classes W_ν, with index ν less than ϵ_0.

Kreisel had introduced intuitionist theories of transfinitely iterated inductive definitions in the mimeographed Stanford seminar reports (1963); these theories were viewed as codifying principles that might be used in constructive consistency proofs for subsystems of analysis. (The whole effort had been prompted by Clifford Spector's consistency proof of classical analysis using bar recursive functionals of finite type.) It was only the work of Feferman and Friedman that established significant connections between subsystems of analysis and classical theories of inductive definition, making a crucial step towards answering the major problem posed in (Kreisel 1968, p. 352): reduce $(\Sigma^1_2\text{-AC})$ to a suitable constructive theory of inductive definitions, which would provide, as Kreisel put it then, "a solution of Hilbert's problem for the subsystem of analysis ... Σ^1_2-AC". The systematic context I just sketched for this part of Friedman's work is completely missing in Simpson's account, i.e., it is not explained why the results of *Iterated inductive definitions and Σ^1_2-AC* have anything to do with iterated inductive definitions. As to the further development, the classical theories for i.d. classes turned out to be reducible to intuitionist theories for accessible i.d. classes. This allowed, in particular, a satisfactory solution of Kreisel's open problem: $(\Sigma^1_2\text{-AC})$ is reducible to $(\text{ID})_{<\epsilon_0}(O)$, the intuitionist theory of constructive number classes with index less than ϵ_0. (These further results were obtained in (Buchholz et al. 1981).)

There are two other papers of Friedman's that are strongly related to these foundational investigations. The first, (Friedman 1969), is concerned with the theory (BI_{pr}), i.e., the subsystem with the schema of bar induction

BI $\text{WF}(R) \Rightarrow \text{TI}(R, S)$

for arbitrary formulas S, but only primitive recursive binary relations R; $\text{WF}(R)$ expresses the well-foundedness of R and $\text{TI}(R, S)$ the transfinite induction principle along R for S. (Howard showed that the full schema BI, for arbitrary definable R, is equivalent to CA.) (BI_{pr}) was the strongest system, genuinely impredicative, for which a constructive consistency proof (relative to accepted intuitionist principles) was presented in Kreisel's survey. It was clearly intermediate between (Σ_1^1-AC) and (Π_1^1-CA)$\restriction$; Friedman established that the latter theory actually proves the existence of an ω-model for (BI$_{pr}$). (Later work showed the proof theoretic equivalence of (BI$_{pr}$) to the classical theory of non-iterated i.d. classes.) The second paper, (Friedman 1978), is not discussed in Simpson's survey. It gives a surprisingly simple way of extending Gödel and Gentzen's double-negation translation to show that a variety of classical theories are conservative over their intuitionist versions—not just for negative arithmetic, but also for Π_2^0-formulas. Consequently, the theories have the same class of provably recursive functions. The theories to which Friedman applied his technique include elementary number theory (PA), (CA), finite type theory, and ZF set theory. In (Feferman and Sieg 1981a), Friedman's method was adapted to systems of ramified analysis and theories of iterated inductive definitions. Its full generality and beautiful simplicity were brought out by (Leivant 1985).

Hilbert and Bernays's presentation of mathematical analysis in (a version of) (CA) can be given quite readily in (Π_1^1-CA)$\restriction$. The slow progress in obtaining constructive consistency proofs for weak second-order (and higher-order) extensions of elementary number theory was accompanied by strictly mathematical work; the latter had the aim of establishing the mathematical significance of those extensions and made use of work in the constructivist tradition, such as Bishop's. By the mid-seventies, through final efforts of Takeuti, Feferman, and Friedman, it was clear that mathematical analysis could be carried out in conservative extensions of number theory. Friedman's seminal papers are *Set theoretic foundations for constructive analysis* and *A strong conservative extension of Peano arithmetic*. The first paper introduced a fragment of Zermelo set theory with intuitionist logic that serves as a framework for constructive analysis and extends (HA) conservatively. In the second paper Friedman used a similar theory expanded by Bishop's principle of limited omniscience (i.e.. the statement that every sequence of 0's and 1's is either everywhere different from 0 or somewhere equal to 0) to develop classical analysis; the theory is shown to be a conservative extension of elementary number theory (PA). It is the actual development of parts of mathematics in subsystems of analysis—with a methodological twist—that is most prominent in Simpson's survey. It is here that, in Simpson's view, "Friedman's insights have been particularly influential and indeed decisive for later developments" (p. 137). Friedman's opening statement in *Some systems of second order arithmetic and their use* (Friedman 1975) formulates most clearly the central issue: "The questions underlying the work presented here on subsystems of second order arithmetic are the following. What are the proper axioms to use in carrying out proofs of particular theorems, or bodies of theorems, in mathematics? What are those formal systems which isolate the essential principles needed to prove them?"

The methodological twist demands establishing the necessity of using a certain set-existence principle for the proof of a mathematical theorem (provable from the principle). The strategy—show that the principle is a consequence of the theorem—is familiar from investigations of the axiom of choice in set theory and was already used in the context of analysis by Dedekind in *Stetigkeit und irrationale Zahlen* (1872): he showed in Section 7 of his booklet that the continuity principle (postulating the topological completeness of the reals) is equivalent to central theorems of analysis. In any event, this intriguing theme is played with beautiful and surprising variations in Friedman's work on subsystems; his initial successes and results of Simpson's gave substance to the enterprise of "reverse mathematics". The list of theorems equivalent to the main set-existence principles of four subsystems that emerged over the years as central is quite impressive (and it has been extended further in significant ways by Friedman, Simpson, and students of Simpson). Let me describe the systems very briefly and present two paradigmatic equivalences for each. There is, first of all, the base theory (RCA_0), consisting of recursive comprehension and the induction principle for Σ_1^0-formulas. To it one adds principles (of increasing strength): WKL (weak König's lemma), ACA (arithmetic comprehension), ATR (arithmetic transfinite recursion), and Π_1^1-CA. (Let me call the resulting subsystems **A**, **B**, **C**, and **D**.) Over (RCA_0) we have the following groups of equivalences:

WKL is equivalent to (i) the sequential Heine-Borel theorem (every covering of the closed unit interval by a sequence of open intervals has a finite subcovering), and to (ii) the completeness theorem for predicate logic.

ACA is equivalent to (i) every bounded sequence of reals has a least upper bound, and to (ii) König's lemma.

ATR is equivalent to (i) any two well-orderings of the natural numbers are comparable, and to (ii) the perfect set theorem.

Π_1^1-CA is equivalent to (i) every bounded, arithmetically definable set of reals has a least upper bound, and to (ii) the perfect kernel theorem.

These examples are all found in Friedman's early work; in addition Simpson's paper lists many newer and most interesting results, including some in algebra. Note that the theories **A**, **B**, **C**, and **D** are in effect theories with restricted induction.

Simpson remarks that "the whole point of Reverse Mathematics is to prove ordinary mathematical theorems using only the weakest possible set existence principles" (p. 150). So why focus exclusively on second-order theories? In many publications, most recently in his (1988a), Feferman has urged the use of standard or not-so-standard (type) theories for a more direct formalization of mathematical practice and then the establishment of their conservativeness over foundationally significant theories. Indeed, Friedman emphasized in his Vancouver paper, after raising the underlying questions above: "Ultimately, answers to these questions will require use of systems that are not subsystems of second order arithmetic, but have variables ranging over objects such as sets of sets of natural numbers. Such systems would be needed in order to formalize directly theorems about continuous functions on the reals, or measurable sets of reals." Then he goes on to point out, however, that (i) the language of second-order arithmetic is "basic among the possible languages relevant to the formalization of mathematics", and (ii) the most important systems not formalized in that language are conservative over those that are. Thus, he claims, "the systematic study reported here of subsystems of second order arithmetic is a necessary and important step in answering the underlying questions". (All quotes are from p. 235.) Simpson takes the quasi-empirical success of reverse mathematics as revealing the significance of these systems; but note the tension between this evaluation and the significance of theorems of intermediate strength such as the Paris-Harrington version of Ramsey's theorem or forms of Kruskal's theorem. So what really is the mathematical significance of these systems, in particular of A, B, C, and D, if they are taken as "a natural vehicle for the formal axiomatic study of ordinary mathematics"? This question does not seem to have an obvious answer, in spite of the remarkable body of results. Is it thought in any way to be comparable to the paradigm of "formal axiomatic study of ordinary mathematics", namely, the use of important structural notions like group or field?

That brings me to two critical remarks on Simpson's presentation. To re-emphasize, Friedman's most significant mathematical and metamathematical results concerning subsystems of analysis are described with admirable—and characteristic—clarity. The discussion of the general import of the work, however, does not have the same perspicuity. Let me explain why by raising and answering one question for each of the two natural categories under which Friedman's work on subsystems falls and under which it is presented by Simpson.

(I) Wherein lies the general interest of the metamathematical work, in particular of (4)? For Simpson, it is important primarily because it brought about "a fairly profound methodological shift in the study of subsystems of Z_2", i.e., second-order arithmetic; a shift from proof theoretic to model theoretic methods. One might very well doubt whether such a shift took place, and the evidence Simpson adduces (p. 142) is certainly not persuasive. One might also disagree whether such a shift is always desirable, and the remarks Simpson makes about "the usual disadvantages of syntax as against semantics" (p. 141) and about the reasons why "a number of researchers ... continue to shun semantical methods in favor of syntactical ones" (p. 143) are not incisive. For a balanced discussion of the methodological issues, the reader should consult the appendix of Feferman's paper (1988a) that was mentioned previously. But be that as it may, for me the conservative extension results of Friedman's are significant in the context of the foundational investigations I sketched above. And in that context the proof theoretic methods used by Feferman and Sieg, for example, to reprove Friedman's results, have the advantage of being directly extendible to finite type theories; they are also flexible enough to allow the investigation of restricted subsystems and of (weak) fragments of arithmetic. Cf. their paper (1981b) and Sieg's (1985a).

(II) Wherein lies the philosophical significance of the mathematical work? For Simpson, reverse mathematics has a moral: "For many key theorems of ordinary mathematics, there is a weakest natural subsystem of Z_2 in which the given theorem is provable. Furthermore, this weakest natural system often turns out to be one of the systems RCA, WKL, ACA, ATR, and Π_1^1-CA. These conclusions have obvious significance for any philosophical investigation of the role of set existence axioms in ordinary mathematics" (p. 148). (This holds, indeed by the same mathematical evidence, for the restricted systems (RCA$_0$), **A**, **B**, **C**, and **D**.) The reader is referred to the discussion of reductionist programs in Section 5 of the paper, where such programs are seen as investigating the question of how much of mathematical practice can be developed using "some restricted set of 'acceptable' principles" (p. 152). The systems **A** and **C** are related to finitism and predicativism, respectively, via suitable conservative extension results: **A** is conservative over primitive recursive arithmetic, **C** is conservative for Π_1^1-sentences over Feferman's theory IR for predicative mathematics. The remark can actually be expanded to **B** and **D**, as **B** is conservative for Π_2^0-formulas over intuitionist number theory (HA) and **D** for the same class of formulas over $(ID)_{<\omega}(O)$, the intuitionist theory for the finite constructive number classes. **B** and **D** are consequently reducible to theories based on intuitionistically acceptable principles. This provides a coherent perspective bringing out the complementary character of mathematical and metamathematical work on subsystems—work that aims for reductions of significant parts of mathematical practice to distinctive foundational positions. Such a scheme guided, in the form of a modified Hilbert program formulated by Paul Bernays, much of the work in proof theory. But as I did not see an obvious answer to the question, "What is the mathematical significance of the systems **A**, **B**, **C**, and **D**, when taken as vehicles for the formal axiomatic study of ordinary mathematics?" so I do not see an obvious answer to the question, "What is the philosophical significance of the corresponding systems PRA, HA, IR, and $(ID)_{<\omega}(O)$, when taken as formal expressions of foundational positions?" There is ample room for reflection on this question; the work of Friedman's reported here provides rich and crucial data.

One final remark: Simpson considers the finitist reductionist program (understood in his special sense described above) as Hilbert's program. This is inaccurate. Hilbert did not propose to redo all of mathematics with only finitist principles, but rather to justify—via finitist consistency proofs—the use of strong classical theories sufficient for the direct formalization of mathematical practice. If this particular reductionist program should be adorned with a name, then it seems appropriate to attach Kronecker's to it. Recall that on Hilbert's view the principles accepted by Kronecker coincided essentially with finitist ones, and Kronecker certainly insisted on using just those. Indeed, it would be highly interesting and quite possibly mathematically rewarding, if parts of Kronecker's work were to be analyzed within restricted axiomatic frameworks. Theories weaker than A have mathematical interest as Simpson points out by referring to joint work with Smith in their paper (1986). This indicates a direction for reverse mathematics that may join fortuitously with recent developments in "feasible mathematics".

II.10

Relative consistency and accessible domains*

> ... weil Nichts in der Mathematik
> gefährlicher ist, als ohne genügenden
> Beweis Existenzen anzunehmen...
> Dedekind[1]

1 Introduction

The goal of Hilbert's program — to give consistency proofs for analysis and set theory within finitist mathematics — is unattainable; the program is dead. The mathematical instrument, however, that Hilbert invented for attaining his programmatic aim is remarkably well: *proof theory* has obtained important results and pursues fascinating logical questions; its concepts and techniques are fundamental for the mechanical search and transformation of proofs; and I believe that it will contribute to the solution of classical mathematical problems.[2] Nevertheless, we may ask ourselves, whether the results of proof theory are significant for the foundational concerns that motivated Hilbert's program and, more generally, for a reflective examination of the nature of mathematics.

The results I alluded to establish the consistency of *classical* theories relative to *constructive* ones and give in particular a constructive foundation to mathematical analysis. They have been obtained in the pursuit

---

*This chapter was originally published in *Synthese* **84**, 1990, pp. 259–297, an essay that in turn was a much-expanded version of *Relative Konsistenz*—written in German and published in (Börger 1987). That collection of papers was dedicated to the memory of Dieter Rödding, my first logic teacher.

The text of the Synthese paper is essentially unchanged, except for the incorporation of some of the (still numerous) footnotes. In the meantime much illuminating historical research has been carried out and many significant mathematical results have been obtained. Some of these developments are reflected in four papers I have since published and that are most closely related to central issues in this essay; (Sieg 1999), (Sieg 2002), (Sieg and Ravaglia 2005), and (Sieg and Schlimm 2005).

Translations in this chapter are my own, unless texts are taken explicitly from English editions. In the notes, some quotations that are not central to my arguments are given only in the original German.

[1] From Dedekind's letter to Lipschitz of 27 July 1876, published in (Dedekind 1932, p. 477).
[2] Finally, there are real beginnings; see (Luckhardt 1989).

of a *reductive program* that provides a coherent scheme for metamathematical work and is best interpreted as a far-reaching generalization of Hilbert's program. For philosophers these definite mathematical results (should) present a profound challenge. To take it on means to explicate the reductionist point of constructive relative consistency proofs; the latter are to secure, after all, classical theories on the basis of more elementary, more evident ones. I take steps towards analyzing the precise character of such implicitly epistemological reductions and thus towards answering the narrow part of the above question. But these steps get their direction from a particular view on the question's wider part.

As background for that view, I point to striking developments within mathematics, namely to the emergence of set theoretic foundations, particularly for analysis, and to the rise of modern axiomatics with a distinctive structuralist perspective. These two developments overlap, and so do the problems related to them. Indeed, they came already to the fore in Dedekind's work and in the controversy surrounding it.[3] They were furthered by Hilbert's contributions to algebraic number theory and the foundations of geometry; the difficult issues connected with them prompted his foundational concerns during the late 1890s. Hilbert's program, though formulated only in the 1920s, can be traced to this earlier "problematic". I argue that it was meant to mediate between broad foundational conceptions and to address related, but quite specific methodological problems. An example of the latter is the use of "abstract" (analytic) means in proofs of "concrete" (number theoretic) results: the program — in its instrumentalist formulation — attempts to exploit the formalizability of mathematical theories for a systematic and philosophically decisive solution.

This instrumentalist aspect, as a matter of fact equivalent to the program's consistency formulation, has been overemphasized in the literature and leaves unaccounted-for critical features of Hilbert's thought. The historical part of this chapter brings into focus such neglected features and sets the stage for an analysis of proof theoretic reductions as *structural* ones. The philosophical significance of relative consistency results is viewed in terms of the *objective underpinnings* of theories to which reductions are (to be) achieved.[4] The elements of *accessible domains* that provide such

---

[3]Dedekind played a significant role in the development of nineteenth-century mathematics. As far as our century is concerned I mention his influence on Hilbert and Emmy Noether, thus on Bourbaki's conceptions. An illuminating analysis of Dedekind's work is given in (Stein 1988); the major influences on Bourbaki are documented in (Dieudonné 1970). In (Zassenhaus 1975) one finds on p. 448 the remark: "... we can see in Dedekind more than in any other single man or woman the founder of the conceptual method of mathematical theorization in our century. The new generation of mathematicians ... after the First World War realized in full detail Dedekind's self-confessed desire for conceptual clarity not only in the foundations of number theory, ring theory and algebra, but on a much broader front, in all mathematical disciplines."

[4]The reduced theories have to be mathematically significant. Indeed, the consistency program has been accompanied from its inception by work intended to show that the theories permit the formal development of substantial parts of mathematics.

underpinnings have a unique build-up through basic operations from distinguished objects; the theories formulate principles that are evident — given an understanding of the build-up and a minimalist delimitation of the domain. But note that (i) the objects in accessible domains need not be constructive in any traditional sense: certain segments of the cumulative hierarchy will be seen to be accessible, and (ii) the restriction of logical principles used is not central: the theories of interest turn out to be such that the consistency of their classical versions is established easily relative to their intuitionist versions (by finitist arguments).

Even in mathematical practice relative consistency proofs are prompted by epistemological concerns. One wants to guarantee the coherence of a complex (new) theory in terms of comprehended notions and does so frequently by devising suitable models. This general goal is pursued, e.g. when Euclidean models for non-Euclidean geometries are given. Proof theoretic reductions have two special features: (i) they focus on the deductive apparatus of theories, and (ii) they are carried out within theories that have to measure up to restrictive epistemological principles. The latter are traditionally of a more or less narrow "constructivist" character. In broadening the range of theories to "quasi-constructive" ones and concentrating on one central feature, namely accessibility, we will be able to evaluate their (relative) epistemological merits. And in this way, it seems to me, we can gain a deepened understanding of what is *characteristic of and possibly problematic* in classical mathematics and of what is *characteristic of and taken for granted as convincing* in constructive mathematics.

In the current discussion, some do as if an exclusive alternative between platonism and constructivism had emerged from the sustained mathematical and philosophical work on foundations for mathematics; others do as if this work were deeply misguided and did not have any bearing on our understanding of mathematics. Both attitudes prevent us from using the insights (of preeminent mathematicians) that underly such work and the significant results that have been obtained. They also prevent us from turning attention to central tasks; namely, to understand the role of abstract structures in mathematical practice and the function of (restricted) accessibility notions in "foundational" theories or "methodical frames", to use Bernays's terminology. I attempt to give a perspective that includes traditional concerns, but that allows — most importantly — to ask questions transcending traditional boundaries. This perspective is deeply influenced by the writings of Paul Bernays.

2 Mathematical reflections

These are concerned with mathematical analysis and theories in which its practice can be formally represented. So I start out by describing attempts to clarify the very object of analysis and thus, it was assumed, the role of analytic methods in number theory. These attempts came under the headings *arithmetization of analysis* and *axiomatic characterization of the real numbers.* I discuss two kinds of arithmetizations put forward by Dedekind and Kronecker, respectively. Dedekind proceeded axiomatically and sought to secure his characterization by a consistency proof relative to logic broadly conceived, whereas Kronecker insisted on a radical restriction of mathematical objects and methods. (Dedekind's arithmetization of analysis should perhaps be called *set theoretic* and Kronecker's by contrast *strict.*) Hilbert's axiomatization of the real numbers grew directly out of Dedekind's and was the basis for two proposals to overcome at least for analysis the set theoretic difficulties that had been discovered around the turn of the century. The second proposal, when suitably amended by the formalist conception of mathematics, led to Hilbert's program.

2.1 Consistent sets. A systematic arithmetization is to achieve, Dirichlet demanded, that *any* theorem of algebra and higher analysis can be formulated as a theorem about natural numbers.[5] If that had been clearly so, Dirichlet's introduction of analytic methods to prove his famous theorem on arithmetic progressions would have been methodologically innocuous. But in using properties of "continuous magnitudes" to prove facts concerning natural numbers, he pushed aside a traditional, partly epistemologically motivated boundary.[6] Dirichlet himself remarks: "The method I employ seems to me to merit attention above all by the connection it establishes between the infinitesimal analysis and the higher arithmetic . . ."[7] In another paper that explores further uses of analytic methods in number theory he writes: ". . . I have been led to investigate a large number of questions concerning numbers from an entirely new point of view, that attaches itself to the principles of infinitesimal analysis and to the remarkable properties of a class of infinite series and infinite products . . ." The significance of these methodological innovations can be fathomed from remarks such as Kummer's, who compares them in his eulogy on Dirichlet to Descartes's "applications of analysis to geometry", or Klein's, who stated that they gave "direction to the entire further development of number theory".

The essays of Dedekind and Kronecker[8] seek an arithmetization satisfying Dirichlet's demand, but proceed in radically different ways. Kronecker admits as objects of analysis only natural numbers and constructs from them (in now well-known ways) integers and rationals. Even algebraic reals are introduced, since they can be isolated effectively as roots of algebraic equations. The general notion of irrational number, however, is rejected in consequence of two restrictive methodological requirements to which mathematical considerations have to conform: (i) concepts must be decidable, and (ii) existence proofs must be carried out in such a way that they present

---
[5]That is reported in the preface to (Dedekind 1888).
[6]I allude, of course, to Gauss's attitude; compare (Sieg 1984, p. 162).
[7](Dirichlet 1838, p. 360), respectively (Dirichlet 1839/40, p. 411).
[8]Dedekind's relevant papers are the essays (1872c) and (1888); his letters to Lipschitz and Weber are also of considerable interest and were published in (Dedekind 1932). As to Kronecker I refer to his (1887) and Hensel's introduction to (Kronecker 1901).

objects of the appropriate kind. For Kronecker there can be no infinite mathematical objects, and geometry is banned from analysis even as a motivating factor. Clearly, this procedure is strictly arithmetic, and Kronecker believes that following it analysis can be re-obtained. In (Kronecker 1887) we read:

I believe that we shall succeed in the future to "arithmetize" the whole content of all these mathematical disciplines [including analysis and algebra]; i.e. to base it [the whole content] on the concept of number taken in its most narrow sense ...

Kronecker did prove, to his great pleasure, Dirichlet's theorem on arithmetic progressions satisfying his restrictive conditions.[9] But it is difficult for me to judge to what extent Kronecker pursued a program of developing (parts of) analysis systematically. In any event, such a program is not chimerical: from mathematical work during the last decade it has emerged that a good deal of analysis and algebra can indeed be done in conservative extensions of primitive recursive arithmetic.[10] Finally, let me mention that Kronecker begins the paper by hinting at his philosophical position — through quoting Gauss on the epistemologically special character of the laws for natural numbers; only these laws, in contrast to those of geometry, carry the complete conviction of their necessity and thus of their absolute truth.

Dedekind, a student of Gauss, emphasized already in his Habilitationsvortrag of 1854 a quite different and equally significant aspect of mathematical experience; namely, the introduction and use of new concepts to grasp composite phenomena that are being governed by the old notions only with great difficulty.[11] Referring to this earlier talk, Dedekind asserts in the preface to his (1888) that most of the great and fruitful advances in mathematics have been made in exactly this way. He gives, in contrast to Kronecker, a general definition of reals: cuts are explicitly motivated in geometric terms, and infinite sets of natural numbers are used as respectable mathematical objects. Kronecker's methodological restrictions are opposed by him, in particular the decidability of concepts; he believes that it is determined independently of our knowledge, whether an object does or does not fall under a concept. In this way Dedekind defends general features of his work in the foundations of analysis and in algebraic number theory.[12] But,

---

[9] (Kronecker 1901, p. 11).

[10] It is most plausible that such work would be enriched by paying attention to Kronecker's. For references to the contemporary mathematical work see (Simpson 1988). As PRA is certainly a part of number theory unproblematic even for Kronecker, this work can be seen as a partial realization of "Kronecker's program" (and not, as it is done by Simpson, of Hilbert's).

[11] Dedekind mentions that Gauss approved of the "Absicht" of his talk. Kneser reports in his *Leopold Kronecker, Jahresbericht der DMV*, 33, 1925, that Dedekind referred often to a remark of Gauss that (for a particular number theoretic problem) notions are more important than notations. In pointing to the "Gaussian roots" of Dedekind's and Kronecker's so strikingly different positions, I want to emphasize already here that they can (and should) be viewed as complementary.

[12] Kronecker spurned Dedekind's algebraic conceptions. See, e.g., the note on p. 336 of his (1886) and Dedekind's gentle rejoinder in, what else, a footnote of his (1888): "but to enter into a discussion [of such restrictions] seems to be called for only when the distinguished

the reader may ask, how does Dedekind secure the existence of mathematical objects? To answer this question I examine Dedekind's considerations for real and natural numbers. The principles underlying the definition of cuts are for us set theoretic ones, for Dedekind they belong to logic[13]: they allow — as Dedekind prefers to express it — the creation of new numbers, such that their system has "the same completeness or ... the same continuity as the straight line". Dedekind emphasizes in a letter to Lipschitz that the *stetige Vollständigkeit* (continuous completeness) is essential for a scientific foundation of the arithmetic of real numbers, as it relieves us of the necessity to assume in analysis *existences without sufficient proof*. Indeed, it provides the answer to his own rhetorical question:

How shall we recognize the admissible existence assumptions and distinguish them from the countless inadmissible ones ... ? Is this to depend only on the success, on the accidental discovery of an internal contradiction?[14]

Dedekind is considering assumptions that concern the existence of individual real numbers; such assumptions are not needed, when we are investigating a complete system — ein denkbar vollständigstes Größen-Gebiet. By way of contrast, and in defense against the remark that all of his considerations are already contained in Euclid's *Elements*, he notices that such a complete system is not underlying the classical work. The definition of proportionality is applied only to those (incommensurable) magnitudes that occur already in Euclid's system and whose existence is evident for good reasons. And he argues in this letter of 1876 and later in the preface to *Was sind und was sollen die Zahlen?* that the algebraic reals form already a model of Euclid's presentation. For Euclid, Dedekind argues, that was sufficient, but it would not suffice, if arithmetic were to be founded on the very concept of number as proportionality of magnitudes.[15]

---

mathematician will have published his reasons for the necessity or even just the advisability of these restrictions." Kronecker expressed his views quite drastically in letters; for example in a letter to Lipschitz of 7 August 1883 he writes: "Bei dieser Gelegenheit habe ich das lange gesuchte Fundament meiner ganzen Formentheorie gefunden, welches gewissermassen "die Arithmetisierung der Algebra" — nach der ich ja das Streben meines mathematischen Lebens gerichtet habe — vollendet, und welches zugleich mir mit Evidenz zeigt, dass auch umgekehrt die Arithmetik dieser "Association der Formen" nicht entbehren kann, dass sie ohne deren Hülfe nur auf Irrwege geräth oder sich Gedankengespinste macht, die wie die Dedekindschen, die wahre Natur der Sache mehr zu verhüllen als zu klären geeignet sind." (Lipschitz 1986, pp. 181–182).

[13] Why then "arithmetization"? Dedekind views cuts as "purely arithmetical phenomena"; see the preface to (1872c) or (1888), where Dedekind talks directly about the "rein arithmetische Erscheinung des Schnitts". In the latter work he immediately goes on to pronounce arithmetic as a part of logic: "By calling arithmetic (algebra, analysis) only a part of logic I express already that I consider the concept of number as completely independent of our ideas or intuitions of space and time, that I view it rather as an immediate outflow from the pure laws of thought." (Dedekind 1932, p. 335). The next three references in this paragraph are to (Dedekind 1932), namely, p. 321, p. 472, and p. 477, respectively.

[14] Letter to Lipschitz of 27 July 1876; in (Dedekind 1932, p. 477).

[15] (Dedekind 1932, pp. 477–8, in particular top of p. 478). — Added for this volume: the precise definition of Dedekind's model is found in note 10 of the Introduction to this volume.

The question as to the existence of particular reals has thus been shifted to the question as to the existence of their complete system. If we interpret the essay on continuity in the light of considerations in *Was sind und was sollen die Zahlen?* and Dedekind's letter to Keferstein, we can describe Dedekind's procedure in a schematic way as follows. Both essays present first of all informal analyses of basic notions, namely of continuity by means of cuts (of points on the straight line and rationals, respectively) and of natural number by means of the components *system, distinguished object 1*, and *successor operation*. These analyses lead with compelling directness to the definitions of a complete, ordered field and of a simply infinite system. Then — in our terminology — models for these axiom systems are given. In *Stetigkeit und irrationale Zahlen* the system of all cuts of rationals is shown to be (topologically) complete and, after the introduction of the arithmetic operations, to satisfy the axioms for an ordered field. The parallel considerations for simply infinite systems in *Was sind und was sollen die Zahlen?* are carried out more explicitly. Dedekind gives in Section 66 of that essay his "proof" of the existence of an infinite system. Such systems contain a simply infinite (sub-) system, as is shown in Section 72.

Dedekind believes to have given purely logical proofs for the existence of these systems and thus to have secured the consistency of the axiomatically characterized notions.[16] With respect to simply infinite systems he writes to Keferstein in a letter of 27 February 1890:

After the essential nature of the simply infinite system, whose abstract type is the number sequence N, had been recognized in my analysis . . . , the question arose: does such a system *exist* at all in the realm of our ideas? Without a logical proof of existence it would always remain doubtful whether the notion of such a system might not perhaps contain internal contradictions. Hence the need for such a proof (articles 66 and 72 of my essay).[17]

I emphasize that Dedekind views these considerations not as specific for the foundational context of the essays analysed here, but rather as paradigmatic for a general mathematical procedure, when abstract, axiomatically characterized notions are to be introduced. That is unequivocally clear, e.g. from his discussions of ideals in (Dedekind 1877), where he draws direct parallels to the steps taken here.[18] The particular constructions leading to the general concept of real number provide an arithmetization of analysis: they proceed, as Dedekind believes, solely within logic and thus purely arithmetically (cf. footnote 13). Their specific logical character implies almost trivially that Dirichlet's demand is satisfied; any analytic statement

---

[16]That such a proof is intended also in (Dedekind 1872c) is most strongly supported by the discussion in (Dedekind 1888, p. 338). — The Fregean critique of Dedekind in section 139 of *Grundgesetze der Arithmetik*, vol. II, is quite misguided. For a deeper understanding of Dedekind's views on creation (Schöpfung) of mathematical objects see also his letter of 24 January 1888 to H. Weber in (Dedekind 1932, p. 489). That, incidentally, anticipates and resolves Benacerraf's dilemma in *What numbers could not be*.

[17]In (van Heijenoort 1967, p. 101). The essay Dedekind refers to is (Dedekind 1888).

[18](Dedekind 1877, pp. 268–269; in particular the long footnote on p. 269).

can be viewed as (a complicated way of making) a statement concerning natural numbers. But Dedekind states, that it is nothing meritorious "to actually carry out this tiresome rewriting (mühselige Umschreibung) and to insist on using and recognizing only the natural numbers".

The very beginnings of the Hilbertian program can be traced back to these foundational problems in general and to Dedekind's proposed solution in particular. Hilbert turned his attention to them, as he recognized the devastating effect on Dedekind's essays of observations that Cantor communicated to him in letters, dated 26 September and 2 October 1897.[19] Cantor remarks there that he was led "many years ago" to the necessity of distinguishing two kinds of totalities (multiplicities, systems): namely, *absolutely infinite* and *completed* ones. Multiplicities of the first kind are called *inconsistent* in his famous letter to Dedekind of 28 July 1899, and those of the second kind *consistent*. Only consistent multiplicities are viewed as sets, i.e. proper objects of set theory. This distinction is to avoid, and does so in a trivial way, the contradictions that arise from assuming that the multiplicity of all things (all cardinals, or all ordinals) forms a set.

In 1899 Hilbert writes *Über den Zahlbegriff*, his first paper addressing foundational issues of analysis. He intends — never too modest about aims — to rescue the set theoretic arithmetization of analysis from the Cantorian difficulties. To this end he gives a categorical axiomatization of the real numbers following Dedekind's work in *Stetigkeit und irrationale Zahlen*. He claims that its consistency can be proved by a "suitable modification of familiar methods"[20] and remarks that such a proof constitutes "the proof for the existence of the totality of real numbers or — in the terminology of G. Cantor — the proof of the fact that the system of real numbers is a consistent (completed) set". In his subsequent Paris address Hilbert goes even further, claiming that the existence of Cantor's higher number classes and of the alephs can also be proved. Cantor, by contrast, insists in a letter to Dedekind, written on 28 August 1899, that even finite multiplicities cannot be proved to be consistent. The fact of their consistency is a simple, unprovable truth — "the axiom of arithmetic"; and the fact of the consistency of those multiplicities that have an aleph as their cardinal number is in exactly the same way an axiom, the "axiom of the extended transfinite arithmetic".[21]

Hilbert recognized soon that his problem, even for the real numbers, was not as easily solved as he had thought. Bernays writes in his (1935a)

---

[19]In particular in Section 66 of (Dedekind 1888). That is clear from Cantor's response of 15 November 1899 to a letter of Hilbert's (presumably not preserved). Cantor's letter is published in (Purkert and Ilgauds 1987, p. 154). See also remark A in Section 2.3.

[20](Hilbert 1900b, p. 261). The German original is: "Um die Widerspruchsfreiheit der aufgestellten Axiome zu beweisen, bedarf es nur einer geeigneten Modifikation bekannter Schlußmethoden." (Bernays 1935a) reports on pp. 198–199 in very similar words, but with a mysterious addition: "Zur Durchführung des Nachweises gedachte Hilbert mit einer geeigneten Modifikation der in der Theorie der reellen Zahlen angewandten Methoden auszukommen."

[21](Cantor 1932, pp. 447–448).

on p. 199, "When addressing the problem [of proving the above consistency claims] in greater detail, the considerable difficulties of this task emerged." It is the realization, I assume, that distinctly new principles have to be accepted; principles that cannot be pushed into the background as "logical" ones.[22] Dedekind's arithmetization of analysis has not been achieved without "mixing in foreign conceptions" after all;[23] a rewriting, however tiresome, of analytic arguments in purely number theoretic terms is seemingly not always possible.

2.2 Consistent theories. In his address to the International Congress of Mathematicians, Heidelberg 1904, Hilbert examines again and systematically various attempts of providing foundations for analysis, in particular Cantor's. The critical attitude towards Cantor that was implicit in *Über den Zahlbegriff* is made explicit here. Hilbert accuses Cantor of not giving a rigorous criterion for distinguishing consistent from inconsistent multiplicities; he thinks that Cantor's conception on this point "still leaves latitude for subjective judgment and therefore affords no objective certainty". He suggests again that consistency proofs for suitable axiomatizations provide an appropriate remedy, but proposes a radically new method of giving such proofs: develop logic (still vaguely conceived) together with analysis in a common frame, so that proofs can be viewed as *finite* mathematical objects; then show that such formal proofs cannot lead to a contradiction. Here we have, seemingly in very rough outline, Hilbert's program as developed in the 1920s; but notice that the point of consistency proofs is still to guarantee the existence of sets, and that a reflection on the mathematical means admissible in such proofs is lacking completely. Before describing the later program, let me mention that this address and *Über den Zahlbegriff* are squarely directed against Kronecker. In his Heidelberg address Hilbert claims that he has refuted Kronecker's standpoint — by partially embracing it, as I hasten to add. I will explain below that this is by no means paradoxical. Indeed, a genuine methodological shift had been made; Bernays remarks that Hilbert started, clearly before giving this address, "to do battle with Kronecker with his own weapons of finiteness by means of a modified conception of mathematics".[24]

---

[22] This general concern comes out in Husserl's notes on a lecture that Hilbert gave to the Göttingen Mathematical Society in 1901 and, in very similar terms, in (Hilbert *1904, p. 266); Husserl's notes are quoted in full in (Wang 1987, p. 53).

[23] Dedekind points out emphatically, e.g., in the letter to Lipschitz (1932, p. 470) and in the introduction to (1888), that his constructions do not appeal anywhere to "fremdartige Vorstellungen"; he has in mind appeals to geometric ones.—Bernays has again and again made the point that a "restlose strikte Arithmetisierung" cannot be achieved. In (Bernays 1941) one finds on p. 152 the remark: "... one can say — and that is certainly the essence of the finitist and intuitionist critique of the usual mathematical methods — that the arithmetization of geometry in analysis and set theory is not a complete one." It is through the powerset of the set of natural numbers that our geometric conception of the continuum is connected to our elementary conception of number; e.g. in his (1978).

[24] The quotation is taken from a longer remark of Bernays in (Reid 1970). It is preceded by: "Under the influence of the discovery of the antinomies in set theory, Hilbert temporarily

There are a number of general tendencies that influenced the Heidelberg address and the further development towards Hilbert's program. First of all, the *radicalization of the axiomatic method*; by that I mean the insight that the linguistic representation of a theory can be viewed as separable from its content or its intended interpretation. That was clear to Dedekind, was explicitly used by Wiener, and brought to perfection by Hilbert in his *Grundlagen der Geometrie*.[25] Secondly, the *instrumentalist view of (strong mathematical) theories*; the earliest explicit formulation I know of is due to Borel discussing the value of abstract, set theoretic arguments from a Kroneckerian perspective.

> One may wonder what is the real value of these [set theoretic] arguments that I do not regard as absolutely valid but that still lead ultimately to effective results. In fact, it seems that if they were completely devoid of value, they could not lead to anything... This, I believe, would be too harsh. They have a value analogous to certain theories in mathematical physics, through which we do not claim to express reality but rather to have a guide that aids us, by analogy, in predicting new phenomena, which must then be verified.

Can one systematically explore, Borel asks, the sense of such arguments. His answer is this:

> It would require considerable research to learn what is the real and precise sense that can be attributed to arguments of this sort. Such research would be useless, or at least it would require more effort than it would be worth. How these overly abstract arguments are related to the concrete becomes clear when the need is felt.[26]

To grapple with this problem clearly one has to use, thirdly, the *strict formalization of logic* that had been achieved by Frege (Peano, and Russell/ Whitehead). That is a moment not yet appreciated in Hilbert's Heidelberg address, where one finds a discussion of logical consequence (Folgerung)

---

thought that Kronecker had probably been right there. (That is, right in insisting on restricted methods.) But soon he changed his mind. Now it became his goal, one might say, to do battle with..."

[25] Dedekind describes, on p. 479 of (1932), such a separation before making the claim that the algebraic reals form a model of the Euclidean development. For a penetrating discussion of the general development see (Guillaume 1985). Such a separation appears to us as banal, but it certainly was not around the turn of the century, as the Frege-Hilbert controversy amply illustrates. — Added for this volume: Cf. note 15.

[26] (Baire, Borel, Hadamard and Lebesgue 1905, p. 273). A striking, but different suggestion along these lines was made already in (Cantor 1879, p. 173): "If, as is assumed here [i.e. from a restrictive position], only the natural numbers are real and all others just relational forms, then it can be required that the proofs of theorems in analysis are checked as to their "number-theoretic content" and that every gap that is discovered is filled according to the principles of arithmetic. The feasibility of such a supplementation is viewed as the true touchstone for the genuineness and complete rigor of those proofs. It is not to be denied that in this way the foundations of many theorems can be perfected and that also other methodological improvements in various branches of analysis can be effected. Adherence to the principles justified from this viewpoint, it is believed, secures against any kind of absurdities or mistakes." This is in a way closer to Hilbert's belief that finitist statements must admit a finitist proof. That belief is implicitly alluded to in (Bernays 1941, p. 151): "The hope that the finitist standpoint (in its original sense) could suffice for all of proof theory was brought about by the fact that the proof theoretic problems could be formulated from that point of view."

Relative consistency and accessible domains 309

quite uninformed by this crucial aspect of Frege's work. Hilbert succeeded to join these tendencies into a sharply focused program with a very special mathematical and philosophical perspective.

The *modified conception of mathematics* underlying the formulation of the program is characterized by Hilbert in the 1920s most pointedly and polemically: classical mathematics is a *formula game* that allows "to express the whole thought content of mathematics in a uniform way"; its consistency has to be established within finitist mathematics, however. Finitist mathematics is taken to be a philosophically unproblematic part of number theory and, in addition, to coincide with the part of mathematics accepted by Kronecker and Brouwer.[27] Not every formula of this "game" has a meaning but only those that correspond to finitist statements, i.e. universal sentences of the kind of Fermat's Theorem. For a precise description of the role of consistency proofs let **P** be a formal theory that allows the representation of classical mathematical practice and let **F** formulate the principles of finitist mathematics. The consistency of **P** is in **F** equivalent to the reflection principle

$$(\forall x)(\Pr(x, \text{'}s\text{'}) \Rightarrow s).$$

Pr is the finitistically formulated proof predicate for **P**, s a finitist statement, and 's' the corresponding formula in the language of **P**. A consistency proof in **F** was programmatically sought; it would show, because of the above equivalence, that the mere technical apparatus **P** can serve reliably as an instrument for the proof of finitist statements. After all, the consistency proof would allow to transform any P-derivation of 's' into a finitist proof of s (and thus give a quite systematic answer to Borel's question). Hilbert believed that consistency proofs would settle foundational problems — once and for all and by purely mathematical means. Bernays judged in (1922b, p. 19):

The great advantage of Hilbert's method is precisely this: the problems and difficulties that present themselves in the foundations of mathematics can be transferred from the epistemological-philosophical to the properly mathematical domain.

Because of Gödel's incompleteness theorems this advantage proved to be illusory, at least when finitist mathematics is contained in **P**:[28] for such Ps the Second Incompleteness Theorem just states that their consistency cannot be established by means formalizable in **P**. The radical restriction of what was "properly mathematical" had to be given up; a modification of the program was formulated and has been pursued successfully for parts of analysis.[29] The crucial tasks of this general reductive program are: (i) find

---

[27] See remark B in Section 2.3 below.

[28] And that is a more than plausible assumption for those Ps Hilbert wanted to investigate and that contain elementary number theory. Consider the practice of finitist mathematics, for example in volume I of *Grundlagen der Mathematik*, the explicit remarks on p. 42 of that book, but also the analyses given by (Kreisel 1965) and (Tait 1981).

[29] Hilbert and Bernays, Gentzen, Lorenzen, Schütte, Kreisel, Feferman, Tait, and many other logicians and mathematicians have contributed; for detailed references to the literature see (Buchholz, Feferman, Pohlers and Sieg 1981) or (Sieg 1985b).

an appropriate formal theory **P*** for a significant part of classical mathematical practice, (ii) formulate a "corresponding" constructive theory **F***, and (iii) prove in **F*** the partial reflection principle for **P***, i.e.

$$\Pr{}^*(d, \text{'}s\text{'} \Rightarrow s)$$

for each **P***-derivation d. $\Pr{}^*$ is here the proof-predicate of **P*** and s an element of some class F of formulas. The provability of the partial-reflection principle implies the consistency of **P*** relative to **F***. (For the theories considered here, this result entails that **P*** is conservative over **F*** for all formulas in F.) Gödel and Gentzen's consistency proof of classical number theory relative to Heyting's formalization of intuitionist number theory was the first contribution to the reductive program; as a matter of fact, their result made that program at all plausible.

I do not intend to sketch the development of proof theory and, consequently, I will comment only on some central results concerning theories for the mathematical continuum. Second-order arithmetic was taken by Hilbert and Bernays as the formal framework for analysis. The essential set theoretic principles are the comprehension principle

$$(\exists X)(\forall y)(y \in X \Leftrightarrow S(y))$$

and forms of the axiom of choice

$$(\forall x)(\exists Y) S(x, Y) \Rightarrow (\exists Z)(\forall x) S(x, (Z)_x);$$

S is an arbitrary formula of the language and may in particular contain set quantifiers. These general principles are impredicative, as the sets X and Z whose existence is postulated are characterized by reference to all sets (of natural numbers). Subsystems of second-order arithmetic can be defined by restricting S to particular classes of formulas. The subsystems that have been proved consistent contain for example the comprehension principle for Π^1_1- and Δ^1_2-formulas; the latter have the shape $(\forall X) R$, respectively, are provably equivalent to formulas of the shape $(\forall X)(\exists Z) R$ and $(\exists Z)(\forall X) T$, where R and T are purely arithmetic.[30] These particular subsystems are of direct mathematical interest, as analysis can be formalized in them by (slightly) refining the presentation of Hilbert and Bernays in supplement IV of *Grundlagen der Mathematik II*. The proof theoretic investigations have been accompanied by mathematical ones, showing that even weaker subsystems will do. Really surprising refinements have been obtained during the last fifteen years: all of classical analysis can be formalized in conservative extensions of elementary number theory, significant parts also of algebra already in conservative extensions of primitive recursive arithmetic.[31]

---

[30] For details concerning the theories with versions of the axiom of choice see (Feferman and Sieg 1981a). The character of the consistency proofs is indicated below. Foundationally significant results are also described in (Feferman 1988a).

[31] For the discussion of these results and detailed references to the literature see (Simpson 1988) and (Sieg 1988).

Relative consistency and accessible domains 311

These two complexes of results indicate corresponding complexes of problems for future development; namely, (1) to give constructive consistency proofs of stronger subsystems of analysis, first of all for the system with Π_2^1-comprehension, and (2) to find weaker, but mathematically still significant subsystems (whose consistency is easily seen from the finitist standpoint and) whose provably recursive functions are in complexity classes. These are mathematically and logically most fascinating problems.

2.3 Remarks. They are partly of historical, partly of systematic character and concern mostly the axiomatic method underlying Hilbert's program. Though they are intended to ease the transition to the philosophical reflections in the second part of this essay, they are also to defuse some of the widely held misconceptions of the program, e.g., its "crude formalism" or its *ad hoc* character to serve as a "weapon" against Brouwer's intuitionism.

2.3.A Paradoxical background. The concern with consistent sets and the explicit use of Cantorian terminology in *Über den Zahlbegriff* show clearly that Hilbert was informed about the set theoretic difficulties Cantor had found and communicated to Dedekind in the famous letter of 28 July 1899. The recently published earlier letters of Cantor's to Hilbert I mentioned above throw light on this background. (They provide also surprising new information on the early history of the set theoretic paradoxes and on the circumstances surrounding Cantor's letter to Dedekind.) There is no doubt that Hilbert was prompted by these difficulties to think seriously about foundational issues. After all, as I pointed out, he recognized the impact of Cantor's observations on Dedekind's logical foundations of arithmetic presented in *Was sind und was sollen die Zahlen?* Here I just want to recall Dedekind's reaction to the Cantorian problems, reported in a letter of Felix Bernstein to Emmy Noether and published in (Dedekind 1932). Bernstein had visited Dedekind on Cantor's request in the spring of 1897. The express purpose was to find out what Dedekind thought about the paradox of the system of all things; Cantor had informed Dedekind about it already by letter in 1896. Bernstein reports: "Dedekind had not arrived yet at a definite position and told me, that in his reflections he almost arrived at doubts, whether human thinking is completely rational."[32] Strong words from a man as sober and clearheaded as Dedekind.

---

[32](Dedekind 1932, p. 449). Even six years later, in 1903, Dedekind still had such strong doubts that he did not allow a reprinting of his booklet. In 1911, he consented to a republication and wrote in the preface, "Die Bedeutung und teilweise Berechtigung dieser Zweifel verkenne ich auch jetzt nicht. Aber mein Vertrauen in die innere Harmonie unserer Logik ist dadurch nicht erschüttert; ich glaube, daß eine strenge Untersuchung der Schöpferkraft des Geistes, aus bestimmten Elementen ein neues Bestimmtes, ihr System zu erschaffen, das notwendig von jedem dieser Elemente verschieden ist, gewiß dazu führen wird, die Grundlagen meiner Schrift einwandfrei zu gestalten." (Dedekind 1932, p. 343).

2.3.B Assumption. How is it possible to reconcile Hilbert's programmatic formalism with his deep trust in the correctness of classical mathematics? — Most easily, when the formal theories of central significance are complete or deductively closed, as the Hilbertians used to say. This completeness assumption is already found in *Über den Zahlbegriff*. Hilbert writes there that the set of real numbers should be thought of as ". . . a system of things, whose mutual relations are given by the above finite and closed system of axioms, and for which statements have validity, only if they can be deduced from those axioms by means of a finite number of logical inferences". Hilbert talks about a non-formalized axiomatic theory. But if it is adequately represented by a formal theory P, then P must naturally be deductively closed. As a matter of fact, it was believed in the Hilbert school — until Gödel's incompleteness theorems became known — that the formalisms for elementary number theory and analysis were complete. For the purpose of obtaining a completeness proof Hilbert suggested in his Bologna address (1928) to reinterpret finitistically the familiar arguments for the categoricity of the Peano-axioms and of his axioms concerning the real numbers. The assumed completeness and the ensuing harmony of provability and truth help understand how Hilbert could take his radical formalist position, in order to simply bypass the epistemological problems associated with the classical infinite structures.[33]—The finitist mathematical basis was thought to be co-extensive with the part of arithmetic accepted by Kronecker and Brouwer. As to Kronecker, Hilbert mentions in his (1931a): "At about the same time [i.e. at the time of Dedekind's (1888)] . . . Kronecker formulated most clearly a view, and illustrated it by numerous examples, that essentially coincides with our finitist standpoint." The relation to intuitionism is discussed explicitly at a great number of places by Bernays; e.g. (Bernays 1967, p. 502). A particularly concise formulation was given by Johann von Neumann in his (1931, pp. 116–7).

2.3.C Doubts. Two mathematicians with quite different foundational views criticized Hilbert's formalism at exactly this point; i.e. they criticized the assumption that parts of mathematics can be represented (completely) by formal theories. The first of them was Brouwer, the second Zermelo.

Brouwer used in his development of analysis infinite proofs and treated them mathematically as well-founded trees. He wrote with respect to them: "These *mental* mathematical proofs that in general contain infinitely many terms must not be confused with their linguistic accompaniments, which are finite and necessarily inadequate, hence do not belong to mathematics."[34] He added that this remark contains his "main argument against the claims of Hilbert's metamathematics". The well-founded trees of Brouwer's can be viewed as inductively generated sets of sequences of natural

---

[33]Compare (Hilbert *1929, pp. 14–15), (Bernays 1930b, pp. 59–60), and the discussion in Section 3.1.

[34]In (Brouwer 1927, footnote 8, p. 460).

numbers; that is the essential claim of the bar-theorem. In the case of the constructive ordinals the inductive generation proceeds by the following rules: $0 \in \mathbf{O}$, $\alpha \in \mathbf{O} \Rightarrow \alpha' \in \mathbf{O}$, $(\forall n)\alpha_n \in \mathbf{O} \Rightarrow \alpha := \sup \alpha_n \in \mathbf{O}$. Notice, it is the bar-theorem together with the continuity principle that implies the fan-theorem and thus the properties of the intuitionist continuum so peculiar from a classical point of view, e.g., the uniform continuity of all real-valued functions on the closed unit interval.

Also Zermelo claimed that finite linguistic means are inadequate to capture the nature of mathematics and mathematical proof. In a brief note he argued: "Complexes of signs are not, as some assume, the true subject matter of mathematics, but rather *conceptually ideal relations* between the elements of a conceptually posited *infinite manifold*. And our systems of signs are only *imperfect* and *auxiliary means* of our *finite* mind, changing from case to case, in order to master at least in stepwise approximation the infinite, that we cannot survey *directly* and *intuitively*."[35] Zermelo suggested using an infinitary logic to overcome finitist restrictions. The concept of well-foundedness is fundamental for Zermelo's infinitary logic as well, but in an unrestricted set theoretic framework. Zermelo's investigations of infinitary systems can be found in (Zermelo 1935).

2.3.D Reduction. Ironically, the constructive consistency proofs of impredicative theories, mentioned at the end of Section 1.2, use infinitary logical calculi; but the syntactic objects constituting them (namely, formulas and derivations) are treated in harmony with intuitionist principles. The theories $\mathbf{F}^*$ — in which the infinitary calculi are investigated and to which the impredicative theories are reduced — are extensions of intuitionist number theory by definition and proof principles for constructive ordinals or other accessible i.d. [inductively defined] classes of natural numbers.[36] The above process of inductive generation for constructive ordinals can be expressed by an arithmetical formula $A(X, x)$. The two crucial principles are

(O1) $(\forall x)(A(\mathbf{O}, x) \Rightarrow \mathbf{O}(x))$, and

(O2) $(\forall x)(A(\mathrm{F}, x) \Rightarrow \mathrm{F}(x)) \Rightarrow (\forall x)(\mathbf{O}(x) \Rightarrow \mathrm{F}(x))$.

The former expresses the closure principle for $\mathbf{O}$, the latter the appropriate induction schema for any formula F. These principles and corresponding ones for other inductively defined classes are correct from an intuitionist point of view; the theories $\mathbf{F}^*$ are based on intuitionist logic. Because of

---

[35](Zermelo 1931, p. 85).

[36]See (Buchholz et al. 1981) and for an informal introduction the second part of (Sieg 1984).—*Accessible* or *deterministic* i.d. classes are distinguished by the fact that all their elements have unique construction trees. If the construction trees for all elements of an i.d. class are finite, we say that the class is given by a finitary inductive definition. For a detailed discussion of these notions see (Feferman and Sieg 1981a, pp. 22–23), (Feferman 1982), and (Feferman 1989).

these facts we can claim that the consistency of the impredicative theories has been established relative to constructive theories.

3 Philosophical reflections

The reductions of impredicative subsystems of analysis to intuitionist theories of higher number classes or other distinguished inductively defined classes are certainly significant results; were not all impredicative definitions supposed to contain vicious circles? The question is, nevertheless, what has been achieved in a general, philosophical way. Gödel remarked that giving a constructive consistency proof for classical mathematics means "to replace its axioms [i.e. those of classical mathematics] about abstract entities of an objective Platonic realm by insights about the given operations of our mind".[37] This pregnant formulation gives a most dramatic philosophical meaning to such proofs; it seems to me to be mistaken, however, in its radical opposition of classical and constructive mathematics and even in the very characterization of their subject matters. I prefer to formulate the task of such proofs as follows: they are to relate two aspects of mathematical experience; namely, the impression that mathematics has to do with abstract objects arranged in structures that are independent of us, and the conviction that the principles for some structures are evident, because we can grasp the build-up of their elements. I will argue that this is indeed central to the mediating task of the (modified) Hilbert program. The starting point of my argument is a reanalysis of the reductive goals of the original program; that will lead to the notion of "structural reduction" and to questions concerning its epistemological point.

3.1 Structural reduction. The description of Hilbert's program in Section 1.2 brings out, appropriately, the goal of justifying the instrumentalist use of classical theories for the proof of true finitist statements; it captures also important features of Hilbert's approach in a natural way, for example his concern with "Methodenreinheit" and the method of ideal elements. And yet, it truncates the program by leaving out essential and problematic considerations. Hilbert and Bernays both argue for a more direct mathematical significance of consistency proofs: such proofs are viewed as the last desideratum in justifying the existential supposition of infinite structures made by modern axiomatic theories.[38] It is clearly this concern that links the program to Hilbert's first foundational investigations and to Dedekind's attempted consistency proofs. Dedekind considers consistency proofs also as a last desideratum, but there seems to be a decisive difference

---

[37](Reid 1970, p. 218); compare also Gödel's remarks in (Wang 1974).

[38]"Existential supposition" is to correspond to the term "existentielle Setzung" that is used by Hilbert and Bernays as a quasitechnical term. The problem pointed to is presented as a central one in *Grundlagen der Mathematik I*; see the summary of the discussion on p. 19 of that work. As to the role of the reflection principle, compare the informative remarks on pp. 43–44.

as to the nature of theories: for him the theories (of natural and real numbers) are not just formal systems with some instrumentalist use. On the contrary, they are contentually motivated, have a materially founded necessity, and mathematical efficacy. They play an important epistemological role by giving us a conceptual grasp of composite mathematical as well as physical phenomena; Dedekind claims, for example, that it is only the theory of real numbers that enables us "to develop the conception of continuous space to a definite one".[39]

None of these points are lost in the considerations of Hilbert and Bernays. The contentual motivation of axiom systems, for example, plays a crucial role for them, as is clear from the very first chapter of *Grundlagen der Mathematik I* where the relation between contentual and formal axiomatics ("inhaltliche", respectively "formale Axiomatik") and its relevance for our knowledge is being discussed. "Formal axiomatics", they explain, "requires contentual axiomatics as a necessary supplement; it is only the latter that guides us in the selection of formalisms and moreover provides directions for applying an already given formal theory to an objective domain."[40] The basic conviction is that the contentual axiomatic theories are fully formalizable; formalisms, according to Hilbert (1927), provide "a picture of the whole science". Bernays (1930b) discusses the completeness problem in detail and conjectures that elementary number theory is complete. Though there is "a wide field of considerable problems", Bernays claims, "this 'problematic' is not an objection against the standpoint taken by us". He continues, arguing as it were against the doubts of Brouwer and Zermelo:

We only have to realize that the [syntactic] formalism of statements and proofs we use to represent our conceptions does not coincide with the [mathematical] formalism of the structure we intend in our thinking. The [syntactic] formalism suffices to

---

[39](Dedekind 1932, p. 340).—The underlying general position is persuasively presented in (Dedekind 1854). Dedekind viewed it as distinctive for the sciences (not just the natural ones) to strive for "characteristic" and "efficacious" basic notions; the latter are needed for the formulation of general truths. The truths themselves have, in turn, an effect on the formation of basic notions: they may have been too narrow or too wide, they may require a change so that they can "extend their efficacy and range to a greater domain". Dedekind continues, and that just cannot be adequately translated: "Dieses Drehen und Wenden der Definitionen, den aufgefundenen Gesetzen und Wahrheiten zuliebe, in denen sie eine Rolle spielen, bildet die größte Kunst des Systematikers." In mathematics we encounter the same phenomenon, e.g. when extending the definition of functions to greater number domains. In contrast to other sciences, however, mathematics does not leave any room for arbitrariness in how to extend definitions. Here the extensions follow with "compelling necessity", if one applies the principle that "laws, that emerged from the initial definitions and that are characteristic for the notions denoted by them, are viewed as generally valid; then these laws in turn become the source of the generalized definitions . . ." What a marvelous general description of his own later work in algebra (in particular the introduction of ideals) and in his foundational papers, the guiding idea of which is formulated clearly on p. 434 of this very essay.

[40] *Grundlagen der Mathematik I*, p. 2. — I translated by "an objective domain" the phrase "ein Gebiet der Tatsächlichkeit".—Similar remarks can be found in earlier, pre-Gödel papers; see especially the comprehensive and deeply philosophical (Bernays 1930b).

formulate our ideas of infinite manifolds and to draw the logical consequences from them, but in general it cannot combinatorially generate the manifold as it were out of itself.[41]

The close, but not too intimate connection between intended structure and syntactic formalism is to be exploited as the crucial means of reduction. This idea is captured in papers by Bernays through a mathematical image. (The papers are separated by almost fifty years; I emphasize this fact to point out that the remarks are not incidental, but touch the core of the strategy.) The first observation, from 1922, follows a discussion of Hilbert's *Grundlagen der Geometrie*.

Thus the axiomatic treatment of geometry amounts to this: one abstracts from geometry, given as the science of spatial figures, the purely mathematical component of knowledge [Erkenntnis]; the latter is then investigated separately all by itself. The spatial relations are *projected* as it were into the sphere of the mathematically abstract, where the structure of their interconnection presents itself as an object of purely mathematical thinking and is subjected to a manner of investigation focused exclusively on logical connections.

What is said here for geometry is stated for arithmetic in (Bernays 1922b) and for theories in general in (Bernays 1970), where a sketch of Hilbert's program is supplemented by a clear formulation of the epistemological significance of such "projections".

In taking the deductive structure of a formalized theory ... as an object of investigation the [contentual] theory is *projected* as it were into the number theoretic domain. The number theoretic structure thus obtained is in general essentially different from the structure intended by the [contentual] theory. But it [the number theoretic structure] can serve to recognize the consistency of the theory from a standpoint that is more elementary than the assumption of the intended structure.

Recalling that — according to Hilbert — the axiomatic method applies in identical ways to different domains, these projections have a uniform character. Thus Hilbert's program can be seen to seek a *uniform structural reduction*: intended structures are projected through their assumed complete formalizations into the properly mathematical domain (of Kronecker's and Brouwer's), i.e., finitist mathematics. The equivalence of consistency and satisfiability was claimed or at least conjectured;[42] consequently, it seemed that the existence of intended structures would be secured by the mathematical solution of the purely combinatorial consistency problem.

---

[41](Bernays 1930b, p. 59). The words in brackets were added by me to make the translation as clear as the German original.—The next longer quotations are taken from (Bernays 1922a, p. 96), and (Bernays 1970, p. 186).

[42]In (Bernays 1930b, p. 21), one finds the following phrase: "It is for this reason necessary to prove for every axiomatic theory the *satisfiability*, i.e., the consistency of its axioms." Compare also *Grundlagen der Mathematik I*, p. 19.—Mints and Friedman have shown that Gödel's completeness theorem for predicate logic can be established in a conservative extension of primitive recursive arithmetic. (That is an obvious improvement of Bernays's proof of the completeness theorem in elementary number theory.) The result is of considerable interest in this connection, as it justifies the equivalence of consistency and satisfiability from a finitist point of view — at least for formal first-order theories.

Relative consistency and accessible domains 317

The principles used in the solution were of course to be finitist; the epistemological gain of such reductions is described in *Grundlagen der Mathematik I*:

> Formal axiomatics, too, requires for the checking of deductions and the proof of consistency in any case certain evidences, but with the crucial difference [when compared to contentual axiomatics] that this evidence does not rest on a special epistemological relation to the particular domain, but rather is one and the same for any axiomatics; this evidence is the primitive manner of recognizing truths that is a prerequisite for any theoretical investigation whatsoever.[43]

This reconstruction of the intent of Hilbert's program is supported most explicitly by (Bernays 1922b) and (Bernays 1930b). Let me focus briefly on the earlier paper, not to report on all its detailed points, but rather to depict the structure of its argumentation. The problem faced by the program is seen in the following way. In providing a rigorous foundation for arithmetic (taken in a wide sense to include analysis and set theory) one proceeds axiomatically and starts out with the assumption of a system of objects satisfying certain structural conditions. But in the assumption of such a system "lies something so-to-speak transcendental for mathematics, and the question arises, which principled position is to be taken [towards that assumption]". Bernays considers two "natural positions". The first, attributed to Frege and Russell, attempts to prove the consistency of arithmetic by purely logical means; this attempt is judged to be a failure. The second position is seen in counterpoint to the logical foundations of arithmetic: As one does not succeed in establishing the logical necessity of the mathematical transcendental assumptions, one asks oneself, is it not possible "to simply do without them". Thus one attempts a constructive foundation, replacing existential assumptions by construction postulates; that is the second position and is associated with Kronecker, Poincaré, Brouwer, and Weyl. The methodological restrictions to which this position leads are viewed as unsatisfactory, as one is forced "to give up the most successful, most elegant, and most proven methods only because one does not have a foundation for them from a particular standpoint". Hilbert takes from these foundational positions what is "positively fruitful": from the first, the strict formalization of mathematical reasoning; from the second, the emphasis on constructions. Hilbert does not want to give up the constructive tendency, but — on the contrary — emphasizes it in the strongest possible terms. The program, as described in Section 1.2, is taken as the tool for an alternative constructive foundation of all of classical mathematics.

It is not the case — as is so often claimed — that the difficult philosophical problems brought out by the axiomatic method and the associated structural view of mathematics were not seen. They motivated the enterprise and were seen perfectly clearly; however, it was hoped, perhaps too naively, to either avoid them directly in a systematic-mathematical development

---

[43] *Grundlagen der Mathematik I*, p. 2. The parenthetical remark is mine.—Bernays uses the term "primitive Erkenntnisweise" that I tried to capture by the somewhat unwieldy phrase "primitive manner of recognizing truths".

(by presenting appropriate models) or to solve them in the case of "fundamental" structures on the finitist basis. In any case, a so-to-speak absolute epistemological reduction was envisioned. These radical, philosophically motivated aspirations of Hilbert's program were blocked by Gödel's incompleteness theorems: according to the first theorem it is not possible, even in the case of natural numbers, to exclude systematically all contentual considerations concerning the intended structure; the second theorem implies that formal theories can be used at most as vehicles for partial structural reductions to strengthenings of the finitist basis. Bernays wrote in the epilogue to his (1930b, p. 61):

> On the whole the situation is like this: Hilbert's proof theory — together with the discovery of the formalizability of mathematical theories — has opened a rich field of research, but the epistemological views that were taken for granted at its inception have become problematic.

At the inception of Hilbert's program, it seems, the epistemological views had not been dogmatically and unshakably fixed. As will be pointed out in the next section, Hilbert's original position had to be and was extended; in addition, to judge from (Bernays 1922b), the focus on finitist mathematics was viewed as part of an "Ansatz" to the solution of a problem. Having formulated the question as to a principled position towards the transcendental assumptions underlying the axiomatic foundations of arithmetic (see above), Bernays remarks, p. 11:

> Under this perspective[44] we are going to try, whether it is not possible to give a foundation to these transcendental assumptions in such a way that only primitive intuitive knowledge (primitive anschauliche Erkenntnisse) is used.

Viewing the philosophical position in this more experimental spirit, we can complement the metamathematical reductive program by a philosophical one that addresses two central issues: (i) what is the nature and the role of the reduced structures? and (ii) what is the special character of the theories to which they are reduced? As to the latter issue, our greater metamathematical experience allows us to point to perhaps significant general features.

3.2 Accessible domains. The reductive program I described in Section 1.2. has been pursued successfully. I think there can be no reasonable doubt that (meta)mathematically and, prima facie, also philosophically significant solutions have been obtained. As to the mathematical results it can be observed:

> A considerable portion of classical mathematical practice, including all of analysis, can be carried out in a small corner of Cantor's paradise that is consistent relative to the constructive principles formalized in intuitionist number theory. This is not trivial, if one bears in mind that in particular for analysis non-constructive principles seemed to be necessary.

---

[44] of taking into account the tendency of the exact sciences to use as far as possible only the most primitive "Erkenntnismittel". That does not mean, as Bernays emphasizes, to deny any other, stronger form of intuitive evidence.

Relative consistency and accessible domains 319

The metamathematical results concerning the relative consistency of impredicative theories speak also for themselves.

The constructive principles formalized in intuitionist theories for i.d. classes[45] allow us to recognize the relative consistency of certain impredicative theories. This is again not trivial, if one takes into account that any impredicative principle, from a broad constructive point of view, seemed to contain vicious circles.[46]

These relative consistency results provide material for critical philosophical analysis. After all, they raise implicitly the traditional question: "What is the (special) evidence of the mathematical principles used in (these) consistency proofs?"—The intuitionist theories for i.d. classes formulate complex principles that are recognized by classical and constructivist mathematicians alike. On the one hand they are more elementary than the principles used in their set theoretic justification, but on the other hand they cannot be given a direct (primitive) intuitive foundation. For a philosophical analysis that attempts to clarify extensions of the finitist standpoint and to explicate — relative to them — the epistemological significance of these particular results some clear and concrete tasks can be formulated.

At the very beginning of the development of Hilbert's program one finds an extension not of, but rather towards the finitist standpoint. Originally, Hilbert intended to make do with a mathematical basis that did not even include the "Allgemeinbegriff der Ziffer": all mathematical knowledge (Erkenntnis) was to be reduced to primitive formal evidence.[47] This extremely restricted undertaking was given up quickly: how could the central goal of the program, consistency, be formulated within its framework? A "finitist standpoint" that is to serve as the basis for Hilbert's investigations cannot be founded on the intuition of concretely given objects; it rather has to correspond to a standpoint, as Bernays explained, "where one already reflects on the general characteristics of intuitive objects".[48] A first task presents itself.

---

[45] i.e., very special ones: higher number classes and, more generally, accessible i.d. classes.

[46] That point is clearly and forcefully made in Gödel's (1944); see pp. 455–456 in (Benacerraf and Putnam 1983).

[47] This is reported in (Bernays 1946, p. 91); an example of the form of this quite primitive evidence can be found l.c. p. 89, but compare also (Bernays 1961, p. 169). As to the historical point see (Bernays 1967, p. 500): "At the time of his Zürich lecture Hilbert tended to restrict the methods of proof theoretic reasoning to the most primitive evidence. The apparent needs of proof theory induced him to adopt successively those suppositions that constitute what he then called the 'finite Einstellung'."

[48] From (Bernays 1930b, p. 40). The context is this: "Diese Heranziehung der Vorstellung des Endlichen [used from the finitist standpoint] gehört freilich nicht mehr zu demjenigen, was von der anschaulichen Evidenz notwendig in das logische Schließen eingeht. Sie entspricht vielmehr einem Standpunkt, bei dem man bereits auf die allgemeinen Charakterzüge der anschaulichen Objekte *reflektiert*."—This is a clearer and more promising starting point for an analysis than the one offered through Hilbert's own characterization in (van Heijenoort 1967, p. 376). Important investigations, in addition to those of Bernays, have been contributed by i.a. Kreisel, Parsons, and Tait. See (Tait 1981) and the references to the literature given there.

(I) Analyze this reflection for the natural numbers (and the elements of other accessible i.d. classes given by finitary inductive definitions) and investigate whether and how induction and recursion principles can be based on it.

Without attempting to summarize the extended (and subtle) discussion in (Bernays 1930b) I want to point to one feature that is crucial in it and important for my considerations. For Bernays the natural numbers (as ordinals) are the simplest formal objects; they are obtained by formal abstraction and are representable by concrete objects, numerals. This representation has a very special characteristic: the representing things contain ("enthalten") the essential properties of the represented things in such a way that relations between the latter objects obtain between the former and can be ascertained by considering those.[49] It is this special characteristic that has to be given up when extending the finitist standpoint: symbols are no longer carrying their meaning on their face, as they cannot exhibit their build-up.[50] For the consistency proofs mentioned in Remark 2.3.D above one uses accessible i.d. classes of natural numbers; numerals for the elements of such a class are now understood as denoting infinite objects, namely the unique construction trees associated effectively with the elements. So we have as a generalization of (I) — to begin with — the task:

(II) Extend the reflection to constructive ordinals and the elements of other accessible i.d. classes and investigate whether and how induction and recursion principles can be based on it.

One delicate question has not been taken into account here. For the consistency proofs of strong impredicative theories the definition of i.d. classes has to be iterated uniformly; that means the branching in the well-founded construction trees is not only taken over natural numbers, but also over

---

However, in this systematic context (and also for a general discussion of feasibility) I should point out that weakenings of the finitist standpoint are of real interest; a penetrating investigation is carried out by R. Gandy in his (1982). Notice that the type-token problematic has to be faced already from weakened positions.—The crux of the additional problematic was compressed by Bernays into one sentence: "Wollen wir ... die Ordnungszahlen als eindeutige Objekte, frei von allen unwesentlichen Zutaten haben, so müssen wir jeweils das bloße Schema der betreffenden Wiederholungsfigur als Objekt nehmen, was eine sehr hohe Abstraktion erfordert." (Bernays 1930b, pp. 31–32). It is for these formal objects that the "Gedankenexperimente" are carried out, that play such an important role in *Grundlagen der Mathematik I* (p. 32) for characterizing finitist considerations.

[49]This is found on pp. 31–32, in particular in footnote 4. The general problematic is also discussed in (Bernays 1935b, pp. 69–71). Compare the previous footnote to recognize that an isomorphic representation by a particular, physically realized object is not intended. The uniform character of the generation and the local structure of the schematic "iteration figure" are important.

[50]Gödel described in his (1958) a standpoint that extends the finitist one and that is appropriate for the consistency proofs for number theory given by Gentzen and Gödel himself. The starting point of this proposal are considerations of Bernays — e.g. in his (1935b) and (1941) — concerning the question, "In what way does intuitionism go beyond finitism?" Bernays's answer is, "Through its abstract notion of consequence." And it is this abstract concept that is to be partially captured by the computable functionals of finite type.—Note that the specifically finitist character of mathematical objects requires, according to Gödel, that they are "finite space-time configurations whose nature is irrelevant except for equality and difference". This seems to conflict with Bernays's analysis pointed to in the previous two footnotes.

other already obtained i.d. classes. These trees are of much greater complexity. For example, it is no longer possible — as it is in the case of constructive ordinals — to generate effectively arbitrary finite subtrees; that has to be done now through procedures that are effective relative to already obtained number classes. Thus we have to modify (II).

(III) Extend the reflection to uniformly iterated accessible i.d. classes, in particular to the higher constructive number classes.

Buchholz and Pohlers used systems of ordinal notations in their investigations of theories for i.d. classes. It is clearly in the tradition of Gentzen and Schütte to use for consistency proofs the principle of transfinite induction along suitable ordinals (represented through effective notation systems). But in parallel to proving the consistency of formal theories by such means, Gentzen wrote in a letter to Bernays of 3 March 1936, one has to pursue a complementary task, namely "... to carry out investigations with the goal of making the validity of transfinite induction constructively intelligible for higher and higher limit numbers". It is only through such investigations that the philosophical point of consistency proofs can be made, namely, to secure a theory by reliable ("sichere") means. Gentzen's task is included in (III), since the systems of notations used in Buchholz and Pohlers's work are generated as accessible i.d. classes, and their well-ordering is recognized through the proof principle for these i.d. classes. These systems of notations were quite complicated, but in their latest and conceptually best form they are given by clauses of the same character as those for higher number classes (Buchholz 1990). I mentioned earlier the consistency problem for the subsystem of analysis with Π_2^1-comprehension; this is not only a mathematical problem, but also an open conceptual problem, as new "constructive" objects are needed for a satisfactory "constructive" solution.[51]

The number classes provide special cases in which generating procedures allow us to grasp the intrinsic build-up of mathematical objects.[52] Such an understanding is a fundamental and objective source of our knowledge of mathematical principles for the structures or domains constituted by those objects: is it not the case that the definition and proof principles follow directly this comprehended build-up? Clearly, we have to complement an analysis of this source — as requested in (I)–(III) — by formulating (the reasons for the choice of) suitable deductive frames in which the mathematical principles are embedded. Thus, there are substantial questions concerning the language, logic, and the exact formulation of schematic principles. But notice that for the concerns here these questions are not of primary importance. For example, the restriction to intuitionist logic is rather insignificant: the double-negation translation used by Gödel and

---

[51] Work of Pohlers and his students to extend the method of local predicativity make it most likely that a close connection to set theory (in particular the study of large cardinals and the fine structure of the constructible universe **L**) is emerging.

[52] By calling their build-up "intrinsic" I point again to the parallelism between the generating procedure and the structure of the intended object; compare the case under discussion here, for example, with that of the computable functionals of finite type.

Gentzen to prove the consistency of classical relative to intuitionist arithmetic can be extended to a variety of theories to yield relative consistency results. Indeed, Friedman showed for arithmetic, finite type theories, and Zermelo-Fraenkel set theory that the classical theories are Π_2^0-conservative over their intuitionist version. Using Friedman's strikingly simple techniques Feferman and Sieg (1981a) established such conservation results for some subsystems of analysis and also for the theories of iterated inductive definitions.[53] In the latter case it is the *further* restriction to accessible i.d. classes that is (technically difficult and) conceptually significant.

Disregarding the traditional constructive traits of the objects considered up to now we can extend the basic accessibility conception from i.d. classes of natural numbers to broader domains. A comprehensive framework for the "inductive or rule-governed generation" of mathematical objects is given in (Aczel 1977); it is indeed so general that it encompasses finitary i.d. classes, higher number classes, the set theoretic model of Feferman's theory T_0 of explicit mathematics and of other constructive theories (like Martin-Löf's), but also segments of the cumulative hierarchy. Clearly, not all of Aczel's i.d. classes have the distinctive feature of accessible i.d. classes; those whose elements do have unique associated well-founded "construction" trees are called deterministic and, here again, accessible. Segments of the cumulative hierarchy — that contain some ordinals $(0, \omega,$ or large cardinals) and are closed under the powerset, union, and replacement operations — are in this sense accessible: the uniquely determined transitive closure of their elements are "construction" trees.[54] Here, as above, we have the task of explicating (the difficulties in) our understanding of generation procedures. After all, for accessibility to have any cognitive significance such an understanding has to be assumed. The latter is in the present case relatively unproblematic, if we restrict attention to hereditarily finite sets; then we have an understanding of the combinatorial generation procedures and, in particular, of forming arbitrary subcollections. Indeed, ZF^-, i.e. Zermelo-Fraenkel set theory without the axiom of infinity, is equivalent to elementary number theory. The powerset operation is the critical generating principle; its strength when applied to infinite sets is highlighted by the fact that ZF without the powerset axiom is equivalent to second-order arithmetic.[55] But if we do assume an understanding of the set theoretic

---

[53]The theory of arithmetic properties and ramified systems were shown to be Π_2^0-conservative over their intuitionist versions in (Feferman and Sieg 1981a, pp. 57–59).—The generality of Friedman's techniques was brought out in (Leivant 1985).

[54]Here is the basis for $\in$-induction and recursion. It seems to me that in this context the discussion and results concerning Fraenkel's *Axiom of Restriction* would be quite pertinent.

[55]Using powerset one obtains not the elements of a subclass from a given set, but rather all subclasses in one fell swoop. It is this utter generality that creates a difficulty even when the given set is that of the natural numbers; see the comprehensive discussion of Bernays's views in (Müller 1981). The difficulty is very roughly this: in terms of the basic operations one does not have "prior" access to all elements of the powerset, unless one chooses a second-order formulation of replacement. That would allow the joining of arbitrary subcollections, but "arbitrary subcollection" has then to be understood in whatever sense the second-order

generation procedure for a segment of the cumulative hierarchy, then it is indeed the case that the axioms of ZF⁻ together with a suitable axiom of infinity "force themselves upon us as being true" — in Gödel's famous phrase; they just formulate the principles underlying the "construction" of the objects in this segment of the hierarchy.[56] In summary, we have a wealth of accessible domains, and it seems that we can understand the pertinent mathematical principles quasi-constructively, as we grasp the build-up of the objects constituting such structures.

3.3 Contrasts. The "ontological status" of mathematical objects has not been discussed. The reason is this: I agree with the subtle considerations of Bernays in the essay *Mathematische Existenz und Widerspruchsfreiheit* and suggest only one amendation, namely to distinguish the "methodical frames" (methodische Rahmen) by having their objects constitute accessible domains. The contrast between "platonist" and "constructivist" tendencies is then not localized in the stark opposition formulated by Gödel; it comes to light rather in refined distinctions concerning the admissibility of operations, of their iteration, and of deductive principles. In this way, it seems to me, methodical frames are not only distinguishable from each other, but also epistemologically differentiated from "abstract" theories formulated within particular frames. I want to focus on this differentiation now and contrast the quasi-constructive aspect of mathematical experience I have been analyzing to — what I suggest to call — its "conceptual" aspect. The latter aspect is most important for mathematical practice and understanding, but also for the sophisticated uses of mathematics in physics; it is quite independent of methodical frames.

As a first step let us consider Dedekind's way of comprehending the accessible domain of natural numbers. The informal analysis underlying *Was sind und was sollen die Zahlen?* described in his letter to Keferstein, (Dedekind 1890a), starts out with the question:

What are the mutually independent fundamental properties of the sequence N, that is, those properties that are not derivable from one another but from which all others follow? And how should we divest these properties of their specifically arithmetic character so that they are subsumed under more general notions and under activities of the understanding *without* which no thinking is possible at all but *with* which a foundation is provided for the reliability and completeness of proofs and for the construction of consistent notions and definitions?

---

variables are interpreted. — A focus on definable subsets leads to the ramified hierarchy, to Gödel's constructible sets, and to the consideration of subsystems of ZF. The investigations concerning subsystems of analysis can be turned into investigations of natural subsystems of set theory. That was done by G. Jäger. His work is presented in (Jäger 1986).

[56]This reason for accepting the axioms of ZF seems to be (at least) consonant with Gödel's analysis in his (1947) and does not rest on the strong Platonism in the later supplement of the paper. The conceptual kernel of this analysis goes back to Zermelo's penetrating paper of 1930. A discussion of the rich literature on the "iterative conception of set" is clearly not possible here. — Notice that the length of iteration is partly determined through the adopted axiom of infinity built into the base clause of the i.d. definition.

One is quickly led to infinite systems that contain a distinguished element 1 and are closed under a successor operation φ. Dedekind notes that such systems may contain non-standard "intruders" and that their exclusion from N was for him "one of the most difficult points" in his analysis; "its mastery required lengthy reflection".[57] The notion of chain allows him to give "an unambiguous conceptual foundation to the distinction between the [standard] elements n and the [nonstandard] t". By means of this notion he captures the informal understanding that the natural numbers are just those objects that are obtained from 1 by finite iteration of φ, or rather the objects arising from any simply infinite system "by entirely disregarding the special nature of its elements, and retaining only their distinguishability and considering exclusively those relations that obtain between them through the ordering mapping φ".[58] He continues: "Taking into account this freeing of the elements from every other content (abstraction), we can justifiably call the [natural] numbers a free creation of the human mind." How startlingly close is this final view of natural numbers to that arrived at by Bernays through "formal abstraction"!

For Dedekind the considerations (concerning the existence of infinite systems) guarantee that the notion "simply infinite system" does not contain an internal contradiction.[59] The "purely logical" and presumably reliable foundation did not, of course, allow this goal to be reached. In Section 1.1, I emphasized methodological parallels between Dedekind's treatment of the natural and real numbers; here I want to bring out a most striking difference. We just saw that Dedekind's analysis of natural numbers is based on a clear understanding of their accessibility through the successor operation. This understanding allows the distinction between standard objects and "intruders" and motivates directly the axioms for simply infinite systems. Given the build-up of the objects in their domains, it is quite obvious that any two simply infinite systems have to be isomorphic via a unique isomorphism. By way of contrast, consider the axioms for dense linear ordering without endpoints; their countable models are all isomorphic, but Cantor's back-and-forth argument for this fact exploits broad structural conditions and not the local build-up of objects. The last observation gives also the reason why these axioms do not have an "intended model": it is the accessibility of objects via operations not just the categoricity of a theory

---

[57]The minimalist understanding is taken for granted when it is claimed that the induction principle is evident from the finitist point of view. Thus, even this seemingly most elementary explanation of induction leaves us with a certain "impredicativity". "The same holds", Parsons rightly argues, "for other domains of objects obtained by iteration of operations yielding new objects, beginning with certain initial ones. It seems that the impredicativity will lose its significance only from points of view that leave it mysterious why mathematical induction is evident." (Parsons 1983, pp. 135–136).

[58]Section 73 of (Dedekind 1888).

[59]What is so astonishing in every rereading of Dedekind's essay is the conceptual clarity, the elegance and generality of its mathematical development. As to the latter, it really contains the general method of making monotone inductive definitions explicit. The treatment of recursive definitions is easily extendible; that it has to be restricted, in effect, to accessible i.d. classes is noted. Compare (Feferman and Sieg 1981a, footnotes 2 and 4 on p. 75).

that gives us such a model. Similar remarks apply to the reals, as the isomorphism between any two models of the axioms for complete, ordered fields is based on the topological completeness requirement, not any build-up of their elements. (That requirement guarantees the continuous extendibility of any isomorphism between their respective rationals.)

This point is perhaps brought out even more clearly by a classical theorem of Pontrjagin's, stating that connected, locally compact topological fields are either isomorphic to the reals, the complex numbers, or the quaternions. For this case Bourbaki's description, that the "individuality" of the objects in the classical structures is induced by the superposition of structural conditions, is so wonderfully apt; having presented the principal structures (order, algebraic, topological) he continues:

> Farther along we come finally to the theories properly called particular. In these the elements of the sets under consideration, which, in the general structures have remained entirely indeterminate, obtain a more definitely characterized individuality. At this point we merge with the theories of classical mathematics, the analysis of functions of real or complex variable, differential geometry, algebraic geometry, theory of numbers. But they have no longer their former autonomy; they have become crossroads, where several more general mathematical structures meet and react upon one another.[60]

Here we are dealing with abstract notions, distilled from mathematical practice for the purpose of comprehending complex connections, of making analogies precise, and to obtain a more profound understanding; it is in this way that the axiomatic method teaches us, as Bourbaki expressed it in Dedekind's spirit (l.c., p. 223),

> to look for the deep-lying reasons for such a discovery [that two, or several, quite distinct theories lend each other "unexpected support"], to find the common ideas of these theories, buried under the accumulation of details properly belonging to each of them, to bring these ideas forward and to put them in their proper light.

Notions like group, field, topological space, differentiable manifold fall into this category, and (relative) consistency proofs have here indeed the task of establishing the consistency of abstract notions relative to accessible domains. In Bourbaki's enterprise one might see this as being done relative to (a segment of) the cumulative hierarchy. But note, this consideration cuts across traditional divisions, as it pertains not only to notions of classical mathematics, but also to some of constructive mathematics. A prime example of the latter is that of a choice sequence introduced by Brouwer into intuitionist mathematics to capture the essence of the continuum; the consistency proof of the theory of choice sequences relative to the theory of (non-iterated) inductive definitions can be viewed as fulfilling exactly the above task.[61] The restriction of admissible operations (and deductive principles) can lead to the rejection of abstract notions; that

---

[60](Bourbaki 1950, p. 229). The natural numbers are not obtained at a crossroad.

[61](Kreisel and Troelstra 1970). It is really just the theory of the second constructive number class that is needed.

comes most poignantly to the surface in the philosophical dispute between Kronecker and Dedekind, but also in Bishop's derisive view of Brouwer's choice sequences. Bishop is not only scornful of the "metaphysical speculation" underlying the notion of choice sequence, but he also views the resulting mathematics as "bizarre". (Bishop 1967, p. 6)

4 Concluding remarks

The conceptual aspect of mathematical experience and its profound function in mathematics has been neglected almost completely in the logico-philosophical literature on the foundations of mathematics.[62] Abstract notions have been important for the internal development of mathematics, but also for sophisticated applications of mathematics in physics and other sciences to organize our experience of the world. It seems to me to be absolutely crucial to gain genuine insight into this dual role, if we want to bring into harmony, as we certainly should, philosophical reflections on mathematics with those on the sciences.

Results of mathematical logic do not give precise answers to large philosophical questions; but they can force us to think through philosophical positions. Broad philosophical considerations do not provide "foundations" for mathematics; but they can bring us to raise mathematical problems. We shall advance our understanding of mathematics only if we continue to develop the dialectic of mathematical investigation and philosophical reflection; a dialectic that has to be informed by crucial features of the historical development of its subject. In Brecht's *Galileo Galilei* one finds the remark:

A main reason for the poverty of the sciences is most often imagined wealth. It is not their aim to open a door to infinite wisdom, but rather to set bounds to infinite misunderstanding.[63]

What is said here for the sciences holds equally for mathematical logic and philosophy.

---

[62]The exception are papers of Bernays: there it is absolutely central.

[63]The German text is: "Eine Hauptursache der Armut der Wissenschaften ist meist eingebildeter Reichtum. Es ist nicht ihr Ziel, der unendlichen Weisheit eine Tür zu öffnen, sondern eine Grenze zu setzen dem unendlichen Irrtum." In its "application" to philosophy it mirrors (for me) the views of the man who influenced so deeply Dedekind, Kronecker, Hilbert, ... ; they are reported in (Kummer 1860, p. 340): "Er [Dirichlet] pflegte von der Philosophie zu sagen, es sei ein wesentlicher Mangel derselben, dass sie keine ungelösten Probleme habe wie die Mathematik, dass sie sich also keiner bestimmten Grenze bewusst sei, innerhalb deren sie die Wahrheit wirklich erforscht habe und über welche hinaus sie sich vorläufig bescheiden müsse, nichts zu wissen. Je grössere Ansprüche auf Allwissenheit die Philosophie machte, desto weniger vollkommen klar erkannte Wahrheit glaubte er ihr zutrauen zu dürfen, da er aus eigener Erfahrung in dem Gebiete seiner Wissenschaft wusste, wie schwer die Erkenntnis der Wahrheit ist, und welche Mühe und Arbeit es kostet, dieselbe auch nur einen Schritt weiter zu führen."

III Philosophical horizons

A brief guide

The last essay in the previous part, (II.10), introduced accessible domains and described their importance for relative consistency proofs: if such proofs are to provide an epistemologically significant (structural) reduction, then the objects constituting these domains must be accessible to our imagination and the mathematical principles for them must be evident to our understanding. The first essay of this part, (III.1), *Aspects of mathematical experience*, contrasts the "constructive" aspect of mathematical experience (underlying accessibility) with a conceptual one; the latter reflects the use of abstract notions that are introduced in order to make precise (in what respect) quite different areas of mathematics are "analogous".

Essay (II.10) formulates also specific tasks for philosophical analysis that focus on the *deterministic inductive generation* of the objects constituting accessible domains. One of these tasks is taken up in essay (III.2) (*Beyond Hilbert's reach?*), after a description has been given of the evolution of Hilbert's programs and their connection to the radical transformation of mathematics in the 19th century. The starting point is (Bernays 1930b), where Bernays tries to characterize mathematical knowledge in a principled way; crucial is mathematical abstraction emphasizing "the structural moments of an object, i.e., the way it is composed from parts, and taking them exclusively into consideration". That leads in complex, and sometimes conflicting, ways to formal objects, in particular to finite ordinals that make up a very special accessible domain.

The first essay in this part addresses issues that were kept in the background when taking for granted that the incompleteness theorems apply to *all* formal theories (satisfying standard conditions). It is the analysis of computability, and in particular Turing's analysis, that allows an adequate characterization of formality. It opened for Gödel, in his Gibbs lecture, a way of arguing for the superiority of the human mind over finite machines; it opened for Turing new ways of intellectual experimentation; it opens for us ways of uncovering capacities of the mathematical mind. That is at the heart of the last essay, *Searching for proofs*: what shapes mathematical arguments into proofs that are intelligible to us, and what allows us to find proofs efficiently? – One part of the answer leads us back to Dedekind.

Paul Bernays

III.1

Aspects of mathematical experience[1]

1

In the current discussion on philosophy of mathematics some do as if systematic foundational work supported an exclusive alternative between Platonism and Constructivism; others do as if such mathematical and logical research were deeply misguided and had no bearing on our understanding of mathematics. Both attitudes prevent us from grasping insights that underlie such work and from appreciating significant results that have been obtained. In consequence, they keep us from turning attention to the task of understanding the role of *accessible domains for foundational theories* and that of *abstract structures for mathematical practice*.

This twofold task derives from a probing perspective that takes seriously traditional epistemological concerns, but that does not respect time-honored boundaries drawn for philosophical convenience. It will be approached mainly through work that has been done during the last seventy years on versions of Hilbert's Program. Such an avenue may be surprising, because the stand that was taken in the foundational discussion by Hilbert, Bernays, and their collaborators is widely perceived as extremely narrow and technical. So I will give in section 2 a revisionary description of Hilbert's Program and sketch in section 3 some results that have been obtained within a *general reductive program*.

A prerequisite for Hilbert's Program is the effective or formal presentation of mathematical thought. Gödel took his incompleteness theorems as refuting any form of "pure formalism", in particular the variety (he thought to be) underlying Hilbert's Program. The discussion of Gödel's reflections on this issue, in section 4, will lead me to focus on two aspects of mathematical experience. The first is the *quasi-constructive aspect*, and it has to do with accessible domains; the second is the *conceptual aspect*, and it deals with axiomatically characterized abstract structures. These two aspects are

---

[1]My considerations are based on two papers of mine: the first is (Sieg 1990a) and the second is (Sieg 1994a). [The first is reprinted in this volume as (II.10).] Here, I focus squarely on broader strategic points and rely for details concerning the relevant (meta) mathematical results, historical connections, and conceptual analyses on those earlier papers.

discussed in sections 5 and 6. In the seventh and final section I come back to the question of "mechanizing" mathematical thought and contrast Turing's views with Gödel's.

2

Are the results of contemporary proof theory significant for the foundational concerns that motivated Hilbert's Program, and are those concerns connected to insightful reflections on the nature of mathematics? — Before we can assess answers to either question, we have to be clear about the specific foundational concerns and the general character of solutions proposed by the program. The broad background is provided by striking developments in 19th century mathematics, namely the emergence of set theory, the discovery of set theoretic foundations for analysis, and the rise of modern axiomatics with a distinctive structuralist bent. These developments came to the fore in Cantor's and Dedekind's work. Some of the difficult issues connected with them were seen by Cantor, others were made explicit by Dedekind and by Kronecker (when criticizing Dedekind); and they clearly prompted Hilbert's foundational studies in the late 1890's. Dedekind and Kronecker were both deeply influenced by Dirichlet; their divergent development of algebraic number theory together with their general reflections pinpoint the central issues most clearly. Hilbert's Program was formulated only in the early twenties, but it evolved out of this earlier "problematic".

The program was to mediate between the opposing foundational views represented by Dedekind and Kronecker, and it was to address a methodological problem, to wit, the use of 'abstract' analytic means in proofs of 'concrete' number theoretic results (employed first by Dirichlet). The expressibility of parts of classical mathematics in axiomatic systems P was the basic datum for conceiving of *consistency proofs* for **P** as a possibly convincing approach. The basic datum, after such **P**'s had been sharpened to formal theories, allowed Hilbert to think of classical mathematics *programmatically* as a formula game and, thus, of the consistency problem as a syntactic one. In this way Hilbert side-stepped the philosophical problems associated with the content of **P** and turned to restrictive demands on consistency proofs. He required that they be given in finitist mathematics, believed to be a philosophically unproblematic part of number theory and to coincide with the mathematics accepted by Kronecker and Brouwer.

To describe the role of consistency proofs in greater detail, let **P** be a formal theory in which the practice of classical mathematics can be represented, and let **F** formulate the principles of finitist mathematics. The formal character of **P** allowed Hilbert to express in the language of **F** the proof-predicate Pr for **P** and thus **P**'s consistency; Hilbert expected that an elementary proof of this elementary statement could be found! The consistency of **P** can actually be shown in **F** to be equivalent to the reflection principle

$(\forall x)(\Pr(x, \text{'}s\text{'}) \to s);$

where s is a finitist statement, and 's' is the corresponding formula in the language of **P**. A consistency proof in **F** would show, because of this equivalence, that the technical apparatus **P** can serve as a reliable instrument for establishing true finitist statements; after all, it would allow the transformation of any **P**-derivation of 's' into an **F**-proof of s. Finitist consistency proofs would thus resolve the methodological problem mentioned above, guaranteeing *Methodenreinheit* in a systematic manner. *And yet*, describing the program in this way truncates it by leaving out essential and problematic considerations.

Hilbert and Bernays attributed to consistency proofs a further philosophical significance: such proofs were thought to provide the last desideratum for *justifying* the *existential supposition* of infinite structures made by modern axiomatic theories.[2] This issue links the program of the twenties to Hilbert's first foundational investigations. At the core of the strategic considerations was the perceived close connection between (truth in) mathematical structures and (provability in) syntactic formalisms; this connection was to be exploited as a crucial means of "reduction". Bernays expressed the central idea repeatedly through a mathematical image (in papers that span close to fifty years). In (1922a) he observed after a discussion of Hilbert's *Grundlagen der Geometrie*:

Thus the axiomatic treatment of geometry amounts to this: one abstracts from geometry, given as the science of spatial figures, the purely mathematical component of knowledge [Erkenntnis]; the latter is then investigated separately all by itself. The spatial relations are *projected* as it were into the sphere of the mathematically abstract, where the structure of their interconnection presents itself as an object of purely mathematical thinking and is subjected to a manner of investigation focused exclusively on logical connections. (p. 96)

What is said here for geometry was stated in (Bernays 1922b) for arithmetic and in (Bernays 1970) for formal theories in general: "Taking the deductive structure of a formalized theory . . . as an object of investigation, the [contentual] theory is *projected* as it were into the number-theoretic domain." The number-theoretic structure obtained in this way, Bernays emphasized then, will usually be different from the structure intended by the theory in essential ways. And yet, the projection has a point, because ". . . [the number-theoretic structure] can serve to recognize the consistency of the theory from a standpoint that is more elementary than the assumption of the intended structure." Since Hilbert saw the axiomatic method as applying in identical ways to different domains, these projections are epistemologically uniform. That is explicitly described in *Grundlagen der Mathematik I*:

---

[2]"Existential supposition" is to correspond to the term "*existentielle Setzung*" that is used by Hilbert and Bernays as a quasi-technical term. The problem pointed to is presented as a central one in *Grundlagen der Mathematik I*; see p. 19 of that work.

Formal axiomatics, too, requires for the checking of deductions and the proof of consistency in any case certain evidences, but with the crucial difference [when compared to contentual axiomatics] that this evidence does not rest on a special epistemological relation to the particular domain, but rather is one and the same for any axiomatics; this evidence is the primitive manner of recognizing truths that is a prerequisite for any theoretical investigation whatsoever.[3]

Hilbert's Program is thus seen to aim for *uniform structural reductions*: arbitrary mathematical structures are projected through their presumably "complete" formalizations into the properly mathematical domain of finitist mathematics. As the equivalence of consistency and satisfiability was assumed, or at least conjectured, the existence of such structures seemed to be secured by solving the finitist consistency problem.

It is often claimed that the difficult philosophical problems inherent in the axiomatic method and the associated structuralist view of mathematics were not addressed in the Hilbert school. That is incorrect; those problems were seen clearly and indeed motivated the enterprise. But Hilbert and Bernays hoped, perhaps too naively, either to avoid them in a systematic-mathematical development by appropriate interpretations or to solve them for fundamental structures by finitist arguments. In any case, they envisioned an absolute reduction to a basis that was viewed by them, to repeat, as "the prerequisite for any theoretical investigation whatsoever". I assume, it is this clear reductive and philosophically motivated goal (to be reached by purely mathematical means) that made Hilbert's Program attractive; even Gödel admitted in his 1938 lecture at Zilsel's: "If the original Hilbert program could have been carried out, that would have been without any doubt of enormous epistemological value. The following requirements would both have been satisfied: (*A*) Mathematics would have been reduced to a very small part of itself . . . . (*B*) Everything would really have been reduced to a concrete basis, on which everyone must be able to agree." Note that (*B*) paraphrases Hilbert's characterization of the finitist basis in his (1926).

3

Gödel's Incompleteness Theorems blocked, however, the radical aspiration of Hilbert's Program. If the program was to be pursued in *some* form, the sharp restriction of the "properly mathematical" domain had to be given up; Gödel's second theorem implies, after all, that structural reductions even for arithmetic can be obtained only by strengthening the finitist basis. A suitable modification of the program has been pursued with remarkable success.[4] The crucial tasks of this *general reductive program* are: (1) find a

---

[3]*Grundlagen der Mathematik I*, p. 2. The parenthetical remark is mine. — Hilbert and Bernays use the term "*primitive Erkenntnisweise*" which I tried to capture by the somewhat unwieldy phrase "primitive manner of recognizing truths".

[4]Hilbert and Bernays, Ackermann, von Neumann, Herbrand, Gödel, Gentzen, Schütte, Kreisel, Feferman, Tait, Takeuti, and many others contributed. For references and detailed

Aspects of mathematical experience

formal theory $\mathbf{P}^*$ for a significant part of classical mathematical practice, (2) formulate an unmistakably constructive theory $\mathbf{F}^*$, and (3) prove in $\mathbf{F}^*$ the *partial reflection principle* for $\mathbf{P}^*$, i.e.,

$$\mathrm{Pr}^*(d,\text{'}s\text{'}) \to s$$

for each $\mathbf{P}^*$-derivation d. Pr^* is the proof-predicate of $\mathbf{P}^*$, and s is an element of some class of formulas in the language of $\mathbf{F}^*$. The provability of this partial reflection principle implies the consistency of $\mathbf{P}^*$ *relative* to $\mathbf{F}^*$. Clearly, for such a result to be of foundational significance, $\mathbf{F}^*$ must be philosophically distinguished. The first contributions to the reductive program were the proofs given by Gödel, respectively Bernays and Gentzen, who established independently the consistency of classical arithmetic relative to its intuitionist version; as a matter of fact, this result made the modification of Hilbert's Program at all plausible.

As I do not intend to sketch the development of proof theory, I will only comment on some central results concerning analysis, i.e., concerning theories for the mathematical continuum. Hilbert and Bernays considered analysis as the touchstone for the feasibility of the reductive program and took second order arithmetic as the framework for its formal development and its metamathematical investigation. In contemporary presentations the essential set theoretic principles are the comprehension principle

$$(\exists X)(\forall y)(y \in X \leftrightarrow S(y))$$

and forms of the axiom of choice

$$(\forall x)(\exists Y)S(x,Y) \leftrightarrow (\exists Z)(\forall x)S(x,(Z)_x);$$

here S is an *arbitrary* second order formula.[5] The principles in this general form are impredicative, as the sets X and Z whose existence is postulated are characterized by reference to *all* sets of natural numbers — if S contains set quantifiers. Subsystems of second order arithmetic are defined by restricting S to particular classes of formulas; and subsystems that have been proved consistent contain, for example, the impredicative comprehension principle for Π^1_1- and Δ^1_2-formulas.

These subsystems are of genuine mathematical interest, since analysis can be formalized in them by refining the presentation in Supplement IV of *Grundlagen der Mathematik II*. (This presentation goes actually back to lectures of Hilbert's starting with those given in the winter term 1917/18, when Bernays had just started to work with him on foundational matters.) Really surprising refinements have been obtained during the last twenty

---

discussions, in particular on the consistency proofs for impredicative theories, see (Buchholz, Feferman, Pohlers and Sieg 1981), (Sieg 1984), (Feferman 1988a), (Pohlers 1989), or (Rathjen 1991).

[5]The lower case variables range over natural numbers, the upper case variables over sets of natural numbers. In the axiom of choice, $y \in (Z)_x$ is understood as $\langle y, x \rangle \in Z$ and $\langle, \rangle$ is a pairing function.

years[6]: all of analysis can be formalized in conservative extensions of number theory (containing the comprehension principle for arithmetic formulas with set parameters); significant parts of analysis and of algebra can be developed already in conservative extensions of primitive recursive arithmetic, which is arguably the exact formal frame for finitist mathematics.[7] The further mathematical investigations, showing that ever weaker subsystems allow the formalization of at least significant parts of analysis, have been complemented by proof theoretic reductions of ever stronger subsystems of analysis to constructive theories. However, the treatment of full second order arithmetic is still an open issue: even the subsystem with Π_2^1-comprehension presents a formidable obstacle.

Before discussing the character of relative consistency proofs for impredicative theories, I want to recall that Brouwer's mathematical universe was richer than assumed in Göttingen. In his development of analysis Brouwer used infinite proofs and treated them mathematically as well-founded trees. Such trees can be viewed as inductively generated sets of sequences of natural numbers. For constructive ordinals the generation proceeds in a similar manner according to the rules $0 \in \mathbf{O}$, $\alpha \in \mathbf{O} \to \alpha' \in \mathbf{O}$, and $(\forall n)\alpha_n \in \mathbf{O} \to \alpha := \sup \alpha_n \in \mathbf{O}$. With respect to the infinite proofs Brouwer wrote in his (1927): "These *mental* mathematical proofs that in general contain infinitely many terms must not be confused with their linguistic accompaniments, which are finite and necessarily inadequate, hence do not belong to mathematics." (Footnote 8, p. 460) He added that this remark contains his "main argument against the claims of Hilbert's metamathematics".

The relative consistency proofs for impredicative theories I alluded to, ironically use infinitary logical calculi; the syntactic objects constituting them, i.e., infinitary formulas and derivations, are treated as well-founded trees in harmony with intuitionist principles. The theories $\mathbf{F}^*$, in which the infinitary calculi are investigated and to which the impredicative theories are reduced, are extensions of intuitionist number theory by definition and proof principles for constructive ordinals or other i.d. [inductively defined] classes of natural numbers. As the process of inductive generation for constructive ordinals can be expressed by an arithmetic formula $A(X,x)$, the two principles are in this case

(O1) $(\forall x)(A(\mathbf{O}, x) \to \mathbf{O}(x))$, and

(O2) $(\forall x)(A(F, x) \to F(x)) \to (\forall x)(\mathbf{O}(x) \to F(x))$.

These principles are correct from an intuitionist point of view.

There is no doubt that (meta) mathematically and prima facie also philosophically significant results have been obtained. As to the mathematical

---

[6] For references to the rich literature see (Feferman 1977), (Feferman 1988a), (Simpson 1988) and (Sieg 1990b).

[7] This is the basis for Simpson's version of Hilbert's Program that should better be called "Kronecker's Program"; see (Simpson 1988) and (Sieg 1990b). [The latter review is reprinted here as II.9.]

Aspects of mathematical experience 335

results it can be observed: a considerable portion of classical mathematical practice, including all of analysis, can be carried out in a small corner of Cantor's paradise that is consistent relative to the constructive principles formalized in intuitionist number theory. And this is not trivial, if one bears in mind that strong non-constructive principles seemed to be necessary for analysis. As to the metamathematical results it can be noted: the constructive principles formalized in intuitionist theories for special i.d. classes allow us to recognize the relative consistency of some impredicative theories. This is again not trivial, if one takes into account that any impredicative principle, from a broad constructive point of view, seemed to contain vicious circles.

The relative consistency proofs provide material for critical philosophical investigations. After all, they press on us the question, "What is the special evidence of the mathematical principles used in these proofs?" The principles for special i.d. classes are recognized by classical and constructivist mathematicians alike: they are more elementary than the principles used in their set theoretic justification, but they cannot be given a direct intuitive foundation. In section 5 I will formulate some tasks for an analysis that attempts to clarify the *objective underpinnings* for extensions of the finitist standpoint and to explicate, relative to them, the epistemological significance of particular results.

4

The task of assessing the epistemological significance of particular proof theoretic results was briefly taken up by Gödel: in his lecture at Zilsel's Seminar of 29 January 1938, he investigated several ways of extending the finitist basis and the possibility of proving the consistency of arithmetic and analysis on that basis. The lecture extended the considerations of a talk Gödel had given in Cambridge on 30 December 1933. In both lectures he was sympathetic to a reductive program of the sort I sketched; cf. the discussion in (Sieg and Parsons 1995, II.4 above). However, this line of research was not pursued and the underlying "methodische Einstellung" was not adopted by Gödel.[8] He tried later, most explicitly in his Gibbs Lecture of 1951, to use the incompleteness theorems as a starting point for an argument in favor of Platonism.[9] Central features of Gödel's argument are, first, the fact that formal theories are being investigated and, second, the belief that the concept of formality had been captured adequately through Turing's analysis. The first point is also important for the very formulation of Hilbert's Program, and the second is crucial for the generality of the incompleteness theorems — used for the program's refutation!

The insistence on the *effective* presentation or the *formal* nature of theories had been motivated by epistemological concerns; and it is quite clear

---

[8] Indeed, not properly appreciated.
[9] To understand this development in Gödel's views is most important, particularly in light of the critical remarks on Platonism made in his 1933 lecture quoted below.

that a restriction on our cognitive, more particularly mathematical, capacities had been intended. For this reason it is surprising that some of the logical pioneers interpreted the incompleteness and undecidability results in a quite dramatic way. Post, for example, emphasized in 1936 that these theorems exemplify "a fundamental discovery in the limitations of the mathematizing power of Homo Sapiens"; a few years later he remarked with respect to the same results:

Like the classical unsolvability proofs, these proofs are of unsolvability by means of given instruments. What is new is that in the recent case these instruments, in effect, seem to be the only instruments at man's disposal. (Post 1944, p. 310)

Turing's work provided for Gödel "a precise and unquestionably adequate definition of the general concept of formal system"; consequently, the incompleteness theorems hold for *arbitrary* formal systems (satisfying the usual conditions). Yet in contrast to Post, Gödel did not see them as establishing "any bounds for the powers of human reason, but rather for the potentialities of pure formalism in mathematics".[10]

In his Gibbs Lecture Gödel argued that human reason goes beyond the bounds for formalism in mathematics. To begin with he stated, if mathematics is viewed as a body of propositions that "hold in an absolute sense", then the incompleteness theorems bring to light the fact that mathematics is not exhaustible by a mechanical enumeration of its theorems. Already the first theorem yields for any consistent formal system **P**, containing a modicum of number theory, a simple arithmetic sentence that is independent of **P**. But he emphasized that the second theorem makes particularly evident the phenomenon of inexhaustibility.

For it makes it impossible that someone should set up a certain well-defined system of axioms and rules and consistently make the following assertion about it: all of these axioms and rules I perceive (with mathematical certitude) to be correct, and moreover I believe that they contain all of mathematics. (1951, pp. 5–6)

If someone claims this he contradicts himself, because recognizing the correctness of all the axioms and rules means recognizing the consistency of the system. Thus, a mathematical insight has been gained that does not follow from the axioms.

To explain the meaning of this situation Gödel distinguished between "objective" and "subjective" mathematics: objective mathematics is the body of all *true* mathematical propositions, subjective mathematics is that of all *humanly provable* ones. There clearly cannot be complete formal systems

---

[10](1964, pp. 72–73). As a footnote to this remark Gödel discussed in (1972) a "philosophical error in Turing's work". Gödel claimed that Turing intended to show that "mental procedures cannot go beyond mechanical procedures" and pointed to page 136 (in (Davis 1965)) of Turing's *On Computable Numbers*, where a very brief argument is to show that the number of states of mind that need be taken into account for a computation is finite. As mechanical procedures, not mental procedures in general, are analyzed there, I do not see a philosophical error in Turing's work, but rather in Gödel's interpretation.

Aspects of mathematical experience

for objective mathematics. For subjective mathematics the existence of a finite procedure yielding all its evident axioms cannot be excluded. But if there were such a procedure, then we could not be certain that all of the generated axioms are correct, and — as far as mathematics is concerned — the human mind would be equivalent to a Turing machine. Furthermore, there would exist simple arithmetical problems that could not be decided by a mathematical proof intelligible to the human mind. Calling such problems *absolutely undecidable* Gödel thus established: *either mathematics is inexhaustible in the sense that its evident axioms cannot be generated by a finite procedure* **or** *(in case there is a procedure generating the axioms of subjective mathematics) there are absolutely undecidable arithmetic problems.* (1951, p. 7)

This disjunction is of "great philosophical interest" to Gödel; not surprisingly, because he rejects the second alternative and explicates the first in the following way: ". . . that is to say, the human mind (even within the realm of pure mathematics) infinitely surpasses the powers of any finite machine." Gödel's elucidation of this remark invokes his Platonism; already in (1933b, p. 50) he had claimed that the axioms of set theory, "if interpreted as meaningful statements, necessarily presuppose a kind of Platonism". But at that time he added the relative clause "which cannot satisfy any critical mind and which does not even produce the conviction that they [the axioms of set theory] are consistent".

I would go too far afield, if I presented the reasons why I do not find Gödel's considerations for Platonism convincing. In any event, my criticism does not start with his treatment of set theory, but at the point where he contrasts the objects of finitist and intuitionist mathematics in his *Dialectica* paper of 1958. (The basic considerations go back to 1938 and 1941.) According to Gödel, finitist mathematical objects are required to be "finite space-time configurations whose nature is irrelevant except for equality and difference"; furthermore, in proofs of propositions concerning them one uses only insights that derive from the combinatorial space-time properties of sign combinations representing them.[11] These remarks, though consonant with Hilbert's very early views, stand in sharp conflict with Bernays's position to which Gödel appealed in his Dialectica paper. Bernays stressed already in 1930 the uniform character of the generation of natural numbers, the local structure of the schematic "iteration figure", and the need to "*reflect* on the general features of intuitive objects". Indeed, our *understanding* of natural numbers as being generated in a uniform way allows us to grasp *laws* concerning them. This observation is also correct for more general inductively defined classes, and it points to the first of two critical aspects of mathematical experience I want to describe now.

---

[11](Gödel 1958), in (Gödel 1995, p. 240). It is informative to compare this statement with the incorrect translation on p. 241 and, most significantly, with the corresponding remark in (Gödel 1972, p. 273). In the latter Gödel expanded "insights that derive from" by "a reflection upon" in this remark.

5

If one takes seriously the reformulation of the first alternative in Gödel's disjunction, one should try to see ways in which the human mind goes beyond the limits of mechanical computors. Gödel suggested in (1972) that there may be humanly effective, but non-mechanical mental procedures; yet even the most specific of his proposals, he admitted then, "would require a substantial advance in our understanding of the basic concepts of mathematics". That proposal concerned extensions of the cumulative hierarchy or, rather, of Zermelo Fraenkel set theory by axioms of infinity. The problem of extending what I call accessible domains is not special to the case of set theory: there are completely analogous issues, e.g., for the theory of primitive recursive functionals and for the theory of constructive ordinals.

Accessible domains comprise elements that are inductively and uniquely generated. They are most familiar from mathematics and logic: the natural numbers, the formulas of first order logic, the constructive ordinals, and the sets in segments of the cumulative hierarchy are generated in this way and form accessible domains. The generating procedures allow us in all these cases to grasp the build-up of the objects and to recognize mathematical principles for the domains constituted by just them. For it is the case, I suppose, that the definition and proof principles for such domains follow directly from the comprehended build-up.

A broad framework for the inductive generation of mathematical objects is described by Aczel (1977). It is indeed so general that it encompasses all the examples I mentioned, and allows us to compare and explicate the difficulties (in our understanding) of generating procedures. This echoes considerations of Gödel's in his 1933 lecture in Cambridge, when discussing varieties of constructive mathematics as follows:

> ... it is certainly true that there are different notions of constructivity and, accordingly, different layers of intuitionistic or constructive mathematics. As we ascend in the series of these layers, we are drawing nearer to ordinary non-constructive mathematics, and at the same time the methods of proof and construction are becoming less satisfactory and less convincing. (p. 51)

Let us continue, I suggest, the ascent to classical mathematics and investigate, in what way the methods of proof and construction are becoming "less satisfactory and less convincing"; let us consider, in particular, (extensions of) Zermelo Fraenkel set theory! It seems that, *if* we understand the generating procedure for a segment of the cumulative hierarchy, then it is the case that the axioms of ZF^*[12] together with a suitable axiom of infinity "force themselves upon us as being true" (in Gödel's famous phrase). They formulate, after all, the principles underlying the "construction" of the objects in the segment; this reason for accepting the axioms is consonant with Gödel's analysis in *What is Cantor's continuum problem?*[13] and

---

[12] ZF^* denotes ZF set theory without the axiom of infinity.
[13] The conceptual kernel of the analysis goes back to Zermelo's penetrating (1930).

Aspects of mathematical experience 339

does not rest on the Platonism advocated in the later supplement of the paper.

By broadening the range of foundational theories for relative consistency proofs from constructive to "quasi-constructive" ones and concentrating on one central feature of objects in the intended domains, namely accessibility, we can understand better what is characteristic of and considered as problematic in classical mathematics, and what is characteristic of and taken for granted as convincing in constructive mathematics. I want to raise issues concerning the second conjunct and start at a very elementary level. A finitist standpoint that is to serve as the basis for Hilbert's Program cannot be founded on just the intuition of concretely given objects, but has to incorporate *reflection* as Bernays explained in (1930b).[14] Thus a first task presents itself.

(I) Analyze Bernays's reflection for the natural numbers (and elements of other accessible i.d. classes given by finitary inductive definitions) and investigate, whether and how induction- and recursion principles can be based on it.

For Bernays, the natural numbers are the simplest formal objects that can be (partially) represented by concrete objects. That representation has a special feature: the representing things, numerals, contain the essential properties of the represented things in such a way that relations between the latter objects obtain between the former and can be ascertained by considering those. This feature has to be given up when we extend the finitist standpoint; symbols are no longer carrying their meaning on their face, as they cannot exhibit their intended build-up. Numerals for the elements of accessible i.d. classes, for example, are understood as denoting infinite objects, namely the unique construction trees associated effectively with the elements. So we generalize (I) to a second task:

(II) Extend the reflection to constructive ordinals and elements of other accessible i.d. classes and investigate, whether and how induction and recursion principles can be based on it.

For the consistency proofs of impredicative theories the definition of i.d. classes has to be iterated; that means, branchings in the well-founded construction trees are not only taken over natural numbers, but also over already obtained i.d. classes. These trees are of much greater complexity. Thus, modifying (II) we have a third task.

(III) Extend the reflection to iterated accessible i.d. classes, in particular to the higher constructive number classes.[15]

The reflective analyses have to be complemented by reasoned choices of deductive frames in which the mathematical principles are embedded. Thus, there are substantial questions concerning the language, logic, and the exact formulation of schematic principles; but these questions are of

---

[14] Cf. the discussion of Bernays's views at the end of section 4.

[15] As I am presenting only broad strategic considerations, I do not discuss the use of systems of ordinal notations in the work of Gentzen, Schütte, Feferman, e.a.; cf. (Sieg 1990b, p. 281).

only secondary importance for my concerns here. The restriction to intuitionist logic, for example, is rather insignificant, as the consistency proofs for classical arithmetic relative to intuitionist arithmetic can be extended to a variety of theories. Indeed, Friedman showed for arithmetic, finite type theories, and Zermelo Fraenkel set theory that the classical theories are Π_2^0-conservative over their intuitionist versions. With Friedman's strikingly simple techniques such results were also established for some subsystems of analysis and for the theories of iterated inductive definitions.[16] In the latter case it is the *further* restriction to accessible i.d. classes that is technically difficult and conceptually significant.

6

We have a wealth of accessible domains and seem to understand the pertinent mathematical principles, because we grasp the build-up of the objects constituting these domains. I did not discuss at all the ontological status of mathematical objects, as I agree with the subtle considerations of Bernays in his essay *Mathematische Existenz und Widerspruchsfreiheit* and suggest only one emendation: the objects of "methodical frames" (methodische Rahmen) should constitute accessible domains. In this way methodical frames may be epistemologically differentiated from each other and from "abstract" theories formulated within particular frames. I want to focus on this latter differentiation and contrast now the *quasi-constructive* aspect of mathematical experience (I sketched in the previous section) with its *conceptual* aspect.

In my paper *Relative consistency and accessible domains* (cf. note 1) I pointed out methodological parallels between Dedekind's treatment of natural and real numbers; here I want to emphasize a striking difference. Dedekind's analysis of natural numbers is based on a clear understanding of their accessibility through the successor operation. Given the build-up of objects in their domains, it is quite obvious that any two simply infinite systems have to be isomorphic, indeed, via a unique mapping. By way of contrast consider the axioms for dense linear orderings without endpoints; their countable models are all isomorphic, but Cantor's back-and-forth argument for this fact exploits the density condition and the nonexistence of endpoints, not any build-up of objects. This observation provides also the reason, why these axioms do not have an *intended model*: the accessibility of objects via operations gives us such models, not the categoricity of a theory. Similar remarks apply to the reals, as the isomorphism between any two models of the axioms for complete, ordered fields is based on the topological completeness requirement, not any build-up of their elements. The crucial point is illustrated even more clearly by a classical

---

[16]In (Feferman and Sieg 1981a, pp. 57–59); the subsystems that were shown to be Π_2^0-conservative over their intuitionist versions include the theory of arithmetic properties and ramified systems.

Aspects of mathematical experience 341

theorem of Pontrjagin's stating that connected, locally-compact topological fields are either isomorphic to the reals, the complex numbers, or the quaternions. For this case Bourbaki's description, that the individuality of the objects in the classical structures is induced by the superposition of structural conditions, is so wonderfully apt; having presented the principal structures (order, algebraic, topological) he continues:

> Farther along we come finally to the theories properly called particular. In these the elements of the sets under consideration, which, in the general structures have remained entirely indeterminate, obtain a more definitely characterized individuality. At this point we merge with the theories of classical mathematics, the analysis of functions of real or complex variable, differential geometry, algebraic geometry, theory of numbers. But they have no longer their former autonomy; they have become crossroads, where several more general mathematical structures meet and react upon one another. (Bourbaki 1950, p. 229)

The general structures fall under *abstract notions* that are distilled from mathematical practice for the purpose of comprehending complex connections (in the case of the reals, connections to geometry), of making analogies between different theories precise, and thus to obtain a more profound understanding. Notions like group, field, topological space, differentiable manifold are abstract in this sense, and *relative consistency proofs* have here indeed the *task of establishing the consistency of these notions relative to accessible domains*. Bourbaki's enterprise might be seen as being pursued relative to (a segment of) the cumulative hierarchy. The abstract, structural concepts are properly and in full generality investigated in category theory; Groethendieck's introduction of universes and MacLane's distinction between small and large categories can be viewed as attempts to establish the consistency of the general theory relative to extended segments of the cumulative hierarchy.[17] — These broad considerations pertain not only to notions of classical mathematics, but apply also to notions distinctive for constructive mathematics. A prime example is the (abstract, axiomatically characterized) concept of a choice sequence that was introduced by Brouwer into intuitionist mathematics in order to capture the essence of the continuum. Kreisel and Troelstra's consistency proof for the theory of choice sequences relative to the theory of O can be viewed as fulfilling exactly the above reductive task.

The conceptual aspect of mathematical experience and its profound function in mathematics has been entirely neglected in the logico-philosophical literature on the foundations of mathematics, except in the writings of Bernays. Among the major contributors to the foundational discussion in our century, it was Bernays who steered clear of divisionary formulations and emphasized the complementary character of seemingly conflicting aspects of mathematical experience (and philosophical positions). We have been discussing, implicitly and in his spirit, a redirected

---

[17] To review in this context the earlier discussion on the foundations of category theory seems very much worthwhile; cf. for example (Feferman 1969).

Hilbert Program searching for *structural reductions of abstract concepts to accessible domains*. Such structural reductions are most significant for any methodical frame: the traditional contrast between "Platonist" and "constructivist" tendencies in mathematics comes to light in refined distinctions concerning the admissibility of operations, of their iteration, and of deductive principles considered as fundamental for a particular frame.[18]

7

The sharpening of axiomatic theories to formal ones was motivated by normative epistemological demands: checking of proofs ought to be done in a radically intersubjective way and ought to involve only operations similar to those used by a human computor when carrying out an arithmetic calculation. Turing analyzed the processes underlying such operations and formulated a notion of computability by means of his machines; that was in 1936. In a paper written about ten years later and entitled *Intelligent Machinery*, he stated what really is *the* problem of cognitive psychology:

> If the untrained infant's mind is to become an intelligent one, it must acquire both discipline and initiative. ... But discipline is certainly not enough in itself to produce intelligence. That which is required in addition we call initiative. This statement will have to serve as a definition. Our task is to discover the nature of this residue as it occurs in man, and to try and copy it in machines. (Turing 1948a, p. 21)

The task of copying may be difficult, and Gödel would argue that it is impossible for mathematical thinking. But before we can start copying, we have to discover at least partially the nature of the residue, and we are led back to the questions: What are essential aspects of mathematical experience? Are they mechanizable?

I tried to give a very tentative and partial answer to the first question. As far as the second question is concerned, I don't have even a conjecture on how it will be answered. To come closer to an answer, we should investigate aspects of mathematical experience vigorously: by historical case studies, theoretical analyses, psychological experimentation, and by machine simulation. That the latter is still a real issue is counter to Turing's expectations. In 1947 he expressed this view:

> As regards mathematical philosophy, since the machines will be doing more and more mathematics themselves, the centre of gravity of the human interest will be driven further and further into philosophical questions of what can in principle be done etc. (Turing 1947, p. 122)

Even now, machines don't do much mathematics themselves — when doing mathematics includes: finding intelligible proofs of given theorems,

---

[18]Abstract notions have been important for the internal development of mathematics and for sophisticated applications in the sciences to organize our experience of the world. It seems to me to be absolutely crucial to gain insight into this dual role — to bring into harmony philosophical reflections on mathematics with those on the sciences.

Aspects of mathematical experience 343

introducing appropriate defined notions, formulating motivated conjectures, discovering new abstract concepts, and recognizing new axioms for accessible domains. For the first, relatively easy question, calculi that were developed in the Hilbert school provide the necessary logical framework. And with respect to the other issues Hilbert, I assume, would have been very optimistic; he claimed in 1927:

The formula game that Brouwer so deprecates has, besides its mathematical value, an important general philosophical significance. For this formula game is carried out according to definite rules in which the technique of our thinking is expressed. These rules form a closed system that can be discovered and definitively stated. The fundamental idea of my proof theory is none other than to describe the activity of our understanding, to make a protocol of the rules according to which our thinking actually proceeds.

And he added, "If any totality of observations and phenomena deserves to be made the object of serious and thorough investigation, it is this one. . . ." This remark of Hilbert's is undoubtedly correct (and independent from his claims for proof theory).

III.2

Beyond Hilbert's reach?*

> ... historical reflection serves, in the end,
> to shape the present and the future.
> Ernst Cassirer[1]

Abstract. Work in the foundations of mathematics should provide systematic frameworks for important parts of the practice of mathematics, and the frameworks should be grounded in conceptual analyses that reflect central aspects of mathematical experience. The Hilbert School of the 1920s used suitable frameworks to formalize (parts of) mathematics and provided conceptual analyses. However, its analyses were mostly restricted to finitist mathematics, the programmatic basis for proving the consistency of frameworks and, thus, their instrumental usefulness. Is the broader foundational quest beyond Hilbert's reach? The answer to this question seems simple: "Yes and No." It is "Yes" if we focus exclusively on Hilbert's finitism; it is "No" if we take into account the more sweeping scope of Hilbert and Bernays's foundational thinking. The evident limitations of Hilbert's "formalism" have been pointed out all too frequently; in contrast, I will trace connections of Hilbert's work, beginning in the late nineteenth century, to contemporary work in mathematical logic. Bernays's reflective philosophical investigations play a significant role in reinforcing these connections. My paper pursues two complementary goals, namely, to describe a global, integrating perspective for foundational work and to formulate some more local, focused problems for mathematical work.

1 What is at issue?

It is a fact of intellectual history, perhaps a curious one, but nonetheless a fact, that the *Grundlagenstreit* of the 1920s colors even now our perspectives on the foundations of mathematics and beyond. In those early debates, we find dramatically formulated stances, and we tend to interpret them as being substantively and starkly opposed to each other. Minimal historical awareness should have undermined that tendency a long time ago, as the

---

*To Howard Stein — friend and teacher.
[1]In (Paetzold 1995, p. 112). "Immer wieder schärft er seinen Zuhörern ein, daß alle historische Betrachtung letztlich—symbolisch—im Dienste der Gestaltung der Gegenwart und der Zukunft steht."

finitist program, first formulated in Hilbert's Leipzig talk of 22 September 1922, was explicitly intended to mediate between constructivist and logicist set theoretic positions. Weyl recognized that point in papers[2] from 1925 and 1928, and even Brouwer's polemical *Intuitionistic reflections on formalism*, published in 1928, contain these remarks:

> The disagreement over which is correct, the formalistic way of founding mathematics anew or the intuitionistic way of reconstructing it, will vanish, and the choice between the two activities be reduced to a matter of taste, as soon as the following insights[3] . . . are generally accepted. The acceptance of these insights is only a question of time, since they are the results of pure reflection and hence contain no disputable element, so that anyone who has once understood them must accept them.[4]

From Hilbert's perspective there was every principled reason to view the mathematical substance of Weyl's and Brouwer's foundational work as part of broadly conceived axiomatic investigations; in lecture notes from the summer term of 1920 he had already emphasized:

> But what these two researchers [Weyl and Brouwer] have achieved in terms of positive and fruitful results through their investigations on the foundations of mathematics, that fits very well with the axiomatic method, indeed, it is exactly in the spirit of this method. For it is being investigated, how a part of analysis can be delimited by a narrower system of assumptions.[5]

One is tempted to think that the *Grundlagenstreit* could have given way to a calmer discussion; but it did not.[6] I will not try to disentangle aspects

---

[2]See (Weyl 1925, p. 540 ff), also (Weyl 1927, pp. 482–84).

[3]Brouwer lists four basic insights. The first two are constitutive of Hilbert's proof theoretic program, and Brouwer emphasizes that they have been "understood and accepted in the formalistic literature", not without claiming that they have been taken over—without proper acknowledgement—from intuitionism. The first insight concerns the distinction between the construction of formalist mathematics and the intuitive (contentual) metamathematics concerning this construction; the second insight points to the problematic character of the law of excluded middle (lem). The remaining two insights are formulated straightforwardly and with great clarity, namely, that the lem is identical with the claim that every mathematical problem is solvable, and that consistency does not guarantee correctness. The fourth insight can be reformulated as stating that consistency does not provide a contentual justification of formalistic mathematics. The third point would have been disputed, and the fourth was in this general formulation undoubtedly clear to Hilbert through his early work on non-Euclidean geometries. Brouwer made also a very specific claim concerning the lem as part of the fourth insight, namely, that its correctness can be justified only by the lem itself. This claim was taken back in (Brouwer 1953, p. 14, fn. 1); see the editor's addition to footnote 8 in (Brouwer 1927) on p. 460 of (van Heijenoort 1967). The substance of the claim had been refuted already earlier by the Gödel-Gentzen reduction of classical to intuitionist arithmetic.

[4](Brouwer 1928, p. 490).

[5](Hilbert *1920b, p. 34). "Was aber diese beiden Forscher [Weyl and Brouwer] in ihren Untersuchungen über die Grundlagen der Mathematik an Positivem und Fruchtbarem leisten, das fügt sich der axiomatischen Methode durchaus ein und ist gerade im Sinne dieser Methode. Denn es wird hier untersucht wie sich ein Teil der Analysis durch ein gewisses engeres System von Voraussetzungen abgrenzen lässt."

[6]It should be noted, however, that some literally identified finitism and intuitionism—before Gödel's and Gentzen's result. I am thinking of Bernays, von Neumann, and Herbrand

of personality, professional aspirations, or ideological judgments. Rather, I intend to describe the broad foundational context and attempt to understand what the programmatic constructivist Hilbert defended against his flamboyant fellow constructivist Brouwer, how he tried to do so, and how he got there.

The logico-philosophical community has focused on Hilbert's finitist means for securing "classical" mathematics and on the epistemological distinctiveness of those means—as viewed in the twenties; and I think that a deepened mathematical and philosophical analysis of finitism remains an important issue. However, we have not been equally concerned with the substance of what Hilbert strove to secure—over a lifetime. And that is not *classical* mathematics as it evolved until the nineteenth century, but rather *modern* mathematics as it resulted from a radical transformation during the second half of that century. Howard Stein called it a transformation "so profound that it is not too much to call it a second birth of the subject"; he argued, and I agree, that it was effected mainly by the work of Gauss, Dirichlet, Riemann, and Dedekind.[7]

Hilbert was intimately connected to this part of mathematical tradition (in Göttingen), but also to a second significant aspect. I am alluding to the free use of mathematical concepts in, and indeed their invention or free creation for, applications in the sciences. (I should point out that Weyl, in 1928, viewed intuitionist mathematics as inadequate for the sciences!) In the introductory remarks of his Paris Lecture Hilbert described most vividly the rich interplay of mathematical thought and experience. Discussing the central importance of problems for mathematics, he commented on their origins as follows:

Surely the first and oldest problems in every branch of mathematics spring from experience and are suggested by the world of external phenomena. . . . But, in the further development of a branch of mathematics, the human mind, encouraged by the success of its solutions, becomes conscious of its independence. It evolves from itself alone . . . by means of logical combination, generalization, specialization, . . . and appears then itself as the real questioner. . . . while the creative power of pure reason is at work, the outer world again comes into play, forces upon us new questions from actual experience, opens up new branches of mathematics. . . . And it seems to me that the numerous and surprising analogies and the apparent harmony which the mathematician so often perceives in the questions, methods, and ideas of the various branches of his science, have their origin in this ever-recurring interplay between thought and experience.[8]

One basic condition has to be met, however, in order to safeguard creative freedom within mathematics and within contexts of applications: the

---

"within" the Hilbert school, but also of Carnap and Fraenkel. As to the former, I am alluding to (Carnap 1930), in particular, pp. 309–310; as to the latter, let me quote the ironic (but historically inaccurate) remark in his (1930, p. 294), "Wie mir scheint, hat *Brouwer* den größten Erfolg für seine Anschauungen dadurch erzielt, daß er als Anhänger seiner Ausgangsposition—*Hilbert* gewonnen hat!"

[7](Stein 1988, p. 238).
[8](Hilbert 1900a, p. 1098).

introduced notions must be consistent. That was clearly expressed in the Paris Lecture, and Hilbert reiterated this view four years later in his Heidelberg talk. About the underlying *creative principle* he wrote then: ". . . in its freest use [that principle] justifies us in forming ever new notions with the sole restriction that we avoid a contradiction."[9] Thus, the central methodological issue is, how we can rationally assess whether this restriction has been met.

The methodological issue was more concrete and limited already at this point, as Hilbert sought to establish the consistency of axiom systems, for example, in 1900 for the real numbers and in 1904 for the natural numbers. Such a proof was to ensure the existence of the set or, in Cantor's terminology, of the consistent multiplicity of the real and natural numbers. The issue can be traced back to Dedekind and is, according to Bernays (and in harmony with my earlier remarks), most closely connected to the "transformation the methodological approach of mathematics underwent towards the end of the nineteenth century".[10] One characteristic feature of that transformation is the emergence of *existential axiomatics* described in the first part of my paper. That part is entitled "Logical Models" and examines Dedekind's and Hilbert's attempts to secure the consistency of analysis by logical means, before 1900. The second part, "Direct Proofs", presents in detail Hilbert's attempt to solve the consistency problem for elementary arithmetic in 1904, Poincaré's critical objections, and the impact of *Principia Mathematica* in Göttingen. "Proof Theoretic Strategies" is the title of the third part; here finitist mathematics moves to center stage, and the methodological perspective for Hilbert's proof theory is described. An analysis of the informal ideas underlying this approach leads to the fourth part, "Accessible Domains", and serves as the motivating background for a programmatic formulation of *reductive structuralism*. My goal there is twofold, namely, to describe a global, integrating perspective of foundational work on the one hand, and to formulate some more local, focused problems for mathematical work on the other hand.[11]

---

[9](Hilbert 1905a, p. 136). The German text refers to principle I of mathematical thought: "In I. kommt das schöpferische Prinzip zum Ausdruck, das uns im freiesten Gebrauch zu immer neuen Begriffsbildungen berechtigt mit der einzigen Beschränkung der Vermeidung eines Widerspruchs."

[10](Bernays 1930b, p. 17). There Bernays locates first, in a most perspicuous way, the philosophical questions concerning mathematics. "Diese Fragen philosophischen Charakters haben eine besondere Dringlichkeit erhalten seit der Wandlung, welche die methodische Einstellung der Mathematik gegen Ende des 19. Jahrhunderts erfuhr." Then he describes the characteristic features of this transformation, namely, the advance of set theory, the emergence of existential axiomatics, and the forging of close connections between logic and mathematics.

[11]The considerations in this paper have profited from critical reactions to a number of earlier presentations, namely, at the Boolos Conference at Notre Dame (16 April 1998), the Steinfest at the University of Chicago (23 May 1999), the Hilbert Workshop at the Sorbonne (26 May 2000), and the Annual Meeting of the Association for Symbolic Logic in Urbana (3 June 2000). Special thanks for helpful criticism go to Bernd Buldt, Michael Detlefsen, Jacques Dubucs, Sol Feferman, Carl Posy, Howard Stein, and Bill Tait.

2 Logical models

Hilbert attempted to secure analysis from contradictions at the close of the nineteenth century. His formulation of a theory for real numbers in 1899 was inspired by Dedekind's and is distinctly modern. Recall that Kronecker, a mere decade earlier, had still been trying to banish the general notion of irrational number from mathematics; and Hilbert's lecture notes from the period between 1894 and 1899 show how difficult it was for him to obtain a proper perspective on the notion of number (*Zahlbegriff*). In the end, Hilbert associated all the central foundational issues with the *axiomatic method*. To proceed axiomatically means for Hilbert to think with consciousness, but also with critical awareness. The method allows the rigorous investigation of independence and completeness issues, and it is needed for securing, completely and logically, the content of our knowledge.[12] Already Dedekind had most explicitly aimed at grounding—by logic—our arithmetical knowledge!

2.1 Existence.

A rather direct interpretation of the essay *Was sind und was sollen die Zahlen?* and of the later explanatory letter to Keferstein shows that Dedekind strove to give a consistency proof relative to a logic that allowed the construction of models, here, for simply infinite systems. In other words, a logical proof of the existence of such a system was to secure that the very notion did not contain an "internal contradiction". Dedekind wrote to Keferstein:

After the essential nature of the simply infinite system, whose abstract type is the number sequence N, had been recognized in my analysis . . . , the question arose: does such a system *exist* at all in the realm of our ideas? Without a logical proof of existence, it would always remain doubtful whether the notion of such a system might not perhaps contain internal contradictions. Hence the need for such a proof (articles 66 and 72 of my essay).[13]

These observations can be extended in a natural way to cuts and complete ordered fields, treated in the earlier essay *Stetigkeit und irrationale Zahlen*. Dedekind viewed his broad methodological considerations not as specific for the foundational context of these essays, but rather as paradigmatic for the sound introduction of axiomatically characterized notions.[14]

In his proof of the existence of a simply infinite system, Dedekind had used however the (in)famous "system of all objects of my thinking". Hilbert learned from Cantor as early as 1897 that this gave rise to a contradiction and, thus, undermined the logical basis of Dedekind's essay. The very title of Hilbert's historically sweeping lectures from 1894, *Quadratur des*

---

[12] The full German text is: "Trotz des hohen pädagogischen Wertes der genetischen Methode verdient doch zur endgültigen Darstellung und völligen logischen Sicherung des Inhaltes unserer Erkenntnis die axiomatische Methode den Vorzug."

[13] (Dedekind 1890a, p. 101). The essay Dedekind refers to is obviously (Dedekind 1888).

[14] See (Dedekind 1877, p. 268–69), in particular the long footnote on p. 269.

Kreises, was consequently expanded in 1897 to *Zahlbegriff und Quadratur des Kreises*. Hilbert emphasized there the importance of the "fixation of the (real) number concept", and he defined the reals as fundamental sequences taking for granted the natural numbers. Two years later, when finishing *Grundlagen der Geometrie*, Hilbert formulated axioms for the reals including the completeness axiom in yet another version of these lectures on the quadrature of the circle. His paper *Über den Zahlbegriff* was completed for publication on 12 October 1899 and summarized these early investigations; Hilbert had presented it already on 19 September 1899 to the Munich meeting of the German Association of Mathematicians.[15]

A neglected, but most significant link to Dedekind should be noted. Hilbert followed Dedekind in formulating the central axiomatic conditions for real numbers, as well as in setting up the very framework by assuming the existence of a system of things satisfying the conditions, "We think a system of things . . . " (Wir denken ein System von Dingen . . .). He proceeded methodologically in exactly the same way for *Grundlagen der Geometrie*, where the existential framework for the axioms of geometry is introduced by "We think three different systems of things . . . " (Wir denken drei verschiedene Systeme von Dingen . . .). Thus, as in Dedekind's case, there is an explicit existential assumption that has to be secured or discharged in some way. To emphasize this crucial aspect of Hilbert's method, both Hilbert and Bernays called it *existential axiomatics* (existentielle Axiomatik). In the case of geometry Hilbert discharged the existential assumption for (parts of) the axiom system by appropriate analytic models. But how could the problem be addressed for analysis? How could the existence of the system of reals be secured? In his answer, Hilbert referred to Cantor's distinction between consistent and inconsistent multiplicities that presented the former as the proper objects of set theory. Hilbert was critical of this distinction. His critical attitude, only implicit here, was made explicit in the Heidelberg talk of 1904 where he claimed that Cantor's conception "leaves latitude for subjective judgement and therefore affords no objective certainty".

2.2 A partial syntactic turn.   Dedekind had given a logical existence proof of a simply infinite system in order to guarantee that the very notion of such a system does not contain "internal contradictions". Hilbert recast consistency as a syntactic property of axiom systems, demanding that no contradiction be provable from the axioms in a finite number of steps.[16] That allowed him to attack the problem of arriving at a consistent multiplicity of real numbers in *Über den Zahlbegriff* and his Paris lecture from a new viewpoint: the existence of sets is to be guaranteed by *consistency*

---

[15]In (Peckhaus 1990) the reader will find an informative, complementary discussion on pp. 29–33.

[16]He had done so also in section 9 of *Grundlagen der Geometrie*, but established consistency there semantically—by an "inductive" argument—on pp. 19–20; see note 12 of my (1999). Notice that Hilbert did not specify in either of these works the character of the "steps".

proofs for appropriate axiom systems. Hilbert had shifted from consistent multiplicities to consistent theories[17] and suggested to give objective content to Cantor's notion through consistency proofs for theories. In the Paris lecture he thought that a *direct proof* should be possible:

> I am convinced that it must be possible to find a direct proof for the consistency of the arithmetical axioms [proposed in *Über den Zahlbegriff* for the reals] by means of a careful study and suitable modification of the known methods of reasoning in the theory of irrational numbers.[18]

It is quite obscure from the published papers I referred to what would constitute a *direct* proof. A reasoned, though by no means unproblematic conjecture can be based on earlier lecture notes. Hilbert thought that the construction of the reals, and also of the natural numbers, could be given directly and be exploited as a blueprint for a Dedekindian consistency proof. This seems to be supported also by the (one-sidedly preserved) correspondence with Cantor at the time of the Munich and Paris talks. Cantor insisted in his letters on two points: (i) the consistency even of finite multiplicities (i.e., the existence of finite sets) has to be postulated, and (ii) Dedekind's considerations are fundamentally flawed. I conjecture Hilbert believed, despite the Cantorian admonitions, that Dedekind's logicism with suitable restrictions might after all provide the means for a principled consistency proof.[19]

Hilbert changed his views dramatically after Zermelo and Russell discovered their elementary contradiction, a contradiction that had according to his own testimony "a catastrophic effect in the mathematical world".[20] It had undoubtedly a catastrophic effect on Hilbert himself: Bernays reported to Constance Reid that Hilbert believed at the time, even if only for a very brief period, that Kronecker might have been right in demanding a radical restriction of mathematical notions and methods. In lecture notes from the summer of 1904, just before the Heidelberg talk in August of that year, one finds these illuminating and revealing remarks on Dedekind and Kronecker:

> He [Dedekind] arrived at the opinion that the standpoint of viewing the integers as obvious cannot be sustained; he recognized that the difficulties Kronecker saw in the

---

[17] See (Bernays 1976a, p. 46); footnote 11 makes an explicit terminological recommendation with regard to "Konsistenz": "Es mag hier angeregt sein, diesen von Cantor speziell in bezug auf Mengenbildungen gebrauchten Ausdruck allgemein mit Bezug auf irgendwelche theoretischen Ansätze zu verwenden."

[18] (Hilbert 1900a, p. 1104). This is reemphasized in (Bernays 1935a, pp. 198–99): "Zur Durchführung des Nachweises gedachte Hilbert mit einer geeigneten Modifikation der in der Theorie der reellen Zahlen angewandten Methoden auszukommen."

[19] Cantor attended the Munich meeting and met with Hilbert. Cantor's views are carefully presented in his letter to Hilbert that was written on 27 January 1900. (The letter is contained in the Hilbert Nachlaß in Göttingen; Cod. Ms. Hilbert 54: 18.)

[20] (Hilbert 1927, p. 169). "Insbesondere war es ein von Zermelo und Russell gefundener Widerspruch, dessen Bekanntwerden in der mathematischen Welt geradezu von katastrophaler Wirkung war." Some indication of related activities in Göttingen from 1902 through 1904 is found in (Peckhaus 1990, p. 57).

definition of irrationals arise already for integers; furthermore, if they are removed here, they disappear there. This work [*Was sind und was sollen die Zahlen?*] was epochal, but it did not yet provide something definitive, certain difficulties remain. These difficulties are connected, as with the definition of the irrationals, above all to the concept of the infinite.[21]

These difficulties were plainly stated at the very beginning of the Heidelberg lecture, where Hilbert described in detail alternative foundational approaches and remarked about Dedekind:

> R. Dedekind clearly recognized the mathematical difficulties encountered when a foundation is sought for the notion of number; for the first time he offered a construction of the theory of integers, and in fact an extremely sagacious one. However, I would call his method *transcendental* insofar as in proving the existence of the infinite he follows a method that, though its fundamental idea is used also by philosophers, I cannot recognize as practicable or secure because it employs the notion of the totality of all objects, which involves an unavoidable contradiction.[22]

What could be done?—Hilbert shifted, first of all, his efforts from the theory of real numbers to that of integers; he proposed, secondly, to give a genuinely direct proof of the existence of "the smallest infinite", and that was to be done by establishing the consistency of an axiom system that reflected Dedekind's conditions for a simply infinite system.

3 Direct proofs

The elementary Zermelo-Russell paradox had convinced Hilbert, as we saw, that there *was* a problem with his earlier considerations and that difficulties had to be faced at a more fundamental level. In the Heidelberg Lecture, Hilbert reasserted most strongly his view that the problems for the reals are resolved once matters are resolved for the natural numbers.

> The existence of the totality of real numbers can be demonstrated in a way similar to that in which the existence of the smallest infinite can be proved; in fact, the axioms for real numbers as I have set them up ... can be expressed by precisely such formulas as the axioms hitherto assumed.... the axioms for the totality of real numbers do not differ qualitatively in any respect from, say, the axioms necessary for the definition of the integers. In the recognition of this fact lies, I believe, the real refutation of the conception of arithmetic associated with L. Kronecker.[23]

Hilbert actually claimed, "In the same way we can show that the fundamental notions of Cantor's set theory, in particular Cantor's alephs, have a

---

[21](Hilbert *1904, p. 166). "Er [Dedekind] drang zu der Ansicht durch, dass der Standpunkt mit der Selbstverständlichkeit der ganzen Zahlen nicht aufrecht zu erhalten ist; er erkannte, dass die Schwierigkeiten, die Kronecker bei der Definition der irrationalen Zahlen sah, schon bei den ganzen Zahlen auftreten und dass, wenn sie hier beseitigt sind, sie auch dort wegfallen. Diese Arbeit [*Was sind und was sollen die Zahlen?*] war epochemachend, aber sie lieferte doch noch nichts definitives, es bleiben gewisse Schwierigkeiten übrig. Diese bestehen hier, wie bei der Definition der irrationalen Zahlen, vor allem im Begriff des Unendlichen."

[22](Hilbert 1905a, pp. 130–31).

[23](Hilbert 1905a, pp. 137–38).

Beyond Hilbert's reach?

consistent existence." Let us come back to the more modest goal of establishing the existence of the smallest infinite.

3.1 Turning further. In section 2.2, I described the partial syntactic turn Hilbert had taken by recasting consistency as a syntactic notion. However, he had neither specified inference steps nor had he presented other than semantic arguments. That was remedied here at least in a broad programmatic sense: he developed logic and arithmetic simultaneously and inferred the consistency of the joint system from elementary syntactic observations. The methodological starting-point was formulated in this way:

Arithmetic is often considered to be a part of logic, and the traditional fundamental logical notions are usually presupposed when it is a question of establishing a foundation for arithmetic. If we observe attentively, however, we realize that in the traditional exposition of the laws of logic certain fundamental arithmetic notions are already used, for example, the notion of set and, to some extent, also the notion of number. Thus we find ourselves turning in a circle, and that is why a partly simultaneous development of the laws of logic and of arithmetic is required if paradoxes are to be avoided.[24]

The theory Hilbert proposed consists of axioms for identity and Dedekind's requirements for a simply infinite system, except that induction is not explicitly formulated; in modern notation:

(1) $x = x$

(2) $x = y \;\&\; A(x) \to A(y)$

(3) $x' = y' \to x = y$

(4) $x' \neq 1$

The rules, extracted from Hilbert's description of "consequence", are modus ponens and a substitution rule that allows the replacement of variables by arbitrary sign combinations. Other modes of logical inferences are mentioned later, but neither formally stated nor incorporated into the consistency proof. The idea of the consistency proof is this: formulate a property P and show by induction on derivations that all provable equations have P. The property Hilbert considers is homogeneity: an equation $a = b$ is called *homogeneous* if and only if a and b have the same number of symbol occurrences; it is easily seen that all equations derivable from axioms (1)-(3) are indeed homogeneous. But a contradiction can be obtained only by establishing an unnegated instance of (4) from (1)-(3). Such an instance is necessarily inhomogeneous and, consequently, not provable.

Hilbert saw his considerations as answering, for the first time, the earlier call for a direct proof. He commented:

---

[24](Hilbert 1905a, p. 131).

The considerations just sketched constitute the *first case* [my emphasis, WS] in which a direct proof of consistency has been successfully carried out for axioms, whereas the method of a suitable specialization, or of the construction of examples, which is otherwise customary for such proofs—in geometry in particular—necessarily fails here.[25]

Hilbert had emphasized the need to develop logic and mathematics simultaneously, but the actual work had significant shortcomings: there is no calculus for sentential logic, there is no proper treatment of quantification, and induction is neither rigorously formulated nor incorporated into the argument. In sum, there *is* an important shift from "semantic" arguments to a "syntactic" one, but the set-up is utterly incomplete as a formal framework for arithmetic.

3.2 Critical analysis. The presumed foundational import of Hilbert's talk was not left unchallenged. On account of the inductive character of the consistency proof Poincaré criticized Hilbert's considerations severely; this critique is well-known and absolutely to the point. Toward the end of Hilbert's paper there is a peculiar "uncertainty" that reveals underlying methodological problems; they too were pointed out by Poincaré.[26] It becomes very clear how penetrating Poincaré's considerations were when one reads in parallel the 1935 remarks on Hilbert's Heidelberg talk by Bernays: they give a précis of Poincaré's critique. We should keep in mind, however, that the existence of sets was the central issue and was to be guaranteed by the consistency of an appropriate axiom system, a viewpoint shared explicitly and strongly by Poincaré. "If therefore", Poincaré wrote, "we have a system of postulates, and if we can demonstrate that these postulates imply no contradction, we shall have the right to consider them as representing the definition of one of the notions entering therein."[27] In any event, as to the critical aspect Bernays wrote,

... the systematic standpoint of Hilbert's proof theory is not yet fully and clearly developed. Some places indicate that Hilbert wants to avoid the intuitive conception of number and replace it by its axiomatic introduction. Such a procedure would lead to a circle in the proof theoretic considerations.[28]

---

[25](Hilbert 1905a, p. 135).

[26]On pp. 1042–43 in (Ewald 1996). This is part of Poincaré's review of contemporaneous investigations of the foundations of mathematics. The critical, but also sympathetic discussion of Hilbert is mainly found in (Poincaré 1906, pp. 1038–46). Brouwer pointed to Poincaré, when elaborating on the first insight in his *Intuitionistic reflections on formalism*, and suggested that this insight had been "strongly prepared" by him.

[27](Poincaré 1905, p. 1026). Having discussed Mill's view of (mathematical) existence and calling it "inadmissible", Poincaré writes in the immediately preceding paragraph: "Mathematics is independent of the existence of material objects; in mathematics the word 'exist' can have only one meaning; it means free from contradiction. ... in defining a thing, we affirm that the definition implies no contradiction."

[28](Bernays 1935a, p. 200). "Außerdem ist auch der methodische Standpunkt der Hilbertschen Beweistheorie in dem Heidelberger Vortrag noch nicht zur vollen Deutlichkeit entwickelt. Einige Stellen deuten darauf hin, daß Hilbert die anschauliche Zahlvorstellung vermeiden und durch die axiomatische Einführung des Zahlbegriffs ersetzen will. Ein solches Verfahren würde in den beweistheoretischen Überlegungen einen Zirkel ergeben."

Bernays emphasized also that Hilbert had not articulated distinctions central to the later finitist program:

> Also, the viewpoint of restricting the contentual application of the forms of existential and general judgments is not yet put forth explicitly and completely.[29]

Hilbert's own views of his objective accomplishments were formulated in lectures from the summer term of 1905.[30] These lectures contain additional technical details, but point out basic shortcomings as well; Hilbert bemoans the unsatisfactory state of logic, in particular, the state of quantification theory. Hilbert had a distinctive approach already in the Heidelberg lecture, clearly recognized and applauded by Poincaré, as Hilbert's "all" ranged only over the limited domain of combinations of thought-objects, not as Russell's over everything whatsoever; see (Poincaré 1905, p. 1040).

3.3 Proper formalisms. During the period from 1905 to 1917 Hilbert gave almost annually lectures on the foundations of mathematics, but these lectures did neither break new ground, nor did they return to a proof theoretic study;[31] another approach opened up, however. Around 1913 Hilbert started to become familiar with some of Russell's writings. The official lecture notes from the winter term 1914–15 contain brief remarks about type theory, and the notes from a student, serendipitously preserved in the Institute for Advanced Study at Princeton, more extended ones. There was even some correspondence between Hilbert and Russell, reported in appendix B of (Sieg 1999). A number of relevant talks on the foundations of mathematics were given in Göttingen during this period by Behmann, Bernstein, Hilbert, and Zermelo; most significantly, Hilbert directed Behmann's dissertation of 1918, *Die Antinomie der transfiniten Zahl und ihre Auflösung durch die Theorie von Russell und Whitehead*. A detailed description of this work is found in (Mancosu 1999a), narrowing the real gap in our historical understanding of the details of the Russellian influence on Hilbert.[32] How strongly Russell influenced Hilbert has been clear from the notes for his course on *Set Theory* (summer term 1917) and

---

[29](Bernays 1935a, p. 200). "Auch wird der Gesichtspunkt der Beschränkung in der inhaltlichen Anwendung der Formen des existentialen und des allgemeinen Urteils noch nicht ausdrücklich und restlos zur Geltung gebracht."

[30] These lectures are discussed in detail by Peckhaus in (1990, 1994a, 1994b, 1995); a broad philosophical perspective is also provided by Hallett in (1994, 1995).

[31] That is supported, in a general way, by (Bernays 1935a). However, in Bernays's description there is a peculiar "smoothing" of the developments between 1904–05 and 1917–22: Bernays does not mention that Hilbert gave lectures on the foundations of mathematics during that period; the 1917–18 lectures are not hinted at. Thus, he effectively creates the impression that the period is one of inactivity; e.g., on p. 200 one finds: "In diesem vorläufigen Stadium hat Hilbert seine Untersuchungen über die Grundlagen der Arithmetik für lange Zeit unterbrochen. Ihre Wiederaufnahme finden wir angekündigt in dem 1917 gehaltenen Vortrag 'Axiomatisches Denken'." This impression is reinforced by the footnote attached to the first sentence in this quote, where Bernays points to the work of others who pursued the research direction stimulated by (Hilbert 1905a).

[32] The list of talks is found on pp. 304–5 of (Mancosu 1999a). Behmann is viewed by Mancosu as a central player and as the (indirect) source for some of Hilbert's views, i.e., as a conduit for Russellian views.

his Zürich talk *Axiomatisches Denken* given on 11 September 1917; they reveal renewed logicist tendencies in Hilbert's work. Hilbert wrote in the essay on which his talk had been based:

> The examination of consistency is an unavoidable task; thus, it seems to be necessary to axiomatize logic itself and to *show that number theory as well as set theory are just parts of logic.*

If we try to achieve such a reduction to logic, Hilbert said at the very end of the set theory notes, "we are facing one of the most difficult problems of mathematics".[33]

Russell and Whitehead provided not only the stimulus for this programmatic redirection, but also powerful technical tools.[34] The latter were ingeniously adapted and mathematically analyzed in the winter term 1917–18, when Hilbert offered lectures under the title *Prinzipien der Mathematik* with the assistance of Paul Bernays. Hilbert had finally a proper formalism for the development of mathematics: a language for capturing the logical form of informal statements and a calculus for representing the structure of logical arguments. The presentation is carried through with focus, elegance, and directness. The logical work is complemented by real metamathematical considerations; the latter are certainly inspired by the (perspective underlying the) work that had been done at the turn of the century on the foundations of geometry. These beautifully written, detailed notes include all the basic material that is contained in the 1928 book by Hilbert and Ackermann; thus, they are the real beginning of modern *mathematical* logic. For my purpose the main points can be summarized as follows: (i) there is a full development of (the syntax and semantics for) sentential, monadic, first-order logic, and ramified analysis, (ii) independence and completeness problems are formulated and partly solved, and (iii) theories are always presented with appropriate domains or, more precisely, many-sorted structures. The last point brings out in this setting the crucial aspect of existential axiomatics that had been so important in Hilbert's early investigations; see section 2.1.

Absolutely no proof theoretic considerations are presented in these notes, though consistency is a real issue. The consistency of pure logic is examined, and both sentential and first-order logic are semantically shown to be consistent, the latter by considering a one-element domain. A footnote warns the reader, however, not to overestimate the significance of this result for first-order logic, because "[i]t does not give us a guarantee that the system of provable formulas remains free of contradictions after the

---

[33]That is not at all reflected in Bernays's presentation in (1935a); the logicist tendency is suppressed and the "ungelöste Problematik" of the consistency of *Principia Mathematica* is emphasized immediately; see p. 201.

[34]How much direct continuity is there between *Principia Mathematica* and *Prinzipien der Mathematik?* That remains an important question for detailed investigation.

Beyond Hilbert's reach?

symbolic introduction of contentually correct assumption."[35] After all, these assumptions may force the domain to be infinite.

At the beginning of 1920, having abandoned for good reasons the logicist route and responding in part to the contemporaneous investigations of Brouwer and Weyl, Hilbert and Bernays pursued a radically constructive redevelopment of arithmetic. This took up a recurring Kroneckerian theme in Hilbert's foundational reflections. However, it was realized quickly that this could not provide a foundation for classical forms of reasoning, as the law of excluded middle does not hold for constructively understood quantified statements. Having recognized this fact, Hilbert and Bernays mentioned Brouwer for the first time when closing with: "This consideration helps us to gain an understanding for the sense of the paradoxical claim, made recently by Brouwer, that for infinite systems the law of the excluded middle (the 'tertium non datur') loses its validity."[36]

To us it may seem as if Hilbert had available all the mathematical and logical means for the formulation of *the* program. Yet it took some more time before he had gained the appropriate methodological perspective, and before finitist mathematics and proof theory emerged in a programmatically coherent alignment.

4 Proof theoretic strategies

Hilbert had argued for a theory of proofs in his 1904 Heidelberg talk; he had mentioned it also in his 1917 Zürich talk, but without any programmatic direction. The suggestion was finally taken up again in the summer semester of 1920: the notes from that term contain a consistency proof for a restricted part of elementary number theory. Indeed, it is (almost exactly) the system of the Heidelberg lecture.[37]

4.1 Turning further, ctd. The syntactic turn in treating the consistency problem, from choosing a syntactic formulation of consistency to developing logic and arithmetic simultaneously, is pursued further in these notes. The description of the system of elementary arithmetic is given in a more coherent way and is evidently informed by the logical work of the

---

[35](Hilbert *1917/18, p. 156). "Man darf dieses Ergebnis in seiner Bedeutung nicht überschätzen. Wir haben ja damit noch keine Gewähr, dass bei der symbolischen Einführung von inhaltlich einwandfreien Voraussetzungen das System der beweisbaren Formeln widerspruchsfrei bleibt."

[36]See (Sieg 1999, pp. 23–27), for more details on the attempted radical constructive development.

[37]The language of this fragment of arithmetic consists of variables $a, b, \ldots$, nonlogical constants $1, +$, and all numerals; $=$ and $\neq$ are the only relation symbols, and $\to$ is the sole logical symbol. The axioms are:
$1 = 1$
$a = b \to a + 1 = b + 1$
$a + 1 = b + 1 \to a = b$
$a = b \to (a = c \to b = c)$
$a + 1 \neq 1$
As to inference rules we have modus ponens and a substitution rule for numerals.

prior years. Attention is paid to the mathematical means used in the proof theoretic arguments, but the formalism that is being investigated is semi-constructive; cf. the end of this subsection. The formalism is almost exactly that of the Heidelberg lecture, but the argument for its consistency is quite different, mainly through the introduction and use of the notion "kürzbar":

> If one considers a proof with respect to a particular concrete property it has, then it is possible that the removal of some formulae in this proof still leaves us with a proof that has that particular property. In this case we are going to say that the proof is *kürzbar* with respect to the given property.[38]

Hilbert establishes three lemmata: the first claims that a theorem can contain at most two occurrences of →, the second asserts that no statement of the form (A → B) → C can be proved, and the third expresses that a formula $a = b$ is provable only if a and b are the same term. To recognize the distinctive character of the arguments, let me look at the proof of the first lemma. Hilbert proceeds indirectly and assumes that there is a theorem with at least three occurrences of →. Without loss of generality he further assumes that the theorem has a proof that is not "kürzbar" (with regard to the property of having an end formula with at least three occurrences of the →). The theorem cannot be an axiom, as axioms contain at most two occurrences of the →. Thus, it must have been obtained by modus ponens. The major premise of that inference contains at least one more occurrence of → than its conclusion, i.e., the given theorem; we have consequently a contradiction to the "Nicht-Kürzbarkeit" of the given proof.

On its surface, Hilbert's new proof does not use the induction principle. It is structured in analogy to the standard proof of the fact that $\sqrt{2}$ is not rational. Hilbert frequently asserted, not only here, that consistency proofs should be of the same character as the proof of the irrationality of $\sqrt{2}$.[39] The analogy plays even on the double meaning of "kürzbar"; on the one

---

[38](Hilbert *1920b, p. 38). "Betrachtet man einen Beweis in Hinsicht auf eine bestimmte, konkret aufweisbare Eigenschaft, welche er besitzt, so kann es sein, dass, nach Wegstreichung einiger Formeln in diesem Beweise noch immer ein Beweis (. . .) übrig bleibt, welcher auch noch jene Eigenschaft besitzt. In diesem Fall wollen wir sagen, dass der Beweis sich in bezug auf die betreffende Eigenschaft kürzen lässt."

[39] On pages 7a–8a of (Hilbert *1921/22) one finds the remark: "Diese Aufgabe [to show that it is impossible to derive in a given calculus certain formulas like $1 \neq 1$] liegt grundsätzlich ebenso im Bereich der anschaulichen Betrachtung wie etwa die Aufgabe des Beweises, dass es unmöglich ist, zwei Zahlzeichen a, b zu finden, welche in der Beziehung $a^2 = 2b^2$ stehen. Hier soll gezeigt werden, dass sich nicht zwei Zahlzeichen von einer gewissen Beschaffenheit angeben lassen. Entsprechend kommt es für uns darauf an zu zeigen, dass sich nicht ein *Beweis* von einer bestimmten Beschaffenheit angeben lässt. Ein formalisierter Beweis ist aber, ebenso wie ein Zahlzeichen, ein konkreter und überblickbarer Gegenstand. Er is (wenigstens grundsätzlich) von Anfang bis Ende mitteilbar. Auch die verlangte Eigenschaft der Endformel, z.B. dass sie "$1 \neq 1$" lautet, ist eine konkret feststellbare Eigenschaft des Beweises." In the lecture notes from the following year this view is expressed again as follows (p. 33): "Hier kommt es zur Geltung, dass die Beweise, wenn sie auch inhaltlich sich im Transfiniten bewegen, doch, als Gegenstände genommen und formalisiert, von finiter Struktur sind. Aus diesem Grunde ist die Behauptung, dass aus bestimmten Aussagen nicht zwei Formeln A, ¬A bewiesen werden können, methodisch gleichzustellen mit inhaltlichen Behauptungen der anschaulichen Zahlentheorie, wie z.B. der, dass man nicht zwei Zahlzeichen a, b finden kann, für welche

Beyond Hilbert's reach? 359

hand, "kürzbar" means, when applied to proofs, "can be shortened", but on the other hand it also applies to fractions and then means "(a common factor) can be canceled". Recall that the standard argument proceeds also indirectly, assuming that $\sqrt{2}$ is rational, i.e., equals p/q, $q \neq 0$; without loss of generality it is then assumed further that p/q is not "kürzbar". In his 1922 publication, based on lectures he had given in the spring and summer of 1921 in Kopenhagen and Hamburg (and submitted for publication not before November of 1921), Hilbert makes explicit the strategic point of the modified argument:

Poincaré's objection, that the principle of complete induction cannot be proved but by complete induction, has been refuted by my theory.[40]

Is this to be taken in the strong sense that induction is not used at all? Or is it to be understood, perhaps, just in the weaker sense that a special procedure is being used—a procedure based on the construction and deconstruction of numerals and that, by its very nature, is different from the induction principle?[41] From a mathematical point of view Hilbert used the least number principle in an elementary form, namely, applied to a purely existential statement. The work in this first of Hilbert's foundational articles of the twenties is evidently transitional. It does have a major problem in not recognizing clearly necessary metamathematical means, but also in

---

$a^2 = 2b^2$ gilt." That is also asserted in publications, for example, in *On the Infinite*, p. 383 of (van Heijenoort 1967).

In (Bernays 1935a, p. 76), one finds this remark about the character of consistency proofs: "Diese Unmöglichkeitsbehauptung, um deren Beweis es sich hier handelt, hat die gleiche Struktur wie z.B. die Behauptung, daß es unmöglich ist, die Gleichung $a^2 = 2b^2$ durch zwei ganze Zahlen a und b zu erfüllen."

[40](Hilbert 1922, p. 161). "Sein [Poincarés] Einwand, dieses Prinzip [der vollständigen Induktion] könnte nicht anders als selbst durch vollständige Induktion bewiesen werden, ist durch meine Theorie widerlegt."

[41]Bernays, in his contemporaneous paper (1922b), acknowledges explicitly that a form of induction has to be used; indeed, the writing of (Bernays 1922b) preceded the 1921–22 lectures, where induction is discussed (in particular on p. 57) and used for the development of finitist arithmetic. Reflecting on the proof of the commutativity of addition, Hilbert and Bernays write (on pp. 56–57): "Bei diesem Beweis wenden wir eine Art von vollständiger Induktion an, die aber auch, in der Form, wie sie hier gebraucht wird, ganz dem Standpunkt unserer anschaulichen Betrachtungsweise entspricht. Das Beweisverfahren kommt auf einen *Abbau der Zahlzeichen* hinaus, d.h. wir benutzen die Tatsache, dass die Zahlzeichen, ebenso wie sie durch Zusammensetzung von 1 und + aufgebaut sind, sich auch umgekehrt durch Wegnahme von 1 und + abbauen lassen müssen." Bernays discusses matters in a very similar manner in his (1935a, p. 203): "Was ferner die methodische Einstellung betrifft, welche Hilbert seiner Beweistheorie zugrunde legt und welche er an Hand der anschaulichen Zahlentheorie erläutert, so liegt darin—ungeachtet der Stellungnahme Hilberts gegen Kronecker—eine weitgehende Annäherung an den Standpunkt Kroneckers vor. Eine solche besteht insbesondere in der Anwendung des anschaulichen Begriffes der Ziffer und ferner darin, daß die anschauliche Form der vollständigen Induktion, d.h. die Schlußweise, welche sich auf die anschauliche Vorstellung von dem 'Aufbau' der Ziffern gründet, als einsichtig und keiner weiteren Zurückführung bedürftig anerkannt wird. Indem Hilbert sich zur Annahme dieser methodischen Voraussetzung entschloß, wurde auch der Grund der Einwendungen behoben, welche seinerzeit Poincaré gegen Hilberts Unternehmen in dem Heidelberger Vortrag gerichtet hatte."

not fixing appropriately the very logic of the formal system to be investigated: Hilbert tried to keep it constructive by using, for example, negation only in a restricted way.[42]

4.2 Principled formulation.
Proof theoretic considerations were pursued with novel metamathematical means and with a principled foundational perspective in the lectures from the winter term 1921–22. For the first time, Hilbert and Bernays used the terms *finitist mathematics* and *Hilbert's proof theory* and made explicit the domain of mathematical (finitist) objects appealed to in proof theoretic investigations. They pointed out:

> We have to extend the domain of objects to be considered; i.e., we have to apply our intuitive considerations also to figures that are not number signs. Thus, we have good reason to distance ourselves from the earlier dominant principle according to which each theorem of pure mathematics is in the end a statement concerning integers.

With a jibe at such distinguished mathematicians as Dirichlet and Dedekind, they continued, "This principle was viewed as expressing a fundamental methodological insight, but it has to be given up as a prejudice."[43] After all, formulas and proofs of formal theories are the direct object of proof theoretic investigation, and appropriate definition and proof principles (analogous to those for numbers) have to be used.

Hilbert proved the consistency of a fragment of number theory with *full classical sentential logic* and free variable statements. The very elaborate and detailed proof given in the notes was sketched in Hilbert's Leipzig talk of 22 September 1922.[44] A treatment of quantifiers is indicated there, and genuine transformations of formal proofs are used to carry out the argument. Equally striking is the underlying idea that expresses in a novel, precise way the nineteenth-century methodological maxim that elementary statements should be provable by elementary means. *Elementary* statements are those formulas, also called *numeric*, that are built up solely from $=$, $\neq$, numerals, and sentential logical connectives.

The central steps of the proof theoretic argument are described easily: (i) formal proofs with a numeric end formula are transformed into configurations that are not necessarily proofs, but consist only of numeric formulas; (ii) formulas in these configurations are all effectively brought into

---

[42]For a more detailed discussion of this "transitional stage", see (Sieg 1999, pp. 26–27 and appendix A). The "proper" response to Poincaré is formulated in Hilbert's second Hamburg talk, (Hilbert 1927, p. 473) see also the remarks by Bernays quoted at the end of the preceding note.

[43](Hilbert *1921/22, p. 4a). "Wir müssen den Bereich der betrachteten Gegenstände erweitern, d.h. wir müssen unsere anschaulichen Überlegungen auch auf andere Figuren als Zahlzeichen anwenden. Wir sehen uns somit veranlasst, von dem früher herrschenden Grundsatz abzugehen, wonach jeder Satz der reinen Mathematik letzten Endes in einer Aussage über ganze Zahlen bestehen sollte. Dieses Prinzip, in welchem man eine grundlegende methodische Erkenntnis erblickt hat, müssen wir jetzt als Vorurteil preisgeben."

[44]The text of the lecture was submitted to *Mathematische Annalen* on 29 September 1922, and published in 1923 as *Die logischen Grundlagen der Mathematik*.

Beyond Hilbert's reach?

normal forms; (iii) the resulting normal form statements are all recognized to be "true". Given a formal proof of $0 \neq 0$, the transformations leave the end formula unchanged. From (iii) and the fact that $0 \neq 0$ is not true, it follows that $0 \neq 0$ is not provable. Clearly, these considerations are preliminary in the sense that they concern a theory that is part of finitist mathematics and thus need not be secured by a consistency proof. The next step is crucial with regard to the real issue of securing parts of mathematics that properly extend finitist mathematics.

Hilbert treats quantifiers with the τ-function, the dual of the later ϵ-operator; τ associates with every predicate $A(a)$ a particular object $\tau_a(A(a))$ or simply τA. The *transfinite axiom* $A(\tau A) \to A(a)$ expresses, according to Hilbert, "if a predicate A holds for the object τA, then it holds for all objects a". The τ-operator allows the definition of the quantifiers:

$(\forall a)A(a) \leftrightarrow A(\tau A)$

$(\exists a)A(a) \leftrightarrow A(\tau(\neg A))$.

Hilbert extends the proof theoretic considerations to the "first and simplest case" of going beyond the finitist system. The technique used will become the ϵ-substitution method, allowing the elimination of quantifiers from proofs of quantifier-free statements. Thus, finitist proof theory is given not only its principled formulation, but also its guiding idea (reflection principle)[45] and a dual version of its technical tool (ϵ-calculus). Bernays writes in 1935: "With the presentation of proof theory as given in the Leipzig talk the principled form of its structure had been reached."[46] Ackermann's thesis, published in 1925, is a direct continuation of Hilbert's paper.

4.3 Uniform projection. Instead of pursuing the all-too-well-known sequence that starts with Hilbert and Bernays, goes through Ackermann, von Neumann, Herbrand, and then ends with Gödel, I turn to the question: What is the informal idea underlying the proof theoretic work, including the very idea of formalizing mathematical theories? An answer to this question is found most directly in papers by Bernays from 1922 and in the related lecture notes from 1921–22 and 1922–23.[47] As we saw, Hilbert's existential axiomatics assumed always a system of objects satisfying the axiomatic conditions, and Bernays remarked:

---

[45] Already in the lecture notes of (*1921/22) we find, on p. 4a, this remark, after a discussion of the "incorrect application of the law of the excluded middle": "Wir sehen also, dass für den Zweck einer strengen Begründung der Mathematik die üblichen Schlussweisen der Analysis in der Tat nicht als logisch selbstverständlich übernommen werden dürfen. Vielmehr ist es gerade erst die Aufgabe für die Begründung, zu erkennen, warum und in wieweit die Anwendung der transfiniten Schlussweisen, so wie sie in der Analysis und in der (axiomatisch begründeten) Mengenlehre geschieht, stets richtige Resultate liefert."

[46] (Bernays 1935a, p. 204). "Mit der Gestaltung der Beweistheorie, die uns in dem Leipziger Vortrag entgegentritt, war die grundsätzliche Form ihrer Anlage erreicht." Bernays gives on that page also a summary of the crucial features of the proof.

[47] The (*1921/22) lectures, p. 7a, emphasize the methodological point of formalization (and its relation to *existential axiomatics*) as follows:

In the assumption of such a system with particular structural properties lies something so-to-speak transcendental for mathematics, and the question arises which principled position with respect to it should be taken.

An intuitive grasp of the completed sequence of natural numbers, for example, or of the manifold of real numbers should not be excluded outright. However, taking into account tendencies in the exact sciences, Bernays suggested a different strategic direction, namely, to try "whether it is not possible to give a foundation to these transcendental assumptions in such a way that only primitive intuitive knowledge is used". This suggestion is supplemented by a wonderful image of how to exploit the formalizability of axiomatic theories for this goal: their formalization serves to *project* the associated structures uniformly into the proper mathematical, finitist domain. Even fifty years later Bernays used that image and emphasized the epistemological significance of such projections:

In taking the deductive structure of a formalized theory ... as an object of investigation the [contentual] theory is projected as it were into the number theoretic domain. The number theoretic structure thus obtained is in general essentially different from the structure intended by the [contentual] theory. But it [the number theoretic structure] can serve to recognize the consistency of the theory from a standpoint that is more elementary than the assumption of the intended structure.[48]

The reader may consider this image as merely playful or as genuinely helpful. I choose the latter view, because the substantive point can be recast, and was recast by Bernays in 1930, as an explication of Hilbert's existential axiomatics that reveals a thoroughly structuralist perspective. "Structuralist" is here to be taken in the modern philosophical sense as described so masterfully in (Parsons 1990) and discussed extensively in (Shapiro 1996). Parsons states there that views of this kind can be traced back to the end of the nineteenth century, but attributes clear general statements only to Bernays in 1950 and to Quine somewhat later.[49] However, Bernays points already in 1930 to a characteristic aspect of the

---

"Diesen Formalismus können wir nun zum Gegenstand einer anschaulichen Betrachtung machen, und damit eröffnet sich uns die Möglichkeit einer strengen Begründung der Mathematik.

Denn das Problem der Widerspruchsfreiheit, welches ja die grundsätzlichen Schwierigkeiten bot, erhält durch den neuen Standpunkt eine ganz konkrete Fassung. Es handelt sich nicht mehr darum, ein System von unendlich vielen Dingen mit bestimmten Verknüpfungseigenschaften (eine stetige Mannigfaltigkeit von gewisser Art) als logisch möglich zu erweisen, sondern es kommt nur darauf an einzusehen, dass es unmöglich ist, aus den (in Formeln aufgeschriebenen) Axiomen nach den Regeln des logischen Kalküls gewisse Formeln wie z.B. $1 \neq 1$ abzuleiten."

[48](Bernays 1970, p. 186). The same image is used in the almost contemporaneous correspondence with Gödel; Bernays, after describing how intuitionism considers proofs as proper objects of mathematics, remarks in his letter of 16 March 1972: "Gewiss macht auch die Hilbert'sche Metamathematik die mathematischen Beweise zum Gegenstand, aber doch nur, nachdem sie diese durch die Formalisierung gleichsam in die mathematische Gegenständlichkeit projiziert hat."

[49]Parsons refers to (Bernays 1950) and states with regard to Quine: "Quine is generally most explicit when speaking of natural numbers. For a very explicit general statement, however,

newer mathematics and describes the subject repeatedly as the study of structures (e.g., on p. 32). He presents concisely the standard account of if-then-ism or deductivism, and raises—as the starting point of his systematic philosophical investigations—the vacuity issue for that account. He takes this problematic as arising from two moments of modern axiomatics, namely, (i) the purely hypothetical connection between axioms and theorems, abstracting from the content and truth of the axioms, and (ii) the existential formulation of mathematical theories, assuming a given and from the very beginning fixed system of things and relations pertaining to them.[50] Let me present Bernays's discussion (on pp. 20–21) of the central point in greater detail.

Given the perspective on modern axiomatics I just sketched, the axioms and theorems of an axiomatic theory are statements that concern the relations occurring in them, and the relations pertain to the things of an assumed system. The knowledge provided by a proof of a theorem (Lehrsatz) L from axioms $A_1, \ldots, A_k$ consists in the determination (Feststellung) that, if the statements $A_1, \ldots, A_k$ hold for the relations, then the statement L also holds for these relations. Here we have, as Bernays puts it, a general theorem on relations, i.e., a theorem of pure logic: the results of an axiomatic theory present themselves as theorems of logic. However, these theorems have significance only if the axiomatic conditions can be satisfied at all:

If such a satisfying structure is unthinkable, i.e., logically impossible, then the axiom system does not lead to any theory at all, and the only logically meaningful statement concerning the system [of axioms] is thus the determination

---

see *Ontological Relativity and Other Essays*, (Columbia University Press, New York, 1969), pp. 43–45."

[50] The first moment is beautifully formulated on pp. 3–4 of the (*1921/22) lecture notes (and takes up the theme of the Paris Lecture quoted above): "Auf diese Weise bildete sich die Einsicht heraus, dass das Wesentliche an der axiomatischen Methode nicht in der Gewinnung einer absoluten Sicherheit besteht, die auf logischem Wege von den Axiomen auf die Lehrsätze übertragen wird, sondern darin, dass die Untersuchung der logischen Zusammenhänge von der Frage der sachlichen Wahrheit abgesondert wird.

Unter diesem Gesichtspunkt stellt sich die Methode des axiomatischen Aufbaues einer Theorie dar als ein Verfahren der Abbildung eines Wissensgebietes auf ein Fachwerk von Begriffen, welche so geschieht, dass den Gegenständen des Wissensgebietes die Begriffe und den Aussagen über die Gegenstände die logischen Beziehungen zwischen den Begriffen entsprechen.

Durch diese Abbildung wird die Untersuchung von der konkreten Wirklichkeit ganz losgelöst. Die Theorie hat mit den realen Objekten und mit dem anschaulichen Inhalt der Erkenntnis gar nichts mehr zu tun; sie ist ein reines Gedankengebilde, von dem man nicht sagen kann, dass es wahr oder falsch ist. Dennoch hat dieses Fachwerk von Begriffen eine Bedeutung für die Erkenntnis der Wirklichkeit, weil es eine mögliche Form von wirklichen Zusammenhängen darstellt. Die Aufgabe der Mathematik ist es, solche Begriffsfachwerke logisch zu entwickeln, sei es, dass man von der Erfahrung her oder durch systematische Spekulation auf sie geführt wird.

Hier erhebt sich nun die Frage, ob denn jedes beliebige Fachwerk ein Abbild wirklicher Zusammenhänge sein kann. Eine Bedingung ist dafür jedenfalls notwendig: Die Sätze der Theorie dürfen einander nicht widersprechen, das heisst, die Theorie muss in sich möglich sein, somit entsteht das *Problem der Widerspruchsfreiheit*."

(Feststellung) of the contradiction following from the axioms. For this reason there is for every axiomatic theory the requirement of a proof of the *satisfiability*, i.e., the *consistency* of its axioms.[51]

Bernays observes further that such proofs are usually given by providing arithmetical models, unless one can get by with the construction of finite ones. Thus, Bernays has retraced Hilbert's motivation for a consistency proof or, perhaps better, isolated the methodological core of his considerations. What is surprising after Herbrand's and Gödel's dissertations is what seems to be an unapologetic identification of satisfiability and consistency. In an unpublished note (found in the appendix), Bernays describes this connection properly and in harmony with the principles guiding proof theoretic investigations. In any event, Bernays formulates here first a position of (what Parsons calls) *eliminative structuralism* in a concise way. That position is complemented by principled, philosophical reflections and programmatic, mathematical efforts to obtain finitist consistency proofs. It is this additional reflective perspective that allows us to see Hilbert and Bernays's structuralism as being of the noneliminative variety; see the beginning of section 5.2.

We saw how Hilbert and Bernays tried to exploit the special epistemological status of finitist mathematics for consistency proofs. After the discovery of Gödel's Second Incompleteness Theorem, however, the fundamental status of finitist mathematics had to be given up, and finitist considerations had to be expanded by considerations in stronger theories. To avoid a threatening vicious circle, as in (Hilbert 1905a), these stronger theories had to be constructively motivated. The first consistency proof satisfying such an informal demand was obtained independently by Gödel and Gentzen and established the consistency of classical arithmetic relative to its intuitionist version: the latter theory was indeed based on an extended constructive viewpoint, ironically, the intuitionist one. (Bernays 1954) called this sharpened axiomatics (verschärfte Axiomatik) and formulated as a minimal requirement that "the objects [making up the intended model of the theory] are not taken from a domain that is thought as being already given, but are rather constituted by generative processes".[52] There is no

---

[51](Bernays 1930b, p. 21). "Ist eine solche Erfüllung undenkbar, d.h. logisch unmöglich, so führt das Axiomensystem zu gar keiner Theorie, und die einzige logisch belangvolle Aussage über das System [von Axiomen, WS] ist dann die Feststellung des aus den Axiomen sich ergebenden Widerspruchs. Aus diesem Grunde besteht für jede axiomatische Theorie die Erforderlichkeit eines Nachweises der *Erfüllbarkeit*, d.h. der *Widerspruchsfreiheit* ihrer Axiome." It should be evident from my earlier discussion that Dedekind saw both of these moments very clearly and, consequently, formulated the consistency problem most appropriately and sharply. He tried to resolve it by model theoretic considerations within logic.

[52](Bernays 1954, pp. 11–12). "Die Mindest-Anforderung an eine verschärfte Axiomatik ist die, dass die Gegenstände nicht einem als vorgängig gedachten Bereich entnommen werden, sondern durch Erzeugungsprozesse konstituiert werden." Bernays continues with a methodologically important remark: "Es kann aber dabei die Meinung sein, dass durch diese Erzeugungsprozesse der Umkreis der Gegenstände determiniert ist; bei dieser Auffassung erhält das *tertium non datur* seine Motivierung. In der Tat kann Offenheit eines Bereiches in zweierlei Sinn verstanden werden, einmal nur so, dass die Konstruktionsprozesse über jeden

Beyond Hilbert's reach?

indication in Bernays's (1954) or in his later writings what kind of generative processes should be considered, and why that particular feature of domains should play a distinctive, foundational role. These two issues are at the center of the considerations in the next section.

5 Accessible domains

When Gödel considered Platonism still as a doctrine "which cannot justify any critical mind and which does not even produce the conviction that they [the axioms of set theory] are consistent", he analyzed also different layers of constructive mathematics in most informative ways. The lowest layer, identified with finitist mathematics, had one important characteristic:

The application of the notion "all" or "any" is to be restricted to those infinite totalities for which we can give a finite procedure for generating all their elements (as we can, e.g., for the totality of integers by the process of forming the next greater integer and as we cannot, e.g., for the totality of all properties of integers).[53]

Directly associated with the procedure for generating the integers are the principles of proof by induction and definition by recursion. There is no further analysis of this direct association; the principles are simply taken to have a high degree of evidence. Can one go beyond such a brief, purely descriptive account and, perhaps, extend the considerations to other classes of mathematical objects?

5.1 Finitist objects and processes. Gödel considered the totality of integers as just one *example* of totalities whose elements are generated by a finite procedure. That a greater class of such totalities has directly associated principles had been emphasized already by Poincaré (1905). After a discussion of the induction principle for natural numbers, he remarked:

I did not mean to say, as has been supposed, that all mathematical reasonings can be reduced to an application of this principle. Examining these reasonings closely, we should see applied there many analogous principles, presenting the same essential characteristics. In this category of principles, that of complete induction is only the simplest of all and this is why I have chosen it as a type. (p. 1025)

Modern expositions and critical examinations of Hilbert's considerations, for example, those of Parsons and Tait, focus on natural numbers. As a

---

einzelnen Gegenstand hinausführen, und andererseits in dem Sinne, dass der resultierende Bereich überhaupt nicht eine mathematisch bestimmte Mannigfaltigkeit darstellt. Je nachdem die Zahlenreihe in dem erstgenannten oder in dem zweiten Sinne aufgefasst wird, hat man die Anerkennung des *tertium non datur* in bezug auf die Zahlen oder den intuitionistischen Standpunkt. Bei dem finiten Standpunkt kommt noch die Anforderung hinzu, dass die Überlegungen an Hand der Betrachtung von endlichen Konfigurationen verlaufen, somit insbesondere Annahmen in der Form allgemeiner Sätze ausgeschlossen werden."

[53](Gödel 1933b, p. 51). The reflections in this paper are continued most directly in Gödel's *Lecture at Zilsel's* (1938); the latter notes contain in particular a detailed analysis of Gentzen's first consistency proof for elementary number theory.

matter of fact, so did Bernays (1930b), but he viewed the case of numbers as paradigmatic and embedded it into broader reflections on the nature of mathematical knowledge; the latter was to be captured in a principled way, independently of the current inventory of mathematical disciplines. Bernays viewed a *certain kind of abstraction* as distinctive for the nature of mathematical thought:

> This abstraction may be called formal or mathematical abstraction. It consists in emphasizing the structural moments of an object, i.e., the way it is composed from parts, and taking them exclusively into consideration; 'object' is here to be understood in the broadest sense. Accordingly, mathematical knowledge can be defined as knowledge based on the structural consideration of objects.[54]

The crucial questions are undoubtedly, what is the extension of *object*, and what kind of objects can be considered or viewed *structurally*. The second question implicitly concerns the boundary between mathematical knowledge secured by intuition (Anschauung), respectively obtained by thinking (Denken) and systematic extrapolation; it is to this question that Bernays turned.

Bernays's analysis of intuitive mathematical knowledge attempts to balance, uneasily, the philosophical demand for intuitive concreteness and the mathematical need for formal abstractness. The tension comes to the fore in first taking formal abstraction as the characteristic feature of intuitive mathematical knowledge, and in then claiming that it is naturally bound to finiteness and finds a principled delimitation only when facing the infinite.[55] Precisely this coextensiveness of finite and intuitive—and thus the sharp differentiation of the intuitive from the non-intuitive—was

---

[54](Bernays 1930b, p. 23). The German text: "Diese Abstraktion, welche als die *formale* oder *mathematische Abstraktion* bezeichnet werden möge, besteht darin, daß von einem Gegenstand—'Gegenstand' hier im weitesten Sinne genommen—die strukturellen Momente, d.h. die Art der Zusammensetzung aus Bestandteilen hervorgekehrt und ausschließlich in Betracht gezogen wird. Man kann demnach als mathematische Erkenntnis eine solche definieren, welche auf der strukturellen Betrachtung von Gegenständen beruht."

[55]The first feature is expressed most clearly on p. 30 of (Bernays 1930b): "Als das Charakteristische an der mathematischen Erkenntnisweise haben wir die formale Abstraktion, d.h. die Einstellung auf die strukturelle Seite der Gegenstände festgestellt und damit das Feld des Mathematischen in grundsätzlicher Weise abgegrenzt." That is supplemented most forcefully on p. 40: "Die wesentliche Gebundenheit der formalen Abstraktion an das Moment der Endlichkeit macht sich insbesondere dadurch geltend, daß bei den Betrachtungen von Gesamtheiten und von Figuren die Eigenschaft der Endlichkeit für den Standpunkt der anschaulichen Evidenz gar kein besonderes beschränkendes Merkmal bildet. Die Beschränkung auf das Endliche wird von diesem Standpunkt aus ganz ohne weiteres, sozusagen *stillschweigend* vollzogen. Wir brauchen hier keine besondere Definition der Endlichkeit, denn die Endlichkeit der Objekte versteht sich für die formale Abstraktion ganz von selbst." The second aspect is emphasized on pages 38 and 39, where Bernays argues that formal abstraction helps us to transcend the limits of our "faktischen" or "wirklichen Vorstellungskraft": "An solche Grenzen für die Möglichkeit der Verwirklichung kehrt sich aber die anschauliche [sic!] Abstraktion nicht. Denn diese Grenzen sind vom Standpunkt der formalen Betrachtung zufällig. Die formale Abstraktion findet sozusagen keine frühere Stelle für eine prinzipielle Abgrenzung als bei dem Unterschied des Endlichen und Unendlichen." Bernays continues in the next paragraph: "Dieser Unterschied ist in der Tat ein grundsätzlicher. Wenn wir uns genauer besinnen, wie denn überhaupt eine

Beyond Hilbert's reach? 367

questioned by Bernays himself in the Nachtrag to (1930b), when arguing that the epistemological considerations underlying his paper should be revised in light of Gödel's results:

Of course, the positive remarks, in particular those bringing out the mathematical element in logic and those highlighting elementary arithmetical evidence, are hardly in need of revision. It seems, however, that the sharp differentiation between the intuitive and non-intuitive, as used in treating the problem of the infinite, cannot be carried through this strictly. In this respect then, the view on the formation of mathematical ideas has to be worked out in further detail.[56]

Bernays refers to his later essays in *Abhandlungen zur Philosophie der Mathematik* as containing considerations to address this fundamental issue. His arguments of 1930 provide, it seems to me, an excellent starting point, as their detailed examination uncovers revealing difficulties. Thus, I will focus on them without relating them at this occasion to the broader philosophical framework in which they have their systematic place. That framework is deeply influenced by Kant, Fries, and Nelson, with Bernays keeping however a distinctive critical distance. The curious reader may consult Bernays's papers (1928b, 1928a, 1930a, 1937); to recognize that some of the issues discussed below are parallel to (still unresolved) problems in Kant's philosophy of arithmetic, see in particular the writings of Parsons, e.g., (1980), (1982), (1984), and (1994).

The arguments that support the uneasy balancing act between philosophically motivated concreteness and mathematically necessitated abstractness and generality are at crucial places strained. That is most evident, when formal abstraction is supposed to help us in going beyond the limits of our real power of representation (our "faktische Vorstellungskraft" or "tatsächliches Vorstellungsvermögen"); here, *intuitiveness of objects is to be secured by intuitiveness of processes generating them*. A similar step from objects to processes is taken when Bernays argues next that formal abstraction is essentially bound to the "moment of finiteness". Indeed, Bernays claims, finiteness is not at all a restrictive feature of objects from the standpoint of intuitive evidence: the finiteness of objects is obvious for formal abstraction ("die Endlichkeit der Objekte versteht sich für die

---

unendliche Mannigfaltigkeit als solche charakterisiert sein kann, so finden wir, daß dieses gar nicht nach der Art einer anschaulichen Aufweisung möglich ist, sondern nur auf dem Wege der Behauptung (bzw. der Annahme oder der Feststellung) einer gesetzlichen Beziehung. Unendliche Mannigfaltigkeiten sind uns demnach nur durch das *Denken* zugänglich. Dieses Denken ist zwar auch eine Art des Vorstellens, aber es wird dadurch nicht die Mannigfaltigkeit als Gegenstand vorgestellt, sondern es werden Bedingungen vorgestellt, denen eine Mannigfaltigkeit genügt (bzw. zu genügen hat)."

[56](Bernays 1976a, p. 61). The German text: "Freilich, die positiven Ausführungen, insbesondere die Aufweisung des mathematischen Elementes in der Logik und die Herausstellung der elementaren arithmetischen Evidenz, bedürfen wohl kaum der Revision. Jedoch, die scharfe Unterscheidung des Anschaulichen und des Nicht-Anschaulichen, wie sie bei der Behandlung des Problems des Unendlichen angewandt wird, ist anscheinend nicht so strikt durchführbar, und die Betrachtung der mathematischen Ideenbildung bedarf wohl in dieser Hinsicht noch der näheren Ausarbeitung."

formale Abstraktion ganz von selbst"). Why should that be? The answer to this question is given by Bernays paradigmatically for the case of numbers and appeals to their introduction as the "simplest formal objects" by iteration of a successor operation. This intuitive-structural introduction of numbers is appropriate, Bernays claims, only for finite numbers, as repetition is from the standpoint of "intuitive-formal considerations" *eo ipso* finite repetition. In short, and in parallel to the above italicized claim, *finiteness of these objects is to be secured by finiteness of the underlying generative process.*[57]

Finally, according to Bernays, it is the intuitive representation of the finite (die anschauliche Vorstellung des Endlichen) that provides the justification (Erkenntnisgrund) for the principle of complete induction and for the admissibility of recursive definitions, both in their elementary forms. Such a representation of the finite is thus explicitly used when reflecting on general characteristics of intuitive objects, and it is a crucial presupposition for the proof theoretic approach.

> The *intuitive representation of the finite* [my emphasis, WS] is forced on us, as soon as a formalism is turned into an object of investigation, thus especially in the systematic theory of logical inferences. This brings out that finiteness is an essential moment of any formalism whatsoever.[58]

Bernays's analysis is consequently also basic for other domains whose elements are generated in elementary ways, especially for the domains of syntactic objects needed in proof theoretic investigations. Indeed, he continues by claiming that the limits of formalisms coincide with those of the general representability of intuitive combinations. (Die Grenzen des Formalismus sind aber keine anderen als die der Vorstellbarkeit überhaupt von anschaulichen Zusammensetzungen.)

How do these considerations compare with Hilbert and Bernays's earlier ones? Should the objects obtained through such elementary generation satisfy the demand articulated in the notes from 1921–22, namely, that "the figures we take as objects must be completely surveyable and only discrete determinations are to be considered for them"? Surveyability was

---

[57]Two aspects are finite: the number of repetitions and what Bernays calls the "iteration figure" that formally represents the generating steps of the elementary operation. Bernays discusses on pp. 30–32 the introduction of natural numbers as the simplest formal objects; the argument for the first italicized claim is presented on pp. 38–39, that for the second claim on p. 40. See also note 54. The importance of the "iterativistic tendency" was emphasized by Hand (1989, 1990); see also section 3 of (Zach 1998). The suggestion, however, to base a nonstandard semantics for numerical statements on this tendency runs into difficulty when trying to account for the meaningfulness of statements concerning syntactic objects, and that is crucial for proof theoretic investigations. It is an interesting suggestion, if one takes as the "explicit (and only) goal" of the finitist viewpoint "to give an account of truth for (a fragment of) arithmetic which is secure", as claimed in (Zach 1998, p. 44).

[58](Bernays 1930b, p. 40). The German text: "Zwangsmäßig aber stellt sich die *anschauliche Endlichkeitsvorstellung* ein [my emphasis, WS], sobald man einen Formalismus selbst zum Gegenstand der Betrachtung macht, insbesondere also in der systematischen Theorie der logischen Schlüsse. Es kommt hiermit zum Ausdruck, daß die Endlichkeit ein wesentliches Moment an den Gebilden eines jeden Formalismus ist."

Beyond Hilbert's reach? 369

then thought to insure that "our claims and considerations have the same reliability and clarity (Handgreiflichkeit) as in intuitive number theory". Against the backdrop of the generation of numerals we have here the same tension as in the considerations by Bernays, just replacing "surveyability" with "intuitive representability". In order to ground mathematical principles for finitist objects, the elementary and uniform generation of figures has to be appealed to—leading to *purely formal objects* of appropriate abstractness; in order to ground philosophical reflections on the primitive intuitive character of finitist mathematical knowledge, focus is shifted to the surveyability or intuitive representability of individual mathematical objects. Indeed, Bernays claims that we are free to represent (repräsentieren) the purely formal objects by concrete objects (e.g., numbers by numerals) in such a way that these representing concrete objects are intuitable and contain in their structure the essential properties of the represented objects, so that "the relations—to be investigated and holding—between the represented objects are found also between the representatives and can be ascertained by considering the representatives alone".[59]

The deep conflict that is apparent in this intricate discussion is not resolved by argument, but by *fiat*: numbers and other purely formal objects just are intuitively given—via representing concrete objects. It is most interesting to observe that Bernays (1935b) contemplates narrower and admittedly vague boundaries for what is intuitive and distinguishes between numbers that are *reachable* (zugänglich) and those that are not. He does so in a critical discussion of intuitionism and intuitive evidence, viewing as reachable those numbers that do not outstrip our actual power of representation (Vorstellungskraft).[60] He suggests also a way of extending mathematical knowledge from reachable numbers to unreachable ones by the *general method of analogy* (die allgemeine Methode der Analogie), i.e., by extending the relations that can be verified for the former numbers to the latter. However one may want to interpret this, it seems clear that

---

[59]This is formulated in (Bernays 1930b, fn. 4 on p. 32). The reader should note that "represent" is here actually translating "repräsentieren"; earlier on and later on it is the translation for "vorstellen". The full German text of the note is: "Der Philosoph ist geneigt, dieses Verhältnis der Repräsentation als einen Bedeutungszusammenhang anzusprechen. Man hat aber zu beachten, daß gegenüber dem gewöhnlichen Verhältnis von Wort und Bedeutung hier der wesentliche Unterschied besteht, daß der repräsentierende Gegenstand in seiner Beschaffenheit die wesentlichen Eigenschaften des repräsentierten Objektes enthält, so daß die zu untersuchenden Beziehungen der repräsentierten Objekte sich auch an den Repräsentanten vorfinden und durch die Betrachtung der Repräsentanten selbst festgestellt werden können."

[60]The discussion is found on p. 70. (Parsons 1982, p. 496) makes a related distinction for Kant's philosophy of mathematics. He distinguishes between weak and strong intuitability as follows: "An object is strongly intuitable if it can be intuited, i.e., if it can be an object of intuition. An object is weakly intuitable if it can be represented in intuition without itself being intuitable. This notion is vague because we have not said what is meant by 'representing' an object in intuition. However, representation of abstract objects by concrete objects, or by objects relatively closer to the concrete, is a pervasive phenomenon and of great importance for understanding abstract objects."

finitist mathematics is not secured by intuitive evidence alone. For an adequate conceptual analysis of finitist mathematics one has to go beyond (the representation of) finiteness, admit rather abstract means for capturing the arbitrary finite iteration of elementary steps, and grant in the end for potentially infinite domains what Bernays asserts for actually infinite ones, namely, that they can be characterized only by way of a lawful relation (gesetzliche Beziehung). However, the distinctive feature of domains with generated elements is that their lawful relation is not just assumed or claimed, but rooted in our understanding of the underlying generative process; that understanding allows us also to recognize induction principles for proofs and recursion principles for functions.

Bernays's broad *informal considerations* leading up to the natural numbers as *unique* (eindeutige) and *purely formal objects* (distinct from formal objects, i.e. types, that allow different concrete instantiations by tokens) are very appealing and, in a deep sense, similar to those of Dedekind, Helmholtz, and Kronecker.[61] However, only Dedekind, who reported in the letter to Keferstein on the informal reflections underlying his work in (1888) took the further step to a sharp and completely novel mathematical formulation. The latter makes crucial use of infinite sets and in particular of *simply infinite systems*. Following Dedekind, but avoiding infinite sets, Zermelo (1909) presented an analysis based on finite sets and "*simply finite systems*". A central question is, *can Zermelo's considerations provide the mathematical basis for a detailed conceptual analysis of natural numbers?* (Bernays's natural numbers as purely formal objects might be obtained then by Tait's "Dedekind abstraction" applied to simply finite systems. Zermelo's work and its connections to that of others is described by Parsons (1987).)

5.2 I.d. classes and abstract notions. Hilbert and Bernays's structuralism, when joined with their finitist methodological reflections, is really a structuralism of Parsons's *noneliminative* variety: it accepts basic, potentially infinite domains of constructed objects, in particular, of natural numbers and syntactic figures constituting formalisms.[62] I propose to call their structuralism and extensions thereof *reductive*, because of the special justificatory role the basic structures play. This is in analogy to "reductive proof theory". The systematic connection will become clear, I trust, from the following considerations that concern the challenging question, how to extend the preliminary and not unproblematic reflections of section 5.1 to appropriate infinitary configurations. Let me make this challenge concrete by

---

[61] I refer to Kronecker's *Über den Zahlbegriff* and Helmholtz's *Zählen und Messen, erkenntnistheoretisch betrachtet*; both papers were published in the Zeller-Festschrift, Leipzig 1887. Dedekind refers to these two papers when remarking in the first note to the *Vorwort* of his (1888): "Das Erscheinen dieser Abhandlungen ist die Veranlassung, welche mich bewogen hat, nun auch mit meiner, in mancher Beziehung ähnlichen, aber durch ihre Begründung doch wesentlich verschiedenen Auffassung hervorzutreten, die ich mir seit vielen Jahren und ohne jede Beeinflussung von irgendwelcher Seite gebildet habe."

[62] (Parsons 1990, section 8).

describing one paradigmatic result of reductive proof theory. The domains of constructed objects are here the higher constructive number classes.

Brouwer (1927) considered infinite proofs[63] and treated them as well-founded trees, i.e., as constructive ordinals of the second number class **O**. The latter are inductively generated according to the following clauses:

0 is in **O**;

if a is in **O**, then the successor of a is in **O**;

if f is a function from **N** to **O** and, for all n in **N**, $f(n)$ is in **O**, then the supremum of the $f(n)$ is also in **O**.

Even higher number classes were inductively defined by Brouwer (and by Hilbert in *Über das Unendliche*); the trees branch over **N** and over previously obtained number classes. These are quite complex i.d. classes, but acceptable to at least some constructivists, among them Bishop, Lorenzen, Myhill, but also Martin-Löf. Iterated i.d. classes were at the center of the foundational investigations in the Stanford Report and much subsequent work; see (Feferman 1981). Church and Kleene had formulated already in the mid-thirties recursive analogues of the higher constructive number classes by requiring that the function f in the third defining clause be (partial) recursive. The elements of **O** can be pictured as infinitary trees that are uniformly and effectively generated; indeed, arbitrary finite subtrees can be effectively determined.

With that specification of constructive function it is quite direct to formulate proof and definition principles for the finite constructive number classes in the language of elementary number theory expanded by predicate symbols for the number classes. The resulting theory, based on intuitionist logic, is denoted by $ID^i(\mathbf{O})_{<\omega}$. The paradigmatic result I want to discuss briefly reduces the impredicative subsystem of classical analysis $(\Pi_1^1\text{-CA})_0$ to the theory $ID^i(\mathbf{O})_{<\omega}$.[64] This reduction is pleasing for two reasons, especially, if one is affected by the implicit irony: (i) $(\Pi_1^1\text{-CA})_0$ suffices as a comprehensive formal framework for the development of mathematical analysis presented in the fourth supplement of Hilbert and Bernays's *Grundlagen der Mathematik II*, and (ii) Brouwer's constructive number classes provide the objective underpinnings for proving the consistency of a blatantly impredicative classical theory. There is a great deal of contemporary work in proof theory that extends this kind of result mainly

---

[63] Brouwer added in a famous footnote: "These mental mathematical proofs that in general contain infinitely many terms must not be confused with their linguistic accompaniments, which are finite and necessarily inadequate, hence do not belong to mathematics." Brouwer claimed that this remark contains his "main argument against the claims of Hilbert's metamathematics".

[64] The result is obtained from work by Tait or by Feferman (in the Buffalo volume, (Kino, Myhill and Vesley 1970) and my Stanford dissertation of 1977; Tait's work goes back to early 1967 and is reported in (1968b). For the detailed exposition of this and related results see (Buchholz et al. 1981) and in particular my chapter "Inductive definitions, constructive ordinals, and normal derivations", pp. 143–87 of that volume.

by providing ordinal analyses for stronger theories. However, attention has been shifted from subsystems of analysis to a more uniform setting of subsystems of set theory, and the systems of ordinal notations needed for the proof theoretic analysis are connected rather directly with large cardinals in set theory.[65]

Aczel (1977) presented a very general notion of i.d. classes, that is, of classes given by inductive definitions in the broadest sense. All the examples I mentioned (the elementary i.d. classes of terms, formulas, and proofs constituting a formal theory, the Brouwerian constructive ordinals) fall under Aczel's notion. Indeed, Aczel's notion is so general that it encompasses also segments of the cumulative hierarchy of sets. These i.d. classes are in Aczel's terminology deterministic and, thus, guarantee the unique generation of the objects falling under them. Given an understanding of the uniform generation steps, the resulting processes allow us to understand the build-up of objects and to recognize proof and definition principles for the domains constituted by them. I call such i.d. classes *accessible domains*[66] and would like to see an abstract mathematical description that highlights their distinctive features. Joyal and Moerdijk's book *Algebraic Set Theory* and the subsequent paper (2000b) by Moerdijk and Palmgren take, it seems, interesting steps that characterize some classes of accessible domains from the perspective of category theory. Here then is the general question, namely, *can one give a category theoretic characterization of accessible domains*?

The crucial task for Hilbert's proof theory is to insure the consistency of the "idea of the infinite totality of numbers and of number sets". That is formulated in (Bernays 1930b) in accord with the historical development sketched in sections 2 through 4, and it required then the use of finitist methods. The task is taken up again in (Bernays 1935b) with a broadened methodological perspective. The assumptions of totalities underlying mathematical theories are called *Platonist*; the condition that restricts their use, as well as the application of the principle of analogy, is described as follows:

The assumptions we are dealing with amount to representations of totalities and to the principle of analogy or of the permanence of laws. And the condition restricting the application of these leading ideas is none other than the consistency of the consequences that can be drawn from those basic assumptions.[67]

Accessible domains have a foundational role in providing the means for consistency proofs, whether syntactic, proof theoretic, or semantic,

---

[65] For discussions of this part of advanced proof theory, see (Jäger 1986), (Pohlers 1989), (Rathjen 1995), and (Buchholz 2002).

[66] A detailed presentation of the central features of Aczel's i.d. classes is found in (Feferman and Sieg 1981a, p. 18–25).

[67] (Bernays 1935b, p. 75). The German text: "Die Annahmen, um die es sich dabei handelt, laufen hinaus auf Vorstellungen von Gesamtheiten und auf das Prinzip der Analogie oder der Permanenz der Gesetze. Und die Bedingung, welche die Anwendung dieser Leitgedanken einschränkt, ist keine andere als die der Widerspruchsfreiheit der Folgerungen, die sich aus jenen zugrundegelegten Annahmen ergeben."

model-theoretic ones. Detailed mathematical and philosophical analyses of accessible domains will allow us to make informative distinctions that concern (constructive) generating operations and their (transfinite) iteration, but also fundamental deductive principles. This leads to a rather natural question, namely, *can theories for accessible domains be given in such a form that their classical versions are uniformly reducible to their intuitionist variants?* As I consider suitable segments of the cumulative hierarchy as accessible domains, the investigations of axioms of infinity (i.e., large cardinal assumptions) are of deep conceptual interest and increasingly connected to proof theoretic work, as I mentioned above. This is a part of set theory, where wide-ranging Platonist assumptions in Bernays's sense are being made, and where their consequences on more concrete mathematical problems are being explored. This latter point was emphasized already by Gödel and has been pursued most vigorously by H. Friedman.

Accessible domains reflect the constructive or, if you wish, *quasi-constructive aspect* of mathematical experience; abstract notions like groups, fields, topological spaces reflect its *conceptual aspect*. These two aspects should be contrasted rather sharply. Accessible domains allow us to formulate correct fundamental principles, whereas abstract notions are distilled from mathematical practice to make precise analogies between different areas; that is done for the purpose of comprehending complex connections and obtaining a more profound understanding.[68] I have stressed (1990a, p. 284–5) the broad significance of this distinction; in addition, I argued specifically that the notion of a complete ordered field (characterizing its model, the reals, up to isomorphism) is an abstract one. How different this case is from that of a simply infinite system is revealed by an analysis of the categoricity proofs: in the one case the desired isomorphism follows the build-up of the objects in the domain, whereas in the other the topological completeness has to be appealed to. Thus, there are two complementary important tasks: to analyze the principles for accessible domains and to establish the consistency of abstract notions relative to (theories for) appropriate domains.[69]

Let me return to Hilbert. As philosophers, mathematicians, and scientists we should explore Hilbert's broad insights into the complex workings of mathematics instead of keeping him shackled to a narrow foundational position that was taken for programmatic reasons in the twenties. Hilbert's

---

[68] (Bernays 1930b, p. 44) makes a related, but certainly not "identical" distinction, referring in a footnote to Fries; he distinguishes between "dem elementar-mathematischen Standpunkt und einem darüber hinausgehenden systematischen Standpunkt". This more systematic standpoint covers not only analysis, but also set theory, whereas for me set theory with its iterative conception falls under the quasi-constructive aspect of mathematical experience.

[69] I also pointed out (1990a) that this second task (as well as the first one) cuts across the traditional foundational divides: if there is an abstract notion in intuitionist mathematics, it is that of a choice sequence introduced by Brouwer to capture the essence of the continuum; the consistency proof of the theory of choice sequences relative to the theory of the second constructive number class $ID_1^i(O)$ can be viewed as fulfilling exactly this task. The proof was given by (Kreisel and Troelstra 1970).

particular proposal for mediating between constructivist and classical positions did not work out. However, the reductive program that emerged from it provides, in my view, an important perspective on mathematical experience. *Reductive structuralism* allows us to connect, in a prima facie coherent way, developments in (the foundations of) mathematics and more directly philosophical studies; it helps us to gain a better understanding of the distinctive character of modern mathematics and its role in our broader intellectual enterprises, in particular its role in the sciences.[70] Hilbert's modernized self can be taken as arguing for creative freedom along two dimensions: *constructions* and *abstract concepts*; the former call for abstract analysis, the latter for constructed models.

6 Concluding remark

I want to end with a most appropriate comment of Stein's on Hilbert; it mirrors the remark I quoted from Hilbert's Paris Lecture in section 1. After complaining gently about Hilbert's insistence (in his later foundational investigations) that the statements of ordinary mathematics are meaningless and only finitist statements have meaning, Stein points out:

Hilbert certainly never abandoned the view that mathematics is an organon for the sciences . . . ; and he surely did not think that physics is meaningless, or its discourse a play with "blind" symbols. His point is, I think, this rather: that the mathematical logos has no responsibility to any imposed *standard* of meaning: not to Kantian or Brouwerian "intuition", not to finite or effective decidability, not to anyone's metaphysical standards for "ontology"; its *sole* "formal" or "legal" responsibility is to be consistent (of course, it has also what one might call a "moral" or "aesthetic" responsibility: to be useful, or interesting, or beautiful; but to this it cannot be constrained—poetry is not produced through censorship).[71]

The mathematical (in particular, proof theoretic) and philosophical challenge is, of course, to analyze on what basis we can live up to the responsibility of being consistent.

Appendix

This appendix consists of a note by Bernays that is found in the Hilbert Nachlaß, cod. 685:9, 2 and was written, presumably, between 1925 and 1928. It is entitled *Existenz und Widerspruchsfreiheit*. The note is preceded by a brief note on the finitist standpoint (containing only well-known observations) and followed by a note analyzing the criticism of axiomatic set theory by Skolem and von Neumann. For our purposes the latter note is of interest only by the way in which Bernays explicates "consistency of the

---

[70]See (Weyl 1925, pp. 540ff) and (Weyl 1927, p. 482–84), in particular p. 484 where the earlier ideas are reexpressed.
[71](Stein 1988, p. 255).

countable infinite". The consistency proof of arithmetic (including the transfinite axioms for the epsilon operator) establishes the consistency of the countable infinite in the following sense: "An axiom system has been recognized as consistent that cannot be satisfied by a finite system of objects." Clearly, this topic is taken up in a very illuminating way in (Bernays 1950).

Existence and consistency. The claim: "Existence = consistency" can only refer to a system *as a whole*. *Within* an axiomatic system the axioms decide about the existence of objects.

If, for a system as a whole, consistency is to be synonymous with existence, then the proof of consistency must consist in an exhibition [of a model].

(All consistency proofs up to now have been either direct exhibitions or indirect ones by reduction; in the latter case a certain other system is already taken as existent. *Frege* has defended with particular emphasis the view that any proof of consistency has to be given by the actual presentation of a system of objects.)

In proof theory, laying a new foundation of arithmetic, consistency proofs are *not* given by exhibition. From this foundational standpoint it does not hold any longer that existence equals consistency. Indeed, it is not the opinion that the possibility of an infinite system is to be proved, rather it is only to be shown that *operating with such a system* does not lead to contradictions in mathematical reasoning.

Existenz und Widerspruchsfreiheit. Die Behauptung: "Existenz = Widerspruchsfreiheit" kann sich immer nur auf ein System *als Ganzes* beziehen. *Innerhalb* eines axiomatischen Systems wird über die Existenz von Dingen durch die Axiome entschieden.

Soll für ein System als Ganzes die Widerspruchsfreiheit mit der Existenz gleichbedeutend sein, so muss der Beweis der Widerspruchsfreiheit in einer Aufweisung bestehen.

(Alle bisherigen Wf.-Beweise sind auch entweder direkte Aufweisungen oder indirekte durch Zurückführung, wobei dann ein gewisses anderes System schon als existent angenommen wird. *Frege* hat bes. nachdrücklich den Standpunkt vertreten, dass der Nachweis der Widerspruchsfreiheit durch wirkliche Aufzeigung eines Systems von Dingen geschehen müsse.)

In der neuen Grundlegung der Arithmetik durch die Beweistheorie geschieht der Wf.-Beweis *nicht* durch eine Aufweisung. Vom Standpunkt dieser Begründung gilt auch nicht mehr, dass Existenz = Widerspruchsfreiheit ist. In der Tat ist ja auch gar nicht die Meinung, dass die Möglichkeit eines unendlichen Systems erwiesen werden soll, vielmehr soll nur gezeigt werden, dass das *Operieren mit einem solchen System* beim Schliessen nicht zu Widersprüchen führt.

III.3

Searching for proofs (and uncovering capacities of the mathematical mind)*

Abstract. What is it that shapes mathematical arguments into proofs that are intelligible to us, and what is it that allows us to find proofs efficiently? — This is the informal question I intend to address by investigating, on the one hand, the abstract ways of the axiomatic method in modern mathematics and, on the other hand, the concrete ways of proof construction suggested by modern proof theory. These theoretical investigations are complemented by experimentation with the proof search algorithm AProS. It searches for natural deduction proofs in pure logic; it can be extended directly to cover elementary parts of set theory and to find abstract proofs of Gödel's incompleteness theorems. The subtle interaction between understanding and reasoning, i.e., between *introducing concepts* and *proving theorems*, is crucial. It suggests principles for structuring proofs conceptually and brings out the dynamic role of *leading ideas*. Hilbert's work provides a perspective that allows us to weave these strands into a fascinating intellectual fabric and to connect, in novel and surprising ways, classical themes with deep contemporary problems. The connections reach from proof theory through computer science and cognitive psychology to the philosophy of mathematics and all the way back.

1 Historical perspective

It is definitely counter to the standard view of Hilbert's formalist perspective on mathematics that I associate his work with uncovering aspects of the mathematical mind; I hope you will see that he played indeed a pivotal role. He was deeply influenced by Dedekind and Kronecker; he connected these extraordinary mathematicians of the 19th century to two equally remarkable logicians of the 20th century, Gödel and Turing. The character of that connection is determined by Hilbert's focus on the *axiomatic method* and the associated *consistency problem*. What a remarkable path it is: emerging from the radical transformation of mathematics in

---

*This essay is dedicated to Grigori Mints on the occasion of his 70th birthday. Over the course of many years we have been discussing the fruitfulness of searching directly for natural deduction proofs. He and his Russian colleagues took already in 1965 a systematic and important step for propositional logic; see the co-authored paper (Shanin, et al. 1965), but also (Mints 1969) and the description of further work in (Maslov, Mints, and Orevkov 1983).

the second half of the 19th century and leading to the dramatic development of metamathematics in the second half of the 20th century.

Examining that path allows us to appreciate Hilbert's perspective on the wide-open mathematical landscape. It also enriches our perspective on his metamathematical work.[1] Some of Hilbert's considerations are, however, not well integrated into contemporary investigations. In particular, the *cognitive* side of proof theory has been neglected, and I intend to pursue it in this essay. It was most strongly, but perhaps somewhat misleadingly, expressed in Hilbert's Hamburg talk of 1927. He starts with a general remark about the "formula game" criticized by Brouwer:

> The formula game ... has, besides its mathematical value, an important general philosophical significance. For this formula game is carried out according to certain definite rules, in which the technique of our thinking is expressed. These rules form a closed system that can be discovered and definitively stated.

Then he continues with a provocative statement about the cognitive goal of proof theoretic investigations.

> The fundamental idea of my proof theory is none other than to describe the activity of our understanding, to make a protocol of the rules according to which our thinking actually proceeds.[2]

It is clear to us, and it was clear to Hilbert, that mathematical thinking does not proceed in the strictly regimented ways imposed by an austere formal theory. Though formal rigor is crucial, it is not sufficient to shape proofs intelligibly or to discover them efficiently, even in pure logic. Recalling the principle that mathematics should solve problems "by a minimum of blind calculation and a maximum of guiding thought", I will investigate the subtle interaction between understanding and reasoning, i.e., between *introducing concepts* and *proving theorems*. That suggests principles for structuring proofs conceptually and brings out the dynamic role of *leading ideas*.[3]

---

[1] In spite of the demise of the finitist program, proof theoretic work has been continued successfully along at least two dimensions. There is, first of all, the ever more refined formalization of mathematics with the novel mathematical end of extracting information from proofs. Formalizing mathematics was originally viewed as the basis for a mathematical treatment of foundational problems and, in particular, for obtaining consistency results. Gödel's theorems shifted the focus from absolute finitist to relative consistency proofs with the philosophical end of comparing foundational frameworks; that is the second dimension of continuing proof theoretic work. These two dimensions are represented by "proof mining" initiated by Kreisel and "reductive proof theory" pursued since Gödel and Gentzen's consistency proof of classical relative to intuitionist number theory.

[2] (Hilbert 1927) in (van Heijenoort 1967, p. 475).

[3] The way in which I am pursuing matters is programmatically related to Wang's perspective in his (1970). In that paper Wang discusses, on p. 106, "the project of mechanizing mathematical arguments". The results that have been obtained so far, Wang asserts, are only "theoretical" ones, "which do not establish the strong conclusion that mathematical reasoning (or even a major part of it) is mechanical in nature". But the unestablished strong conclusion challenges us to address in novel ways "the perennial problem about mind and machine" — by dealing with mathematical activity in a systematic way. Wang continues: "Even though

In some sense, the development toward proof theory began in late 1917 when Hilbert gave a talk in Zürich, entitled *Axiomatisches Denken*. The talk was deeply rooted in the past and pointed decisively to the future. Hilbert suggested, in particular,

... we must — that is my conviction — take the concept of the specifically mathematical proof as an object of investigation, just as the astronomer has to consider the movement of his position, the physicist must study the theory of his apparatus, and the philosopher criticizes reason itself.

Hilbert recognized, in the very next sentence, that "the execution of this program is at present, to be sure, still an unsolved problem". Ironically, solving this problem was just a step in solving the most pressing issue with modern abstract mathematics as it had emerged in the second half of the 19th century. This development of mathematics is complemented by and connected to the dramatic expansion of logic facilitating steps toward full formalization.[4] Hilbert clearly hoped to address the issue he had already articulated in his Paris address of 1900 and had stated prominently as the second in his famous list of problems:

... I wish to designate the following as the most important among the numerous questions which can be asked with regard to the axioms [of arithmetic]: *To prove that they are not contradictory, that is, that a finite number of logical steps based upon them can never lead to contradictory results.*

As to the axioms of arithmetic, Hilbert points to his paper *Über den Zahlbegriff* delivered at the Munich meeting of the German Association of Mathematicians in September of 1899. The title alone indicates already its intellectual context: twelve years earlier, Kronecker had published a well-known paper with the very same title and had sketched a way of introducing irrational numbers without accepting the general notion. It is precisely to the general concept that Hilbert wants to give a proper foundation — using the axiomatic method and following Dedekind who represents most strikingly the development toward greater abstractness in mathematics.

---

what is demanded is not mechanical simulation, the task requires a close examination of how mathematics is done in order to determine how informal methods can be replaced by mechanizable procedures and how the speed of computers can be employed to compensate for their inflexibility. The field is wide open, and like all good things, it is not easy. But one does expect and look for pleasant surprises in this enterprise which requires a novel combination of psychology, logic, mathematics and computer technology." Surprisingly, there is still no unified interdisciplinary approach; but see Appendix C below with the title "Confluence?".

[4]The deepest philosophical connection between the mathematical and logical developments is indicated by the fact that both Dedekind and Frege considered the concept of a "function" to be central; it is a dramatic break from traditional metaphysics. Cf. Cassirer's *Substanzbegriff und Funktionsbegriff*.

2 Abstract concepts

Howard Stein analyzed philosophical aspects of the 19th century expansion and transformation of mathematics I just alluded to.[5] Underlying these developments is for him the rediscovery of a *capacity of the human mind* that had been first discovered by the Greeks between the 6th and 4th century B.C.:

> The expansion [of mathematics in the 19th century] was effected by the very same capacity of thought that the Greeks discovered; but in the process, something new was learned about the nature of that capacity — what it is, and what it is not. I believe that what has been learned, when properly understood, constitutes one of the greatest advances in philosophy — although it, too, like the advance in mathematics itself, has a close relation to ancient ideas.[6]

The deep connections and striking differences between the two discoveries can be examined by comparing Eudoxos's theory of proportion with Dedekind's and Hilbert's theory of real numbers. Fundamental for articulating this difference is Dedekind's notion of *system* that is also used by Hilbert.

2.1 Systems.

When discussing Kronecker's demand that proofs be constructive and that notions be decidable, Stein writes:

> I think the issue concerns definitions rather more crucially than proofs; but let me say, borrowing a usage from Plato, that it concerns the mathematical *logos*, in the sense both of 'discourse' generally, and of definition — i.e., the formation of concepts — in particular. (p. 251)

Logos refers to definitions not only as abbreviatory devices, but also as providing a frame for discourse, here the discourse concerning irrational numbers. Indeed, the frame is provided by a *structural* definition that concerns systems and that imposes relations between their elements. This methodological perspective shapes Dedekind's mathematical and foundational work, and Hilbert clearly stands in this Dedekindian tradition. The

---

[5] Stein did so in his marvelous paper (Stein 1988). The key words of its title (logos, logic, and logistiké) structure the systematic progression of my essay that was presented as the Howard Stein Lecture at the University of Chicago on 15 May 2008; Part 2 is a discussion of logos, Part 3 of logic, and Part 4 of logistiké. Improved versions of that talk were presented on 8 October 2008 to a workshop on "Mathematics between the Natural Sciences and the Humanities" held in Göttingen, on 28 December 2008 to the Symposium on "Hilbert's Place in the Foundations and Philosophy of Mathematics" at the meeting of the American Philosophical Association in Philadelphia, on 27 February 2009 in the series "Formal Methods in the Humanities" at Stanford University, and on 16 April 2009 to the conference on "The Fundamental Idea of Proof Theory" in Paris. I am grateful to many remarks from the various audiences. The final version of this essay was influenced by very helpful comments from two anonymous referees and Sol Feferman. — Dawn McLaughlin prepared the LaTeX version of this document; many thanks to her for her meticulous attention to detail.

[6] (Stein 1988, pp. 238–239). Stein continues: "I also believe that, when properly understood, this philosophical advance should conduce to a certain modesty: one of the things we should have learned in the course of it is how much we do *not* yet understand about the nature of mathematics." — I could not agree more.

structural definitions of Euclidean space in Hilbert's (1899a) and of real numbers in his (1900b) start out with, *We think three systems of things...*, respectively with *We think a system of things; we call these things numbers and denote them by a, b, c... We think these numbers in certain mutual relations, the precise and complete description of which is given by the following axioms: ...*[7] The last sentence is followed by the conditions characterizing real numbers, i.e., those of Dedekind's (1872c), except that continuity is postulated in a different, though deeply related way (see below). Hilbert and Bernays called this way of giving a structural definition, or formulating a mathematical theory, *existential axiomatics*.

The introduction of concepts "rendered necessary by the frequent recurrence of complex phenomena, which could be controlled only with difficulty by the old ones" is praised by Dedekind as the engine of progress in mathematics and other sciences.[8] The definition of *continuity* or *completeness* in his (1872c) is to be viewed in this light. The underlying complex phenomena are related to orderings. Dedekind emphasizes transitivity and density as central properties of an ordered system O, and adds the feature that every element in O generates a cut; a *cut* of O is simply a partition of O into two non-empty parts A and B, such that all the elements of A are smaller than all the elements of B. Two different interpretations are presented for these principles, namely, the rational numbers with the ordinary less-than relation and the geometric line with the to-the-right-of relation. On account of this fact the ordering phenomena for the rationals and the geometric line are viewed as *analogous*. Finally, the *continuity principle* is the converse of the last condition: every cut of the ordered system is produced by exactly one element. For Dedekind this principle expresses the essence of continuity and holds for the geometric line.[9]

In order to capture continuity arithmetically and to define a system of real numbers, Dedekind turns the analogy between the rationals and the geometric line into a *real correspondence* by embedding the rationals into the line (after having fixed an origin and a unit). This makes clear that the system of rationals is not continuous, and it motivates considering cuts of rationals as arithmetic counterparts to geometric points. Dedekind shows the system of these cuts to be an ordered field that is also continuous or complete.[10] The completeness of the system, its non-extendibility,

---

[7]The German texts are: "*Wir denken drei Systeme von Dingen...*, respectively *Wir denken ein System von Dingen; wir nennen diese Dinge Zahlen und bezeichnen sie mit a, b, c... Wir denken diese Zahlen in gewissen gegenseitigen Beziehungen, deren genaue und vollständige Beschreibung durch die folgenden Axiome geschieht: ...*"

[8](Dedekind 1888, p. vi).

[9]Dedekind remarks on p. 11 of (1872c): "Die Annahme dieser Eigenschaft der Linie ist nichts als ein Axiom, durch welches wir erst der Linie ihre Stetigkeit zuerkennen, durch welches wir die Stetigkeit in die Linie hineindenken." Then he continues that the "really existent" space may or may not be continuous and that — even if it were not continuous — we could make it continuous in thought. On p. vii of (1888) he discusses a model for Euclid's *Elements* that is everywhere discontinuous.

[10]My interpretation of these considerations reflects Dedekind's methodological practice that is tangible in (1872c) and perfectly explicit five years later in his (1877) — with reference back

points to the core of the difference with Eudoxos's definition of proportionality in Book V of Euclid's *Elements*. The ancient definition applies to many different kinds of geometric magnitudes without requiring that their respective systems be complete, as they may be open to new geometric constructions. Hilbert's completeness axiom expresses the condition of non-extendibility most directly as part of the structural definition. As a matter of fact, even in his (1922) Hilbert articulates Dedekind's structural way of thinking of the system of real numbers when describing the *axiomatische Begründungsmethode* for analysis (that is done still before finitist proof theory is given its proper methodological articulation in 1922):

> The continuum of real numbers is a system of things, which are linked to one another by determinate relations, the so-called axioms. In particular, in place of the definition of real numbers by Dedekind cuts, we have the two axioms of continuity, namely, the Archimedean axiom and the so-called completeness axiom. To be sure, Dedekind cuts can then also be used to specify individual real numbers, but they do not provide the definition of the concept of real number. Rather, a real number is conceptually just a thing belonging to our system. . . .
> This standpoint is logically completely unobjectionable, and the only thing that remains to be decided is, whether a system of the requisite sort is thinkable, that is, whether the axioms do not, say, lead to a contradiction.[11]

The axioms serve, of course, also as starting-points for the systematic development of analysis; consistency is to ensure that not too much can be proved, namely, everything. This is one of the crucially important connections to provability. Dedekind also points repeatedly and polemically to the fact that we have finally a proof of $\sqrt{3}\sqrt{2} = \sqrt{6}$ and indicates how analysis can be developed; he shows the continuity principle to be *equivalent* to the basic analytic fact that bounded, monotonically increasing functions have a limit. That is methodology par excellence: The continuity principle is not only sufficient to prove the analytic fact, but indeed necessary.

2.2 Consistency. For both Dedekind and Hilbert, the coherence of their theories for real numbers was central. Dedekind had aimed for, and thought he had achieved in his (1888), "the purely logical construction of the science of numbers and the continuous realm of numbers gained in it."[12]

---

to (1872c). Thus, Noether attributed the "axiomatische Auffassung" to Dedekind in her comments on (1872c). Notice that Dedekind does *not* identify real numbers with cuts of rationals; real numbers are associated with or determined by cuts, but are viewed as *new* objects. That is vigorously expressed in letters to Lipschitz.

[11](Hilbert 1922) in (Ewald 1996, p. 1118). — That is fully in Dedekind's spirit: Hilbert's critical remark about the definition of real numbers as cuts do not apply to Dedekind, as should be clear from my discussion (in the previous note), and the issue of consistency was an explicit part of Dedekind's logicist program.

[12]The systematic build-up of the continuum envisioned in (1872c, pp. 5–6) is carried out in later manuscripts where integers and rationals are introduced as equivalence classes of pairs of natural numbers; they serve as models for subsystems of the axioms for the reals, in a completely modern way. — All of these developments as well as that towards the formulation of simply infinite systems are analyzed in (Sieg and Schlimm 2005).

Within the logical frame of that essay Dedekind defines simply infinite systems and provides also an "example" or "instance". The point of such an instantiation is articulated sharply and forcefully in his famous letter to Keferstein where he asks, whether simply infinite systems "exist at all in the realm of our thoughts". He supports the affirmative answer by a logical existence proof. Without such a proof, he explains, "it would remain doubtful, whether the concept of such a system does not perhaps contain internal contradictions". His *Gedankenwelt*, "the totality S of all things that can be object of my thinking", was crucial for obtaining a simply infinite system.[13]

Cantor recognized Dedekind's *Gedankenwelt* as an inconsistent system and communicated that fact to both Dedekind (in 1896) and to Hilbert (in 1897). When Hilbert formulated arithmetic in his (1900b), he reformulated the problem of instantiating logoi as a quasi-syntactic problem: Show that no contradiction is provable from the axiomatic conditions in a finite number of logical steps. That is, of course, the second problem of his Paris address I discussed in section 1. He took for granted that consistency amounts to mathematical existence and assumed that the ordinary investigations of irrational numbers could be turned into a model theoretic consistency proof within a restricted logicist framework. This was crucial for the arithmetization of analysis and its logicist founding. It should be mentioned that Hilbert in *Grundlagen der Geometrie* also "geometrized" analysis by giving a geometric model via his "Streckenrechnung" for the axioms of arithmetic (with full continuity only in the second edition of the *Grundlagen* volume).

In his lecture (*1920b), Hilbert formulated the principles of Zermelo's set theory (in the language of first-order logic). He considered Zermelo's theory as providing the mathematical objects Dedekind had obtained through logicist principles; Hilbert remarked revealingly:

The theory, which results from developing all the consequences of this axiom system, encompasses all mathematical theories (like number theory, analysis, geometry) in the following sense: the relations that hold between the objects of one of these mathematical disciplines are represented in a completely corresponding way by relations that obtain in a sub-domain of Zermelo's set theory.[14]

---

[13] Let me support, by appeal to authority, the claim that Dedekind's thoughts are not psychological ideas: Frege asserts in his manuscript *Logik* from 1894 that he uses the word "Gedanke" in an unusual way and remarks that "Dedekind's usage agrees with mine". It is worthwhile noting that Frege, in this manuscript, approved of Dedekind's argument for the existence of an infinite system. — Note also that Hilbert formulated his existential axiomatics with the phrase "wir denken", so that the system is undoubtedly an object of our thought, indeed, "ein Gedanke".

[14] (Hilbert *1920b, p. 23). Here is the German text: "Die Theorie, welche sich aus der Entwicklung dieses Axiomensystems in seine Konsequenzen ergibt, schliesst alle mathematischen Theorien (wie Zahlentheorie, Analysis, Geometrie) in sich in dem Sinne, dass die Beziehungen, welche sich zwischen den Gegenständen einer dieser mathematischen Disziplinen finden, vollkommen entsprechend dargestellt werden durch die Beziehungen, welche in einem Teilgebiete der Zermeloschen Mengenlehre stattfinden."

In spite of this perspective, Hilbert reconsidered at the end of the 1920-lecture his earlier attempt (published as (1905a)) to establish by mathematical proof that no contradiction can be proved in formalized elementary number theory. That had raised already then the issue, how proofs can be characterized and subjected to mathematical investigation. It was only after the study of *Principia Mathematica* that Hilbert had a properly general and precise concept of (formal) proof available.

3 Rigorous proofs

Proofs are essential for developing any mathematical subject, vide Euclid in the *Elements* or Dedekind in *Was sind und was sollen die Zahlen?*. In the introduction to his *Grundgesetze der Arithmetik*, Frege distinguished his systematic development from Euclid's by pointing to the list of explicit inference principles for obtaining gapless proofs. As to Dedekind's essay he remarked polemically that no proofs can be found in that work. Dedekind and Hilbert explicated the "science of (natural) number" and "arithmetic (of real numbers)" in similar ways; their theories start from the defining conditions for simply infinite systems, respectively complete ordered fields. Dedekind writes in (1888):

The relations or laws which are derived exclusively from the conditions [for a simply infinite system] and are therefore always the same in all ordered simply infinite systems, ... form the next object of the *science of numbers* or *arithmetic*.[15]

The term "derive" is left informal; hence Frege's critique. Exactly at this point enters logic in the restricted modern sense as dealing with formal methods for correct, truth-preserving inference.

3.1 Natural deductions.

Underlying Dedekind's and Hilbert's descriptions is an abstract concept of *logical consequence*. Hilbert stated in 1891 during a famous stop at a Berlin railway station that in a proper axiomatization of geometry "one must always be able to say 'tables, chairs, beer mugs' instead of 'points, straight lines, planes'." This remark has been taken as claiming that the basic terms must be meaningless, but it is more adequately understood if it is put side by side with a remark of Dedekind's in a letter to Lipschitz written fifteen years earlier: "All technical expressions [can be] replaced by arbitrary, newly invented (up to now meaningless) words; the edifice must not collapse, if it is correctly constructed, and I claim, for example, that my theory of real numbers withstands this test." Thus, logical arguments leading from principles to derived claims cannot be severed by a re-interpretation of the technical expressions or, to put it differently, there are no counterexamples to the arguments.

---

[15](Dedekind 1888, sec. 73). In the letter to Keferstein, on p. 9, Dedekind reiterates this perspective and requires that every claim "must be derived completely abstractly from the logical definition of [the simply infinite system] N".

Dedekind's and Hilbert's presentations are detailed, reveal the logical form of arguments, and reflect features of the mathematical structures. In the very first sentence of the Preface to his (1888), Dedekind programmatically emphasizes that "in science nothing capable of proof should be accepted without proof" and claims that only common sense (gesunder Menschenverstand) is needed to understand his essay. But he recognizes also that many readers will be discouraged, when asked to prove truths that seem obvious and certain by "the long sequence of simple inferences that corresponds to the nature of our step-by-step understanding" (Treppenverstand).[16] Dedekind believes that there are only a few such simple inferences, but he does not explicitly list them. Looking for an expressive formal language and powerful inferential tools, Hilbert moved slowly toward a presentation of proofs in logical calculi. He and his students started in 1913 to learn modern logic by studying *Principia Mathematica*. During the winter term 1917–18 he gave the first course in mathematical logic proper and sketched, toward the end of the term, how to develop analysis in ramified type theory with the axiom of reducibility.[17]

So there is finally (in Göttingen) a way of building up gapless proofs in Frege's sense. However, Hilbert aimed for a framework in which mathematics can be formalized in a natural and direct way. The calculus of *Principia Mathematica* did not lend itself to that task. In the winter term 1921–22 he presented a logical calculus that is especially interesting for sentential logic. He points to the parallelism with his axiomatization of geometry: groups of axioms are introduced there for each concept, and that is done here for each logical connective. Let me formulate the axioms for just conjunction and disjunction:

$A \& B \to A$ $\qquad$ $((A \to C) \& (B \to C)) \to ((A \vee B) \to C)$

$A \& B \to B$ $\qquad$ $A \to (A \vee B)$

$A \to (B \to A \& B)$ $\qquad$ $B \to (A \vee B)$

The simplicity of this calculus and its directness for formalization inspired the work of Gentzen on natural reasoning. It should be pointed out that Bernays had proved the completeness of Russell's calculus in his

---

[16](Dedekind 1888, p. iv). Dedekind continues: "Ich erblicke dagegen gerade in der Möglichkeit, solche Wahrheiten auf andere, einfachere zurückzuführen, mag die Reihe der Schlüsse noch so lang und scheinbar künstlich sein, einen überzeugenden Beweis dafür, daß ihr Besitz oder der Glaube an sie niemals unmittelbar durch innere Anschauung gegeben, sondern immer durch eine mehr oder weniger vollständige Wiederholung der einzelnen Schlüsse erworben ist."

[17]That is usually associated with the book (Hilbert and Ackermann 1928) that was published only in 1928; however, that book takes over the structure and much of the content from these earlier lecture notes. See my paper (1999) and the forthcoming third volume of Hilbert's *Lectures on the Foundations of Mathematics and Physics*. — In the final section of his (2008), Wiedijk lists "three main revolutions" in mathematics: the introduction of *proof* in classical Greece (culminating in Euclid's *Elements*), that of *rigor* in the 19th century, and that of *formal mathematics* in the late 20th and early 21st centuries. The latter revolution, if it is one, took place in the 1920s.

Habilitationsschrift of 1918 and had investigated rule-based variants. The proof theoretic investigations of, essentially, primitive recursive arithmetic in the 1921–22 lectures also led to a tree-presentation of proofs, what Hilbert and Bernays called "the resolution of proofs into proof threads" (die Auflösung von Beweisen in Beweisfäden).[18] The full formulation of the calculus and the articulation of the methodological parallelism to *Grundlagen der Geometrie* are also found in (Hilbert and Bernays 1934, pp. 63–64).

3.2 Strategies. Gentzen formulated *natural deduction calculi* using Hilbert's axiomatic formulation as a starting point and called them calculi of *natural reasoning* (natürliches Schließen); he emphasized that making and discharging assumptions were their distinctive features. Here are the Elimination and Introduction rules for the connectives discussed above and as formulated in (Gentzen 1936); the configurations that are derived with their help are sequents of the form $\Gamma \supset \psi$ with Γ containing all the assumptions on which the proof of ψ depends:

$$\frac{\Gamma \supset A \& B}{\Gamma \supset A} \qquad \frac{\Gamma \supset A \& B}{\Gamma \supset B} \qquad \frac{\Gamma \supset A \vee B \quad \Gamma, A \supset C \quad \Gamma, B \supset C}{\Gamma \supset C}$$

$$\frac{\Gamma \supset A \quad \Gamma \supset B}{\Gamma \supset A \& B} \qquad \frac{\Gamma \supset A}{\Gamma \supset A \vee B} \qquad \frac{\Gamma \supset B}{\Gamma \supset A \vee B}$$

Gentzen and later Prawitz established normalization theorems for proofs in nd calculi.[19] As the calculi are complete, one obtains proof theoretically refined completeness theorems: if ψ is a logical consequence of Γ, then there is a *normal* proof of ψ from Γ. I reformulated the nd calculi as *intercalation calculi*[20] for which these refined completeness theorems can be proved semantically without appealing to a syntactic normalization procedure; see (Sieg and Byrnes 1998) for classical first-order logic as well as (Sieg and Cittadini 2005) for some non-classical logics, in particular, for intuitionist first-order logic.

The refined completeness results and their semantic proofs provide foundations to the systematic search for normal proofs in nd calculi. This is methodologically analogous to the use of completeness results for cut-free sequent calculi and was exploited in the pioneering work of Hao Wang.[21]

---

[18] On account of this background, I assume, Gentzen emphasized in his dissertation and his first consistency proof for elementary number theory the dual character of introduction and elimination rules, but considered making and discharging assumptions as the most important feature of his calculi.

[19] The first version of Gentzen's dissertation was recently discovered by Jan von Plato in the Bernays Nachlass of the ETH in Zürich. It contains a detailed proof of the normalization theorem for intuitionist predicate logic; see (von Plato 2008).

[20] I discovered only recently that Beth in his (1958) employs "intercalate" (on p. 87) when discussing the use of lemmata in the proofs of mathematical theorems.

[21] See the informative and retrospective discussion in his (1984) and, perhaps, also the programmatic (1970). — Cf. also my (2007).

The subformula property of normal and cut-free derivations is fundamental for mechanical search. The ic calculi enforce normality by applying the E-rules only on the left to premises and the I-rules only on the right to the goal. In the first case one really tries to "extract" a goal formula by a *sequence* of E-rules from an assumption in which it is contained as a strictly positive subformula. This feature is distinctive and makes search efficient, but it is in a certain sense just a natural systematization and logical deepening of the familiar forward and backward argumentation. Suitable strategies have been implemented and guide a complete search procedure for first-order logic, called AProS.[22] In Appendix A, I discuss examples of purely logical arguments.[23] The AProS strategies can be extended by E- and I-rules for definitions, so that the meanings of defined notions as well as those of logical connectives can be used to guide search. In this way we have developed quite efficiently the part of elementary set theory concerning Boolean operations, power sets, Cartesian products, etc. In Appendix B, the reader finds two examples of set theoretic arguments.

You might think, that is interesting, but what relevance do these considerations have for finding proofs in more complex parts of mathematics? To answer that question and put it into a broader context, let me first note that the history of such computational perspectives goes back at least to Leibniz, and that it can be illuminated by Poincaré's surprising view of Hilbert's *Grundlagen der Geometrie*. In his review of Hilbert's book, he suggested giving the axioms to a reasoning machine, like Jevons's logical piano, and observing whether all of geometry would be obtained. He wrote that such radical formalization might seem "artificial and childish", were it not for the important question of "completeness":

Is the list of axioms complete, or have some of them escaped us, namely those we use unconsciously? . . . One has to find out whether geometry is a logical consequence of the explicitly stated axioms or, in other words, whether the axioms, when given to the reasoning machine, will make it possible to obtain the sequence of all theorems as output [of the machine].[24]

With respect to a sophisticated logical framework and under the assumption of the finite axiomatizability of mathematics, Poincaré's problem morphed into what Hilbert and others viewed in the 1920s as the *most important problem* of mathematical logic: the decision problem

---

[22]Nd calculi were considered as inappropriate for theorem proving because of the seemingly unlimited branching in a backward search afforded by modus ponens (conditional elimination). The global property of normality for nd proofs could not be directly exploited for a locally determined backward search; hence, the intercalation formulation of natural deduction. The implementation of AProS can be downloaded at http://caae.phil.cmu.edu/projects/apros/

[23]In (Sieg and Field 2005, pp. 334–5), the problem of proving that $\sqrt{2}$ is not rational is formulated as a logical problem, and AProS finds a proof directly; cf. the description of the difficulties of obtaining such a proof in (Wiedijk 2008).

[24](Poincaré 1902b, pp. 252–253).

(*Entscheidungsproblem*) for predicate logic. Its special character was vividly described in a talk Hilbert's student Behmann gave in 1921:

> For the nature of the problem it is of fundamental significance that as auxiliary means ... only the completely mechanical reckoning according to a given prescription [Vorschrift] is admitted, i.e., without any thinking in the proper sense of the word. If one wanted to, one could speak of mechanical or machine-like thinking. (Perhaps it can later even be carried out by a machine.)

Johann von Neumann argued against the positive solvability of the decision problem, in spite of the fact that — as he formulated matters in 1924 — "... we have no idea how to prove the undecidability". It was only twelve years later that Turing provided the idea, i.e., introduced the appropriate concept, for proving the unsolvability of the *Entscheidungsproblem*.

The issue for Turing was, What are the procedures a human being can carry out when mechanically operating as a computer?[25] In his classical paper *On computable numbers with an application to the Entscheidungsproblem*, Turing isolated the basic steps underlying a computer's procedures as the operations of a Turing machine. He then proved: There is no procedure that can be executed by a Turing machine and solves the decision problem. Using the concepts of general recursive and λ-definable functions, Church had also established the undecidability of predicate logic. The core of Church's argument was presented in Supplement II of *Grundlagen der Mathematik, vol. II*. However, it was not only expanded by later considerations due to Church and Kleene, but also deepened by local axiomatic considerations for the concept of a *reckonable function*.[26]

Hilbert and Bernays introduced reckonable functions informally as those number theoretic functions whose values can be determined in a "deductive formalism". They proved that, if the deductive formalism satisfies their *recursiveness conditions*, then the class of reckonable functions is co-extensional with that of the general recursive ones. (The crucial condition requires that the proof relation of the deductive formalism is primitive recursive.) Their concept is one way of capturing the "completely mechanical reckoning according to a given prescription" mentioned in the quotation from Behmann. Indeed, it generalizes Church's informal notion of calculable functions whose values can be determined in a logic and imposes

---

[25] For Turing a "computer" is a human being carrying out a "calculation" and using only minimal cognitive capacities. The limitations of the human sensory apparatus motivate finiteness and locality conditions; Turing's supporting argument is not mathematically precise, and I don't think there is any hope of turning the analysis into a mathematical theorem. What one can do, however, is to exploit it as a starting point for formulating a general concept and establishing a representation theorem; cf. my paper (2008).

[26] I distinguish local from global axiomatics. As an example of the former I discuss in part 4.1 an abstract proof of Gödel's incompleteness theorems. Other examples can be found in Hilbert's 1917-talk in Zürich, but also in contemporary discussions, e.g., Booker's report on L-functions in the Notices of the AMS, p. 1088. Booker remarks that many objects go by the name of L-function and that it is difficult to pin down exactly which ones are. He attributes then to A. Selberg an "axiomatic approach" consisting in "writing down the common properties of the known examples" — as axioms.

the recursiveness condition in order to obtain a mathematically rigorous formulation. For us the questions are of course: Can a machine carry out this mechanical thinking? and, if a universal Turing machine in principle can, What is needed to *copy*, as Turing put it in 1948, aspects of mathematical thinking in such a machine? — Copying requires an original, i.e., that we must have uncovered suitable aspects of the mathematical mind when trying to extend automated proof search from logic to mathematics.

4 Local axiomatics

At the end of his report on *Intelligent Machinery* from 1948, Turing suggested that machines might search for proofs of mathematical theorems in suitable formal systems. It was clear to Turing that one cannot just specify axioms and logical rules, state a theorem, and expect a machine to demonstrate the theorem. For a machine to exhibit the necessary intelligence it must "acquire both discipline and initiative". Discipline would be acquired by becoming (practically) a universal machine; Turing argued that "discipline is certainly not enough in itself to produce intelligence" and continued:

That which is required in addition we call initiative. This statement will have to serve as a definition. Our task is to discover the nature of this residue as it occurs in man, and try and copy it in machines. (p. 21)

The dynamic character of strategies constitutes but a partial and limited copy of human initiative. Nevertheless, local axiomatics that allows the expression of leading ideas together with a hierarchical organization that reflects the conceptual structure of a field can carry us a long way. Hilbert expressed his views in 1919 as follows, arguing against the logicists' view that mathematics consists of tautologies grounded in definitions:

If this view were correct, mathematics would be nothing but an accumulation of logical inferences piled on top of each other. There would be a random concatenation of inferences with logical reasoning as its sole driving force. But in fact there is no question of such arbitrariness; rather we see that the formation of concepts in mathematics is constantly guided by intuition and experience, so that mathematics on the whole forms a non-arbitrary, closed structure.[27]

Hilbert's grouping of the axioms for geometry in his (1899a) had the express purpose of organizing proofs and the subject in a conceptual way: parts of his development are marvelous instances of *local axiomatics*, analyzing which notions and principles are needed for which theorems.

---

[27](Hilbert *1919, p. 5). Here is the German text: "Wäre die dargelegte Ansicht zutreffend, so müsste die Mathematik nichts anderes als eine Anhäufung von übereinander getürmten logischen Schlüssen sein. Es müsste ein wahlloses Aneinanderreihen von Folgerungen stattfinden, bei welchem das logische Schliessen allein die treibende Kraft wäre. Von einer solchen Willkür ist aber tatsächlich keine Rede; vielmehr zeigt sich, dass die Begriffsbildungen in der Mathematik beständig durch Anschauung und Erfahrung geleitet werden, sodass im grossen und ganzen die Mathematik ein willkürfreies, geschlossenes Gebilde darstellt."

4.1 Modern. The idea of local axiomatics can be used for individual mathematical theorems and asks, How can we prove this particular theorem or this particular group of theorems? Hilbert and Bernays used the technique in their *Grundlagen der Mathematik II* also outside a foundational axiomatic context: first for proving Gödel's incompleteness theorems and then, as I indicated at the end of section 3.2, for showing that the functions reckonable in formal deductive systems coincide with the general recursive ones. One crucial task has to be taken on for *local* as well as for *global axiomatics*, namely, isolating what is at the heart of an argument or uncovering its *leading (mathematical) idea*. That was proposed by Saunders MacLane in his Göttingen dissertation (of late 1933) and summarized in his (1935). MacLane emphasized that proofs are not "mere collections of atomic processes, but are rather complex combinations with a highly rational structure". When reviewing in 1979 this early logical work, he ended with the remark, "There remains the real question of the actual structure of mathematical proofs and their strategy. It is a topic long given up by mathematical logicians, but one which still — properly handled — might give us some real insight."[28] That is exactly the topic I am trying to explore.

As an illustration of the general point concerning the "rational structure" of mathematical arguments, I consider briefly the proofs of Gödel's incompleteness theorems. These proofs make use of the connection between the mathematics that is used to present a formal theory and the mathematics that can be formally developed in the theory. Three steps are crucial for obtaining the proofs, steps that go beyond the purely logical strategies and are merged into the search algorithm:

1. *Local axioms*: representability of the core syntactic notions, the diagonal lemma, and the Hilbert & Bernays derivability conditions.

2. *Proof-specific definitions*: formulating instances of existential claims, for example, the Gödel sentence for the first incompleteness theorem.

3. *Leading idea*: moving between object- and meta-theory, expressed by appropriate Elimination and Introduction rules (for example, if a proof of A has been obtained in the object-theory, then one is allowed to introduce the claim 'A is provable' in the meta-theory).

AProS finds the proofs efficiently and directly, even those that did not enter into the analysis of the leading idea, for example, the proof of Löb's theorem. All of this is found in (Sieg and Field 2005).

It has been a long-standing tradition in mathematics to give and to analyze a variety of arguments for the same statement; the fundamental theorems of algebra and arithmetic are well-known examples. In this

---

[28] The first quotation is from MacLane's (1935, p. 130), the second from his (1979, p. 66). The processes by means of which MacLane tries to articulate the "rational structure" of proofs should be examined in greater detail.

way we delimit conceptual contexts, provide contrasting *explanations* for the theorem at hand, and gain a deeper understanding by looking at it in different ways, e.g., from a topological or algebraic perspective.[29] An automated search requires obviously a sharp isolation of local axioms and leading ideas that underlie a proof. Such developments can be integrated into a global framework through a hierarchical organization, and that has been part and parcel of mathematical practice. Hilbert called it *Tieferlegung der Fundamente*!

These broad ideas are currently being explored in order to obtain an automated proof of the Cantor-Bernstein theorem from Zermelo's axioms for set theory.[30] The theorem claims that there is a bijection between two sets, in case there are injections from the first to the second and from the second to the first. The theorem is a crucial part of the investigations concerning the size of sets and guarantees the anti-symmetry of the partial ordering of sets by the "smaller-or-equal-size" relation.[31] We have begun to develop set theory from Zermelo's axioms and use three layers for the conceptual organization of the full proof:

A. Construction of sets, for example, empty set, power set, union, and pairs.
B. Introduction of functions as set theoretic objects.
C. The abstract proof.

The abstract proof is divided in the same schematic way as that of Gödel's theorems and is independent of the set theoretic definition of function. The *local axioms* are lemmata for injective, surjective, and bijective functions as well as a fixed-point theorem. The crucial *proof-specific definition* is that of the bijection claimed to exist in the theorem. Finally, the *leading idea* is simply to exploit the fixed-point property and verify that the defined function is indeed a bijection. — It is noteworthy that the differences between the standard proofs amount to different ways of obtaining the smallest fixed-point of an inductive definition.

---

[29] In the Introduction to the second edition of (Dirichlet 1863) Dedekind emphasized this aspect for the development of a whole branch of mathematics. In the tenth supplement to this edition of Dirichlet's lectures, he presented his general theory of ideals in order, as he put it, "to cast, from a higher standpoint, a new light on the main subject of the whole book". In German, "Endlich habe ich in dieses Supplement eine allgemeine Theorie der Ideale aufgenommen, um auf den Hauptgegenstand des ganzen Buches von einem höheren Standpunkte aus ein neues Licht zu werfen." He continues, "hierbei habe ich mich freilich auf die Darstellung der Grundlagen beschränken müssen, doch hoffe ich, daß das Streben nach charakteristischen Grundbegriffen, welches in anderen Teilen der Mathematik mit so schönem Erfolg gekrönt ist, mir nicht ganz mißglückt sein möge." (Dedekind 1932, pp. 396–7).

[30] My collaborators on this particular part of the AProS Project have been Ian Kash, Tyler Gibson, Michael Warren, and Alex Smith.

[31] On p. 209 of Cantor's (1932) *Gesammelte Abhandlungen*, Zermelo calls this theorem "one of the most important theorems of all of set theory".

4.2 Classical. Shaping a field and its proofs by concepts is classical; so is the deepening of its foundations. That can be beautifully illustrated by the developments in the first two books of Euclid's *Elements* (and the related investigations at the beginning of Book XII). Proposition 47 of Book I, the Pythagorean theorem, is at the center of those developments. The broad mathematical context is given by the *quadrature problem*, i.e., determining the "size" or, in modern terms, the area of geometric figures in terms of squares. The problem is discussed in Book II for polygons. Polygons can be partitioned into triangles that can be transformed individually (by ruler and compass constructions) first into rectangles "of equal area" and then into equal squares.[32] The question is, how can we join these squares to obtain one single square that is equal to the polygon we started out with? It is precisely here that the Pythagorean theorem comes in and provides the most direct way of determining the larger square. The diagram, displayed below, captures the construction and the abstract proof of the theorem. If one views the determination of the larger square as a geometric computation, then the proof straightforwardly verifies its correctness.[33]

For the proof, Euclid has us first construct the squares on the triangle's sides and then make the observation that the extensions of the sides of the smaller squares by the contiguous sides of the original triangle constitute lines. In the next step a crucial auxiliary line is drawn, namely, the line that is perpendicular to the hypotenuse and that passes through the vertex opposite the hypotenuse. This auxiliary line partitions the big square into the black and grey rectangles. Two claims are now considered: the black rectangle is equal to the black square, and the grey rectangle is equal to the

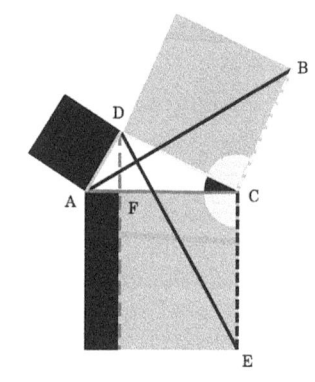

---

[32]Euclid simply calls the geometric figures "equal". This is central and has been pursued throughout the evolution of geometry. In Hilbert's (1899a), a whole chapter is devoted to "Die Lehre von den Flächeninhalten in der Ebene", Chapter IV, making the implicit Euclidean assumptions concerning "area" explicit. See also Hartshorne's book, section 22, *Area in Euclid's geometry*.

[33]Hilbert remarked in his (*1899b), "Wir werden im folgenden häufig Gebrauch von Figuren machen, wir werden uns aber niemals auf sie verlassen. Stets müssen wir dafür sorgen, daß die an einer Figur vorgenommenen Operationen auch rein logisch gültig bleiben. Es kann dies garnicht genug betont werden; im richtigen Gebrauch der Figuren liegt eine Hauptschwierigkeit unserer Untersuchungen." (p. 303) Here I am obviously not so much interested in the (correct) use of diagrams as analyzed by Manders and for which Avigad e.a. have provided an informative formal framework. Manders's analysis led to the assertion that only topological features of diagrams are relevant for and appealed to in Euclidean proofs; the conceptual setting sketched above with its focus on "area" gives a reason for that assertion. There is also important work from the early part of the 20th century by Hans Brandes and, in particular, Paul Mahlo in his 1908 dissertation. This work tries to classify the "Zerlegungsbeweise" of the Pythagorean theorem and should be investigated carefully; (Bernstein 1924) reflects on those dissertations.

grey square. Euclid uses three facts that are readily obtained from earlier propositions: (α) Triangles are equal when they have two equal sides and when the enclosed angles are equal (Proposition I.4); (β) Triangles are equal when they have the same base and when their third vertex lies on the same parallel to that base (Proposition I.37); (γ) A diagonal divides a rectangle into two equal triangles (Proposition I.41).[34]

Here is the proof based on (α) through (γ) for the grey square and the grey rectangle. (The common notions are implicitly appealed to; the argument for the equality of the black square and the black rectangle is analogous.) The triangles ABC and DCE satisfy the conditions of (α) and are thus equal; on account of (β) they are equal to DBC and FCE, respectively. Finally, (γ) ensures the equality of the grey square and grey rectangle. Comparing the structure of this argument to that of the abstract proofs for the incompleteness theorem and the Cantor-Bernstein theorem, we can make the following general observations: (α) through (γ) are used as *local axioms*; the auxiliary line drawn through the vertex opposite of the hypotenuse and perpendicular to it is the central *proof-specific definition*; finally, the *leading idea* is the partitioning of squares and establishing that corresponding parts are equal.

The character of the "deepening of the foundations" is amusingly depicted by anecdotes concerning Hobbes and Newton: Hobbes started with Proposition 47 and was convinced of its truth only after having read its proof and all the (proofs of the) propositions supporting it; Newton, in contrast, started at the beginning and could not understand, why such evident propositions were being established — until he came to the Pythagorean theorem. Less historically, there is also a deeper parallelism with the overall structure of the proof of the Cantor-Bernstein theorem from Zermelo's axioms. The construction of figures like triangles and squares corresponds to A (in the list A–C concerning the Cantor-Bernstein theorem); the congruence criteria for such figures correspond to B; the abstract proof of the geometric theorem, finally, has the same conceptual organization as the set theoretic proof referenced in C.

The abstract proof of the Pythagorean theorem and its deepening are shaped by the mathematical context, here the quadrature problem. I want to end this discussion with two related observations. Recall that the Pythagorean theorem is used in Hippocrates's proof for the quadrature of the lune.[35] This is just one of its uses for solving quadrature problems, but it seems to be very special, as only the case for isosceles triangles is exploited. The crucial auxiliary line divides in half the square over the

---

[34]This is not *exactly* Euclid's proof. Euclid does not appeal to I.37, but just to I.41, which is really a combination of (β) and (γ) and applies directly to the diagram; I.37 is used in the proof of I.41. — A colorful version of the diagram can be found in Byrne's edition of the first six books of the Euclidean *Elements*, London, 1847. [The labeling of points was added by me; WS.]

[35]See the very informative discussion in (Dunham 1990).

hypotenuse, and we have a perfectly symmetric configuration.[36] Here is the first observation, namely, the claim concerning the equality of the rectangles (into which the square over the hypoteneuse is divided) and squares (over the legs) is "necessary", and the proof idea is relatively straightforward. That leads me to the second observation that is speculative and formulated as a question: Isn't it plausible that the Euclidean proof is obtained by generalizing this special one?

Karzel and Kroll, in their *Geschichte der Geometrie seit Hilbert* under the heading "Order and Topology", link the classical Greek considerations back (or rather forward) to modern developments:

In Euclidean geometry triangles and also rectangles take on the role of elementary figures out of which more complex figures are thought to be composed. To these elementary figures one can assign in Euclidean geometry an area in a natural way. If one assumes in addition the axiom of continuity, then one arrives at the concept of an integral when striving to assign an area also to more complex figures.[37]

So we have returned to continuity and to Dedekind.

5 Cognitive aspects

In his (1888) Dedekind refers to his *Habilitationsrede* where he claimed that the need to introduce appropriate notions arises from the fact that our intellectual powers are imperfect. Their limitation leads us to frame the object of a science in different forms and introducing a concept means, in a certain sense, formulating a hypothesis on the inner nature of the science. How well the concept captures this inner nature is determined by its usefulness for the development of the science, and in mathematics that is mainly its usefulness for constructing proofs. Dedekind put the theories from his foundational essays to this test by showing that they allow the direct, stepwise development of analysis and number theory. Thus, Dedekind viewed general concepts and general forms of arguments as tools to overcome, at least partially, the imperfection of our intellectual powers. He remarked:

Essentially, there would be no more science for a man gifted with an unbounded understanding — a man for whom the final conclusions, which we obtain through a long chain of inferences, would be immediately evident truths; and this would be so even if he stood in exactly the same relation to the objects of science as we do. (Ewald 1996, pp. 755–6)

---

[36] In (Aumann 2009, pp. 64–65) knowledge of this geometric fact is attributed to the Babylonians, and it is the one Socrates extracts from the slave boy in Plato's *Meno*.

[37] (Karzel and Kroll 1988, p. 121). The German text is: "In der euklidischen Geometrie spielen neben den Dreiecksflächen noch die Rechtecksflächen ... die Rolle von Elementarflächen, aus denen man sich kompliziertere Flächen zusammengesetzt denkt. Diesen Elementarflächen kann man in der euklidischen Geometrie in natürlicher Weise einen Flächeninhalt zuweisen. Setzt man nunmehr noch das Stetigkeitsaxiom voraus, so gelangt man beim Bemühen, auch komplizierteren Flächen einen Inhalt zuzuweisen, zum Integralbegriff."

The theme of bounded human understanding is sounded also in a remark from (Bernays 1954): "Though for differently built beings there might be a different kind of evidence, it is nevertheless our concern to find out what evidence is for us."[38] Bernays put forth the challenge of finding out what is evidence *for us*, not for some *differently built being*. Turing in his (1936) appealed crucially to *human* cognitive limitations to arrive at his notion of computability. Ten years later Gödel took the success of having given "an absolute definition of an interesting epistemological notion", i.e., of effective calculability, as encouragement to strive for "the same thing" with respect to demonstrability and mathematical definability. That was attempted in his (1946). Reflecting on a possible objection to his concept of ordinal definability, namely, that uncountably many sets are ordinal definable, Gödel considers as plausible the view "that all things conceivable by us are denumerable". Indeed, he thinks that a concept of definability "satisfying the postulate of denumerability" is possible, but "that it would involve some extramathematical element concerning the psychology of the being who deals with mathematics".

Reflections on cognitive limitations motivated also the finitist program's goal of an absolute epistemological reduction. Bernays provides in his (1922b) a view of the program in *statu nascendi* and connects it to the existential axiomatics discussed above in Part 2. When giving a rigorous foundation for arithmetic or analysis one proceeds axiomatically, according to Bernays, and assumes the existence of a system of objects satisfying the structural conditions expressed by the axioms. In the assumption of such a system "lies something transcendental for mathematics, and the question arises, which principled position is to be taken [towards that assumption]". An intuitive grasp of the completed sequence of natural numbers, for example, or even of the manifold of real numbers is not excluded outright. However, taking into account tendencies in the exact sciences, one might try "to give a foundation to these transcendental assumptions in such a way that only primitive intuitive knowledge is used". That is to be done by giving finitist consistency proofs for systems in which significant parts of mathematics can be formalized. The second incompleteness theorem implies, of course, that such an absolute epistemological reduction cannot be achieved. What then is evidence for principles that allow us to step beyond the finitist framework? — Bernays emphasized in his later writings that evidence is acquired by *intellectual experience* and through *experimentation* in an almost Dedekindian spirit. In his (1946) he wrote:

In this way we recognize the necessity of something like intelligence or reason that should not be regarded as a container of [items of] a priori knowledge, but as a

---

[38](Bernays 1954, p. 18). The German text is: "Obwohl es für anders gebildete Wesen eine andere Evidenz geben könnte, so ist jedoch unser Anliegen festzustellen, was Evidenz für uns ist."

mental activity that consists in reacting to given situations with the formation of experimentally applied categories.[39]

This intellectual experimentation in part supports the introduction of concepts to define abstract structures or to characterize accessible domains (obtained by general inductive definitions), and it is in part supported by using these concepts in proofs of central theorems.[40]

I intended to turn attention to those aspects of the mathematical mind that are central, if we want to grasp the subtle connection between reasoning and understanding in mathematics, as well as the role of leading ideas in guiding proofs and of general concepts in providing explanations. Implicitly, I have been arguing for an *expansion of proof theory*: Let us take steps toward a theory that articulates principles for organizing proofs conceptually and for finding them dynamically. A good start is a thorough reconstruction of parts of the rich body of mathematical knowledge that *is* systematic, but is also structured for intelligibility and discovery, when viewed from the right perspective. Such an expanded proof theory should be called *structural* for two reasons. On the one hand one exploits the intricate internal structure of (normal) proofs, and on the other hand one appeals to the notions and principles characterizing mathematical structures. (Cf. also the very tentative parallel remarks in Appendix C.)

When focusing on formal methods and carrying out computations in support of proof search experiments, we have to isolate truly creative elements in proofs and thus come closer to an understanding of the technique of our mathematical thinking, be it mechanical or non-mechanical. Hilbert continued his remarks in (1927) about the formula game as follows:

Thinking, it so happens, parallels speaking and writing: we form statements and place them one behind another. If any totality of observations and phenomena deserves to be made the object of a serious and thorough investigation, it is this one — since, after all, it is part of the task of science to liberate us from arbitrariness, sentiment, and habit . . . (p. 475)

I could not agree more (with the second sentence in this quotation) and share Hilbert's eternal optimism, "Wir müssen wissen! Wir werden wissen!"

---

[39] (Bernays 1946, p. 91). The German text is: "Wir erkennen so die Notwendigkeit von etwas wie Intelligenz oder Vernunft, die man nicht anzusehen hat als Behältnis von Erkenntnissen a priori, sondern als eine geistige Tätigkeit, die darin besteht, auf gegebene Situationen mit der Bildung von versuchsweise angesetzten Kategorien zu reagieren." — Unfortunately, "applied" is not capturing "angesetzten". The latter verb is related to "Ansatz". That noun has no adequate English rendering either, but is used (as in "Hilbertscher Ansatz") to express a particular approach to solving a problem that does however not guarantee a solution.

[40] Andrea Cantini expresses in his recent (2008) a similar perspective, emphasizing also the significance of "geistiges Experimentieren" in Bernays's reflections on mathematics; see pp. 34–37. In the very same volume in which Cantini's article is published, Carlo Cellucci describes a concept of *analytic* proof that incorporates many features of the experimentation both Cantini and I consider as important. However, Cellucci sharply contrasts that concept with that of an *axiomatic* proof. These two notions, it seems to me, stand in opposition only if one attaches to the latter concept a dogmatic foundationalist intention. — In (Sieg 2010b) I have compared Gödel's and Turing's approach to such intellectual experimentation.

Appendices

APros' distinctive feature is its *goal-directed* search for *normal* proofs. It exploits an essential feature of normal proofs, i.e., the division of every branch in their representing tree into an E- and an I-part; see (Prawitz 1965, p. 41). This global property of nd proofs, far from being an obstacle to backward search, makes proof search both strategic and efficient. — Siekmann and Wrightson collected in their two volume *Automated Reasoning* classical papers that contain marvelous discussions of the broad methodology underlying different approaches in the emerging field from the late 1950s to the early 1970s. The papers by Beth, Kanger, Prawitz, Wang and the "Russian School" are of particular interest from my perspective, as we find in them serious attempts of searching for humanly intelligible proofs and of getting the logical framework right before building heuristics into the search. That was perhaps most clearly formulated by Kanger in his (1963, p. 364): "The introduction of heuristics may yield considerable simplifications of a given proof method, but I have the impression that it would be wise to postpone the heuristics until we have a satisfactory method to start with." The work with APros and automated proof search support that view.

A. Purely logical arguments. In the supplement to (Shanin, et al. 1965), one finds five propositional problems and their proofs; APros solves them with just the basic rules whereas in this paper quite complex derived rules are used. I discuss one example in order to illustrate, how dramatically the search is impacted by "slight" reformulations of the problem to be solved or, what amounts to the same thing, by introducing specific heuristics. The problem in (Shanin, et al. 1965) is to derive

$$(\neg(K \to A) \vee (K \to B))$$

from the premises

$$(H \vee \neg(A \;\&\; K)) \text{ and } (H \to (\neg A \vee B)).$$

The pure APros search procedure uses 277 search steps to find a proof of length 77. If one uses in a first step a derived rule to replace positive occurrences of $(\neg X \vee \Delta)$ or $(X \vee \neg \Delta)$ by $(X \to \Delta)$, respectively, $(\Delta \to X)$ then APros uses 273 search steps for a proof of length 87 (having made the replacement in the first premise), 149 search steps for a proof of length 80 (having made the replacement also in the second premise), and finally 9 search steps to find a derivation of length 12 (having made the replacement also in the conclusion).

The Shanin-procedure introduces also instances of the law of excluded middle. In the above problem it does so for the left disjunct of the goal, i.e., it uses the instance $(\neg(K \to A) \vee (K \to A))$. If one adds that instance as an additional premise, then APros takes 108 search steps to obtain a proof of 49 lines. If the conclusion $(\neg(K \to A) \vee (K \to B))$ is replaced by $((K \to A) \to (K \to B))$ then the instance of the law of excluded middle is not used when AProS obtains a proof of length 29 in 23 search steps. If only the goal is reformulated as a conditional, then the same proof is obtained with just 18 search steps.

The replacement step in the last proof amounts to using one of the available rules from (Shanin, et al. 1965) heuristically: if the goal is of the form $(\neg X \vee \Delta)$ or $(X \vee \neg \Delta)$ then prove instead $(X \to \Delta)$, respectively, $(\Delta \to X)$. Such a reformulation of a problem, or equivalently the strategic use of a derived rule, can thus have a dramatic consequence on the search and the resulting derivation. Let me discuss two additional examples and a motivated extension of this heuristic step:

(1) Prove from the premise $P \vee Q$ the disjunction

$$(P \,\&\, Q) \vee (P \,\&\, \neg Q) \vee (\neg P \,\&\, Q).$$

With its basic algorithm AProS uses 202 search steps to find a proof of length 58; however, if the goal is reformulated as the conditional

$$\neg(P \,\&\, Q) \to ((P \,\&\, \neg Q) \vee (\neg P \,\&\, Q))$$

then 28 steps lead to a proof of length 18.

(2) Prove $((P \vee Q) \to (P \vee R)) \to (P \vee (Q \to R))$.

103 steps in the basic search lead to a proof of length 47. If one considers instead

$$((P \vee Q) \to (P \vee R)) \to (\neg P \to (Q \to R))$$

AProS finds a proof of length 14 with 9 search steps.

These quasi-empirical observations can be used to articulate a heuristic for the purely logical search: if one encounters a disjunction $(X \vee \Delta)$ as the goal, prove instead the conditional $(\neg X \to \Delta)$ (and eliminate in the antecedent a double negation, in case X happens to be a negation).

B. Some elementary set theoretic arguments. As I mentioned in sections 3.2 and 4.1, we have been extending the automated search procedure to elementary set theory. Though our goal is different from that of interactive theorem proving, there is a great deal of overlap: the hierarchical organization of the search can be viewed as reflecting and sharpening the interaction of a user with a proof assistant. After all, we start out by analyzing the structure of proofs, formalizing them, and then automating the proof search, i.e., completely eliminating interaction. The case of using computers as proof assistants is made in great detail in Harrison's paper (2008). For the case of automated proof search it is important, if not absolutely essential, that the logical calculus of choice is natural deduction.[41]

---

[41] There have been attempts of using proofs by resolution or other "machine-oriented" procedures as starting points for obtaining natural deduction proofs; Peter Andrews and Frank Pfenning, but also more recently Xiaorong Huang did interesting work in that direction.

Searching for proofs

Natural deduction has been used for proof search in set theory; an informative description is found, for example, in (Bledsoe 1983). Pastre's (1976) dissertation, deeply influenced by Bledsoe's work, is mentioned in Bledsoe's paper. She has continued that early work, and her most recent paper (2007) addresses a variety of elementary set theoretic problems. Similar work, but in the context of the *Theorema* project, was done in (Windsteiger 2001) and (Windsteiger 2003). However, the term natural deduction is used here only in a very loose way: there is no search space that underlies the logical part and guarantees completeness of the search procedure. Rather, the search is guided in both logic and set theory by "natural heuristics" for the use of reduction rules that are not connected to a systematic logical search and, in Pastre's case, do not even allow for any backtracking.

Let me consider a couple of examples of AProS proofs to show how the logical search is extended in a most natural way by exploiting the meaning of defined concepts by appropriate I- and E-rules. That has, in particular, the "side-effect" of articulating in a mathematically sensible way, at which point in the search definitions should be expanded. In each case, the reader should view the proof strategically, i.e., closing the gap between premises and conclusion by use of (inverted) I-rules and motivated E-rules.

Example 1: $a \in b$ proves $a \subseteq \bigcup(b)$

1. $a \in b$ — Premise
2. $u \in a$ — Assumption
3. $(a \in b \,\&\, u \in a)$ — &I 1, 2
4. $(\exists z)(z \in b \,\&\, u \in z)$ — $\exists$I 3
5. $u \in \bigcup(b)$ — Def.I (Union) 4
6. $(u \in a \to u \in \bigcup(b))$ — $\to$I 5
7. $(\forall x)(x \in a \to x \in \bigcup(b))$ — $\forall$I 6
8. $a \subseteq \bigcup(b)$ — Def.I (Subset) 7

Example 2.1: $a \subseteq b$ proves $\wp(a) \subseteq \wp(b)$

1. $a \subseteq b$ — Premise
2. $u \in \wp(a)$ — Assumption
3. $v \in u$ — Assumption
4. $(\forall x)(x \in a \to x \in b)$ — Def.E (Subset) 1
5. $(v \in a \to v \in b)$ — $\forall$E 4
6. $u \subseteq a$ — Def.E (Power Set) 2
7. $(\forall x)(x \in u \to x \in a)$ — Def.E (Subset) 6
8. $(v \in u \to v \in a)$ — $\forall$E 7
9. $v \in a$ — $\to$E 8, 3
10. $v \in b$ — $\to$E 5, 9
11. $(v \in u \to v \in b)$ — $\to$I 10
12. $(\forall x)(x \in u \to x \in b)$ — $\forall$I 11
13. $u \subseteq b$ — Def.I (Subset) 12

14.	$u \in \wp(b)$	Def.I (Power Set) 13
15.	$(u \in \wp(a) \to u \in \wp(b))$	$\to$I 14
16.	$(\forall x)(x \in \wp(a) \to x \in \wp(b))$	$\forall$I 15
17.	$\wp(a) \subseteq \wp(b)$	Def.I (Subset) 16

Example 2.2: This is example 2.1 with the additional premise (lemma):
$$(\forall x)[(x \subseteq a \ \& \ a \subseteq b) \to x \subseteq b]$$

1.	$(\forall x)[(x \subseteq a \ \& \ a \subseteq b) \to x \subseteq b]$	Premise
2.	$a \subseteq b$	Premise
3.	$u \in \wp(a)$	Assumption
4.	$(u \subseteq a \ \& \ a \subseteq b) \to u \subseteq b$	$\forall$E 1
5.	$u \subseteq a$	Def.E (Power Set) 3
6.	$(u \subseteq a \ \& \ a \subseteq b)$	&I 5, 2
7.	$u \subseteq b$	$\to$E 4, 6
8.	$u \in \wp(b)$	Def.I (Power Set) 7
9.	$(u \in \wp(a) \to u \in \wp(b))$	$\to$I 8
10.	$(\forall x)(x \in \wp(a) \to x \in \wp(b))$	$\forall$I 9
11.	$\wp(a) \subseteq \wp(b)$	Def.I (Subset) 10

C. Confluence? The AProS project intends also to throw some empirical light on the cognitive situation. With a number of collaborators I have been developing a web-based introduction to logic, called *Logic & Proofs*; it focuses on the strategically guided construction of proofs and includes dynamic tutoring via the search algorithm AProS. The course is an expansive Learning Laboratory, as students construct arguments in a virtual Proof Lab in which their every move is recorded. It allows the investigation of questions like:

- How do students go about constructing arguments?
- How do particular pedagogical interventions affect their learning?
- How efficient do students get in finding proofs with little backtracking?
- Does the skill of strategically looking for proofs transfer to informal considerations?

The last question hints at a broader and long-term issue I am particularly interested in, namely, to find out whether strategic-logical skills improve the ability of students to understand complex mathematics.

The practical educational aspects are deeply connected to a theoretical issue in cognitive science, namely, the stark opposition of "mental models" (Johnson-Laird) and "mental proofs" (Rips). I do not see an unbridgeable gulf, but consider the two views as complementary. Proofs as diagrams give rise to mental models, and the dynamic features of proof construction I emphasized are promoted by and reflect a broader structural, mathematical context; all of this is helping us to bridge the gap between premises and conclusion. The crucial question for me is: Can we make advances in isolating basic operations of the mind involved in constructing mathematical proofs or, in other words, can we develop a cognitive psychology of proofs that reflect logical and mathematical understanding?

There is deeply relevant work on analogical reasoning, e.g., Dedre Gentner's. In the (2010) manuscript with J. Colhoun they write, "Analogical processes are at the core of relational thinking, a crucial ability that, we suggest, is key to human cognitive prowess and separates us from other intelligent creatures. Our capacity for analogy ensures that every new encounter offers not only its own kernel of knowledge, but a potentially vast set of insights resulting from parallels past and future." Performance in particular tasks is enhanced when analogies, viewed as relational similarities, are strengthened by explicit comparisons and appropriate encodings. It seems that abstraction is here a crucial mental operation and builds on such comparisons. The underlying theoretical model of these investigations (structure mapping) is steeped in the language of the mathematics that evolved in the 19th century, in particular, through Dedekind's work. It was Dedekind who introduced *mappings* between arbitrary systems; he asserted in the strongest terms that without this *capacity of the mind* (to let a thing of one system correspond to a thing of another system) no thinking is possible at all. Modern abstract, structural mathematics, one can argue convincingly, makes analogies between different "structures" precise via appropriate axiomatic formulations. — All of this, so the rich psychological experimental work demonstrates, is important for learning. In the context of more sophisticated mathematics, *Kaminski e.a.* hypothesized (and confirmed) for example recently "that learning a single generic instantiation [i.e., a more abstract example of a structure or concept; WS] . . . may result in better knowledge transfer than learning multiple concrete, contextualized instantiations." (p. 454)

There is a most plausible confluence of mathematical and psychological reflection that would get us closer to a better characterization of the "capacity of the human mind" that was discovered in Greek and rediscovered in 19th century mathematics; according to Stein, as quoted already at the beginning of Part 2, "what has been learned, when properly understood, constitutes one of the greatest advances in philosophy . . ."

Bibliography

The following abbreviations are being used: *DMV* for Deutsche Mathematiker Vereinigung; *JSL* for Journal of Symbolic Logic; *BSL* for Bulletin of Symbolic Logic; *AMS* for American Mathematical Society.

Abrusci, V. M.
1981 "Proof", "theory", and "foundations" in Hilbert's mathematical work from 1885 to 1900, *in* M. L. D. Chiara (ed), *Italian studies in the philosophy of science*, Dordrecht, pp. 453–491.
1987 David Hilbert's 'Vorlesungen' on logic and foundations of mathematics, *in* G. Corsi, C. Mangione and M. Mugnai (eds), *Atti del convegno internazionale di storia della logica, San Gimignano, 1987*, CLUEB, Bologna, pp. 333–338.

Ackermann, W.
1924 Begründung des "tertium non datur" mittels der Hilbertschen Theorie der Widerspruchsfreiheit, *Mathematische Annalen* **93**, 1–36.
1925–26 Letters to Bernays of 25 June 1925 and 31 March 1926; Wissenschaftshistorische Sammlung, ETH Zürich, Bernays Nachlass, HS 975: 96 and 97.
1928 Zum Hilbertschen Aufbau der reellen Zahlen, *Mathematische Annalen* **99**, 118–133. Translated in (van Heijenoort 1967, pp. 493–507).
1940 Zur Widerspruchsfreiheit der Zahlentheorie, *Mathematische Annalen* **117**, 162–194.

Aczel, P.
1977 An introduction to inductive definitions, in (Barwise 1977, pp. 739–782).

Addison, J. S. and Kleene, S. C.
1957 A note of function quantification, Proc. *AMS* **8**, 1002–1006.

Andrews, P.
2005 Some reflections on proof transformations, *in* D. Hutter and W. Stephan (eds), *Mechanizing Mathematical Reasoning*, Vol. 2605 of *Lecture Notes in Computer Science*, Springer, pp. 14–29.

Andrews, P. and Brown, C.
2005 Proving theorems and teaching logic with TPS and ETPS, *BSL* **11**(1), 108–109.

Apt, K. and Marek, W.
1974 Second order arithmetic and related topics, *Annals of Mathematical Logic* **6**, 177–229.

Artmann, B.

2007   Allgemeine Phänomene mathematischen Denkens in den Elementen der Euklidischen Geometrie, *Mitteilungen der DMV* **15**, 165–172.

Aspray, W. and Kitcher, P. (eds)

1988   *History and philosophy of modern mathematics*, Vol. XI of *Minnesota Studies in the Philosophy of Science*, University of Minnesota Press, Minneapolis.

Asquith, P. D. and Kitcher, P. (eds)

1985   *PSA 1984: Proceedings of the 1984 biennial meeting of the Philosophy of Science Association*, Vol. 2, Philosophy of Science Association.

Aumann, G.

2009   *Euklids Erbe — Ein Streifzug durch die Geometrie und ihre Geschichte*; third edition, Darmstadt.

Avigad, J., Dean, E. and Mumma, J.

2009   A formal system for Euclid's *Elements*. *Review of Symbolic Logic* **2**, 700–768.

Avigad, J. and Towsner, H.

2009   Functional interpretation and inductive definitions, *JSL* **74**, 1100–1120.

Avigad, J. and Zach, R.

2007   The Epsilon Calculus, Stanford Encyclopedia of Philosophy (version of July 2007), http://plato.stanford.edu/entries/epsilon-calculus/#6.

Baire, R., Borel, É., Hadamard, J. and Lebesgue, H.

1905   Cinq lettres sur la théorie des ensembles, *Bulletin de la Société Mathématique de France* **33**, 261–273. Translated in (Moore 1982, pp. 311–320).

Barwise, J. (ed)

1977   *Handbook of mathematical logic*, North-Holland.

1981   Infinitary Logics, *in* E. Agazzi (ed) *Modern Logic — A Survey*, D. Reidel, Dordrecht, pp. 93–112.

Beaney, M.

2003   Analysis, *in* E. N. Zalta (ed), *The Stanford Encyclopedia of Philosophy (Summer 2003 Edition)*, Stanford University.

Beeson, M.

1985   *Foundations of constructive mathematics: metamathematical studies*, Springer.

Behmann, H.

1918   *Die Antinomie der transfiniten Zahl und ihre Auflösung durch die Theorie von Russell und Whitehead*, Dissertation, Georg-August-Universität Göttingen.

Behrmann, J.

1976   Biobibliographische Notiz, in (Zilsel 1976, pp. 44–46).

Belna, J.-P.

1996   *La notion de nombre chez Dedekind, Cantor, Frege*, VRIN.

Bibliography

Benacerraf, P.
1965 What numbers could not be, *The Philosophical Review* **74**, 47–73.

Benacerraf, P. and Putnam, H. (eds)
1983 *Philosophy of mathematics*, second edition, Cambridge University Press.

Bernays, P.
1918 *Beiträge zur axiomatischen Behandlung des Logik-Kalküls*, Habilitation, Georg-August-Universität Göttingen.
1922a Hilberts Bedeutung für die Philosophie der Mathematik, *Die Naturwissenschaften* **4**, 93–99. Translated in (Mancosu 1998, pp. 189–197).
1922b Über Hilberts Gedanken zur Grundlegung der Mathematik, *Jahresbericht der DMV* **31**, 10–19. Translated in (Mancosu 1998, pp. 215–222).
1926 Axiomatische Untersuchung des Aussagen-Kalküls der "Principia Mathematica", *Mathematische Zeitschrift* **25**, 305–320.
1927a Probleme der theoretischen Logik. Reprinted in (Bernays 1976a, pp. 1–16).
1927b Zusatz zu Hilberts Vortrag über "Die Grundlagen der Mathematik", *Abhandlungen aus dem mathematischen Seminar der Hamburgischen Universität* **6**, 89–92. Translated in (van Heijenoort 1967, pp. 485–489).
1928a Die Grundbegriffe der reinen Geometrie in ihrem Verhältnis zur Anschauung, *Die Naturwissenschaften* **16**(12), 197–203.
1928b Über Nelsons Stellungsnahme in der Philosophie der Mathematik, *Die Naturwissenschaften* **16**(9), 142–145.
1930a Die Grundgedanken der Fries'schen Philosophie in ihrem Verhältnis zum heutigen Stand der Wissenschaft, *Abhandlungen der Fries'schen Schule, Neue Folge* **5**(2), 97–113.
1930b Die Philosophie der Mathematik und die Hilbertsche Beweistheorie. Reprinted in (Bernays 1976a, pp. 17–61).
1933 Methoden des Nachweises von Widerspruchsfreiheit und ihre Grenzen, in (Saxer 1933, pp. 342–343).
1935a Hilberts Untersuchungen über die Grundlagen der Arithmetik, in (Hilbert 1935, third volume, pp. 196–216).
1935b Über den Platonismus in der Mathematik. Reprinted in (Bernays 1976a, pp. 62–78). Translated in (Benacerraf and Putnam 1983, pp. 258–271).
1937a Grundsätzliche Betrachtungen zur Erkenntnistheorie, *Abhandlungen der Fries'schen Schule, Neue Folge* **6**(3–4), 278–290.
1937b Thesen und Bemerkungen zu den philosophischen Fragen und zur Situation der logisch-mathematischen Grundlagenforschung. Reprinted in (Bernays 1967a, pp. 79–84).
1938 Über die aktuelle Methodenfrage der Hilbertschen Beweistheorie; unpublished manuscript from the Bernays Nachlass; it was presented at *Les entretiens de Zürich* in December 1938 and published in French as (Bernays 1941).
1941 Sur les questions méthodologiques actuelles de la théorie Hilbertienne de la démonstration, *in* F. Gonseth (ed), *Les entretiens de Zürich sur les fondements et la méthode des sciences mathématiques*, Leemann & Co., pp. 144–152. Discussion, pp. 153–161.
1946 Gesichtspunkte zum Problem der Evidenz. Reprinted in (Bernays 1976a, pp. 85–91).
1950 Mathematische Existenz und Widerspruchsfreiheit. Reprinted in (Bernays 1976a, pp. 92–106).

Bernays, P. (continued)

1954 Zur Beurteilung der Situation in der beweistheoretischen Forschung, *Revue internationale de philosophie* **8**, 9–13. Discussion, pp. 15–21.

1961 Zur Rolle der Sprache in erkenntnistheoretischer Hinsicht. Reprinted in (Bernays 1976a, pp. 155–169).

1965 Betrachtungen zum Sequenzenkalkül, *in* A.-T. Tymieniecka and C. Parsons (eds), *Contributions to Logic and Methodology, in honor of J. M. Bochenski*, North-Holland, pp. 1–44.

1967 Hilbert, David, *in* P. Edwards (ed), *Encyclopedia of philosophy*, Vol. 3, Macmillian and Co., pp. 496–504.

1970 Die schematische Korrespondenz und die idealisierten Strukturen. Reprinted in (Bernays 1976a, pp. 176–188).

1976a *Abhandlungen zur Philosophie der Mathematik*, Wissenschaftliche Buchgesellschaft.

1976b A short biography, *in* G. H. Müller (ed), *Sets and classes*, North-Holland, pp. xi–xiii.

1977 *Three Recorded Interviews, given on 25.7.1977, 13.8.1977, 27.8.1977.* Wissenschaftshistorische Sammlung, ETH Zürich, Bernays Nachlass, Cod. Ms. P. Bernays T 1285.

1978 Bemerkungen zu Lorenzens Stellungsnahme in der Philosophie der Mathematik, in (Lorenz 1978, pp. 3–16).

Bernoulli, J.

1685 Parallelismus ratiocinii logici et algebraici, in *Opera*, 1744, vol. I, 213–218.

Bernstein, F.

1919 Die Mengenlehre Georg Cantors und der Finitismus, *Jahresbericht der DMV* **28**, 63–78.

1924 Der Pythagoräische Lehrsatz, *Zeitschrift für Mathematischen und Naturwissenschaftlichen Unterricht* **55**, 204–207.

Beth, E. W.

1958 On machines which prove theorems. Reprinted in (Siekmann and Wrightson 1983, pp. 79–90).

Bledsoe, W. W.

1983 Non-resolution theorem proving, *Artificial Intelligence* **9**, 1–35.

1984 Some automatic proofs in analysis, in (Bledsoe and Loveland 1984, pp. 89–118).

Bledsoe, W. W. and Loveland, D. W. (eds)

1984 *Automated theorem proving: after 25 years*, Vol. 29 of *Contemporary Mathematics*, AMS.

Blumenthal, O.

1935 Lebensgeschichte [David Hilberts], in (Hilbert 1935, volume 3, pp. 388–429).

Bishop, E.

1967 *Foundations of constructive analysis*, McGraw-Hill.

Bolzano, B.

1851 *Paradoxien des Unendlichen*, Radelli & Hille. Translated in part in (Ewald 1996, pp. 249–292).

Boniface, J. and Schappacher, N.

2001 'Sur le concept de nombre en mathématiques' cours inédit de Leopold Kronecker à Berlin (1891), *Revue d'histoire des mathématiques* **7**, 207–275.

Booker, A. R.

2008 Uncovering a new L-Function, *Notices of the AMS* **55**, 1088–1094.

Boolos, G.

1976 On deciding the truth of certain statements involving the notion of consistency, *JSL* **41**, 778–781.

Borel, É.

1905 Quelques remarques sur les principes de la théorie des ensembles, *Mathematische Annalen* **60**, 194–5.

1914 *Leçons sur la théorie des fonctions*, second edition, Gauthier-Villars.

Börger, E. (ed)

1987 *Computation theory and logic*, Vol. 270 of *Lecture Notes in Computer Science*, Springer.

Bourbaki, N.

1950 The architecture of mathematics, *Mathematical Monthly* **57**, 221–232.

Brandes, H.

1907 *Über die axiomatische Einfachheit mit besonderer Berücksichtigung der auf Addition beruhenden Zerlegungsbeweise des pythagoräischen Lehrsatzes*, Dissertation, Halle.

Brouwer, L. E. J.

1918 Begründung der Mengenlehre unabhängig vom logischen Satz vom ausgeschlossenen Dritten, Erster Teil, Allgemeine Mengenlehre, *Verhandelingen der Koninklijke Akademie van Wetenschappen te Amsterdam,* 1e Sectie, deel XII, no. 5, pp. 1–43.

1919a Begründung der Mengenlehre unabhängig vom logischen Satz vom ausgeschlossenen Dritten, Zweiter Teil, Theorie der Punktmengen, *Verhandelingen der Koninklijke Akademie van Wetenschappen te Amsterdam,* 1e Sectie, deel XII, no. 7, pp. 1–33.

1919b Intuitionistische Mengenlehre, *Jahresbericht der* DMV **28**, 203–208. Translated in (Mancosu 1999a, pp. 23–26).

1921 Besitzt jede reelle Zahl eine Dezimalbruchentwicklung?, *Mathematische Annalen* **83**, 201–210. Translated in (Mancosu 1998, pp. 28–34).

1927 Über Definitionsbereiche von Funktionen, *Mathematische Annalen* **97**, 60–75. Translated in (van Heijenoort 1967, pp. 446–463).

1928 Intuitionistische Betrachtungen über den Formalismus, *Koninklijke Akademie van Wetenschappen te Amsterdam*, Proceedings of the section of sciences 31, pp. 374–379. Translated in (van Heijenoort 1967, pp. 490–492).

1953 Points and spaces, *Canadian Journal of Mathematics* **6**, 1–17.

Buchholz, W.

1981a The $\Omega_{\mu+1}$-rule, in (Buchholz, Feferman, Pohlers and Sieg 1981, pp. 188–233).

1981b Ordinal analysis of ID_ν, in (Buchholz, Feferman, Pohlers and Sieg 1981, pp. 234–260).

Buchholz, W. (continued)

1986 A new system of proof-theoretic ordinal functions, *Annals of Pure and Applied Logic* **32**, 195–207.

1990 Proof theory of iterated inductive definitions revisited. Manuscript, essential parts are incorporated in (Buchholz 2002).

2002 Relating ordinals to proofs in a more perspicuous way, in (Sieg, Sommer and Talcott 2002, pp. 37–59).

Buchholz, W., Feferman, S., Pohlers, W. and Sieg, W.

1981 *Iterated inductive definitions and subsystems of analysis: Recent proof-theoretical studies*, Vol. 897 of *Lecture Notes in Mathematics*, Springer.

Burgess, J.

2009 Putting structuralism in its place; manuscript.

Buss, S. R.

1995 On Gödel's theorems on lengths of proofs II: lower bounds for recognizing k symbol provability, in (Clote and Remmel 1995, pp. 57–90).

Byrnes, J.

1999 *Proof search and normal forms in natural deduction*, Dissertation, Carnegie Mellon University.

Cantini, A.

2008 On formal proofs, in (Lupacchini and Corsi 2008, pp. 29–48).

Cantor, G.

1878 Ein Beitrag zur Mannigfaltigkeitslehre, *Crelles Journal für reine und angewandte Mathematik* **84**, 242–258. Reprinted in (Cantor 1932, pp. 119–133).

1879 Über unendliche, lineare Punktmannigfaltigkeiten, No. 1, *Mathematische Annalen* **15**, 1–7.

1880 —, No. 2, ibid. 17, 355–358.

1882 —, No. 3, ibid. 20, 113–121.

1883a —, No. 4, ibid. 21, 51–58.

1883b —, No. 5, ibid. 21, 545–586.

1883c —, No. 6, ibid. 23, 453–488. This and the preceding five papers are reprinted in (Cantor 1932, pp. 139–246).

1897/9 Letters to Hilbert, in (Purkert and Ilgauds 1987).

1899 Letter to Dedekind. Translated in (van Heijenoort 1967, pp. 13–117).

1932 *Gesammelte Abhandlungen mathematischen und philosophischen Inhalts*, E. Zermelo (ed), Springer.

Carnap, R.

1930 Die Mathematik als Zweig der Logik, *Blätter für Deutsche Philosophie* **4**, 298–310.

1931 Die logizistische Grundlegung der Mathematik, *Erkenntnis* **2**, 91–105. Translated in (Benacerraf and Putnam 1983, pp. 41–52).

Cassirer, E.

1910 *Substanzbegriff und Funktionsbegriff*, Berlin.

Celluci, C.

2008 Why proof? What is a proof?, in (Lupacchini and Corsi 2008, pp. 1–27).

Bibliography

Church, A.
1935 An unsolvable problem of elementary number theory; preliminary report (abstract), *Bulletin of the AMS* **41**, 332–333.
1936 An unsolvable problem of elementary number theory, *American Journal of Mathematics* **58**, 345–363.
1960 The consistency of primitive recursive arithmetic, unpublished notes.

Church, A. and Kleene, S. C.
1937 Formal definitions in the theory of ordinal numbers, *Fundamenta Mathematicae* **28**, 11–21.

Clote, P. and Krajíček, J. (eds)
1993 *Arithmetic, proof theory, and computational complexity*, Oxford logic guides 23, Clarendon Press.

Clote, P. and Remmel, J. B. (eds)
1995 *Feasible mathematics II*, Vol. 13 of *Progress in computer science and applied logic*, Birkhäuser.

Coquand, T.
1995 A semantics of evidence for classical arithmetic, *JSL* **60**, 325–337.

Corry, L.
1996 *Modern algebra and the rise of mathematical structures*, Birkhäuser. (Second revised edition, 2004.)
2004 *Hilbert and the axiomatization of physics (1898–1918): From "Grundlagen der Geometrie" to "Grundlagen der Physik"*, Kluwer.

Davis, M.
1965 *The undecidable*, Raven Press.

Dawson, J. W.
1984 Discussion on the foundations of mathematics, *History and Philosophy of Logic* **5**, 111–129.
1985 The reception of Gödel's incompleteness theorems, in (Asquith and Kitcher 1985, pp. 253–274).
1986 Introductory Note to (Gödel 1931b) in (Gödel 1986, pp. 196–199).
1997 *Logical dilemmas: the life and work of Kurt Gödel*, A. K. Peters. Translated into German by Jakob Kellner as (Dawson 1999).
1999 *Kurt Gödel: Leben und Werk*, Computerkultur XI, Springer.

Dedekind, R.
1854 Über die Einführung neuer Funktionen in der Mathematik, Habilitationsvortrag, in (Dedekind 1932, pp. 428–438). Translated in (Ewald 1996, pp. 754–762).
1871/1872 Stetigkeit und irrationale Zahlen, Cod. Ms. Dedekind III, 17. Printed in (Dugac 1976, pp. 203–209).
1871 Arithmetische Grundlagen, Cod. Ms. Dedekind III, 4, II.
1872/1878 Was sind und was sollen die Zahlen? [erster Entwurf], Cod. Ms. Dedekind III, 1, I. Printed in (Dugac 1976, pp. 293–309).
1872a Die Schöpfung der Null und der negativen ganzen Zahlen, Cod. Ms. Dedekind III, 4, I, pp. 1–4.

Dedekind, R. (continued)

1872b Ganze und rationale Zahlen, Cod. Ms. Dedekind III, 4, I, pp. 5–7.

1872c *Stetigkeit und irrationale Zahlen*, Vieweg. Reprinted in (Dedekind 1932, pp. 315–324). Translated in (Ewald 1996, pp. 765–779).

1877 Sur la théorie des nombres entiers algébriques, *Bulletin des sciences mathématiques et astronomiques* **1**(XI), 2(I), pp. 1–121. Partially reprinted in (Dedekind 1932, pp. 262–296). Translated in (Dedekind 1996).

1887a Was sind und was sollen die Zahlen? [zweiter Entwurf], Cod. Ms. Dedekind III, 1, II.

1887b Was sind und was sollen die Zahlen? [dritter Entwurf], Cod. Ms. Dedekind III, 1, III.

1888 *Was sind und was sollen die Zahlen?*, Vieweg. Reprinted in (Dedekind 1932, pp. 335–391). Translated in (Ewald 1996, pp. 787–833).

189? Die Erweiterung des Zahlbegriffs auf Grund der Reihe der natürlichen Zahlen, Cod. Ms. Dedekind III, 2, I.

1890a Letter to H. Keferstein, Cod. Ms. Dedekind III, I, IV. Printed in (Sinaceur 1974, pp. 270–278). Translated in (van Heijenoort 1967, pp. 98–103).

1890b Über den Begriff des Unendlichen, Cod. Ms. Dedekind III, 1, IV. Submitted to, but not published by, the Mathematische Gesellschaft in Hamburg. Printed in (Sinaceur 1974, pp. 259–269).

1932 *Gesammelte mathematische Werke*, Vol. 3. R. Fricke, E. Noether, and Ö. Ore (eds), Vieweg.

1996 *Theory of algebraic integers*, translated and introduced by J. Stillwell, Cambridge University Press.

Dekker, J. C. E. (ed)

1962 *Recursive function theory*, Vol. 5 of *Proceedings of symposia in pure mathematics*, AMS.

de Mol, L.

2006 Closing the circle: an analysis of Emil Post's early work, *BSL* **12**, 267–289.

Dieudonné, J.

1970 The work of Nicholas Bourbaki, *American Mathematical Monthly* **70**, 134–145.

1985 *Geschichte der Mathematik, 1700–1900*, Vieweg.

Diller, J. and Schütte, K.

1971 Simultane Rekursionen in der Theorie der Funktionale endlicher Typen, *Archiv für mathematische Logik und Grundlagenforschung* **14**, 69–74.

Dirichlet, P. G. L.

1837 Beweis des Satzes, dass jede unbegrenzte arithmetische Progression, deren erstes Glied und Differenz ganze Zahlen ohne gemeinschaftlichen Faktor sind, unendlich viele Primzahlen enthält. Reprinted in (Dirichlet 1889, pp. 357–374).

1838 Sur l'usage des séries infinies dans la théorie des nombres. Reprinted in (Dirichlet 1889, pp. 357–374).

1839/40 Recherches sur diverses applications de l'analyse infinitésimale à la théorie des nombres. Reprinted in (Dirichlet 1889, pp. 411–496).

1863 *Vorlesungen über Zahlentheorie, Hrsg. und mit Zusätzen versehen von R. Dedekind*, Vieweg. 2nd edition 1871; 3rd 1879; 4th 1894.

Bibliography

Dirichlet, P. G. L. (continued)
1889   *Gesammelte Werke I*. L. Kronecker (ed), G. Reimer.
1897   *Gesammelte Werke II*. L. Kronecker and L. Fuchs (eds), G. Reimer.

Dreben, B. and van Heijenoort, J.
1986   Introductory note to (Gödel 1929), in (Gödel 1986, pp. 44–59).

Drucker, T. L.
1985   *Perspectives on the history of mathematical logic*, Birkhäuser.

Dugac, P.
1976   *Richard Dedekind et les fondements des mathématiques*, VRIN.

Dunham, W.
1990   *Journey through genius: The great theorems of mathematics*, Wiley.

Dvořak, J.
1981   *Edgar Zilsel und die Einheit der Erkenntnis*, Löcker.

Ebbinghaus, H.-D.
2007   *Ernst Zermelo: An approach to his life and work*, Springer.

Edwards, H.
1988   Kronecker's place in history, in (Aspray and Kitcher 1988, pp. 139–144).
2009   Kronecker's algorithmic mathematics, *Mathematical Intelligencer* **31**, 8–17.

Ewald, W. B. (ed)
1996   *From Kant to Hilbert: A source book in the foundations of mathematics*, Oxford University Press. Two volumes.

Feferman, S.
1964   Systems of predicative analysis, *JSL* **29**, 1–30. Reprinted in (Hintikka 1969, pp. 95–127).
1969   Set-theoretical foundations of category theory (with an appendix by G. Kreisel), *Reports of the Midwest Category Seminar, III*, Vol. 106 of *Lecture Notes in Mathematics*, Springer, pp. 201—247.
1970   Formal theories for transfinite iterations of generalized inductive definitions and some subsystems of analysis, in (Kino, Myhill and Vesley 1970, pp. 303–325).
1977   Theories of finite type related to mathematical practice, in (Barwise 1977, pp. 913–971).
1978   A more perspicuous formal system for predicativity, in (Lorenz 1978, pp. 68–93).
1979   Constructive theories of functions and classes, *in* M. Boffa, D. van Dalen and K. McAloon (eds), *Logic Colloqium '78*, North-Holland, pp. 159–224.
1981   How we got from there to here, in (Buchholz, Feferman, Pohlers and Sieg 1981, pp. 1–15).
1982   Inductively presented systems and the formalization of metamathematics, in (van Dalen et al. 1982, pp. 95–128).
1985   A theory of variable types, *Revista Colombiana de Matemáticas* **19**, 95–105.
1986   Introductory note to (Gödel 1931c), *in* (Gödel 1986, pp. 208–213).
1988a   Hilbert's program relativized: proof-theoretical and foundational reductions, *JSL* **53**(2), 364–384.

Feferman, S. (continued)

1988b Weyl vindicated: "Das Kontinuum" 70 years later, *Temi e prospettive della logica e della filosophia della scienza contemporanee, vol. I—Logica*, CLUEB, pp. 59–93.

1989 Finitary inductively presented logics, *in* R. Ferro (ed), *Logic Colloqium '88*, North-Holland, pp. 191–220.

1995 Introductory note to (Gödel 1933b), in (Gödel 1995, pp. 36–44).

1998 *In the light of logic*, Oxford University Press.

2003 Introductory note to Gödel's correspondence with Bernays; in (Gödel 2003a, pp. 41–79).

2010 The proof theory of classical and constructive inductive definitions. A forty year saga *in* R. Schindler (ed), *Ways of proof theory*, Ontos Verlag, Frankfurt, pp. 7–30.

Feferman, S. and Sieg, W.

1981a Iterated inductive definitions and subsystems of analysis, in (Buchholz, Feferman, Pohlers and Sieg 1981, pp. 16–77).

1981b Proof-theoretic equivalences between classical and constructive theories for analysis, in (Buchholz, Feferman, Pohlers and Sieg 1981, pp. 78–142).

Feferman, S. and Sieg, W. (eds)

2010 *Proofs, Categories and Computations* — Essays in honor of Grigori Mints; College Publications, London.

Feferman, S. and Strahm, T.

2010 The unfolding of finitist arithmetic, *Review of Symbolic Logic* **3**, 665–689

Ferreira, F.

1994 A feasible theory for analysis, *JSL* **59**, 1001–1011.

Ferreira, F. and Ferreira, G.

2008 The Riemann integral in weak systems of analysis, *Journal of Universal Computer Science* **14**, 908–937.

Ferreira, G.

2006 *Sistemas de Analise Fraca para a Integracao*, Dissertation, Universidade de Lisboa.

Ferreirós, J.

1999 *Labyrinth of thought — A history of set theory and its role in modern mathematics*, Birkhäuser. (Second edition, 2008.)

Ferreirós, J. and Gray, J. (eds)

2006 *The architecture of modern mathematics — Essays in history and philosophy*, Oxford University Press.

Flagg, R. C.

1986 Integrating classical and intuitionistic type theory, *Annals of Pure and Applied Logic* **32**, 27–51.

Flagg, R. C. and Friedman, H.

1986 Epistemic and intuitionistic formal systems, *Annals of Pure and Applied Logic* **32**, 53–60.

Fraenkel, A.

1930 Die heutigen Gegensätze in der Grundlegung der Mathematik, *Erkenntnis* **1**, 286–302.

Frege, G.

1893 *Grundgesetze der Arithmetik, begriffsschriftlich abgeleitet*, Jena.

1969 *Nachgelassene Schriften und wissenschaftlicher Briefwechsel*, H. Hermes, F. Kambartel, and F. Kaulbach (eds), Meiner Verlag.

1980 *Gottlob Freges Briefwechsel*, G. Gabriel, F. Kambartel, C. Thiel (eds), Felix Meiner Verlag.

1984 *Collected papers on mathematics, logic, and philosophy*. B. McGuinness (ed), Oxford University Press.

Frei, G. (ed)

1985 *Der Briefwechsel David Hilbert–Felix Klein (1886–1918)*, Vandenhoeck & Ruprecht.

Friedman, H.

1969 Bar induction and Π_1^1-CA, *JSL* **34**, 353–362.

1970 Iterated inductive definitions and Σ_2^1-AC, in (Kino, Myhill and Vesley 1970, pp. 435–442).

1975 Some systems of second order arithmetic and their use, *in* R. D. James (ed), *Proceedings of the International Congress of Mathematicians*, Vol. 1, Canadian Mathematical Congress, Vancouver, pp. 235–242.

1976 The arithmetic theory of sets and functions I. Mimeographed.

1977 Set theoretic foundations for constructive analysis, *Annals of Mathematics*, ser. 2, **105**, 1–28.

1978 Classically and intuitionistically provably recursive functions, *in* G. H. Müller and D. S. Scott (eds), *Higher Set Theory*, Vol. 669 of *Lecture Notes in Mathematics*, Springer, pp. 21–27.

1980 A strong conservative extension of Peano arithmetic, *in* J. Barwise, H. J. Keisler and K. Kunen (eds), *The Kleene Symposium*, North-Holland, pp. 113–122.

Friedman, H., Simpson, S. G. and Smith, R. L.

1983 Countable algebra and set existence axioms, *Annals of Pure and Applied Logic* **25**(2), 141–183.

Gandy, R. O.

1973 Bertrand Russell, as Mathematician, *Bulletin of the London Mathematical Society* **5**, 342–348.

1980 Church's Thesis and principles for mechanisms, *in* Barwise, Keisler, and Kunen (eds), *The Kleene Symposium*, Amsterdam; pp. 123–148.

1982 Limitations of mathematical knowledge, in (van Dalen, Lascar and Smiley 1982, pp. 129–146).

Garey, M. R. and Johnson, D. S.

1979 *Computers and intractability—a guide to the theory of NP-completeness*, W. H. Freeman.

Gauss, C. F.

1831 Anzeige der Theoria residuorum biquadraticorum, Commentatio secunda, *Göttingische gelehrte Anzeigen, 23 April 1831*. Reprinted in (Gauss 1863–1929, Vol. 2 (1876), pp. 169–78). Translated in (Ewald 1996, pp. 306–313).

1863–1929 *Werke*, Königliche Gesellschaft der Wissenschaften, Göttingen, Leipzig, and Berlin. 12 vols.

Gauss, C. F. and Bessel, F. W.

1880 *Briefwechsel*, W. Engelmann.

Geach, P. and Black, M. (eds)

1977 *Translations from the philosophical writings of Gottlob Frege*, Oxford.

Gentner, D.

1983 Structure-Mapping: A theoretical framework for analogy, *Cognitive Science* **7**, 155–170.

Gentner, D. and Colhoun, J.

2010 Analogical processes in human thinking and learning, *in* B. Glatzeder, V. Goel, and A. von Müller (eds), *On Thinking: Vol. 2. Towards a Theory of Thinking*. Springer, 35–48.

Gentzen, G.

1932 Über die Existenz unabhängiger Axiomensysteme zu unendlichen Satzsystemen, *Mathematische Annalen* **107**, 329–350.

1932/3 "Urdissertation", Wissenschaftshistorische Sammlung, Eidgenössische Technische Hochschule, Zürich, Bernays Nachlass, Ms. ULS. (A detailed description of the manuscript is found in (von Plato 2009b, pp. 675–680)).

1933 Über das Verhältnis zwischen intuitionistischer und klassischer Arithmetik, *Archiv für mathematische Logik und Grundlagenforschung* **16** (1974), 119–132.

1934/5 Untersuchungen über das logische Schließen I, II, *Mathematische Zeitschrift* **39**, 176–210, 405–431.

1935 Der erste Widerspruchsfreiheitsbeweis für die klassische Zahlentheorie, *Archiv für mathematische Logik und Grundlagenforschung* **16** (1974), 97–118.

1936 Die Widerspruchsfreiheit der reinen Zahlentheorie, *Mathematische Annalen* **112**, 493–565. Translated in (Gentzen 1969, pp. 132–213).

1938a Die gegenwärtige Lage in der mathematischen Grundlagenforschung, *Forschung zur Logik und zur Grundlegung der exakten Wissenschaften, Neue Folge 4*, S. Hirzel, pp. 5–18. Translated in (Gentzen 1969, pp. 234–251).

1938b Neue Fassung des Widerspruchsfreiheitsbeweises für die reine Zahlentheorie, *Forschung zur Logik und zur Grundlegung der exakten Wissenschaften, Neue Folge 4*, S. Hirzel, 19–44. Translated in (Gentzen 1969, pp. 252–286).

1943 Beweisbarkeit und Unbeweisbarkeit von Anfangsfällen der transfiniten Induktion in der reinen Zahlentheorie, *Mathematische Annalen* **119**, 140–161. Translated in (Gentzen 1969, pp. 287–308).

1969 *The collected papers of Gerhard Gentzen*, North-Holland. Edited and translated by M. E. Szabo.

Giaquinto, M.

1983 Hilbert's philosophy of mathematics, *British Journal for the Philosophy of Science* **34**, 119–132.

Glivenko, V.

1929 Sur quelques points de la logique de M. Brouwer, *Académie Royale de Belgique, Bulletins de la classe des sciences*, ser. 5, vol. 15, 183–188.

Gödel, K.

1929 *Über die Vollständigkeit des Logikkalküls*, Dissertation, Vienna, in (Gödel 1986, pp. 60–101).

193? [Undecidable Diophantine propositions], in (Gödel 1995, pp. 164–175).

1930a Die Vollständigkeit der Axiome des logischen Funktionenkalküls, in (Gödel 1986, pp. 102–123).

1930b Einige metamathematische Resultate über Entscheidungsdefinitheit und Widerspruchsfreiheit, in (Gödel 1986, pp. 140–143).

1930c Vortrag über Vollständigkeit des Funktionenkalküls, in (Gödel 1995, pp. 16–29).

1931a Besprechung von Hilberts *Die Grundlegung der elementaren Zahlentheorie*, Zentralblatt für Mathematik und ihre Grenzgebiete 1, 260, reprinted in (Gödel 1986, pp. 212–214).

1931b Diskussion zur Grundlegung der Mathematik, in (Gödel 1986, pp. 200–205).

1931c Review of (Hilbert 1931a), in (Gödel 1986, pp. 213–215).

1931d Über formal unentscheidbare Sätze der Principia Mathematica und verwandter Systeme I, in (Gödel 1986, pp. 126–195).

1932a Review of (Carnap 1931), in (Gödel 1986, pp. 243–245).

1932b Review of (Heyting 1931), in (Gödel 1986, pp. 246–247).

1932c Review of (von Neumann 1931), in (Gödel 1986, pp. 248–249).

1932d Über Vollständigkeit und Widerspruchsfreiheit, Mathematisches Kolloquium, dated 22 January 1931, Ergebnisse eines mathematischen Kolloquiums 3, 12–13; reprinted and translated in (Gödel 1986, pp. 234–237).

1933a Eine Interpretation des intuitionistischen Aussagenkalküls, in (Gödel 1986, pp. 300–303).

1933b The present situation in the foundations of mathematics, in (Gödel 1995, pp. 36–53).

1933c Über Unabhängigkeitsbeweise im Aussagenkalkül, in (Gödel 1986, pp. 268–271).

1933d Zur intuitionistischen Arithmetik und Zahlentheorie, in (Gödel 1986, pp. 286–295).

1934 On undecidable propositions of formal mathematical systems, in (Gödel 1986, pp. 346–69).

1936 Über die Länge von Beweisen; in (Gödel 1986, pp. 396–399).

1938 Vortrag bei Zilsel, in (Gödel 1995, pp. 85–113).

1941 In what sense is intuitionistic logic constructive?, in (Gödel 1995, pp. 189–200).

1944 Russell's mathematical logic, in (Gödel 1990, pp. 102–141).

1946 Remarks before the Princeton bicentennial conference on problems in mathematics, in (Gödel 1990, pp. 150–153).

1947 What is Cantor's continuum problem?, in (Gödel 1990, pp. 176–187).

1951 Some basic theorems on the foundations of mathematics and their implications, Gibbs Lecture, in (Gödel 1995, pp. 304–323).

1958 Über eine bisher noch nicht benützte Erweiterung des finiten Standpunktes, *Dialectica* 12, 280–287; revised and expanded as (Gödel 1972). The German original was reprinted and translated in (Gödel 1990, pp. 240–251); the expanded English version is in (Gödel 1990, pp. 271–280).

Gödel, K. (continued)

1964 Postscriptum (to 1934 Princeton Lectures), in (Davis 1965, pp. 71–73) and (Gödel 1986, pp. 369–371).

1972 Some remarks on the undecidability results, in (Gödel 1990, pp. 305–306).

1986 *Collected Works*, Vol. I, Oxford University Press.

1990 *Collected Works*, Vol. II, Oxford University Press.

1995 *Collected Works*, Vol. III, Oxford University Press.

2003a *Collected Works*, Vol. IV, Oxford University Press.

2003b *Collected Works*, Vol. V, Oxford University Press.

Goldfarb, W.

1979 Logic in the twenties: the nature of the quantifier, *JSL* **44**(3), 351–368.

Goodman, N. D.

1973 The arithmetic theory of constructions, in (Mathias and Rogers 1973, pp. 274–298).

1970 A theory of constructions equivalent to arithmetic, in (Kino, Myhill and Vesley 1970, pp. 101–120).

Graham, R., Rothschild, B. and Spencer, J.

1980 *Ramsey theory*, Wiley & Sons.

Gray, J.

1992 The nineteenth-century revolution in mathematical ontology, *in* D. Gillies (ed), *Revolutions in mathematics*, Oxford University Press, pp. 226–248.

Guard, J.

1961 *The independence of transfinite induction up to ω^ω in recursive arithmetic*, Dissertation, Princeton University.

Guillaume, M.

1985 Axiomatik und Logik, in (Dieudonné 1985, pp. 748–882).

Hahn, H., Carnap, R., Gödel, K., Heyting, A., Reidemeister, K., Scholz, A. and von Neumann, J.

1931 Diskussion zur Grundlegung der Mathematik, *Erkenntnis* **2**, 135–151. Translated in (Dawson 1984, pp. 116–128).

Hallett, M.

1989 Physicalism, reductionism and Hilbert, *in* A. D. Irvine (ed), *Physicalism in mathematics*, Kluwer, pp. 183–257.

1994 Hilbert's axiomatic method and the laws of thought, *in* A. George (ed), *Mathematics and mind*, Oxford University Press, pp. 158–200.

1995 Hilbert and logic, *in* R. Marion and R. S. Cohen (eds), *Québec studies in the philosophy of science I*, Kluwer, pp. 135–187.

Hallett, M. and Majer, U. (eds)

2004 *David Hilbert's lectures on the foundations of geometry, 1891–1902*, Springer.

Hamilton, W. R.

1853 *Lectures on quaternions*, Hodges and Smith. Reprinted in (Ewald 1996, pp. 375–425).

Hand, M.

1989 A number in the exponent of an operation, *Synthese* **81**, 243–265.

Bibliography

Hand, M. (continued)
1990 Hilbert's iterativistic tendencies, *History and Philosophy of Logic* **11**, 185–192.

Harrington, L., Morley, M., Scedrov, A. and Simpson, S. G. (eds)
1985 *Harvey Friedman's research in the foundations of mathematics*, North-Holland.

Harrison, J.
2008 Formal proof — theory and practice, *Notices of the AMS* **55**(11), 1395–1406.

Hartmanis, J.
1989 Gödel, von Neumann and the P=?NP problem, *Bulletin of the European Association for Computer Science* **38**, 101–107.

Hartshorne, R.
2000a *Geometry: Euclid and beyond*, Springer.
2000b Teaching geometry according to Euclid, *Notices of the AMS* **47**(4), 460–465.

Heine, E.
1872 Die Elemente der Functionenlehre, *Crelles Journal für die reine und angewandte Mathematik* **74**, 172–188.

Herbrand, J.
1930 *Recherches sur la théorie de la démonstration*, Dissertation, University of Paris.
1931 Sur la non-contradiction de l'arithmétique, *Crelles Journal für die reine und angewandte Mathematik* **166**, 1–8; translated in (Herbrand 1971, pp. 282–298).
1968 *Écrits logiques*, J. van Heijenoort (ed), Presses Universitaires de France.
1971 *Logical writings*, Warren Goldfarb (ed), Harvard University Press.

Hertz, H.
1894 *Die Prinzipien der Mechanik*, Vol. III of *Gesammelte Werke*, Barth.

Heyting, A.
1930a Die formalen Regeln der intuitionistischen Logik, *Sitzungsberichte der Preussischen Akademie der Wissenschaften*, 42–56. Translated in (Mancosu 1998, pp. 311–327).
1930b Die formalen Regeln der intuitionistischen Mathematik, *Sitzungsberichte der Preussischen Akademie der Wissenschaften*, 57–71.
1930c Sur la logique intuitionniste, *Académie royale de Belgique, Bulletins de la classe des sciences* **5**(16), 957–963.
1931 Die intuitionistische Grundlegung der Mathematik, *Erkenntnis* **2**, 106–115. Translated in (Benacerraf and Putnam 1983, pp. 52–61).
1934 *Mathematische Grundlagenforschung. Intuitionismus. Beweistheorie.*, Springer.

Hilbert, D.
Unpublished Lecture Notes of Hilbert's are located in Göttingen in two different places, namely, the Staats- und Universitätsbibliothek and the Mathematisches Institut. The reference year of these notes is preceded by a "*"; their location is indicated by SUB xyz, repectively MI. Many of them are being prepared for publication in *David Hilbert's lectures on the foundations of mathematics and physics, 1891–1933*, Springer.

Hilbert, D. (continued)

*1889/90 Einführung in das Studium der Mathematik, SUB 530.

*1894a Die Grundlagen der Geometrie, SUB 541.

*1894b Die Quadratur des Kreises, SUB 542.

1897a Die Theorie der algebraischen Zahlkörper, *Jahresbericht der* DMV **4**, 175–546. Reprinted in *Gesammelte Abhandlungen*, Vol. 1, pp. 63–363.

*1897b Zahlbegriff und Quadratur des Kreises, SUB 549.

*1898/99 Grundlagen der Euklidischen Geometrie. Lecture Notes by H. von Schaper, MI. Printed in (Hilbert 2004, pp. 302–395).

1899a Grundlagen der Geometrie, in *Festschrift zur Feier der Enthüllung des Gauss-Weber-Denkmals in Göttingen*, Teubner, pp. 1–92.

*1899b Zahlbegriff und Quadratur des Kreises, SUB 549 and also in SUB 557.

1900a Mathematische Probleme, *Nachrichten der Königlichen Gesellschaft der Wissenschaften zu Göttingen*, 253–297. Translated in (Ewald 1996, pp. 1096–1105). Reprinted with additions in *Archiv der Mathematik und Physik* **3**(1), 1901.

1900b Über den Zahlbegriff, *Jahresbericht der* DMV **8**, 180–194. Reprinted in *Grundlagen der Geometrie*, third edition, Leipzig, 1909, pp. 256–262. Translated in (Ewald 1996, pp.1089–1095).

1900c Les principes fondamentaux de la géométrie; translation of (Hilbert 1899a) with some additions, Gauthier-Villars.

1902 Sur les problèmes futurs des mathématiques, *Compte Rendu du Deuxième Congrès International des Mathématiciens*, Gauthier-Villars, pp. 59–114.

*1904 Zahlbegriff und Quadratur des Kreises. Lecture notes by M. Born. MI.

1905a Über die Grundlagen der Logik und der Arithmetik, in *Verhandlungen des Dritten Internationalen Mathematiker-Kongresses*, Teubner, pp. 174–185. Translated in (van Heijenoort 1967, pp. 129–138).

*1905b Logische Prinzipien des mathematischesn Denkens. Lecture notes by Hellinger. MI.

*1908 Prinzipien der Mathematik. MI.

*1910 Elemente und Prinzipienfragen der Mathematik. Lecture notes by R. Courant, MI.

*1913 Elemente und Prinzipien der Mathematik, SUB 559.

*1914/15 Probleme und Prinzipienfragen der Mathematik, SUB 559 contains draft.

*1917 Mengenlehre. Lecture notes by M. Goeb, MI.

*1917/18 Prinzipien der Mathematik. Lecture notes by P. Bernays, MI.

1918 Axiomatisches Denken, *Mathematische Annalen* **78**, 405–415. Translated in (Ewald 1996).

*1919 Natur und mathematisches Erkennen. Lecture notes by P. Bernays, MI. (These notes were edited by D. E. Rowe and published in 1992 by Birkhäuser.)

*1920a Logik-Kalkül. Lecture notes by P. Bernays, MI.

*1920b Probleme der mathematischen Logik. Lecture notes by P. Bernays and M. Schönfinkel, MI.

1921 Natur und mathematisches Erkennen; talk given in Copenhagen on 14 March 1921; SUB 589.

Bibliography

Hilbert, D. (continued)

*1921/22 Grundlagen der Mathematik. Lecture notes by P. Bernays, MI.

1922 Neubegründung der Mathematik, *Abhandlungen aus dem mathematischen Seminar der Hamburgischen Universität* **1**, 157–177; translated in (Ewald 1996, pp. 1117–1134).

*1922/23 Logische Grundlagen der Mathematik. Lecture notes by P. Bernays, SUB 567.

1923 Die logischen Grundlagen der Mathematik, *Mathematische Annalen* **88**, 151–165; translated in (Ewald 1996, pp. 1136–1148).

1926 Über das Unendliche, *Mathematische Annalen* **95**, 161–190. Translated in (van Heijenoort 1967, pp. 367–392).

1927 Die Grundlagen der Mathematik, *Abhandlungen aus dem mathematischen Seminar der Hamburgischen Universität* **6**(1/2), 65–85; translated in (van Heijenoort 1967, pp. 464–479).

1928 Probleme der Grundlegung der Mathematik, *Mathematische Annalen* **102**, 1–9. Reprint, with emendations and additions, of paper with the same title, published in *Atti del Congresso internazionale dei matematici, Bologna 1928*, pp. 135–141.

*1929 Mengenlehre, Lecture Notes by L. Collatz, Staats- und Universitätsbibliothek Hamburg, Signatur: NL Collatz, 82 pages.

1930 Naturerkennen und Logik, *Die Naturwissenschaften* **18**, 959–963; reprinted in (Hilbert 1935, pp. 378–387); translated in (Ewald 1996, pp. 1157–1165).

1931a Die Grundlegung der elementaren Zahlenlehre, *Mathematische Annalen* **104**, 485–494; partially reprinted in (Hilbert 1935, pp. 192–195); translated in (Ewald 1996, pp. 1148–1157).

1931b Beweis des *tertium non datur*, *Nachrichten von der Gesellschaft der Wissenschaften zu Göttingen, Mathematisch-physikalische Klasse*, 120–125.

1935 *Gesammelte Abhandlungen*, Springer. Three volumes.

2004 *David Hilbert's Lectures on the Foundations of Geometry, 1891–1902*; M. Hallet and U. Majer (eds), Springer.

2009 *David Hilbert's Lectures on the Foundations of Physics, 1915–1927*; T. Sauer and U. Majer (eds), Springer.

2013 *David Hilbert's Lectures on the Foundations of Arithmetic and Logic, 1917–1933*; W. Ewald and W. Sieg (eds), Springer.

Hilbert, D. and Ackermann, W.

1928 *Grundzüge der theoretischen Logik*, Springer.

Hilbert, D. and Bernays, P.

1934 *Grundlagen der Mathematik*, Vol. I, Springer. Second edition, 1968, with revisions detailed in foreword by Bernays.

1939 *Grundlagen der Mathematik*, Vol. II, Springer. Second edition, 1970, with revisions detailed in foreword by Bernays.

Hintikka, J. (ed)

1969 *The philosophy of mathematics*, second edition, Oxford University Press.

Huang, X.

1996 Translating machine-generated resolution proofs into nd-proofs at the assertion level, *in* N. Foo and R. Goebel (eds), *PRICAI-96*, Vol. 1114 of *Lecture Notes in Artificial Intelligence*, Springer, 399–410.

Jäger, G.
1986 *Theories for admissible sets—a unifying approach to proof theory*, Bibliopolis.

Johnson-Laird, P. N.
1983 *Mental models*, Harvard University Press.

Joyal, A. and Moerdijk, I.
1995 *Algebraic set theory*, Vol. 220 of *London Mathematical Society Lecture Notes Series*, Cambridge University Press.

Kaminski, J. A., Sloutsky, V. M. and Heckler, A. F.
2008 The advantage of abstract examples in learning mathematics, *Science* **320**, 454–455.

Kanamori, A.
2012 In praise of replacement, *BSL* **18**, 46–90.

Kanger, S.
1963 A simplified proof method for elementary logic. Reprinted in (Siekmann and Wrightson 1983, pp. 364–371).

Karzel, H. and Kroll, H.-J.
1988 *Geschichte der Geometrie seit Hilbert*, Wissenschaftliche Buchgesellschaft.

Keferstein, H.
1890 Über den Begriff der Zahl, *Festschrift der Mathematischen Gesellschaft in Hamburg*, pp. 119–125.

Kennedy, J.
2010 Gödel and "Formalism Freeness"; manuscript, 25 February 2010.

Kino, A., Myhill, J. and Vesley, R. E. (eds)
1970 *Intuitionism and proof theory. Proceedings of the summer conference at Buffalo, N.Y., 1968*, North-Holland.

Kleene, S. C.
1987 Reflections on Church's thesis, *Notre Dame Journal of Formal Logic* **28**, 490–498.

Klein, F.
1926 *Vorlesungen über die Entwicklung der Mathematik im 19. Jahrhundert*, Teil I, Springer.

Kneser, H.
1921/22 Grundlagen der Mathematik. Private notes of (Hilbert *1921/22).
1922/23 Logische Grundlagen der Mathematik. Private notes of (Hilbert *1922/23).

Kohlenbach, U.
2008 *Applied proof theory: Proof interpretations and their use in mathematics*, Springer.

Kolmogorov, A. N.
1932 Zur Deutung der intuitionistischen Logik, *Mathematische Zeitschrift* **35**, 58–65.

Kondô, M.
1938 Sur l'uniformisation des complémentaires analytiques et les ensembles projectifs de la seconde classe, *Japanese Journal of Mathematics* **15**, 197–230.

Kreisel, G.

1951 On the interpretation of non-finitist proofs—Part I, *JSL* **16**, 241–267.
1958a Hilbert's programme, *Dialectica* **12**, 346–372. Reprinted in (Benacerraf and Putnam 1983, pp. 207–238).
1958b Mathematical significance of consistency proofs, *JSL* **23**, 155–182.
1959 Proof by transfinite induction and definition by transfinite induction in quantifier-free systems, *JSL* **24**, 322–323.
1962 Foundations of intuitionistic logic, in (Nagel, Suppes and Tarski 1962, pp. 198–210).
1963 *Stanford report on the foundations of analysis*, Stanford University, Stanford, California. With contributions by W. A. Howard and W. W. Tait.
1965 Mathematical logic, *in* T. L. Saaty (ed), *Lectures on Modern Mathematics*, Vol. III, John Wiley and Sons, pp. 95–195.
1968 Survey of proof theory, *JSL* **33**, 321–388.
1971 Review of (Gentzen 1969), *Journal of Philosophy* **68**, 238–265.
1987 Gödel's excursions into intuitionistic logic, in (Weingartner and Schmetterer 1987, pp. 65–186).

Kreisel, G., Mints, G. and Simpson, S. G.

1975 The use of abstract languages in elementary metamathematics: Some pedagogical examples, *in* R. Parikh (ed), *Logic Colloqium*, Vol. 453 of *Lecture Notes in Mathematics*, Springer, pp. 38–131.

Kreisel, G. and Troelstra, A. S.

1970 Formal systems for some branches of intuitionistic analysis, *Annals of Mathematical Logic* **1**(3), 229–387.

Kronecker, L.

1886 Über einige Anwendungen der Modulsysteme, *Crelles Journal für die reine und angewandte Mathematik* **99**, 329–371. Also in (Kronecker 1899, pp. 145–208).
1887 Über den Zahlbegriff, *Crelles Journal für die reine und angewandte Mathematik* **101**, 337–355. Reprinted in (Kronecker 1899, pp. 251–274). This is an expanded version of the essay with the same name that was published in *Philosophische Aufsätze*, Eduard Zeller zu seinem fünfzigjährigen Doctor-Jubiläum gewidmet, Leipzig 1887. The latter essay is translated in (Ewald 1996, pp. 947–955).
1899 *Werke III*, Teubner.
1901 *Vorlesungen zur Zahlentheorie,* K. Hensel (ed), Teubner.

Kummer, E. E.

1860 Gedächtnisrede auf Gustav Peter Lejeune Dirichlet. Reprinted in (Dirichlet 1897, pp. 311–344).

Leibniz, G. W.

1666 Dissertatio de arte combinatoria, in *Sämtliche Schriften und Briefe*, Vol. 1, VI. Reihe: philosophische Schriften, 165–228.

Leivant, D.

1985 Syntactic translations and provably recursive functions, *JSL* **50**, 682–688.

Lipschitz, R.

1986   *Briefwechsel mit Cantor, Dedekind, Helmholtz, Kronecker, Weierstrass und anderen*, W. Scharlau (ed), Vieweg.

Lorenz, K. (ed)

1978   *Konstruktionen versus Positionen*, DeGruyter.

Lorenzen, P.

1955   *Einführung in die Operative Logik und Mathematik*, Springer.

Luckhardt, H.

1989   Herbrand-Analysen zweier Beweise des Satzes von Roth: polynomiale Anzahlschranken, *JSL* **54**(1), 234–263.

Lupacchini, R. and Corsi, G. (eds)

2008   *Deduction, computation, experiments — Exploring the effectiveness of proof*, Springer Italia.

Lützen, J.

2005   *Mechanistic images in geometric form — Heinrich Hertz's 'Principles of Mechanics'*, Oxford University Press.

Mac Lane, S.

1934   *Abgekürzte Beweise im Logikkalkul*, Dissertation, Göttingen.

1935   A logical analysis of mathematical structure, *The Monist* **45**, 118–130.

1971   Categorical algebra and set-theoretic foundations, in *Axiomatic set theory*, Vol. XIII, Part I of *Proceedings of the symposia in pure mathematics*, pp. 231–240.

1979   A late return to a thesis in logic, *in* I. Kaplansky (ed), *Saunders MacLane — Selected papers*, Springer.

Macrae, N.

1992   *John von Neumann*, Pantheon Books.

Mahlo, P.

1908   *Topologische Untersuchungen über Zerlegung in ebene und sphärische Polygone*, Dissertation, Halle.

Mancosu, P.

1998   *From Brouwer to Hilbert. The debate on the foundations of mathematics in the 1920s*, Oxford University Press.

1999a Between Russell and Hilbert: Behmann on the foundations of mathematics, *BSL* **5**(3), 303–330.

1999b Between Vienna and Berlin: the immediate reception of Gödel's incompleteness theorems, *History and Philosophy of Logic* **20**, 33–45.

Manders, K.

2008   The Euclidean diagram, *in* P. Mancosu (ed), *The philosophy of mathematical practice*, Oxford University Press, pp. 80–133. MS first circulated in 1995.

Martin-Löf, P.

1975   An intuitionistic theory of types: predicative part, in (Rose and Shepherdson 1975, pp. 73–118).

1984   *Intuitionistic type theory*, Bibliopolis.

Maslov, Y., Mints, G. E. and Orevkov, V. P.
1983 Mechanical proof search and the theory of logical deduction in the USSR, in (Siekmann and Wrightson 1983, pp. 29–38).

Mathias, A. R. D. and Rogers, H. (eds)
1973 *Cambridge summer school in mathematical logic*, Vol. 337 of *Lecture Notes in Mathematics*, Springer.

McCarty, D.
1995 The mysteries of Richard Dedekind, in J. Hintikka (ed), *From Dedekind to Gödel: Essays on the development of the foundations of mathematics*, Kluwer, pp. 53–96.

Mehrtens, H.
1979 Das Skelett der modernen Algebra. Zur Bildung mathematischer Begriffe bei Richard Dedekind, in C. Scriba (ed), *Disciplinae Novae*, Vandenhoeck & Ruprecht, pp. 25–43.

Menzler-Trott, E.
2007 *Logic's Lost Genius — The life of Gerhard Gentzen*, American Mathematical Society and London Mathematical Society.
2010 Personal communication, 7 December 2010.

Meyer, F.
1888 R. Dedekind: Was sind und was sollen die Zahlen?, *Jahrbuch über die Fortschritte der Mathematik* **20**, 49–52.

Minkowski, H.
1896 *Die Geometrie der Zahlen*, Teubner. (A second, expanded edition was published posthumously in 1910; Hilbert and Speiser served as editors.)

Mints, G. E.
1969 Variation in the deduction search tactics in sequential calculi, *Seminar in Mathematics V. A. Steklov Mathematical Institute* **4**, 52–59.
1971 Quantifier-free and one-quantifier systems (Russian), *Zapiski Nauchnyk Seminarov LOMI* **20**, 115–133. Translated in *Journal of Soviet Mathematics* **1** (1973), 71–84.
1976 What can be done in PRA?, *Zapiski Nauchuyh Seminarov LOMI* **60**, 93–102. Translated in *Journal of Soviet Mathematics* **14**, 1980, 1487–1492.
1991 Proof theory in the USSR 1925–1969, *JSL* **56**(2), 385–424.

Moerdijk, I. and Palmgren, E.
2000a Type theories, toposes, and constructive set theory: Predicative aspects of AST. Manuscript.
2000b Well-founded trees in categories, *Annals of Pure and Applied Logic* **104**, 189–218.

Moore, G.
1980 Beyond first-order logic: The historical interplay between mathematical logic and axiomatic set theory, *History and Philosophy of Logic* **1**, 95–137.
1982 *Zermelo's axiom of choice*, Springer.
1988 The emergence of first-order logic, in (Aspray and Kitcher 1988, pp. 93–135).
1997 Hilbert and the emergence of modern mathematical logic, *Theoria* **12**(1), 65–90.

Mugnai, M.
2010 Logic and mathematics in the seventeenth century, *History and Philosophy of Logic* **31**, 297–314.

Müller, G. H.
1981 Framing mathematics, *Epistemologia* **4**(1), 253–86.

Nagel, E., Suppes, P. and Tarski, A. (eds)
1962 *Logic, methodology, and philosophy of science. Proceedings of the 1960 International Congress*, Stanford University Press.

Paetzold, H.
1995 *Ernst Cassirer—Von Marburg nach New York*, Wissenschaftliche Buchgesellschaft.

Parkinson, G. H. R.
1966 *Leibniz — Logical papers*, Oxford University Press.

Parsons, C. D.
1970 On a number-theoretic choice schema and its relation to induction, in (Kino, Myhill and Vesley 1970, pp. 459–473).
1972 On n-quantifier induction, *JSL* **37**, 466–482.
1980 Mathematical intuition, *Proceedings of the Aristotelian Society N. S.* **80** (1979–80), 145–68.
1982 Objects and logic, *The Monist* **65**(4), 491–516.
1983 The impredicativity of induction, *in* L. S. Cauman, I. Levi, C. D. Parsons and R. Schwartz (eds), *How many questions? — Essays in honor of Sidney Morgenbesser*, Hackett, pp. 132–153.
1984 Arithmetic and categories, *Topoi* **3**(2), 109–121.
1987 Developing arithmetic in set theory without infinity: Some historical remarks, *History and Philosophy of Logic* **8**, 201–213.
1990 The structuralist view of mathematical objects, *Synthese* **84**, 303–346.
1994 Intuition and number, *in* A. George (ed), *Mathematics and mind*, Oxford University Press, pp. 141–157.
2003 Introductory Note to the Wang correspondence, in (Gödel 2003b, pp. 379–397).
2007 *Mathematical thought and its objects*, Cambridge University Press.

Pastre, D.
1976 Démonstration automatique de théorèmes en théorie des ensembles, Dissertation, University of Paris.
2002 Strong and weak points of the MUSCADET theorem prover — examples from CASC-JC, *AI Communications* **15**, 147–160.
2007 Complementarity of a natural deduction knowledge-based prover and resolution-based provers in automated theorem proving. Manuscript, March 2007, 34 pages.

Peano, G.
1889 *Arithmetices Principia, nova methodo exposita*, Bocca.
1891 Sul concetto di numero, *Rivista di Matematica* **1**, 87–102 and 256–267.

Peckhaus, V.
1990 *Hilbertprogramm und Kritische Philosophie*, Vandenhoeck & Ruprecht.

Peckhaus, V. (continued)
1994a Hilbert's axiomatic programme and philosophy, *in* E. Knobloch and D. E. Rowe (eds), *The history of modern mathematics*, Vol. 3, Academic Press, pp. 91–112.
1994b Logic in transition: The logical calculi of Hilbert (1905) and Zermelo (1908), *in* D. Prawitz and D. Westerstahl (eds), *Logic and philosophy of science in Uppsala*, Kluwer, pp. 311–323.
1995 Hilberts Logik. Von der Axiomatik zur Beweistheorie, *Internationale Zeitschrift für Geschichte und Ethik der Naturwissenschaft, Technik und Medizin*, Vol. 3, 65–86.

Petri, B. and Schappacher, N.
2007 On arithmetization, *in* C. Goldstein, N. Schappacher, J. Schwermer (eds) *The shaping of arithmetic after C. F. Gauss's Disquisitiones Arithmeticae*, Springer, pp. 353–374.

Pfenning, F.
1987 *Proof transformations in higher-order logic*, Dissertation, Carnegie Mellon University.

Pohlers, W.
1981 Proof-theoretical analysis of ID_ν by the method of local predicativity, in (Buchholz, Feferman, Pohlers and Sieg 1981, pp. 261–357).
1989 *Proof theory — An introduction*, Vol. 1407 of *Lecture Notes in Mathematics*, Springer.
2009 *Proof theory — The first step into impredicativity*, Springer.

Poincaré, H.
1902a Du rôle de l'intuition et de la logique en mathématiques, *Compte Rendu du Deuxième Congrès International des Mathématiciens*, Paris, pp. 115–130.
1902b Review of (Hilbert 1899a), *Bulletin des sciences mathématiques* **26**, 249–272.
1905 Les mathématiques et la logique, *Revue de métaphysique et de morale* **13**, 815–835. Translated in (Ewald 1996, pp. 1021–1038).
1906 Les mathématiques et la logique, *Revue de métaphysique et de morale* **14**, 17–34, 294–317. Translated in (Ewald 1996, pp. 1038–1071).

Popper, K.
1976 *Unended quest; an intellectual autobiography*, Open Court.

Post, E.
1936 Finite combinatorial processes, Formulation I, *JSL* **1**, 103–105. Reprinted in (Davis 1965, pp. 289–291).
1944 Recursively enumerable sets of positive integers and their decision problems, *Bulletin of the* AMS **50**, 284–316. Also in (Davis 1965, pp. 305–337).
1947 Recursive unsolvability of a problem of Thue; *JSL* **12**, 1–11.

Prawitz, D.
1960 An improved proof procedure, in (Siekmann and Wrightson 1983, pp. 162–198).
1965 *Natural deduction: A proof-theoretical study*, Stockholm.

Purkert, W. and Ilgauds, H. J.
1987 *Georg Cantor, 1845–1918*, Birkhäuser.

Ramsey, F. P.

1925 The foundations of mathematics, *Proceedings of the London Mathematical Society*, Vol. 25 of *Series 2*, pp. 338–384.

Rathjen, M.

1991 Proof-theoretic analysis of KPM, *Archive of Mathematical Logic* **30**, 377–403.

1995 Recent advances in ordinal analysis, *BSL* **1**(4), 468–485.

2009 The constructive Hilbert program and the limits of Martin-Löf type theory, in S. Lindström, E. Palmgren, K. Segerberg, and V. Stoltenberg-Hansen (eds), *Logicism, intuitionism, and formalism: What has become of them?*, Springer, 2009, pp. 397–433.

Ravaglia, M.

2003 *Explicating the finitist standpoint*, Dissertation, Carnegie Mellon University.

Reid, C.

1970 *Hilbert*, Springer.

Richardson, R. G. D.

1932 Report on the Congress, *Bulletin of the AMS* **38**(11), 769–774.

Rips, L. J.

1994 *The psychology of proof — Deductive reasoning in human thinking*, MIT Press.

Rose, H. E.

1984 *Subrecursion: functions and hierarchies*, Clarendon Press.

Rose, H. E. and Shepherdson, J. C. (eds)

1975 *Logic Colloquium '73*, North-Holland.

Rosser, J. B.

1937 Gödel's theorems for non-constructive logics, *JSL* **2**, 129–137.

Russell, B.

1908 Mathematical logic as based on the theory of types, *American Journal of Mathematics* **30**, 222–262. Reprinted in (van Heijenoort 1967, pp. 150–182).

Saxer, W. (ed)

1933 *Verhandlungen des Internationalen Mathematiker-Kongresses Zürich 1932, II. Band: Sektionsvorträge*, Orell Füssli Verlag.

Schappacher, N.

2005 David Hilbert, Report on algebraic number fields ('Zahlbericht'), 1897, in I. Grattan-Guinness (ed), *Landmark writings in western mathematics: case studies, 1640–1940*, Elsevier, pp. 700–709.

Schlimm, D.

2000 *Richard Dedekind: Axiomatic foundations of mathematics*, Master's thesis, Carnegie Mellon University.

Scholz, E.

1982 Herbart's influence on Bernhard Riemann, *Historia Mathematica* **9**, 413–40.

Schröder, E.

1873 *Lehrbuch der Arithmetik und Algebra*, Teubner.

Bibliography

Schröder-Heister, P.
2002 Resolution and the origins of structural reasoning: early proof-theoretic ideas of Hertz and Gentzen, *BSL* **8**, 246–265.

Schütte, K.
1993 Bemerkungen zur Hilbertschen Beweistheorie, ACTA BORUSSICA V, Beiträge zur ost- und westdeutschen Landeskunde 1991–1995.

Schwichtenberg, H.
1977 Proof theory: Some applications of cut-elimination, in (Barwise 1977, pp. 867–895).

Shanker, S. G. (ed)
1988 *Gödel's theorem in focus*, Croom Helm.

Shanin, N. A., Davydov, G. E., Maslov, S. Yu., Mints, G. E., Orevkov, V. P. and Slisenk, A. O.
1965 An algorithm for a machine search of a natural logical deduction in a propositional calculus. Translated in (Siekmann and Wrightson 1983, pp. 424–483).

Shapiro, S.
1985 *Intensional mathematics*, North-Holland.
1996 *Philosophy of mathematics—Structure and ontology*, Oxford University Press.
2000 *Thinking about mathematics: The philosophy of mathematics*, Oxford University Press.
2006 Computability, Proof, and Open-Texture, *in* A. Olszewski and J. Wolenski (eds), *Church's Thesis after 70 years*, Ontos Verlag, 2006, 420–455.

Shoenfield, J. R.
1967 *Mathematical logic*, Addison-Wesley.

Sieg, W.
1977 *Trees in metamathematics (Theories of inductive definitions and subsystems of analysis)*, Dissertation, Stanford University.
1981 Inductive definitions, constructive ordinals, and normal derivations, in (Buchholz, Feferman, Pohlers and Sieg 1981, pp. 143–187).
1984 Foundations for analysis and proof theory, *Synthese* **60**(2), 159–200. In this volume, pp. 229–261.
1985a Fragments of arithmetic, *Annals of Pure and Applied Logic* **28**, 33–71.
1985b Reductions of theories for analysis, *in* G. Dorn and P. Weingartner (eds), *Foundations of logic and linguistics*, Plenum Press, pp. 199–230. In this volume, pp. 263–280.
1987 Relative Konsistenz, in (Börger 1987, pp. 360–381).
1988 Hilbert's program sixty years later, *JSL* **53**(2), 338–348. In this volume, pp. 281–290.
1990a Relative consistency and accessible domains, *Synthese* **84**, 259–297. Reprinted with modifications in (Ferreirós and Gray 2006, pp. 339–368). In this volume, pp. 299–326.
1990b Review of (Simpson 1985b), *JSL* **55**, 870–874. In this volume, pp. 291–298.
1991 Herbrand analyses, *Archive for mathematical logic* **30**, 409–441.
1994a Eine neue Perspektive für das Hilbertsche Programm, *Dialektik*, pp. 163–180.

Sieg, W. (continued)

1994b Mechanical procedures and mathematical experience, *in* A. George (ed), *Mathematics and mind*, Oxford University Press, pp. 71–117.

1997a Aspects of mathematical experience, *in* E. Agazzi and G. Darvas (eds), *Philosophy of mathematics today*, Kluwer, pp. 195–217. In this volume, pp. 329–343.

1997b Step by recursive step: Church's analysis of effective calculability, *BSL* **3**, 154–180. Reprinted *in* A. Olszewski and J. Wolenski (eds), *Church's Thesis after 70 years*, Ontos Verlag, 2006, pp. 456–490.

1999 Hilbert's programs: 1917–1922, *BSL* **5**, 1–44. In this volume, pp. 91–127.

2000 Reductive structuralism, *Technical Report CMU-PHIL-108*, Carnegie Mellon University, Department of Philosophy.

2002 Beyond Hilbert's reach?, *in* D. Malament (ed), *Reading natural philosophy. Essays in the history and philosophy of science and mathematics*, Open Court, pp. 363–405. Reprinted *in* S. Lindström, E. Palmgren, K. Segerberg, and V. Stoltenberg-Hansen (eds), *Logicism, intuitionism, and formalism: What has become of them?*, Springer, 2009, pp. 449–483. In this volume, pp. 345–375.

2003 Introductory Note to Gödel's correspondence with von Neumann, in (Gödel 2003b, pp. 327–335). In this volume, pp. 145–153.

2005 Only two letters: The correspondence between Herbrand and Gödel, *BSL* **11**(2), 172–184.

2007 On mind & Turing's machines, *Natural Computing* **6**, 187–205.

2008 Church without dogma: axioms for computability, *in* S. B. Cooper, B. Löwe, and A. Sorbi (eds), *New computational paradigms — changing conceptions of what is computable*, Springer, pp. 139–152.

2009a On computability, *in* A. Irvine (ed), *Philosophy of mathematics*, Elsevier, pp. 535–630.

2009b Hilbert's proof theory, *in* D. M. Gabbay and J. Woods (eds), *Handbook of the history of logic*, Elsevier, pp. 321–384.

2010a Searching for proofs (and uncovering capacities of the mathematical mind); in (Feferman and Sieg 2010, pp. 189–215). In this volume, pp. 377–401.

2010b Gödel's philosophical challenge (to Turing); to appear in 2012 *in* J. Copeland, C. Posy and O. Shagrir (eds), *COMPUTABILITY: Gödel, Church, Turing, and Beyond*, MIT Press.

2011 In the shadow of incompleteness: Hilbert and Gentzen; to appear in 2012 in P. Dybjer, S. Lindström, E. Palmgren, and G. Sundholm (eds), *Epistemology versus Ontology: Essays on the Philosophy and Foundations of Mathematics in Honour of Per Martin-Löf*, Springer. In this volume, pp. 155–192.

2012 Normal forms for puzzles: a variant of Turing's Thesis; to appear *in* S. B. Cooper and J. van Leeuwen (eds), *Alan Turing: His work and impact*, Elsevier.

Sieg, W. and Byrnes, J.

1998 Normal natural deduction proofs (in classical logic), *Studia Logica* **60**, 67–106.

Sieg, W. and Cittadini, S.

2005 Normal natural deduction proofs (in non-classical logics), *in* D. Hutter and W. Stephan (eds), *Mechanizing mathematical reasoning*, Vol. 2605 of *Lecture Notes in Computer Science*, Springer, pp. 169–191.

Bibliography

Sieg, W. and Field, C.
2005　Automated search for Gödel's proofs, *Annals of Pure and Applied Logic* **133**, 319–338. Reprinted in (Lupacchini and Corsi 2008, pp. 117–140).

Sieg, W. and Parsons, C. D.
1995　Introductory note to (Gödel 1938), in (Gödel 1995, pp. 62–84). In this volume, pp. 193–213.

Sieg, W. and Ravaglia, M.
2005　David Hilbert and Paul Bernays, *Grundlagen der Mathematik*, in I. Grattan-Guinness (ed), *Landmark writings in western mathematics: case studies, 1640–1940*, Elsevier, pp. 981–999.

Sieg, W. and Schlimm, D.
2005　Dedekind's analysis of number: systems and axioms, *Synthese* **147**, 121–170. In this volume, pp. 35–72.
201?　Dedekind's structuralism: mappings and models. In preparation.

Sieg, W. and Tapp, C.
2012　Introduction to the Undated Draft; the "Undated Draft" is a manuscript most likely written in late 1920/early 1921; in (Hilbert 2012).

Sieg, W., Sommer, R. and Talcott, C. (eds)
2002　*Reflections on the foundations of mathematics: Essays in honor of Solomon Feferman*, Association for Symbolic Logic.

Siekmann, J. and Wrightson, G. (eds)
1983　*Automated Reasoning*, two volumes, Springer.

Simpson, S. G.
1982　Which set existence axioms are needed to prove the Cauchy/Peano theorem for ordinary differential equations?, *JSL* **49**, 783–802.
1985a　Friedman's research on subsystems of second order arithmetic, in (Harrington, Morley, Scedrov, and Simpson 1985, pp. 137–159).
1985b　Reverse mathematics, *in* A. Nerode and R. A. Shore (eds), *Recursion Theory*, Vol. 42 of *Proceedings of Symposia in Pure Mathematics*, AMS, pp. 461–471.
1988　Partial realizations of Hilbert's program, *JSL* **53**(2), 359–363.
1999　*Subsystems of second order arithmetic*, Springer.

Simpson, S. G. and Smith, R. L.
1986　Factorization of polynomials and Σ_1^0-induction, *Annals of Pure and Applied Logic* **31**, 289–306.

Sinaceur, M.-A.
1974　L'infini et les nombres — Commentaires de R. Dedekind à 'Zahlen' — La correspondance avec Keferstein, *Revue d'histoire des sciences* **27**, 251–278.

Skolem, T.
1922　Einige Bemerkungen zur axiomatischen Begründung der Mengenlehre. Translated in (van Heijenoort 1967, pp. 290–301).
1930　Über die Grundlagendiskussionen in der Mathematik, *Den syvende skandinaviske matematikerkongress*, Brogger, pp. 3–21.

Smoryński, C.
1977　The incompleteness theorems, in (Barwise 1977, pp. 821–866).

Spector, C.
1962 Provably recursive functionals of analysis: a consistency proof of analysis by an extension of principles formulated in current intuitionistic mathematics, in (Dekker 1962, pp. 1–27).

Stein, H.
1988 Logos, logic, and logistiké: Some philosophical remarks on nineteenth-century transformation of mathematics, in (Aspray and Kitcher 1988, pp. 238–259).
1990 Eudoxos and Dedekind: On the ancient Greek theory of ratios and its relation to modern mathematics, *Synthese* **84**, 163–211.
2000a Dedekind, Julius Wilhelm Richard, *Routledge Encyclopedia of Philosophy*, Routledge.
2000b Logicism, *Routledge Encyclopedia of Philosophy*, Routledge.

Stillwell, J.
2004 Emil Post and his anticipation of Gödel and Turing, *Mathematics Magazine* **77**, 3–14.

Sundholm, G.
1983 Constructions, proofs and the meaning of logical constants, *Journal of Philosophical Logic* **12**, 151–172.

Tait, W. W.
1965 Functions defined by transfinite recursion, *JSL* **30**, 155–174.
1968a Constructive reasoning, *in* B. van Rootselaar and F. Stall (eds), *Logic, methodology, and philosophy of science III*, North-Holland, pp. 185–199.
1968b Normal derivability in classical logic, *in* J. Barwise (ed), *The syntax and semantics of infinitary languages*, Vol. 72 of *Lecture Notes in Mathematics*, Springer, pp. 204–236.
1970 Applications of the cut-elimination theorem to some subsystems of classical analysis, in (Kino, Myhill and Vesley 1970, pp. 475–488).
1981 Finitism, *Journal of Philosophy* **78**, 524–546.
1986 Critical notice: Charles Parsons' *Mathematics in philosophy*, *Philosophy of Science* **53**, 588–607.
1996 Frege versus Cantor and Dedekind: On the concept of number, *in* M. Schirn (ed), *Frege: Importance and legacy*, DeGruyter, pp. 70–113.
2002 Remarks on finitism, in (Sieg, Sommer and Talcott 2002, pp. 410–419).
2005a Gödel's reformulation of Gentzen's first consistency proof of arithmetic, *BSL* **11**, 225–238.
2005b *The provenance of pure reason: Essays in the philosophy of mathematics and its history*, Oxford University Press.
2010 The substitution method revisited; in (Feferman and Sieg 2010, pp. 231–241).

Takeuti, G.
1975 *Proof theory*, North-Holland.
1978 *Two applications of logic to mathematics*, Princeton University Press.

Tapp, C.
2006 *An den Grenzen des Endlichen: Erkenntnistheoretische, wissenschafts-philosophische und logikhistorische Perspektiven auf das Hilbertprogramm*, Dissertation, Universität München. Electronically available at http://edoc.ub.uni-muenchen.de/6523/.

Bibliography

Tarski, A.
1924 Sur les ensembles finis, *Fundamenta Mathematicae* **6**, 45–95.

Taussky-Todd, O.
1987 Remembrances of Kurt Gödel, in: *Gödel remembered. Salzburg, 10–12 July 1983*, Bibliopolis, pp. 31–41.

Toepell, M.-M.
1986 *Über die Entstehung von David Hilberts "Grundlagen der Geometrie"*, Vandenhoeck & Ruprecht.

Torretti, R.
1984 *Philosophy of geometry from Riemann to Poincaré*, Reidel.

Troelstra, A. S.
1973 *Metamathematical investigation of intuitionistic arithmetic and analysis*, Vol. 344 of *Lecture Notes in Mathematics*, Springer.
1980 Extended bar induction of type zero, *in* J. Barwise, H. J. Keisler and K. Kunen (eds), *The Kleene Symposium*, North-Holland, pp. 277–316.
1986 Introductory note to (Gödel 1933a), in (Gödel 1986, pp. 282–287).
1990 Introductory note to (Gödel 1958, Gödel 1972), in (Gödel 1990, pp. 217–240).
1995 Introductory note to (Gödel 1941), in (Gödel 1995, pp. 186–189).

Troelstra, A. S. and van Dalen, D.
1988 *Constructivism in mathematics*, Vol. I and II, North-Holland.

Turing, A.
1936 On computable numbers, with an application to the *Entscheidungsproblem*, *Proceedings of the London Mathematical Society* **42**, 230–265. Also in (Davis 1965, pp. 116–151).
1947 Lecture to the London Mathematical Society on 20 February 1947, *in* B. E. Carpenter and R. W. Doran (eds), *A. M. Turing's ACE report of 1946 and other papers*, Cambridge University Press, pp. 106–124.
1948a Intelligent machinery. Written in September 1947, submitted to the National Physical Laboratory in 1948, and reprinted in *Machine Intelligence*, 5, Edinburgh, 3–23.
1948b Practical form of type theory, *JSL* **13**(2), 80–94.
1950a Computing machinery and intelligence, *Mind* **59**, 433–460.
1950b The word problem in semi-groups with cancellation, *Annals of Mathematics* **52** (2), 491–505.
1954 Solvable and unsolvable problems, *Science News* **31**, 7–23.

Ulam, S. M.
1958 John von Neumann 1903–1957, *Bulletin of the* AMS **3**(2), 1–49.
1976 *Adventures of a mathematician*, Charles Scribner.

van Dalen, D.
1995 Hermann Weyl's intuitionistic mathematics, *BSL* **1**(2), 145–169.

van Dalen, D., Lascar, D. and Smiley, T. J. (eds)
1982 *Logic Colloqium '80*, North-Holland.

van Heijenoort, J. (ed)
1967 *From Frege to Gödel: A sourcebook of mathematical logic, 1879–1931*, Harvard University Press.

von Helmholtz, H.

1887 Zählen und Messen, *Philosophische Aufsätze Eduard Zeller zu seinem fünfzigjährigen Doktorjubiläum gewidmet*, Fues Verlag, pp. 17–52. Reprinted in *Wissenschaftliche Abhandlungen*, Vol. III, pp. 356–391.

von Neumann, J.

1923 Zur Einführung der transfiniten Zahlen. Reprinted in (von Neumann 1961, pp. 24–33); translated in (van Heijenoort 1967, pp. 347–354).

1925 Eine Axiomatisierung der Mengenlehre. Reprinted in (von Neumann 1961, pp. 34–56); translated in (van Heijenoort 1967, pp. 393–413).

1926 Zur Prüferschen Theorie der idealen Zahlen. Reprinted in (von Neumann 1961, pp. 69–103).

1927 Zur Hilbertschen Beweistheorie. Reprinted in (von Neumann 1961, pp. 256–300).

1928a Die Axiomatisierung der Mengenlehre. Reprinted in (von Neumann 1961, pp. 339–422).

1928b Über die Definition durch transfinite Induktion und verwandte Fragen der allgemeinen Mengenlehre. Reprinted in (von Neumann 1961, pp. 320–338).

1929 Über eine Widerspruchsfreiheitsfrage in der axiomatischen Mengenlehre. Reprinted in (von Neumann 1961, pp. 494–508).

1931 Die formalistische Grundlegung der Mathematik, *Erkenntnis* **2**, 116–121. Translated in (Benacerraf and Putnam 1983, pp. 61–65).

1958 *The computer and the brain, the 1956 Silliman Lecture*, Yale University Press. With preface by Klara von Neumann.

1961 *Collected Works*, Vol. I. A. H. Taub (ed), Pergamon.

von Plato, J.

2008 Gentzen's proof of normalization for natural deduction, *BSL* **14**, 240–257.

2009a Gentzen's logic *in* D. M. Gabbay and J. Woods (eds), *Handbook of the History of Logic*, Vol. 5, Elsevier, 667–721.

2009b Gentzen's INH: a brief summary of its main ideas; personal communication, 24 November 2009, three pages.

2010 Gentzen's logical calculi: Aspects of a work of genius; manuscript, to appear.

Wang, H.

1960 Toward mechanical mathematics, in (Siekmann and Wrightson 1983, pp. 244–264).

1970 On the long-range prospects of automatic theorem-proving, *in* M. Laudet et al. (eds), *Symposium on automated demonstration*, Vol. 125 in *Lecture Notes in Mathematics*, Springer, pp. 101–111.

1974 *From mathematics to philosophy*, Routledge.

1981 Some facts about Kurt Gödel, *JSL* **46**, 653–659.

1984 Computer theorem proving and artificial intelligence, in (Bledsoe and Loveland 1984, pp. 49–70).

1987 *Reflections on Kurt Gödel*, MIT Press.

Weber, H.

1893 Leopold Kronecker, *Jahresbericht der* DMV **2**, 5–31.

Weingartner, P. and Schmetterer, L. (eds)

1987 *Gödel remembered. Salzburg 10–12 July 1983*, Bibliopolis.

Weinstein, S.
1983 The intended interpretation of intuitionistic logic, *Journal of Philosophical Logic* **12**, 261–270.

Weyl, H.
1910 Über die Definitionen der mathematischen Grundbegriffe, *Mathematisch-naturwissenschaftliche Blätter* **7**, 93–95, 109–113.
1918 *Das Kontinuum*, Verlag von Veit, Leipzig. Translated by S. Pollard and T. Bole, Dover, 1987.
1921 Über die neue Grundlagenkrise der Mathematik, *Mathematische Zeitschrift* **10**, 39–79.
1925 Die heutige Erkenntnislage in der Mathematik. Reprinted in volume 2 of Weyl's "Gesammelte Abhandlungen", Springer, 1968, 511–542.
1927 Diskussionsbemerkungen zu dem zweiten Hilbertschen Vortrag über die Grundlagen der Mathematik, *Abhandlungen aus dem mathematischen Seminar der Hamburgischen Universität* **6**, 86–88. Translated in (van Heijenoort 1967, pp. 480–484).
1944 David Hilbert and his mathematical work, *Bulletin of the AMS* **50**, 612–654.

Whitehead, A. N. and Russell, B.
1910 *Principia Mathematica*, Vol. 1, Cambridge University Press.
1912 *Principia Mathematica*, Vol. 2, Cambridge University Press.
1913 *Principia Mathematica*, Vol. 3, Cambridge University Press.

Wiedijk, F.
2008 Formal proof — getting started, *Notices of the* AMS **55**(11), 1408–1414.

Wiener, H.
1891 Über Grundlagen und Aufbau der Geometrie, *Jahresbericht der* DMV **1**, 45–48.

Windsteiger, W.
2001 A set theory prover within *Theorema*, in R. Moreno-Diaz et al. (eds), *Eurocast 2001*, Vol. 2178 of *Lecture Notes in Computer Science*, Springer, pp. 525–539.
2003 An automated prover for set theory in *Theorema*, Omega-Theorema Workshop, 14 pages.

Zach, R.
1998 Numbers and functions in Hilbert's finitism, *Taiwanese Journal for Philosophy and History of Science* **10**, 33–60.
1999 Completeness before Post: Bernays, Hilbert, and the development of propositional logic, *BSL* **5**, 331–366.
2003 The practice of finitism: epsilon calculus and consistency proofs in Hilbert's program, *Synthese* **137**, 211–259.
2004 Hilbert's "Verunglücker Beweis", the first epsilon theorem, and consistency proofs, *History and Philosophy of Logic* **25**, 79–94.

Zassenhaus, H. J.
1975 On the Minkowski-Hilbert dialogue on mathematization, *Canadian Mathematical Bulletin* **18**, 443–461.

Zermelo, E.
1904 Beweis, daß jede Menge wohlgeordnet werden kann, *Mathematische Annalen* **59**, 514–516. Translated in (van Heijenoort 1967, pp. 139–141).

Zermelo, E. (continued)

1908a Neuer Beweis für die Möglichkeit einer Wohlordnung, *Mathematische Annalen* **65**, 107–128. Translated in (van Heijenoort 1967, pp. 183–198).

1908b Über die Grundlagen der Arithmetik, *Atti del IV. Congresso internazionale dei matematici*, Academia dei Lincei, Rome, pp. 8–11.

1908c Untersuchungen über die Grundlagen der Mengenlehre I, *Mathematische Annalen* **65**, 261–281. Translated in (van Heijenoort 1967, pp. 199–215).

1909 Sur les ensembles finis et le principe de l'induction complète, *Acta Mathematica* **32**, 185–193.

1930 Über Grenzzahlen und Mengenbereiche, *Fundamenta Mathematicae* **16**, 29–47. Translated in (Ewald 1996, pp. 1219–1233).

1932 Über Stufen der Quantifikation und die Logik des Unendlichen, *Jahresbericht der* DMV **41**, 85–88.

1935 Grundlagen einer Allgemeinen Theorie der Mathematischen Satzsysteme, *Fundamenta Mathematicae* **25**, 136–146.

Zilsel, E.

1976 *Die sozialen Ursprünge der neuzeitlichen Wissenschaft.* W. Krohn (ed), Suhrkamp.

2000 *The social origins of modern science.* D. Raven, W. Krohn, and R. S. Cohen (eds), Kluwer.

Zucker, J. I.

1971 Iterated inductive definitions, trees, and ordinals, Dissertation, Stanford University.

1973 Iterated inductive definitions, trees, and ordinals, in (Troelstra 1973, chapter VI).

Index

Abstract structure (concept), 17, 26, 61–70, 301, 329, 396
Accessibility, 26, 301, 322, 324, 327, 339–340
Accessible domain, 8, 17–19, 26, 119, 227, 299–326, 327, 329, 338, 340–343, 348, 365–374, 396
Ackermann, 17, 19, 30, 102–103, 107–111, 121–122, 126, 131, 134–136, 139, 141, 143, 146, 156–161, 165, 169–171, 177, 183–184, 199–200, 213, 215–216, 222, 242–243, 266, 271, 285, 332, 356, 361, 385
Analysis, *see* Arithmetic, second-order
Archimedean axiom, 76–77, 382
Arithmetic
    first-order, 30, 93, 102, 104–106, 108–109, 111, 118, 129, 137, 139, 146, 149, 152, 156, 165, 169, 175, 177, 181, 185, 188, 194, 197, 200, 356, 383, 386–387
        primitive recursive (PRA), 19–20, 94, 134, 141–143, 157, 176, 197–200, 213, 216–218, 221–222, 225, 245–246, 250–252, 264, 267–268, 270, 274, 276–277, 287, 289, 292–293, 297, 303, 310, 316, 334, 338, 386, 388
        recursive or finitist, 132, 141, 161, 203, 222, 359
    second-order, 7, 160, 245–246, 263, 267–268, 273–274, 281, 285, 291–292, 294–296, 310, 322
Arithmetization, 6, 11, 22, 33, 37, 55, 79, 84, 94–95, 144, 215, 230–231, 234–235, 239, 264, 282–283, 289, 301–302, 304–307, 383
Axiom of arithmetic (Cantor's), 96, 306
Axiom of choice, 58, 146, 151, 158, 198, 238, 245, 267, 275, 277–278, 291, 295, 310, 333
Axiom of complete induction, 215
Axiom of infinity, 322–323, 338
Axiom of reducibility, 16, 89, 101, 106, 108, 112, 135–136, 237, 385
Axiom of restriction, 322
Axiomatic method, 1, 5–6, 16–17, 23, 33, 37, 71, 73, 75, 82, 87, 94, 98, 100, 103, 106, 237, 265, 308, 311, 316–317, 325, 331–332, 346, 349, 377, 379
Axioms of geometry, 76, 350

Benacerraf, 259, 290, 305, 319
Bernays, 4, 7–8, 14, 16–18, 20, 23–26, 29–31, 71, 75, 82, 86, 89, 91, 93, 96, 101–103, 106, 109, 111, 113–115, 117, 119–124, 129, 131–133, 135–144, 149–150, 155–159, 161–162, 165–167, 169–172, 174–176, 178–179, 182–183, 185, 189–190, 192–195, 197, 209, 211–213, 215–226, 229–230, 235–236, 240–245, 257, 259, 261, 265–267, 271–272, 281, 284–287, 289–290, 291, 294, 297, 301, 306–310, 312, 314–324, 326, 327–328, 329, 331–333, 337, 339–341, 345–346, 348, 350–351, 354–357, 359–375, 381, 385–386, 388, 390, 395–396
Bishop, 231, 233, 258, 268, 272, 287, 294, 326, 371

435

Bolzano, 58, 63–66, 85, 233, 247, 268, 287
Bolzano-Weierstrass theorem, 247, 268, 287
Borel, 238, 247, 272, 275, 295, 308–309
Brouwer, 9, 26, 33, 91, 112, 114, 116–117, 120, 147, 180, 197, 203, 230–231, 239–241, 258–259, 264, 268, 271–273, 287, 309, 311–312, 315–317, 325–326, 330, 334, 341, 343, 346–347, 354, 357, 371–374, 378
Buchholz, 27, 156, 192, 210, 227, 244, 255, 274, 288, 293, 309, 313, 321, 333, 371–372
Buchholz, Feferman, Pohlers, and Sieg, 227, 244, 309

Calculus
  elementary, 139–140, 142
  for predicate logic, 20, 139, 142–144
  for sentential logic, 98, 109, 122, 133, 171, 252, 354, 385
  formal, 101, 116
  (in)finitary, 252, 255, 268–271, 274–275, 313, 334
  infinitesimal, 43
    fundamental theorem of, 287
  logical, 15–16, 88, 98, 103–105, 109, 111–112, 127, 131, 134, 158, 171, 176, 244, 247–248, 313, 334, 385, 398
  of *Principia Mathematica*, 7–9, 15–16, 19–20, 89, 93, 103, 125, 148, 156, 163, 217, 237, 241, 285, 348, 356, 384–385
  sequent, 7, 156, 177, 181, 247–248, 251, 253, 264, 268–270, 386
Canonical isomorphism, 13, 26
Cantor, 6, 17, 63–65, 73–74, 84, 86, 88, 95–96, 106, 231–232, 235–236, 238, 240, 245, 264–266, 282–284, 291, 306–308, 311, 318, 324, 330, 335, 338, 340, 348–352, 383, 391, 393
Church, 20, 122, 151, 208, 222, 273, 371, 388

Completeness
  semantic, 30, 109, 182–183
  syntactic, 159–160, 162, 165–167, 171, 219, 236, 243, 269–272
  topological, 295, 325, 340, 373
Completeness theorem, 109, 137, 144, 215, 275–276, 295, 316, 386
Conceptual (structural) definition, 1, 3–4, 8, 11–14, 17, 25, 48, 380–382
Continuity, 11–12, 44, 52–53, 69–70, 75–76, 78–80, 83, 85, 87, 93–94, 143, 155, 195, 231–233, 240, 246, 287, 295, 304–305, 313, 356, 381–383, 394
Continuum hypothesis, 146, 151
Continuum problem, 73, 136, 165, 284, 338
Cut-elimination, 7, 181, 249, 253, 268, 275
Cut-free, 248–249, 251, 253, 256, 260, 268–269, 386–387

Dedekind, 1, 4, 7–14, 17, 23–25, 31, 33–34, 35–72, 73–75, 77–84, 86–90, 94–96, 99–100, 112, 118, 137, 231–236, 238, 240, 246, 257, 264, 282–283, 287, 291, 295, 299–300, 302–308, 311–312, 314–315, 323–327, 330, 340, 347–353, 360, 364, 370, 377, 379–385, 391, 394–395, 401
Derivability conditions, 150, 218–219, 221, 390
Dirichlet, 9, 36, 41, 44, 52, 54, 56, 72, 94, 118, 232, 264, 302–303, 305, 326, 330, 347, 360, 391

Entscheidungsproblem (decision problem), 20–21, 102, 110–111, 144, 152, 181, 215, 221–222, 387–388
ϵ-calculus, 136, 268, 361
Euclid, 15, 75, 233, 304, 384, 392–393
Existential axiomatics, 14, 17, 37, 71, 74–90, 132, 138, 225, 348, 350, 356, 361–362, 381, 383, 395

Index

Fachwerk (Begriffsfachwerk), 24, 363
Feferman, 27, 92, 145, 161–162, 193, 210, 213, 227, 244, 246–247, 250, 254–255, 257–258, 271, 273–275, 278, 287–288, 292–296, 309–310, 313, 322, 324, 332–334, 339–340, 348, 371–372, 380
Finitism, 6, 91, 117, 150, 194, 219, 297, 320, 345–347
Formalism, 8, 18–20, 22, 115, 123, 139–140, 142–144, 161, 189, 195, 217–218, 220–221, 223–225, 230, 244–245, 286, 311–312, 315–316, 329, 336, 345–346, 354, 356, 358, 368, 388
Frame (methodical, methodological), 4, 18, 25, 69, 139, 222–226, 301, 323, 340, 342
Frege, 6, 9, 14–16, 24, 35, 48, 62–63, 87, 89, 101, 105, 112, 115, 120, 123, 127, 230–231, 233, 236, 248, 265, 305, 308–309, 317, 375, 379, 383–385
Friedman, 150, 202, 227, 247, 250, 258, 267–268, 274–277, 287–288, 291–297, 316, 322, 340, 373

Gauss, 5, 11, 39–40, 43, 72, 79, 232, 239, 264, 302–303, 347
Gentzen, 7, 16, 31, 117, 121–122, 129, 133–134, 155–192, 193–195, 197–198, 207–213, 223–225, 228, 242–244, 247–248, 250–253, 255, 258, 260, 267, 270–272, 286, 294, 309–310, 320–322, 332–333, 339, 346, 364–365, 378, 385–386
Gödel, 7, 17, 19–20, 22–23, 31, 66, 91, 93, 103, 108–110, 115, 117, 119, 121–122, 129, 137, 142, 144, 145–153, 155–156, 158, 160–172, 174, 177–179, 181–187, 190–192, 193–213, 215–219, 221–223, 228, 230, 237, 241, 243, 250, 254, 257–260, 263, 266–267, 272, 275, 281, 285–286, 288, 291, 294, 309–310, 312, 314–316, 318–321, 323, 327, 329–330, 332–333, 335–338, 342, 346, 361–362, 364–365, 367, 373, 377–378, 388, 390–391, 395–396
Gödel-number, 217, 254, 272

Hauptsatz, 181, 248–249, 251, 270
Heine, 48
Heine-Borel theorem, 247, 275, 295
Herbrand, 17, 19, 122, 129, 137, 139, 143, 145, 149–150, 155, 167–170, 196, 201, 213, 215–217, 222, 266, 270–271, 332, 346, 361
Heyting, 147–148, 161, 163, 170, 172–173, 175, 178, 180, 197, 202–204, 223, 258
Hilbert
 before 1904, 29, 31, 33, 75, 81, 86, 88, 96, 98, 100, 113, 115, 132, 236, 238, 240, 265, 284, 307, 348, 350–352, 355, 357
 between 1904 and 1922, 8, 17, 19, 25, 30–31, 80, 91–127, 129, 131, 133, 135–136, 171, 346, 359–361, 382
 after 1922, see Hilbert's program (Hilbert program)
 after 1930, 3, 4, 19, 24, 171–177
Hilbert's program (Hilbert program), 1, 3–27, 31, 71, 91–127, 167, 193–195, 197, 213, 219, 221, 227, 230, 240–244, 247, 249–250, 257–258, 263–264, 266–267, 271, 281–290, 291, 297, 299–300, 302, 307–308, 311–312, 314, 316–317, 318–319, 327, 329–330, 332–335, 339

i.d. class, see Inductive definition (i.d. class)
Ideal/real, 266, 285, 314
Incompleteness theorem, 17, 19–20, 22, 31, 121, 129, 145, 147–151, 155, 158, 162, 164, 167–169, 171, 194, 196, 206, 209, 215–224, 230, 243, 258,

263, 266, 281, 285, 291, 309, 312, 318, 327, 329, 332, 335–336, 364, 377, 388, 390, 393, 395
Inductive definition (i.d. class), 7, 18, 206, 210, 227, 250, 254–257, 263, 272–275, 288–289, 293–294, 313, 319–322, 324–325, 335, 339–340, 370–372, 391, 396
Intuitionism, 8, 91, 117, 147, 149–150, 168, 204, 224, 230, 239, 264, 311–312, 320, 346, 362, 369
Inversion, 43, 249, 270, 278–279

Keferstein, 37, 45, 55–56, 59, 61–63, 69, 81, 95, 234, 305, 323, 349, 370, 383–384
Kleene, 151, 222, 257, 273–274, 371, 388
Kneser, 111, 124, 131, 133–136, 176, 178–179, 190, 303
König's Lemma, 227, 275–278, 295
Kreisel, 206–208, 210–212, 241–242, 244–245, 247, 257, 264, 266, 267, 276–277, 287–289, 292–294, 309, 319, 325, 332, 341, 373, 378
Kronecker, 4–5, 7, 9–11, 31, 34, 35, 37, 45, 66, 71, 73, 75–76, 83, 88, 94, 99–100, 113, 120, 132, 231, 234–235, 238–242, 258, 264–265, 268, 282–284, 297–298, 302–304, 307–309, 312, 316–317, 326, 330, 334, 349, 351–352, 357, 359, 370, 377, 379–380
Kummer, 11, 54, 107, 302, 326

λ-calculus, 207
Lipschitz, 52, 54, 69–70, 75, 80, 94, 233, 299, 302, 304, 307, 382, 384
Logicism, 31, 84, 89, 91, 112, 204, 351

Mapping, 12–13, 23–24, 36, 55–58, 60, 64, 66–68, 71, 79–80, 242, 324, 340, 400
Martin-Löf, 26–27, 155, 207, 322, 371

Mechanical Procedure, 5, 8, 15, 19–22, 336
Minkowski, 5
Mints, 156, 197, 275, 316, 377

Natural deduction, 133–134, 172, 174–175, 181, 377, 384, 386–387, 398–399
Non-Euclidean geometry, 73, 143, 232
Normal derivation, 156, 180, 270–271, 278–279, 371
Normal form, 133–134, 211, 361
Normalization, 156, 174–175, 179, 181, 185, 386
Number theory, elementary, *see* Arithmetic, first-order
Numeric formula, 133, 139, 158, 182, 360

Ordinal notation systems, 6, 224, 256, 272, 288, 321, 339, 372
Ordinal numbers
  constructive, 7, 195, 210, 254, 257–258, 293, 313, 320–321, 334, 338–339, 371–372
  countable, 207
  finite, 10, 327
  transfinite, 199, 252

Paradox(es)
  Richard's, 98–99
  Russell's solution to, 103
  Russell-Zermelo, 98, 352
  set theoretic, 230, 240, 284, 307, 311
Parsons, 27, 38, 91, 129, 145, 164, 190, 193, 197, 200, 213, 222, 319, 324, 335, 362, 364–365, 367, 369–370
Pohlers, 210, 227, 244, 255, 257, 274, 288, 309, 321, 333, 372
Poincaré, 26, 30, 87–89, 97, 100, 115–116, 118–120, 125, 229, 234, 283, 317, 348, 354–355, 359–360, 365, 387
Post, 21–22, 109, 336
Prawitz, 227, 386, 397

Principia Mathematica, 7, 15–16, 19–20, 89, 93, 103, 125, 148, 156, 163, 217, 237, 241, 285, 348, 356, 384–385
Proof theory, 3, 6–7, 14, 16–17, 19, 30–31, 33, 74, 86, 88–89, 91, 93, 97, 101–102, 111, 115–116, 122, 129, 131–144, 145, 147, 155–157, 159–160, 162, 166–168, 170, 177–178, 183, 189–190, 208, 210, 215, 217, 220–221, 223–225, 227, 229–261, 264–272, 276, 281, 284, 286, 288, 291, 297, 299, 308, 310, 318–319, 330, 333, 343, 348, 354, 357, 360–361, 370–372, 375, 377–380, 382, 396
Pythagorean theorem, 392–393

Reduction, 17–18, 21, 26, 37, 85, 89–90, 93, 100–101, 108, 111, 117, 142, 144, 175, 181, 185, 187, 189–190, 197–198, 208, 210–212, 223, 227, 229–231, 235, 241, 244, 250, 255–257, 259–260, 263–280, 288–290, 293, 297, 300–301, 313–314, 316–318, 327, 331–332, 334, 342, 346, 356, 371, 375, 395, 399
Representability, 217–219, 221, 368–369, 390
Reverse mathematics, 69, 227, 275, 287, 291–298
Riemann, 72, 245–246, 347
Riemann integral, 245–246
rigorous proof, 8, 14–17, 384–389
Russell, 7, 16, 31, 33, 89, 98–99, 101, 103, 106, 108, 120, 123, 125–126, 136, 142, 156, 167, 197, 231, 236–238, 241, 265, 284, 308, 317, 351–352, 355–356, 385

Schröder, 10, 103, 178
Schütte, 122, 192, 199, 209, 247, 252, 271, 292, 309, 311, 321, 332, 339

Set theory
  axiomatic, 66, 117, 195, 374
  Cantorian, 236
  emergence of, 300, 330
  foundations, 90, 135, 236
  subsystems of, 164, 292, 323, 372
  von Neumann, 146, 217
  Zermelo, *see* Zermelo
  Zermelo-Fraenkel, 292, 322
Simpson, 227, 268, 275–277, 287, 291, 293–298, 303, 310, 334
Spector, 200, 257, 288, 293
Stein, 22, 26–27, 38, 59, 91–92, 96–97, 300, 345, 347–348, 374, 380, 401
Structural definition, 1, 3–4, 8, 11–14, 17, 25, 380–382
Structuralism (structure, structural, structuralist), 1, 3–4, 7–14, 17–18, 23–27, 29–30, 36–38, 46, 54, 61, 66, 71, 74, 76–78, 80, 90, 101–102, 104, 106, 110, 118, 120–121, 132, 138, 157, 161, 166, 176–177, 187, 220, 227, 240–241, 247–248, 254, 257, 260–261, 265–266, 271, 273, 284, 296, 300–301, 312, 314–318, 320–321, 323–325, 327, 329–332, 337, 341–342, 348, 356, 358, 361–364, 366, 368–370, 374, 377–378, 380–382, 385, 389–390, 393, 395–396, 398, 400–401
Subformula property, 156, 175, 177, 179–180, 249, 251, 268, 270, 387

Tait, 27, 38, 80, 156, 158, 161, 190, 208–210, 213, 247, 252, 254–255, 269, 271, 274, 289, 309, 319, 332, 348, 365, 370–371
τ-calculus, 206
Truth-definition, 250–251, 253, 256, 268, 270
Turing, 8, 19, 21–23, 122, 327, 336–337, 342, 377, 388–389, 395

Verifiability (Verifiable), 139–141, 143–144
von Neumann, 8, 17, 118, 122, 129, 131, 136, 142, 145–152, 155, 157, 159–160, 162–171, 177, 181, 216–217, 228, 242–243, 266, 271, 285, 312, 332, 346, 361, 374, 388

Weber, 36, 39, 52, 54–55, 65, 67, 75, 239, 302, 305
Weyl, 29, 33, 103–104, 106, 108, 110–112, 114–117, 120, 136, 146–147, 210, 246–247, 263, 268, 287, 291, 317, 346–347, 357, 374
Whitehead, 7, 16, 103, 125, 142, 156, 237–238, 308, 355–356

Zermelo, 14, 26, 33, 58, 88–90, 92, 98–99, 122, 135, 165, 230–231, 236–238, 243, 259, 265, 271, 284, 292, 294, 312–313, 315, 322–323, 338, 340, 351–352, 355, 370, 383, 391, 393

www.ingramcontent.com/pod-product-compliance
Ingram Content Group UK Ltd.
Pitfield, Milton Keynes, MK11 3LW, UK
UKHW021328180426
11947UKWH00017B/1515